11
Topics in Organometallic Chemistry

Topics in Organometallic Chemistry

Recently Published and Forthcoming Volumes

Metal Carbons in Organic Synthesis
Volume Editor: K.-H. Dötz
Vol. 13, 2004

Theoretical Aspects of Transition Metal Catalysis
Volume Editor: G. Frenking
Vol. 12, 2005

Ruthenium Catalysts and Fine Chemistry
Volume Editors:
C. Bruneau and P. H. Dixneuf
Vol. 11, 2004

New Aspects of Zirconium Containing Organic Compounds
Volume Editor: I. Marek
Vol. 10, 2004

CVD Precursors
Volume Editor: R. Fischer
Vol. 9, 2005

Metallocenes in Stereoselective Synthesis
Volume Editor: T. Takahashi
Vol. 8, 2004

Transition Metal Arene π-Complexes in Organic Synthesis and Catalysis
Volume Editor: E. P. Kündig
Vol. 7, 2004

Organometallics in Process Chemistry
Volume Editor: R. D. Larsen
Vol. 6, 2004

Organolithiums in Enantioselective Synthesis
Volume Editor: D. M. Hodgson
Vol. 5, 2003

Organometallic Bonding and Reactivity Fundamental Studies
Volume Editor: J. M. Brown, P. Hofmann
Vol. 4, 1999

Activation of Unreactive Bonds and Organic Synthesis
Volume Editor: S. Murai
Vol. 3, 1999

Lanthanides: Chemistry and Use in Organic Synthesis
Volume Editor: S. Kobayashi
Vol. 2, 1999

Alkene Metathesis in Organic Synthesis
Volume Editor: A. Fürstner
Vol. 1, 1998

Ruthenium Catalysts and Fine Chemistry

Volume Editors: C. Bruneau · P.H. Dixneuf

With contributions by
I.W.C.E. Arends · S. Blechert · C. Bruneau · N. Chatani · S.J. Connon · L. Delaude · A. Demonceau · S. Dérien · P.H. Dixneuf · K. Itoh · F. Kakiuchi · T. Kodama · C.-J. Li · B. Marciniec · F. Monnier · H. Nishiyama · A.F. Noels · C. Pretraszuk · R.A. Sheldon · Y. Yamamoto · M. Wang

Springer

The series *Topics in Organometallic Chemistry* presents critical overviews of research results in organometallic chemistry, where new developments are having a significant influence on such diverse areas as organic synthesis, pharmaceutical research, biology, polymer research and materials science. Thus the scope of coverage includes a broad range of topics of pure and applied organometallic chemistry. Coverage is designed for a broad academic and industrial scientific readership starting at the graduate level, who want to be informed about new developments of progress and trends in this increasinly interdisciplinary field. Where appropriate, theoretical and mechanistic aspects are included in order to help the reader understand the underlying principles involved.
The individual volumes are thematic and the contributions are invited by the volumes editors.
In references Topics in Organometallic Chemistry is abbreviated
Top. Organomet. Chem. and is cited as a journal

Springer WWW home page: springeronline.com
Visit the TOMC contents at springerlink.com

ISSN 1436-6002
DOI 10.1007/978-3-540-40027-1

Bibliographic information published by Die Deutsche Bibliothek
Die Deutsche Bibliothek lists this publication in the Deutsche Nationalbibliographie; detailed bibliographic data is available in the Internet at <http://dnb.ddb.de>.

springeronline.com

Originally published by Springer-Verlag Berlin Heidelberg New York in 2004
MyCopy version of the original edition 2004

Typesetting: Fotosatz-Service Köhler GmbH, Würzburg
Production editor: Christiane Messerschmidt, Rheinau
Cover: design & production GmbH, Heidelberg

Printed on acid-free paper 02/3020 – 5 4 3 2 1 0
www.springer.com/mycopy

Volume Editors

Dr. Christian Bruneau
UMR 6509
Université de Rennes 1
(Campus de Beaulieu)
35042 Rennes Cedex
France
christian.bruneau@univ-rennes1.fr

Professor Pierre H. Dixneuf
Institut de Chimie de Rennes
Université de Rennes 1
(Campus de Beaulieu)
35042 Rennes Cedex
France
pierre.dixneuf@univ-rennes1.fr

Editorial Board

Topics in Organometallic Chemistry is also Available Electronically

For all customers who have a standing order to Topics in Organometallic Chemistry, we offer the electronic version via SpringerLink free of charge. Please contact your librarian who can receive a password for free access to the full articles by registration at:

springerlink.com

If you do not have a subscription, you can still view the tables of contents of the volumes and the abstract of each article by going to the SpringerLink Homepage, clicking on "Browse by Online Libraries", then "Chemical Sciences", and finally choose Topics in Organometallic Chemistry.

You will find information about the

- Editorial Board
- Aims and Scope
- Instructions for Authors
- Sample Contribution

at springeronline.com using the search function.

Preface

During the last decade molecular ruthenium catalysts, have provided a variety of novel activation processes leading to powerful new organic synthetic methods, that are not promoted by other metal catalysts. Ruthenium catalysis constitutes an emerging field for the selective preparation of fine chemicals. This is due to the availability of a large number of well-defined and stable ruthenium precatalysts offering several possible oxidation states. They usually tolerate functional groups and have revealed catalytic activities for a wide range of chemical transformations with atom economy. New ruthenium catalysts make possible carbon–carbon, carbon–hydrogen, carbon–heteroatom bond formation and cleavage, and are able to provide non classical activation modes.

The most important discoveries in ruthenium catalysis are highlighted and innovative activation processes, some of which are still controversial, are presented in this volume. They illustrate the usefulness in organic synthesis of specific reactions including carbocyclization, cyclopropanation, olefin metathesis, carbonylation, oxidation, transformation of silicon containing substrates, and show novel reactions operating via vinylidene intermediates, radical processes, inert bonds activation as well as catalysis in water.

This monograph is not intended to provide a comprehensive view of all ruthenium-catalyzed reactions, as it is an explosive growth field. For instance, ruthenium-catalyzed enantioselective hydrogenation, already detailed in several monographs, will not be treated here in spite of its high impact in organic synthesis.

This volume should be helpful to researchers, teachers and students interested in innovative and sustainable chemistry. We are grateful to the experts who have contributed by writing a chapter and we dedicate this volume to all chemists and students who have been the actors in the first steps of this fast developing field.

Rennes, France, March 2004

Christian Bruneau
Pierre H. Dixneuf

Contents

Topics Organomet Chem (2004) 11: 1–44
DOI 10.1007/978-3-540-40027-1

Ruthenium-Catalyzed C–C Bond Formation

Sylvie Dérien (✉) · Florian Monnier · Pierre H. Dixneuf

Institut de Chimie de Rennes, UMR 6509 Université de Rennes 1– CNRS, Organométalliques et Catalyse, Campus de Beaulieu, 35042 Rennes, France
derien@univ-rennes.fr

Abstract Molecular ruthenium catalysts are now currently used to perform selective carbon–carbon bond formation by combination of simple substrates. Their tolerance toward functional groups has allowed the access to high value, multifunctional molecules. It will be shown that ruthenium catalysts allow the coupling of functional alkenes or alkynes with a variety of unsaturated molecules such as alkenes, dienes, alkynes, and diynes. A large range of electron-rich ruthenium or hydridoruthenium complexes are currently used for the formation of cyclic and polycyclic compounds on reaction with substrates containing several unsaturated C–C bonds. Ruthenium complexes have promoted several original activation pathways, such as C–H bond activation, the distribution of carbene from diazoalkanes, and especially their versatility in making a large variety of ruthenacycle intermediates. Besides the applications of ruthenium precatalysts in organic synthesis an important discussion of and mechanisms will be presented.

Keywords Ruthenium catalysts · C–C and C=C bond formation · Alkenes · Alkynes · Allyl ruthenium · Ruthenacycle · Hydroruthenation

1 Introduction

During the last decade, molecular ruthenium catalysts have promoted tremendous developments in organic synthesis methodology and polymer science, and revealed novel activation processes. Ruthenium catalysts have become unavoidable catalysts in enantioselective catalysis, for the production of pharmaceutical intermediates, and they show new hydrogen-transfer processes for the enantioselective reduction of ketones. The high tolerance of ruthenium complexes toward a variety of functional groups and the discovery of innovative, efficient, tunable ruthenium alkylidene catalysts for alkene metathesis have led to the inclusion of alkene metathesis as a very efficient method that is currently modifying synthetic approaches.

Molecular ruthenium catalysts have created a large variety of processes leading to selective C–C bond formation reactions via the combination of several molecules with atom economy. In this direction ruthenium catalysts have promoted reactions that were not previously observed with organic or enzyme catalysts, but especially via activation processes not observed with other metal catalysts.

The objective of this review is to present the most general ruthenium-catalyzed methods for selective C–C bond forming reactions. Particular attention will be paid to the nature of the catalyst and its relevance to the reaction mechanism, rather than to give many examples of applications. However, the C–C bond formation reactions that are the topics of other chapters of this volume will be only briefly indicated but not developed.

2
Coupling Reactions of Two C=C Bonds

Catalyzed C–C bond formation by selective coupling between two C=C bonds gives access to a variety of unsaturated functional compounds. In this area, ruthenium complexes have promoted, in recent years, an impressive development owing to high regioselectivity pathways.

2.1
Dimerization of Functional Alkenes

One of the oldest ruthenium-catalyzed C=C bond coupling reactions deals with the selective dimerization of functionalized alkenes, especially the dimerization of acrylates [1, 2]. It usually involves either an initial hydrometallation process, oxidative coupling, or vinyl C–H bond activation (Scheme 1).

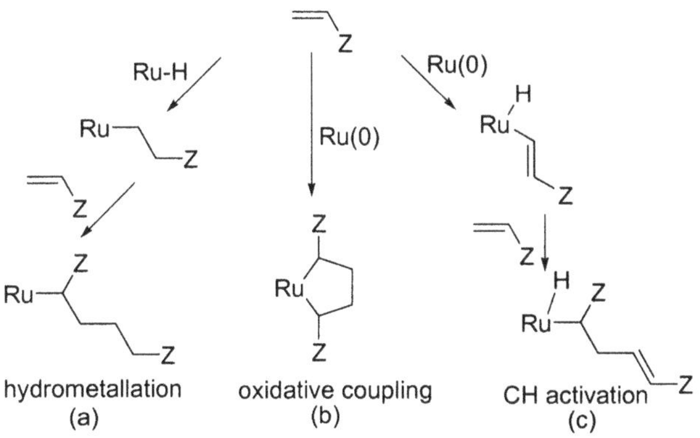

Ru-H
Ru(0)
Ru(0)
hydrometallation (a)
oxidative coupling (b)
CH activation (c)

Scheme 1

For example, the tail-to-tail dimerization of methyl acrylate was catalyzed by ruthenium complexes such as $RuHCl(CO)(P\textit{i}\text{-}Pr_3)_2/CF_3SO_3Ag$ or even $RuCl_3$ and gave dimethyl hexenedioate isomers. Efficient catalytic systems such as $Ru(\eta^6\text{-naphthalene})(COD)/CH_3CN$, where COD is cyclooctadiene, selectively led to the diester **2** in 75% yield [1] (Eq. 1).

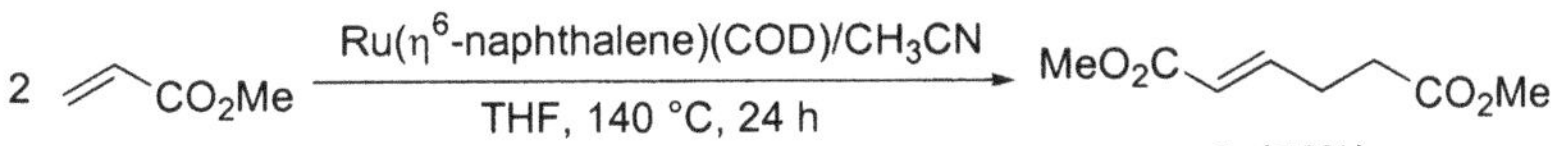

$$2\ \text{CH}_2{=}\text{CHCO}_2\text{Me}\ (\mathbf{1}) \xrightarrow[\text{THF, 140 °C, 24 h}]{\text{Ru}(\eta^6\text{-naphthalene})(\text{COD})/\text{CH}_3\text{CN}} \text{MeO}_2\text{C}\text{CH}{=}\text{CHCH}_2\text{CH}_2\text{CO}_2\text{Me}\ \ \mathbf{2}\ (75\%) \quad (1)$$

The tail-to-tail dimerization of acrolein [3] and acrylonitrile [4, 5] was also obtained, with a lower reactivity and stereoselectivity. However, the dimerization of acrylonitrile was performed under mild conditions in the presence of molecular hydrogen with Ru(COD)(COT), where COT is cyclooctatetraene, [4] (Eq. 2).

2 CH₂=CH–CN (**3**) —[Ru(COD)(COT); H_2, 150 °C, 3h]→ NC–CH=CH–CH₂–CH₂–CN **4** (40% - E/Z : 1/1) (2)

Recently, a selective head-to-tail dimerization of acrylic or α,β-unsaturated carbonyl compounds was performed with Cp*$RuH_3(PCy_3)$ catalyst, where Cp* is pentamethylcyclopentadienyl and Cy is cyclohexyl, and was expected to occur via hydrometallation [6] (Eq. 3).

2 CH₂=CH–CO_2Me (**1**) —[Cp*$RuH_3(PCy_3)$; THF, 80 °C, 24 h]→ MeO_2C–C(=CH₂)–CH₂–CH₂–CO_2Me **5** (91%) (3)

An initial hydrometallation was also invoked in the dimerization of norbornadiene with the catalyst precursor Ru(COD)(COT) to generate pentacyclotetradeca-4,11-diene 7 in very good yield [7] (Eq. 4). A suggested mechanism for the formation of 7 involves olefin insertion into the preformed Ru–H bond and the cleavage of two C–C bonds.

2 **6** —[Ru(COD)(COT); dimethylfumarate, THF, 40 °C, 1 h]→ **7** (96%) (4)

2.2 Cross-Couplings of Alkenes

The mixed coupling of two different alkenes allows the formation of new functional unsaturated products but requires high regioselectivity. A ruthenium hydride complex, generated in situ from the reaction of $RuHCl(CO)(PCy_3)_2$ with $HBF_4.OEt_2$, was found to be an effective catalyst for the hydrovinylation of alkenes [8]. The reaction of styrene with ethylene produced the hydrovinylation compound **10** in 93% yield (Eq. 5). Initial hydrometallation of the alkene and insertion of ethylene seemed to be a plausible mechanism.

CH₂=CH–Ph (**8**) + $CH_2{=}CH_2$ (**9**) —[$RuHCl(CO)(PCy_3)_2/HBF_4.OEt_2$; C_6H_6, 20 °C, 6 h]→ CH₂=CH–CH(CH₃)–Ph **10** (93%) (5)

Activation of vinyl C–H bonds with $RuH_2(CO)(PPh_3)_3$ catalyst has allowed the formal insertion of α,β-unsaturated ketones or esters into the C–H bond of vinylsilanes and led to a regioselective C–C coupling at the β-position [9] (Eq. 6). Activation of the sp^2 C–H bond occurred with the aid of chelation of a coordinating functional group and provided vinylruthenium hydride **14**. Insertion of olefin afforded the tetrasubstituted alkene **13**. The ruthenium activation of a variety of inert C–H bonds has now been performed by Murai [10].

11 + **12** ($Si(OEt)_3$) → [$Ru(H)_2(CO)(PPh_3)_3$, toluene, 135 °C, 0.5 h] → **13** (96%), via **14** (Ru–H) (6)

Ruthenium(0) complexes such as Ru(COD)(COT) catalyze the dehydrohalogenative coupling of vinyl halides with olefins to give substituted conjugated dienes in a Heck-type reaction [11]. Thus, alkenyl halides readily react with activated olefins to produce dienes **16** (Eq. 7). Oxidative addition of vinyl halide, followed by regioselective insertion of an electron-deficient olefin and by β-hydrogen elimination leads to the diene.

15 (Ph–CH=CH–Br) + **1** (CO_2Me) → [Ru(COD)(COT), NEt_3, 100 °C, 4 h] → **16** (79%) (7)

The cross-coupling reaction of vinyl halides with Grignard reagents to provide corresponding alkenes was also promoted by a ruthenium catalyst such as $RuCl_2(PPh_3)_3$ [12].

2.3
Cyclizations

The catalytic intramolecular coupling of two C=C bonds at a ruthenium site leads to cyclization reactions. For example, although generally less reactive than α,ω-diynes or enynes, 1,6-dienes react with $[RuCl_2(COD)]_n$ in 2-propanol, leading to *exo*-methylenecyclopentanes in excellent yields [13] (Eq. 8). The mechanism suggests the formation of the ruthenacyclopentane(hydrido) intermediate **19**.

This reaction applied to diallyllactones allowed the diastereoselective preparation of *exo*-methylene spirolactones [14] (Eq. 9).

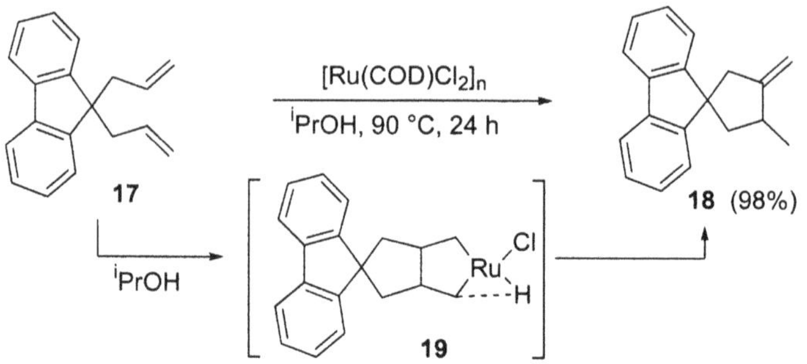

$[Ru(COD)Cl_2]_n$
${}^iPrOH$, 90 °C, 24 h
17
18 (98%)
(8) 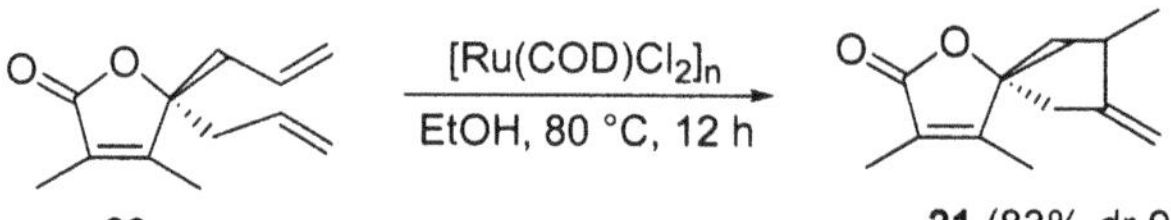
${}^iPrOH$
Ru Cl H
19

$[Ru(COD)Cl_2]_n$
EtOH, 80 °C, 12 h
20
21 (83%-dr 95/5)
(9)

Functionalized *exo*-methylenecyclopentanes can also be obtained by ruthenium-catalyzed intramolecular C–H bond activation [15]. 1-(2-Pyridyl)-, 1-(2-imidazolyl)-, and 1-(2-oxazolyl)-1,5-dienes proceeded in a regiospecific manner to give five-membered ring products (Eq. 10). The proposed mechanism initially involves the activation of the vinylic C–H bond of the exocyclic C=C bond assisted by preliminary coordination of the nitrogen atom, followed by intramolecular insertion of the other C=C bond (see Eq. 6).

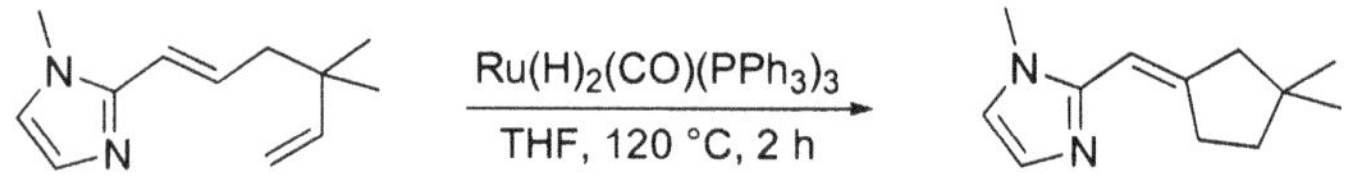

$Ru(H)_2(CO)(PPh_3)_3$
THF, 120 °C, 2 h
22
23 (87%)
(10)

2.4 Carbonylations Involving Two C=C Bonds

When an oxidative coupling or addition takes place in the presence of carbon monoxide, CO insertion occurs leading to ketones. The $Ru_3(CO)_{12}$-catalyzed reaction of alkenylpyridyl or *N*-(2-pyridyl)enamines and ethene performed under an atmosphere of carbon monoxide leads to the selective formation of α,β-unsaturated ketones [16] (Eq. 11). After activation of the vinyl C–H bond, insertion of both carbon monoxide and ethylene takes place to give **25**.

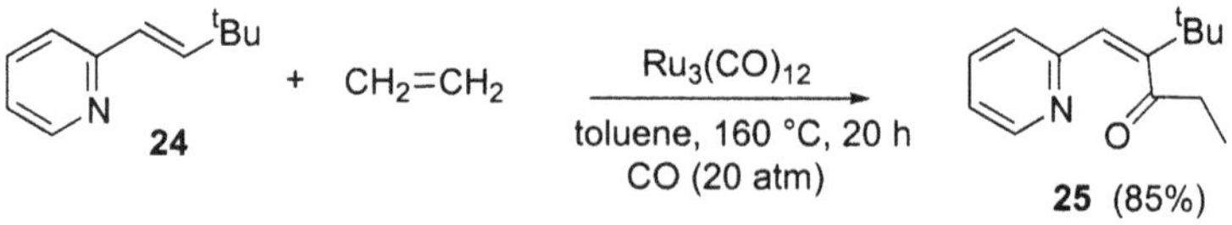

${}^tBu$
24
+ $CH_2{=}CH_2$
$Ru_3(CO)_{12}$
toluene, 160 °C, 20 h
CO (20 atm)
${}^tBu$
25 (85%)
(11)

A related reaction with α,β-unsaturated imines allowed the one-pot synthesis of γ-lactams [17] (Eq. 12).

26 + $CH_2{=}CH_2$ $\xrightarrow[\text{toluene, 150 °C, 15 h; CO (10 atm)}]{Ru_3(CO)_{12}}$ 27 (95%) (12)

Reactions involving carbonylation are detailed in the chapter Selective Carbonylations with Ruthenium Catalysts of this volume.

3 Mixed C=C Bond and 1,3-Diene Coupling Reactions

Ruthenium complexes can promote the catalytic coupling of 1,3-dienes with alkenes, leading to the formation of functionalized dienes, as well as Diels–Alder reaction.

3.1 New Diene Formation from 1,3-Dienes

Functionalized dienes can be obtained by C–C bond formation between 1,3-dienes and alkenes via oxidative coupling with electron-rich ruthenium catalysts but also via insertion into Ru-H and then Ru–C bonds. For example, Ru(COD)(COT) catalyzed the selective codimerization of 1,3-dienes with acrylic compounds to give 3,5-dienoic acid derivatives [18] (Eq. 13). η^4-coordination of 1,3-diene to a hydridoruthenium leads to a π-allylruthenium species to selectively give, after coupling with the C=C bond and isomerization, the functionalized conjugated 1,3-dienes.

28 + 29 (CO_2Et) $\xrightarrow[\text{N-methylpiperidine, 80 °C, 2.5 h}]{\text{Ru(COD)(COT)}}$ 30 (CO_2Et) + 31 (CO_2Et) (13)

30+ 31 (88% - 93/7)

Ru-H; [Ru; CO_2Et; [Ru–H; CO_2Et

A π-allylruthenium complex, formed from 1,3-diene and a preformed Ru–H complex, was also postulated to be an intermediate for the regioselective hydrovinylation of unsymmetrically substituted 1,3-dienes to afford 3-methyl-1,4-dienes as products [19] (Eq. 14). Isomerization of the initially formed 1,4-diene, such as **33**, to the stabler conjugated 1,3-diene did not occur.

32 + $CH_2{=}CH_2$ $\xrightarrow[C_6H_6,\ 20\ °C,\ 6\ h]{RuHCl(CO)(PCy_3)_2/HBF_4.OEt_2}$ 33 (67%) (14)

This reaction was applied to a steroid and was shown to be stereospecific, giving a product with a (S) configuration at carbon 20 (Eq. 15).

34 + $CH_2{=}CH_2$ $\xrightarrow[C_6H_6,\ 20\ °C,\ 7\ h]{RuHCl(CO)(PCy_3)_2/HBF_4.OEt_2}$ 35 (88%) (15)

Functional 1,5-dienes were also synthesized in good yields by ruthenium-catalyzed regioselective codimerization of enol esters with 2-substituted-1,3-butadienes [20] (Eq. 16). A ruthenacycle intermediate formed by oxidative coupling was proposed followed by intracyclic β-hydride elimination. The (Z)-selectivity is thought to result from the configurational inhibition for the β-hydride elimination in the intermediate ruthenacyclopentane.

28 + 36 $\xrightarrow[MeOH,\ 100\ °C,\ 14\ h]{Cp^*RuCl(cod)}$ 37 (95%) (16)

3.2 Diels–Alder and Ene Reactions

The moderate Lewis acidity of ruthenium complexes was used to promote catalytic Diels–Alder reaction of dienes and acrolein derivatives [21–23]. The enantioselective Diels–Alder reaction of methacrolein with dienes was catalyzed with cationic ruthenium complexes containing an arene or cyclopentadienyl (Cp) ligand and a chiral ligand such as phosphinooxazoline, pyridyl-oxazoline, monoxidized 2,2′-bis(diphenylphosphino)-1,1′-binaphthyl (BINPO) or 1,2-bis[bis(pentafluorophenyl)phosphanyloxy]-1,2-diphenylethane (BIPHOP-F). The reaction gave the cycloadduct in high yields with excellent

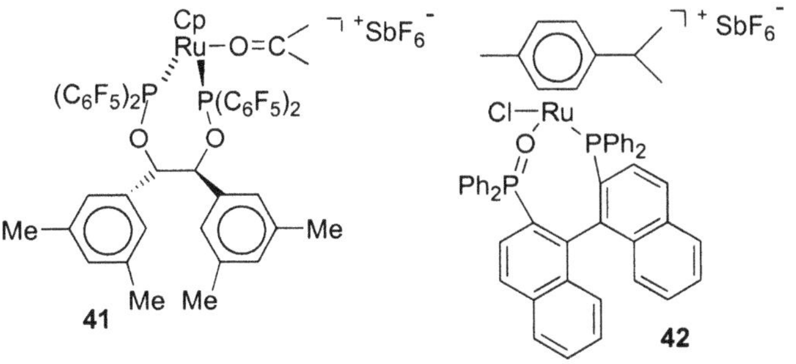

Fig. 1 **41** **42**

exo–endo selectivity (up to 99:1) and enantioselectivity (up to 99%) in several examples, particularly with the complexes [CpRu(acetone)(Me_4BIPHOP-F)] SbF_6 (**41**) [22] and [(*p*-cymene)RuCl(BINPO)](SbF_6)/ $AgSbF_6$ (**42**) [23] (Eq. 17, Fig. 1).

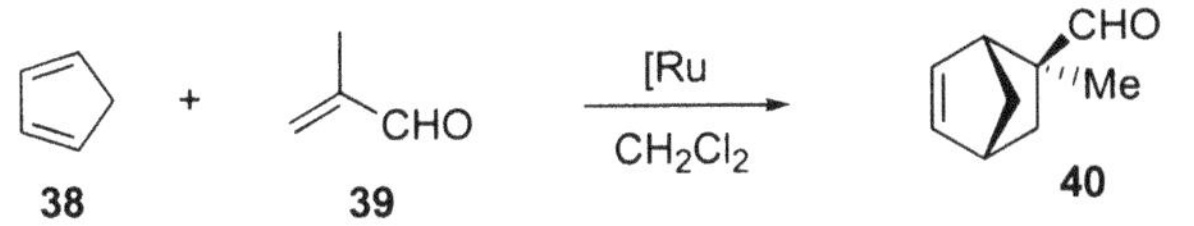

38 **39** **40** (17)

41	- 20°C	87%	exo/endo : 97/3	97% ee (S)
42	- 78°C	100%	exo/endo : 93/7	99% ee
42	- 24°C	100%	exo/endo : 99/1	94% ee

With the complex [(indenyl)Ru(acetone)(Me_4BIPHOP-F)]SbF_6 as catalyst [22], the reaction afforded the exo cycloadduct as the major product for the reaction of acrolein with cyclopentadiene, whereas this noncatalyzed reaction is known to give the endo derivative as the major product. Analogously the catalyst **41** performed the asymmetric 1,3-dipolar addition of nitrones with enals [24].

The ene reaction involving an aldehyde has also been performed with a (salen)ruthenium(II) catalyst [25]. The C–C coupling of an unsaturated carbonyl with aldehydes was also achieved with $RuH_2(PPh_3)_4$ to give an α-methylene-β-hydroxyketone but in that case the reaction proceeds via the preliminary hydrometallation of the double bond [26].

4 Cross-Coupling of 1,3-Diene

Catalyzed oligomerization and co-oligomerization of conjugated dienes have been performed with a wide range of transition-metal complexes. Catalytic cyclodimerizations of conjugated dienes have also been performed selectively [27]. Thus, a catalytic amount of CpRuCl(diene) and $Ag(OSO_2CF_3)$ led to the formation of 1,5-cyclooctadiene, dimethylcyclooctadienes, and 6-methyl-

2,4,7-nonatriene from butadiene, isoprene, or 1,3-pentadiene, respectively (Eq. 18).

2 **43** $Cp^*RuBr(\eta^4$-butadiene)/AgOTf, THF, 20 °C, 10 h → **44** (89%) (18)

The mechanism of the homocoupling of dienes is one of the representative reactions proceeding through a π-allylruthenium intermediate. Indeed, a bis π-allylruthenium complex was produced by oxidative cyclization of two dienes and the coupling of the terminal carbon atoms led to a cationic (diene)(allyl)hydridoruthenium species.

5 Cross-Coupling of a C=C Bond with Allene

The synthesis of unsaturated compounds by C–C bond formation can also be carried out by coupling of alkenes with allenes, intermolecularly or intramolecularly. Thus, 1,3-dienes were selectively obtained by coupling of allenes and vinyl ketones [28–30]. The reaction was catalyzed by the complex CpRuCl(COD) and with $CeCl_3$ as a cocatalyst (Eq. 19). This cocatalyst is expected to decrease the chloride ion concentration to keep the active cationic ruthenium complex coordinatively unsaturated.

45 + **46** CpRu(COD)Cl/$CeCl_3$, DMF, 60 °C, 6 h → **47** (81%) (19)

48

A ruthenacyclopentane **48** has been proposed as an intermediate in this reaction, after coordination of the allene and enone. Exocyclic β-hydride elimination led to the 1,3-dienes. This ruthenacycle possessed a σ-bound ruthenium allyl, allowing nucleophilic additions by alcohols or amines. Alkylative cycloetherification [29] (Eq. 20) and synthesis of pyrrolidine and piperidine [30] were thus achieved.

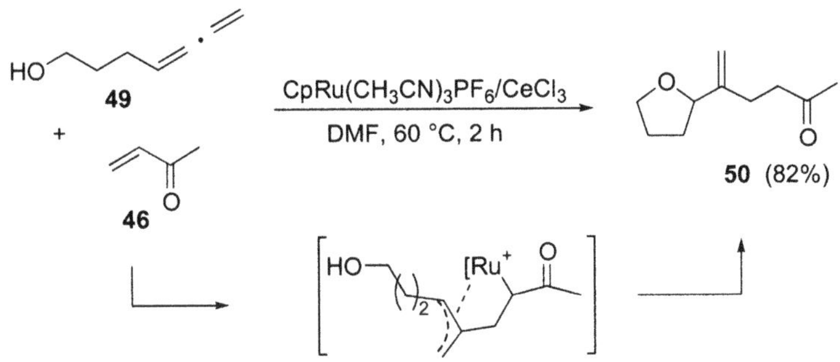

49 + 46 → (CpRu($CH_3CN)_3PF_6/CeCl_3$, DMF, 60 °C, 2 h) → 50 (82%) (20)

This reaction applied to allenylcyclobutanols allowed the synthesis of α-substituted cyclopentanones by ring expansion of the four-membered ring of the π-allylruthenium intermediate [31] (Eq. 21).

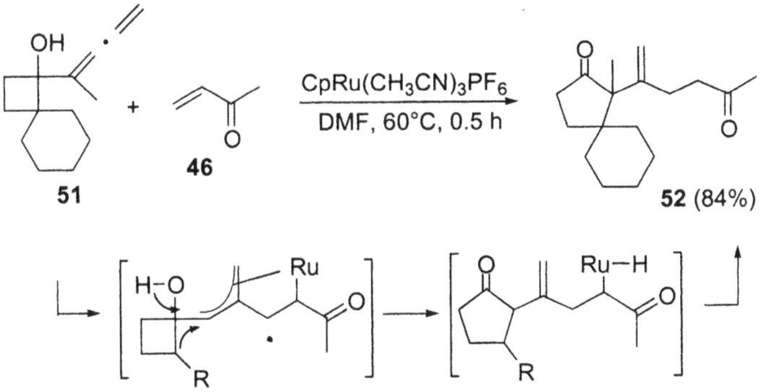

51 + 46 → (CpRu($CH_3CN)_3PF_6$, DMF, 60°C, 0.5 h) → 52 (84%) (21)

The ruthenium-catalyzed cycloisomerization of a variety of δ-enallenes was also achieved, forming cyclic 1,3-dienes or 1,4-dienes depending on the substrates and reaction conditions [32] (Eq. 22). This intramolecular coupling of the C=C bond and allenes can be envisioned by the initial hydrometallation of the allene moiety followed by intramolecular olefin insertion and isomerization.

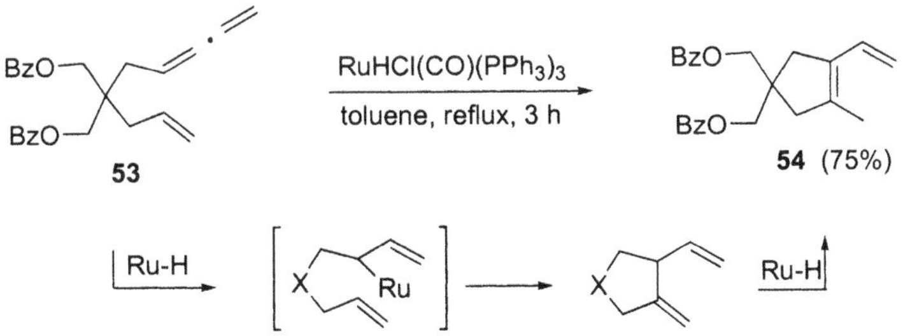

53 → (RuHCl(CO)($PPh_3)_3$, toluene, reflux, 3 h) → 54 (75%) (22)

6
Coupling Reactions of C=C and C≡C Bonds

The excellent coordination properties of alkynes with transition metals led to their use as partners for the coupling with a large variety of unsaturated molecules. Two partners such as alkynes and alkenes can produce various modes of C–C bond formation. Linear or cyclic couplings can occur via different pathways, similar to those reported for two C=C bonds couplings (Scheme 1).

6.1
Linear Intermolecular Couplings Involving Ruthenacycle Intermediates

One of the most reported pathways for C=C and C≡C bonds coupling involves the oxidative coupling and the ruthenacycle intermediate formation. The first ruthenium-catalyzed linear codimerization of disubstituted alkynes and alkenes involved acrylates or acrylamides and selectively produced 1,3-dienes [33] (Eq. 23). The proposed mechanism involves a ruthenacyclopentene via oxidative coupling on the Ru(0) catalyst Ru(COD)(COT). The formation of 1,3-diene results from intracyclic β-hydride elimination, this process taking place only when a favored exocyclic β-elimination is not possible.

PhC≡CPh (**55**) + CO_2Et (**29**) → Ru(COD)(COT), pyridine, 80 °C, 20 h → **56** (93%); intermediate: [Ru, Ph, Ph, H, EWG] (23)

For linear mixed coupling, ruthenium catalysts offer the possibility to develop the potential of the Alder-ene reaction by expanding its scope and selectivity, and the coupling usually takes place with atom economy. This C=C bond/C≡C bond coupling affords a wide variety of linear or branched 1,4-dienes depending on the nature of the substitutions of the starting materials and on the catalyst system based on a Cp ruthenium moiety [34–36]. Thus, pent-1-yne and hex-1-ene at room temperature with CpRuCl(COD)/NH_4PF_6 gave the branched 1,4-diene as the major product [35] (Eq. 24).

57 + **58** → CpRuCl(COD)/NH_4PF_6, DMF/H_2O:4/1, 100 °C, 2 h → **59a** + **59b** (24)

59 (56% - **a**/**b** : 84/16)

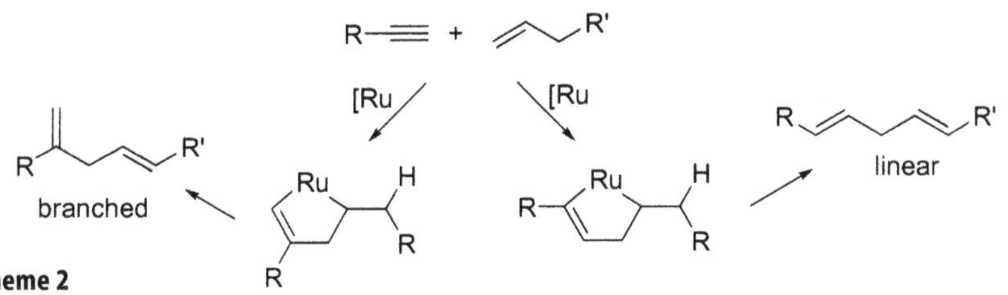

Scheme 2

The presence of two different isomers can be viewed through the competitive ruthenacycle formation, depending on the orientation of the alkyne via oxidative coupling. A β-hydride elimination, which is favored with H exocyclic with respect to intracyclic β-hydride, produces the 1,4-dienes (Scheme 2).

Using the complex $CpRu(CH_3CN)_3PF_6$ as a catalyst, reaction can proceed at room temperature in dimethylformamide (DMF) [36] (Eq. 25).

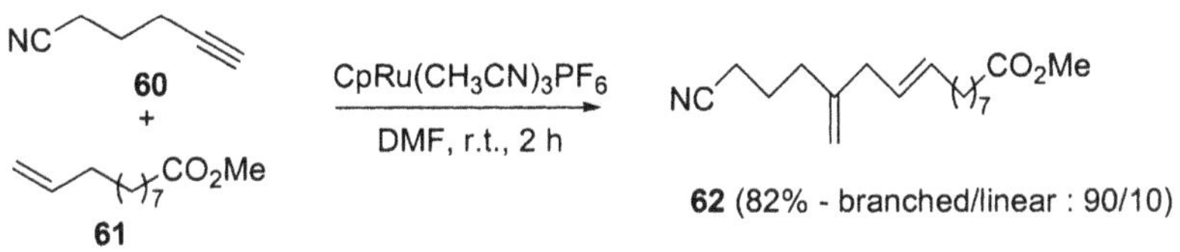

(25)

The regioselective preference for the formation of the branched product can be reversed by an increase of steric hindrance, especially at the propargylic position. Preferential formation of the linear isomer was also observed with 4-hydroxyalkynoates, allowing the synthesis of butenolides via cyclization [37] (Eq. 26).

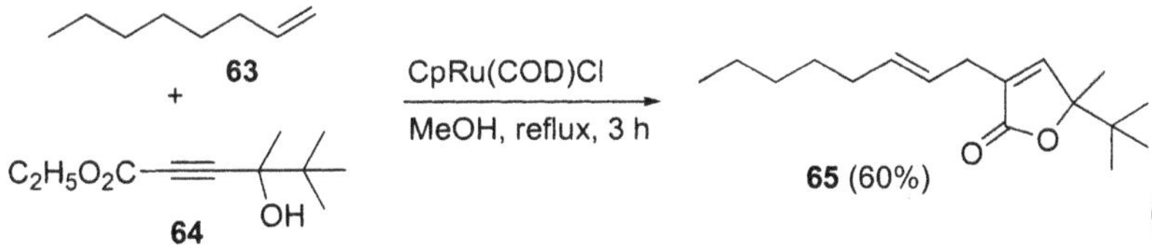

(26)

By contrast, the reaction of silylalkynes and terminal alkenes proceeded with complete control of regioselectivity by the silyl substituent to give only one isomer, similar to the branched isomer [38] (Eq. 27).

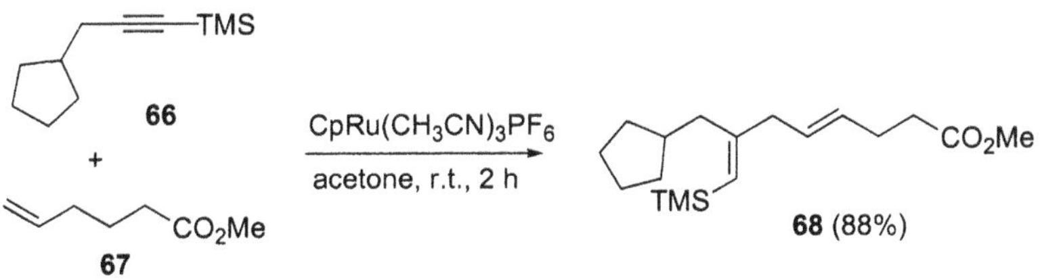

(27)

A similar mechanism, based on a ruthenacyclopentene, can be proposed for the coupling of alkynes and allylic alcohols to lead to γ,δ-unsaturated aldehydes and ketones. When $(C_5H_5)RuCl(COD)$ was used as a catalyst, the ruthenium-catalyzed coupling between alkynes and substituted allylic alcohols afforded γ,δ-unsaturated ketones. The linear isomer was the major product [39] (Eq. 28). Similarly, the linear derivative was also obtained when an allylsilylether or an allylic amide was used in place of the allyl alcohol, leading to 1,4-dienes [40].

C_6H_{13}—≡ **69** + **70** (OH) → CpRuCl(COD)/NH$_4$PF$_6$, DMF/H$_2$O, 100 °C, 2 h → **71a** + **71b** (28)

71 (60% - **a/b** : 70/30)

By using a more sterically hindered and electron-rich catalyst, $(C_5Me_5)RuCl(COD)$, γ,δ-unsaturated acetals and aldehydes have been synthesized from alkynes and allyl alcohol [41] (Eq. 29). It is noteworthy that in this case the branched isomer is the major product. This is due to the bulkiness of the C_5Me_5, with respect to the C_5H_5 ligand, which favors the head-to-tail coupling.

^{t}Bu—≡ **72** + **73** (OH) → Cp*RuCl(COD), neat, r.t., 15 min → ^{t}Bu…CHO **74** (60%) (29)

This reaction applied to propargyl alcohols selectively provided 2-alkoxy-5-methylenetetrahydropyranes in good yields by cyclization of the aldehyde [42] (Eq. 30).

75 + **73** (OH) → Cp*RuCl(COD), allyl alcohol, r.t., 0.5 h → **76** (80%) (30)

6.2 Intermolecular Coupling Involving Hydrometallation or C–H Bond Activation

1,3-Dienes have been synthesized by cross-coupling of alkenes and alkynes involving other types of mechanisms, such as initial hydrometallation or C–H bond activation.

Conjugated dienes were thus selectively obtained by hydrovinylation of alkynes catalyzed by a cationic ruthenium alkylidene complex [43] (Eq. 31). This reaction is thought to be promoted by the ruthenium hydride species resulting from the deprotonation of the δ-methyl group of the metallic precursor, followed by the sequential insertion of alkyne and ethylene into the metal–hydride and metal–vinyl bonds.

Ph—≡—Ph **55** + $H_2C{=}CH_2$ → [$(PCy_3)_2(CO)(Cl)Ru{=}CH\text{-}CH{=}C(CH_3)_2{}^+BF_4^-$, CH_2Cl_2, 75°C, 5 h] → **77** (90%) (31)

Ru-H → [Ph, Ph, Ru, H] →

Another important pathway to generate conjugated dienes from alkyne and alkene with a ruthenium catalyst is based on inert C–H bond activation. The addition of the β-CH bond of conjugated enones to internal alkynes gave conjugated dienones in good yields with the aid of $RuH_2(CO)(PPh_3)_3$ as a catalyst [44] (Eq. 32). The mechanism is initiated by C–H bond cleavage via chelation of the carbonyl group. The cis addition of the Ru–H bond to the alkyne leads to the corresponding conjugated dienones.

78 + Ph—≡—Ph **55** → [$RuH_2(CO)(PPh_3)_3$, toluene, reflux, 1 h] → **79** (71%) (32)

6.3 Intermolecular Coupling with Cycle Formation

The coupling between alkenes and alkynes can also afford cyclization reactions and leads to strained carbocycles. Most of these reactions are performed via a ruthenacycle intermediate leading to [2+2] cycloaddition.

One of the first examples of ruthenium-catalyzed C–C bond formation afforded the synthesis of cyclobutenes, from norbornene derivatives with dimethyl acetylenedicarboxylate, and was reported by Mitsudo and coworkers [45, 46] by using various catalysts such as $RuH_2(CO)[P(p\text{-}C_6H_4F)_3]_3$ or $RuH_2(PPh_3)_4$. More recently, the complex Cp*RuCl(COD) has shown to be an excellent catalyst for the [2+2] cycloaddition of norbornenes with various internal alkynes [45] (Eq. 33) and with a variety of substituted norbornenes and norbornadienes [47]. The ruthenacycle intermediate, formed by oxidative coupling, cannot undergo β-hydride elimination and leads to cyclobutene via a reductive elimination.

80 + Ph—≡—Ph **55** → [Cp*RuCl(COD); NEt_3, 80 °C, 15 h] → **81** (88%) (33)

This method was applied to the synthesis of a new range of rigid linear rods based on the [*n*]ladderanes [48] (Eq. 34).

82 + E—≡—E **83** → [$RuH_2(CO)(PAr_3)_3$; benzene, 50 °C, 5 days; E = CO_2Me] → **84** (52%) (34)

A cyclobutene was recently obtained by a related reaction of dimethyl acetylenedicarboxylate with ethylene in the presence of a cationic ruthenium alkylidene catalyst precursor [43] (Eq. 35).

MeO_2C—≡—CO_2Me **83** + $H_2C{=}CH_2$ → [$(PCy_3)_2(CO)(Cl)Ru{=}CH\text{-}CH{=}C(CH_3)_2{}^+BF_4^-$; CH_2Cl_2, 60 °C] → **85** (70%) (35)

Cyclobutanes are produced by using cyclooctadiene as an olefinic substrate via an unusual [4+2] cycloaddition. The mechanism is postulated to proceed through a ruthenacyclopentene which undergoes an intramolecular insertion reaction of the second C=C bond of the cyclooctadiene [49] (Eq. 36).

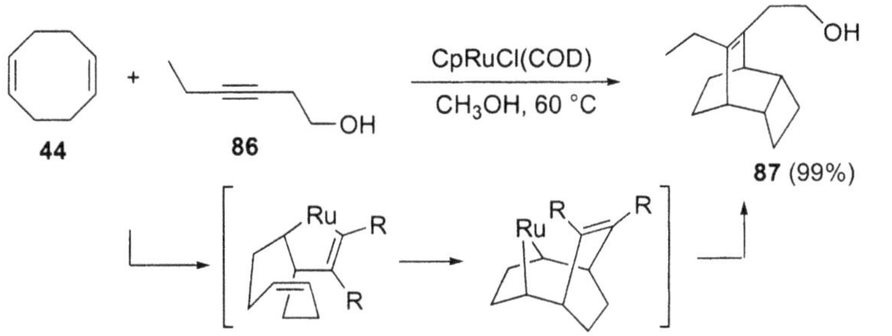

CpRuCl(COD)
CH_3OH, 60 °C

44 86 87 (99%) (36)

Unexpectedly, norbornene derivatives can undergo a novel cyclopropanation reaction with propargyl alcohol in the presence of cationic $[(\eta^5\text{-}Cp)(CH_3CN)_3Ru]^+X^-$ catalysts, which have an electron-withdrawing substituent on the Cp ligand. Cyclopropanation products, *exo*-acetyltricyclooctanes, were obtained in good yields [50] (Eq. 37). The reaction has been shown not to involve the expected allenylidene intermediate but rather to lead to a ruthenacycle intermediate and to a β-hydroxy elimination.

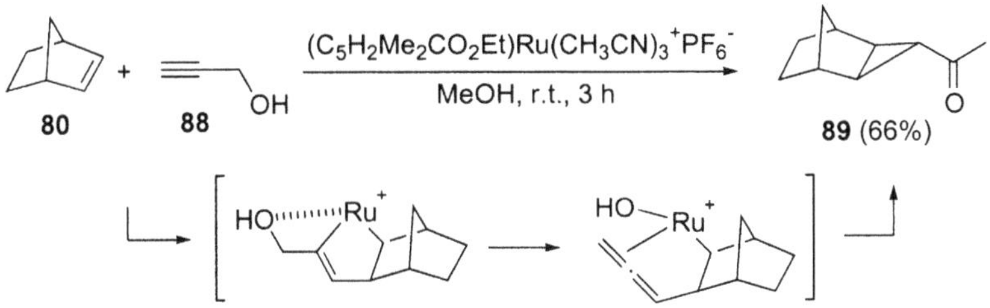

$(C_5H_2Me_2CO_2Et)Ru(CH_3CN)_3{}^+PF_6{}^-$
MeOH, r.t., 3 h

80 88 89 (66%) (37)

Recently, cyclopropane derivatives were produced by a ruthenium-catalyzed cyclopropanation of alkenes using propargylic carboxylates as precursors of vinylcarbenoids [51] (Eq. 38). The key intermediate of this reaction is a vinylcarbene complex generated by nucleophilic attack of the carboxylate to an internal carbon of alkyne activated by the ruthenium complex. Then, a [2+1] cycloaddition between alkenes and carbenoid species affords vinylcyclopropanes.

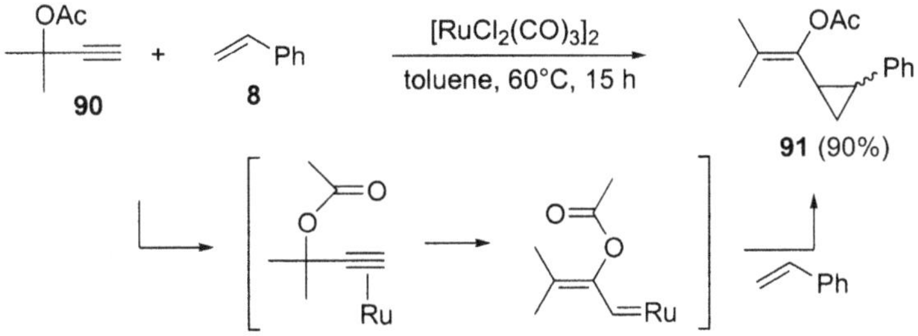

$[RuCl_2(CO)_3]_2$
toluene, 60°C, 15 h

90 8 91 (90%) (38)

Finally, benzenepolycarboxylates were obtained by ruthenium-catalyzed cross-benzannulation of acetylenedicarboxylates with allylic compounds [52] (Eq. 39). A ruthenacyclopentene is postulated to occur via oxidative coupling of one molecule of alkyne with allylic alcohol. Subsequent insertion of another molecule of alkyne gives the corresponding polysubstituted benzene derivatives.

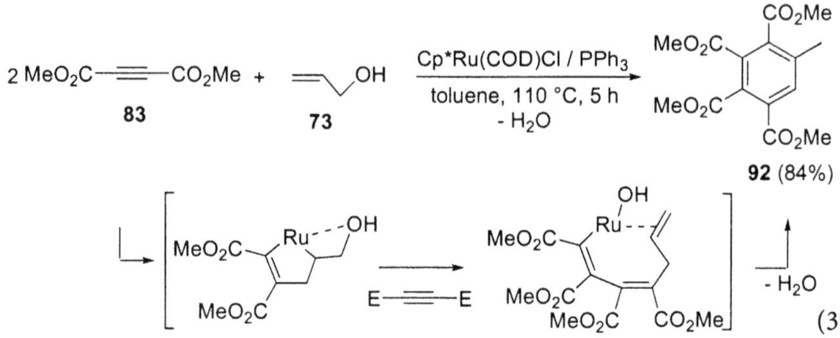

2 MeO_2C—≡—CO_2Me + ═╲╱OH (**83**, **73**) → Cp*Ru(COD)Cl / PPh_3; toluene, 110 °C, 5 h; - H_2O → **92** (84%)

(39)

6.4 Coupling of C=C and C≡C Bonds via Ruthenium Vinylidene Complexes

Ruthenium vinylidene complexes, arising from a 1,2-hydrogen shift of ruthenium alkyne complexes, are reactive intermediates toward nucleophiles owing to the enhanced electrophilicity of the Ru=C carbon atom, and have allowed the coupling between alkynes and alkenes. For example, the system $CpRu(PPh_3)_2Cl/NH_4PF_6$ catalyzed the addition of allylic alcohols to terminal alkynes, yielding β,γ-unsaturated ketones [53, 54], as in the following example [54] (Eq. 40).

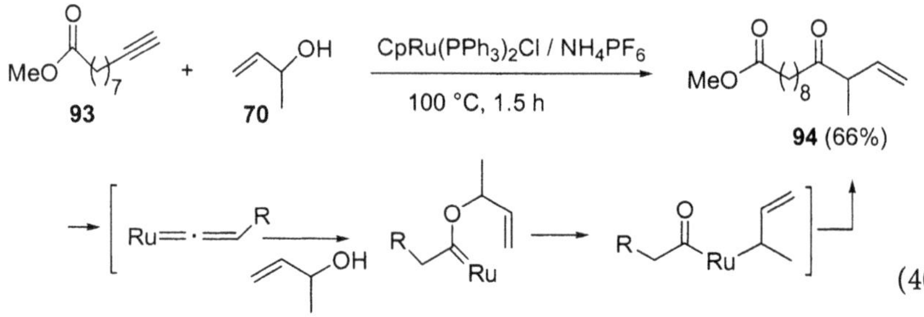

93 + **70** → $CpRu(PPh_3)_2Cl$ / NH_4PF_6; 100 °C, 1.5 h → **94** (66%)

(40)

Carbon nucleophiles can also add to in situ generated vinylidene species. Thus, ruthenium-catalyzed cyclizations of dienylalkynes produced arene derivatives [55] (Eq. 41).

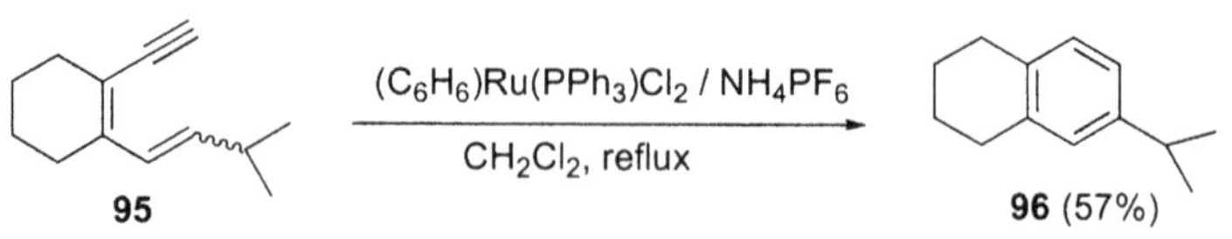

95 → $(C_6H_6)Ru(PPh_3)Cl_2$ / NH_4PF_6; CH_2Cl_2, reflux → **96** (57%)

(41)

Ruthenium vinylidene intermediates have also been proposed in the mechanism of the coupling of unactivated alkenes with terminal alkynes to afford 1,3-dienes as a mixture of two isomers, linear and branched derivatives. The linear one was favored [56] (Eq. 42). The same system has allowed the ruthenium-catalyzed alkenylation of pyridine [57].

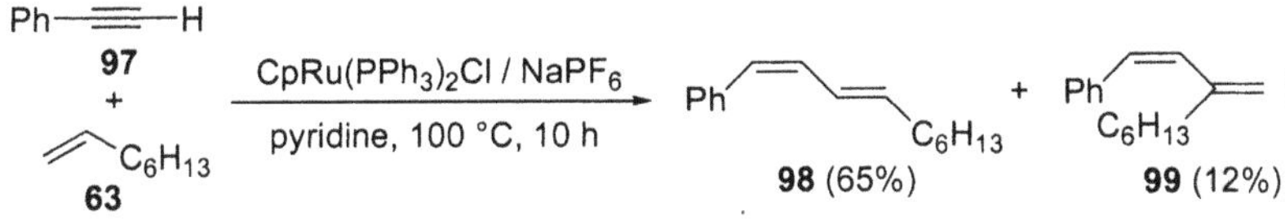

(42)

This mechanism and a large variety of applications are developed in detail in the chapter Ruthenium Vinylidenes and Allenylidenes in Catalysis.

6.5 C=C and C≡C Bond Couplings Involving Heteroatom Additions

The C–C bond formation can also be obtained via a first-step addition of a heteroatom to alkynes. Thus, the reaction of the three components terminal alkyne, water and enone led to 1,5-diketone with atom economy, using the system CpRuCl(COD)/NH_4PF_6 and $In(OSO_2CF_3)_3$ as a cocatalyst [58, 59] (Eq. 43). The mechanism is postulated to proceed by the ruthenium-catalyzed nucleophilic addition of water to alkynes to generate a ruthenium enolate intermediate allowing further insertion of enone and formation of 1,5-diketones after protonation.

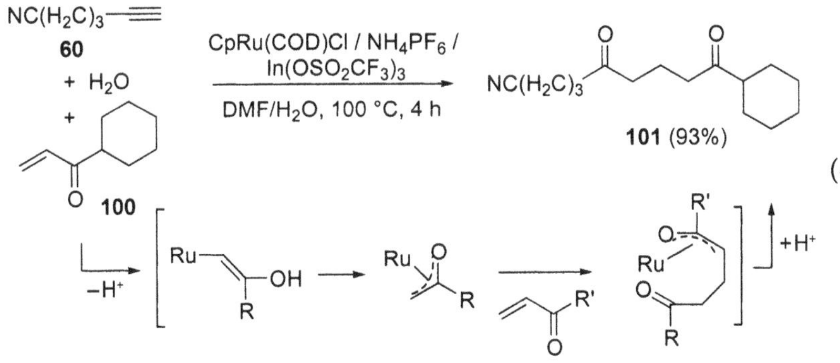

(43)

Enediones were obtained when the same reaction was performed with propargyl alcohols [59] (Eq. 44).

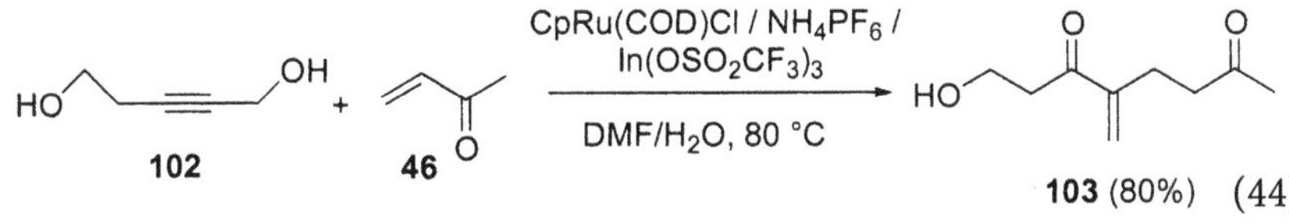

(44)

In both cases, a ruthenacycle intermediate cannot be ruled out. Furthermore, an intramolecular version from yne-enones was carried out and the formation of the products seemed to involve a ruthenacycle intermediate (see Eq. 56).

A related three-component coupling has also been performed with a halide as a nucleophile to lead to (*E*)- or (*Z*)-vinyl halides, depending on the conditions and on the substitution of the alkynes [60–62]. Indeed, (*E*)-vinyl chlorides were preferentially obtained, from a large variety of alkynes and enones, by using CpRuCl(COD) with stannic chloride as a cocatalyst in a polar solvent such as DMF and with ammonium chloride salts [62] (Eq. 45).

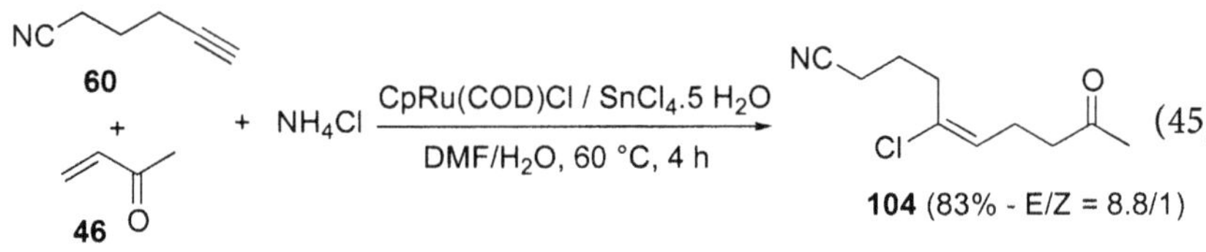

NC **60** + **46** + NH_4Cl —[CpRu(COD)Cl / $SnCl_4.5\ H_2O$; DMF/H_2O, 60 °C, 4 h]→ **104** (83% - E/Z = 8.8/1) (45)

On the other hand, the catalytic system $CpRu(CH_3CN)_3PF_6/SnBr_4$ in the presence of lithium bromide in a less polar solvent such as acetone led to the favored formation of (*Z*)-vinyl bromides. Interestingly, when alkynes with a quaternary propargylic carbon or aryl acetylenes were used, complete selectivity for the (*Z*) isomer was obtained [60] (Eq. 46).

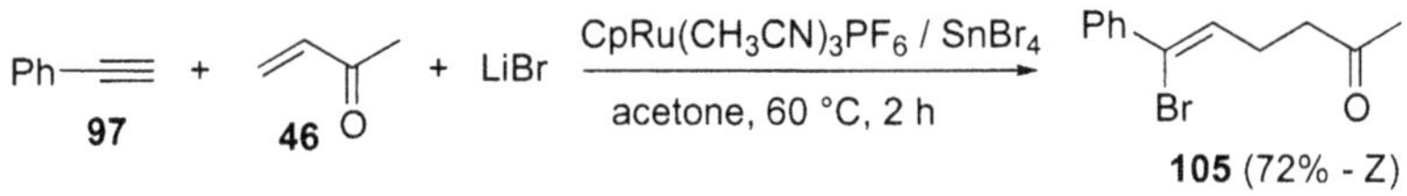

Ph **97** + **46** + LiBr —[$CpRu(CH_3CN)_3PF_6$ / $SnBr_4$; acetone, 60 °C, 2 h]→ **105** (72% - Z) (46)

The *Z* or *E* selectivity may arise from a trans or a cis haloruthenation, depending on the η^2-alkyne ruthenium halide complex, more ionic or more covalent species, this equilibrium being displaced by the reaction conditions (Scheme 3).

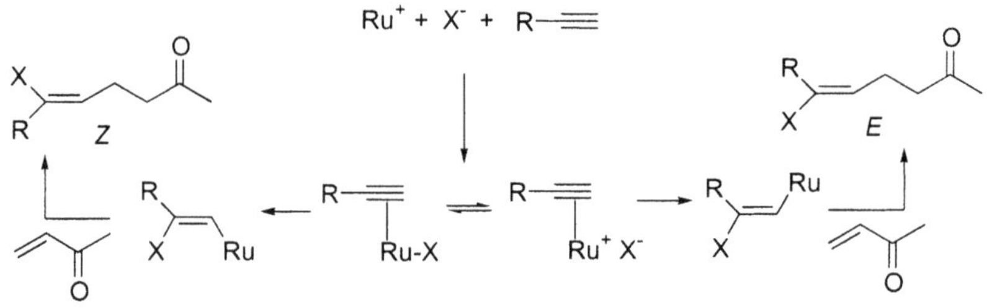

$Ru^+ + X^- + R$—≡

Scheme 3

Access to (*Z*)-vinyl bromides allowed an efficient cyclopentenone synthesis and their application to the formation of cyclopentanoid natural products such as rosaprostol or a selective cox-2 inhibitor [63] (Eq. 47).

A ruthenium-catalyzed four-component combination was also achieved when an aldehyde was added to a mixture of alkyne, enone, and halide anion.

106 + 107 + LiBr — $CpRu(CH_3CN)_3PF_6$ / $SnBr_4$, acetone, 60 °C → 108 (64%)

108 — $CrCl_2$, $NiCl_2$ → → COX-2 inhibitor (47)

New vinyl chlorides or bromides were thus obtained by trapping ruthenium enolate in an aldol reaction [64] (Eq. 48).

60 + $Me_4N^+Cl^-$ + 109 — $CpRu(CH_3CN)_3PF_6/SnCl_4.5\ H_2O$, DMF, 60 °C → 110 (58%) (48)

6.6 Enyne Cycloisomerization

The intramolecular C=C/C≡C bond coupling of enynes generally leads to the formation of conjugated alkenylcycloalkenes. The first ruthenium-catalyzed enyne cycloisomerization of 1,6-enynes into 1-vinylcycloalkenes was reported by Chatani et al. [65], using $[RuCl_2(CO)_3]_2$ as a catalyst, under an atmosphere of CO (Eq. 49). A ruthenacyclopentene has been postulated as an intermediate via oxidative coupling. A conrotary cycloreversion of cyclobutene led to vinylcycloalkene. It is noteworthy that the same product is formed in the presence of the alkene metathesis catalyst $RuCl_2(=CHPh)(PCy_3)_2$ [66].

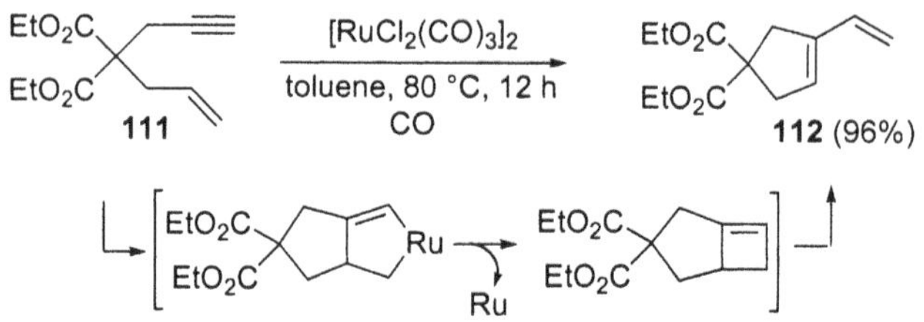

(49)

The reaction of 6,11-dien-1-yne in the presence of the same catalyst gave a tetracyclic compound in 84% yield [67] (Eq. 50). The same ruthenacyclopentene is involved but the reaction is rationalized by the intermediacy of a carbenoid intermediate which undergoes intramolecular cyclopropanation. Alternatively, a polarized η^1-alkyne complex bearing a positive charge at the β-position can also be envisioned.

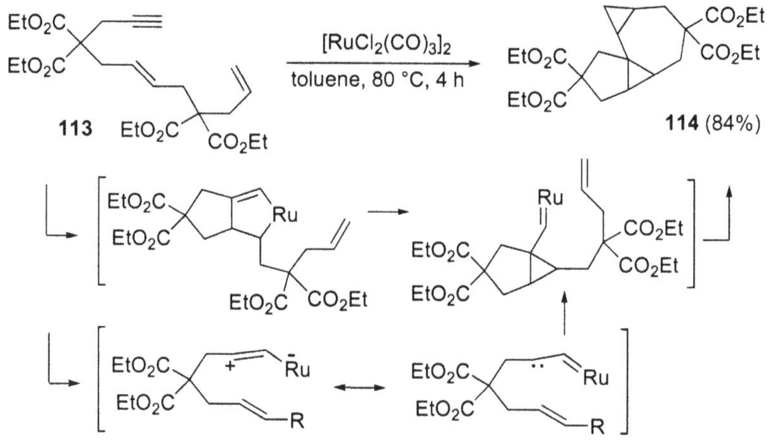

(50)

The catalyst $[RuCl_2(CO)_3]_2$ also promotes the electrophilic activation of the C≡CH bond of ω-arylalk-1-ynes. The intramolecular cycloisomerization takes place with nucleophilic addition of the aryl group to the activated β-carbon of the alkyne bond, thus eliminating a vinylidene intermediate [68].

Besides enyne metathesis [66] (see also the chapter Recent Advances in Alkenes Metathesis in this volume), which generally produces 1-vinylcycloalkenes, ruthenium-catalyzed enyne cycloisomerization can proceed by two major pathways via hydrometallation or a ruthenacycle intermediate. The $RuClH(CO)(PPh_3)_3$ complex catalyzed the cyclization of 1,5- and 1,6-enynes with an electron-withdrawing group on the alkene to give cyclized 1,3-dienes, dialkylidenecyclopentanes (for n=2), or alkylidenecyclopentenes (for n=1) [69, 70] (Eq. 51). Hydroruthenation of the alkyne can give two vinylruthenium complexes which can undergo intramolecular alkene insertion into the Ru–C bond.

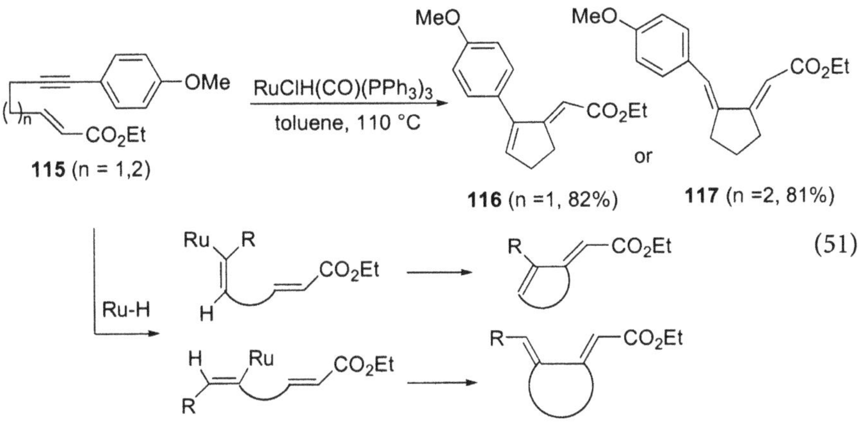

115 (n = 1,2) 116 (n = 1, 82%) 117 (n = 2, 81%)

(51)

This reaction has been applied to the synthesis of carbapenam skeletons [69] (Eq. 52).

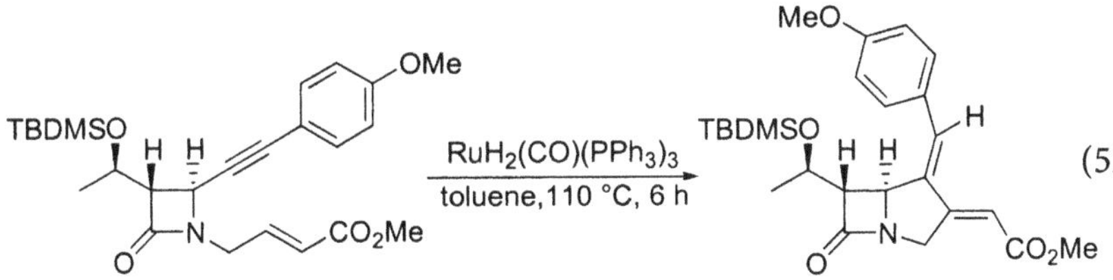

118 119 (63%)

(52)

Cycloisomerization of allyl propargyl ethers takes place under mild conditions when performed with a ruthenium hydride catalyst obtained from Cp*RuCl(COD) in the presence of acetic acid or ethanol. 3,4-Dialkylidenetetrahydrofurans were produced in good yields [71] (Eq. 53). It is noteworthy that the C–H and C–C bonds formed are always syn. This is the result of the cis addition of the Ru–H bond to the C≡C bond.

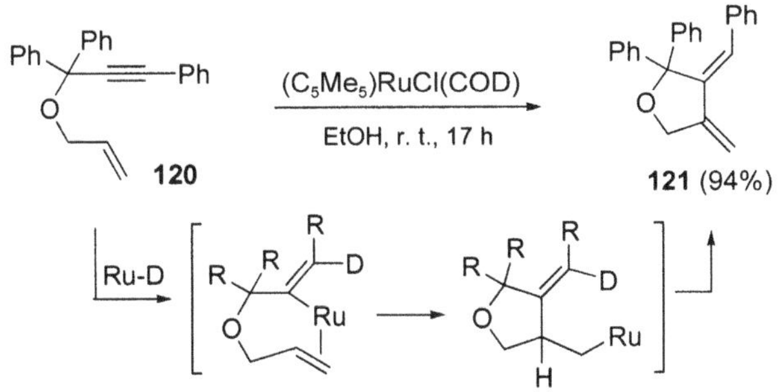

120 121 (94%)

(53)

Ruthenacyclopentenes have also been proposed as intermediates in the intramolecular coupling reaction of the C=C bond and the C≡C bond. Thus, the complex $CpRu(CH_3CN)_3PF_6$ catalyzed the cycloisomerization of a variety of

1,6- and 1,7-enynes to lead to five- or six-membered rings [72,73] (Eq. 54). Similar cyclized products were also obtained by using the complexes $RuCl_3$ or $Ru(AsPh_3)_4Cl_2$ in methanol [74].

MeO_2C MeO_2C **122** — $CpRu(CH_3CN)_3PF_6$, DMF, r. t., 1h → MeO_2C MeO_2C **123** (82%)

[MeO_2C MeO_2C Ru H] (54)

When the double bond of the enyne possesses a cyclopropyl substituent, an intramolecular [5+2] cycloaddition of alkyne and vinylcyclopropane takes place [75, 76]. The ruthenacycle does not undergo β-hydride elimination but a rearrangement of the cyclopropane to produce a ruthenacyclooctadiene. Thus, a variety of bicyclic and tricyclic cycloheptadienes were obtained in good yields [75] (Eq. 55).

MeO_2C MeO_2C H **124** $CH_2OTBDMS$ — $CpRu(CH_3CN)_3PF_6$, acetone, r. t., 6 h → MeO_2C MeO_2C H H $CH_2OTBDMS$ **125** (93%) (55)

[MeO_2C MeO_2C Ru TBDMSO → MeO_2C MeO_2C Ru OTBDMS]

When the double bond is conjugated with a carbonyl group, a different type of product is formed. In the presence of water a hydrative cyclization was performed by reaction of yne-enones, leading to cyclic 1,5-diketones in good yields by using the system $CpRu(CH_3CN)_3PF_6$/camphorsulfonic acid as a catalyst. In anhydrous acetone, pyrans were produced in place of 1,5-diketones [77] (Eq. 56).

$CpRu(CH_3CN)_3PF_6$/CSA, acetone, H_2O, r. t., 12 h → **127** (70%)

$CpRu(CH_3CN)_3PF_6$, acetone, r. t., 4 h → **128** (89%)

126 (56)

When enyne cycloisomerization takes place in the presence of an unsaturated molecule an insertion reaction can occur. Thus, $Ru_3(CO)_{12}$ catalyzes the cycloisomerization of 1,6-enynes under a CO atmosphere to give an insertion of carbon monoxide and the formation of bicyclic cyclopentenones as a catalytic Pauson–Khand reaction [78] (Eq. 57).

129 $Ru_3(CO)_{12}$, dioxane, 160 °C, 20 h, CO (10 atm) → **130** (95%) (57)

The ruthenacyclopentene intermediate can also undergo insertion of ethylene to give a ruthenacycloheptene. Subsequent unexpectedly observed β-hydride elimination occurred and led then to cyclization products with a propenylidene substituent [79] (Eq. 58). Various enynes, with substituents on triple or double bonds, have been cyclized to form carbocyclic and heterocyclic compounds in good yields.

131 + $CH_2{=}CH_2$ Cp*Ru(COD)Cl, toluene, r. t., 22 h → **132** (90%) (58)

Most of the 1,6-enyne cycloisomerizations reported here lead to five-membered rings. However, when the enyne was substituted with a quaternary propargylic center and with an ester group on the triple bond, seven-membered rings were obtained [72, 80] (Eq. 59). In this case, a mechanism via a π-allylruthenium intermediate is proposed.

TBDMSO ... CO$_2$Et (**133**) $\xrightarrow[\text{DMF, r. t., 4 h}]{CpRu(CH_3CN)_3PF_6}$ **134** (83%) (59)

[TBDMSO ... CO$_2$Et, Ru$^+$–H → TBDMSO ... Ru$^+$–H, CO$_2$Et]

Finally, ruthenium-catalyzed carbocyclization by intramolecular reaction of allylsilanes and allylstannanes with alkynes also led to the formation of vinylalkylidenecyclopentanes [81] (Eq. 60). This reaction is catalyzed by $RuCl_3$ or $CpRuCl(PPh_3)_2/NH_4PF_6$ in methanol. The postulated mechanism involves the coordination of the alkyne on the ruthenium center to form an electrophilic η^2-alkyne complex. This complex can thus promote the nucleophilic addition of the allylsilane or stannane double bond.

PhO$_2$S, PhO$_2$S ... SiMe$_3$ (**135**) $\xrightarrow[\text{MeOH, 60 °C, 17 h}]{CpRu(PPh_3)_2Cl / NaPF_6}$ PhO$_2$S, PhO$_2$S (**136** (92%)) (60)

[Ru$^+$, PhO$_2$S, PhO$_2$S, SiMe$_3$]

7
Cross-Coupling of C≡C Bonds and Dienes

The catalytic coupling of C≡C bonds with C=C bonds of 1,3-dienes or 1,2-dienes has been performed. Terminal alkynes reacted with a range of 1,3-dienes in the presence of a catalytic amount of $RuH_2(PBu_3)_4$, $RuH_2(PEt_3)_4$ or $Ru(COD)(COT)/PBu_3$ to give linear conjugated or nonconjugated enynes, with high regioselectivity [82] (Eq. 61). The mechanism is based on the C–H bond activation of terminal alkynes to afford reactive hydrido alkynyl complexes. These intermediates can insert dienes.

$CH_3(CH_2)_5$–≡ **69** + **43** $\xrightarrow[\text{benzene, 80 °C, 2 h}]{RuH_2(PBu_3)_4}$ $CH_3(CH_2)_5$–≡– **137** (88%)

[Ru(H)–≡–R] → [Ru–H …≡–R] (61)

On reaction with hydroxyallenes, a variety of terminal alkynes led selectively to 1,3-disubstituted conjugated enynes in good yields in the presence of the catalyst $RuH_2(PPh_3)_4$ and the ligand 1,1′-bis(di(*p*-methylphenyl)phosphino)ferrocene [83] (Eq. 62). The selectivity may result from the interaction of the allene hydroxyl group with the ruthenium catalyst.

138 + ≡–Ph **97** $\xrightarrow[\text{cyclohexane, reflux, 3 h}]{RuH_2(PPh_3)_4/\text{ dppf}}$ **139** (73%) (62)

8
C≡C/C≡C Bond Coupling

Ruthenium-catalyzed activation of alkynes can lead to the formation of C–C bonds between two C≡C bonds by a variety of pathways.

8.1
Intermolecular Coupling of Alkynes

The first example involves the dimerization of terminal alkynes. It takes place via initial activation of the alkyne C–H bond, but several examples involve a vinylidene intermediate. In most cases, conjugated enynes are obtained by ruthenium-catalyzed tail-to-tail dimerization [84, 85], as in the following example [85] (Eq. 63).

2 Ph–≡ **97** $\xrightarrow[\text{toluene, reflux, 20 h}]{TpRu(PPh_3)_2Cl}$ Ph–≡–CH=CH–Ph **140** (91%) (63)

Butatrienes were formed when *tert*-butylacetylene was used in the presence of $RuH_2(CO)(PPh_3)_3$ [86] (Eq. 64).

2 ^{t}Bu–≡ **72** $\xrightarrow[\text{benzene, 100 °C, 10h}]{Ru(H)_2(CO)(PPh_3)_3}$ ^{t}Bu(H)C=C=C=C(H)^{t}Bu **141** (51%) (64)

These reactions constitute only a few examples of ruthenium vinylidene in catalysis, a topic of increasing importance in organic synthesis that is presented in the chapter Ruthenium Vinylidenes and Allenylidenes in Catalysis of this volume.

Interestingly, the in situ generated ruthenium acetylide complex $C_5Me_5(PPh_3)Ru{-}C{\equiv}CPh$ catalyzed the cross-coupling reaction of terminal and internal alkynes to yield functionalized enynes [87] (Eq. 65). A coordinatively unsaturated enynyl complex is postulated as an intermediate in this mechanism.

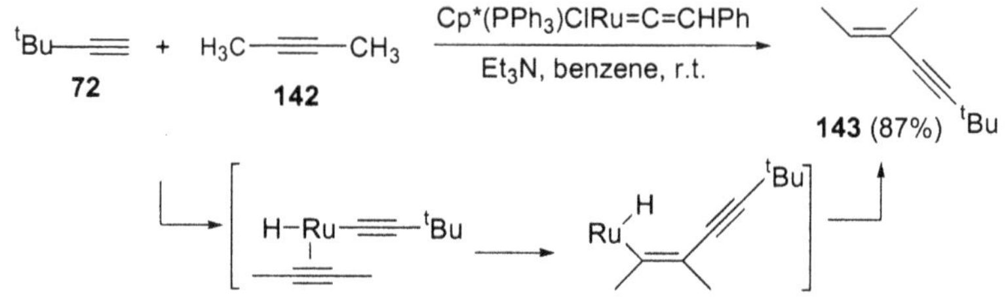

(65)

The linear coupling reaction of acetylene and acrylonitrile afforded 2,4,6-heptatrienenitrile by dimerization of acetylene and insertion of one molecule of acrylonitrile [88] (Eq. 66). The reaction involves the formation of a ruthenacyclopentadiene complex, which also catalyzed the reaction.

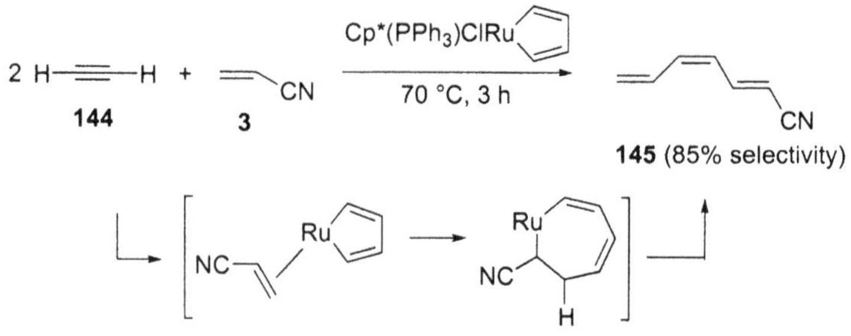

(66)

The precatalyst Cp*RuCl(COD) allowed the head-to-head oxidative dimerization of terminal alkynes and the concomitant 1,4-addition of carboxylic acid to stereoselectively afford 1-acyloxy-1,3-dienes in one step under mild conditions [89] (Eqs. 67, 68). The first step of the reaction consists in the oxidative head-to-head alkyne coupling via the formation of a ruthenacycle intermediate that behaves as a mixed Fischer–Schrock-type biscarbene ruthenium complex, allowing protonation and nucleophilic addition of the carboxylate.

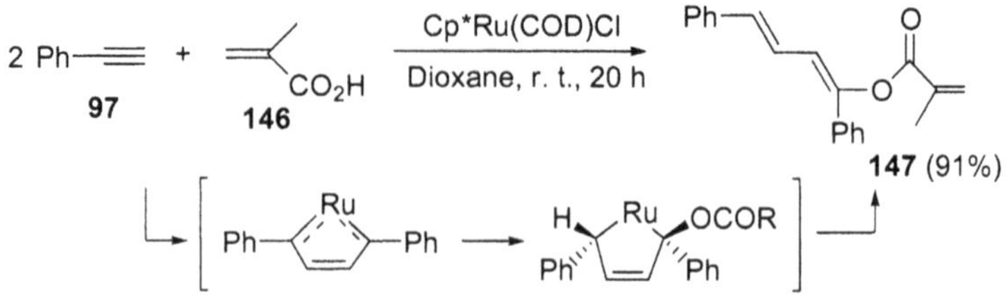

(67)

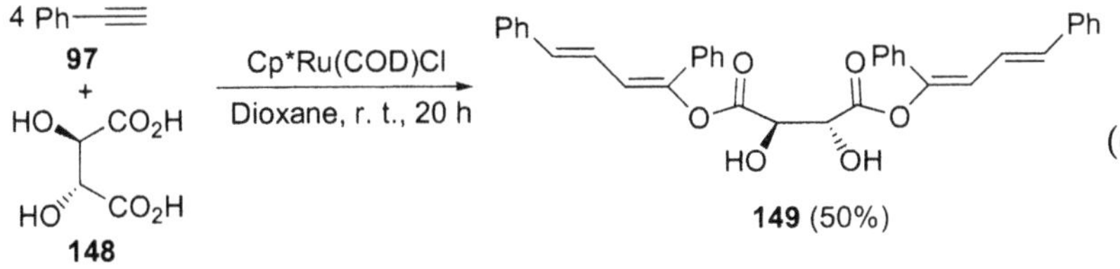

(68)

A similar reaction applied to propargyl alcohols in place of terminal alkynes led to the one-step catalytic head-to-head cyclodimerization of propargyl alcohols and to the formation of alkylidenecyclobutene derivatives [90] (Eq. 69). It was shown that the reaction occurs via cyclobutadieneruthenium and cyclobutenylruthenium intermediates, dehydration and carboxylate addition.

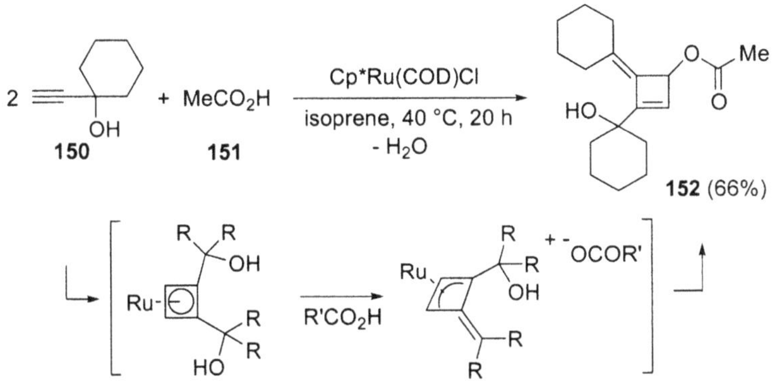

(69)

$CpRu(CH_3CN)_3PF_6$-catalyzed linear dimerization of propargyl alcohols was also carried out in the presence of water and produced conjugated dienones in good yields via a head-to-head oxidative coupling, followed by successive dehydration and addition of water [91] (Eq. 70).

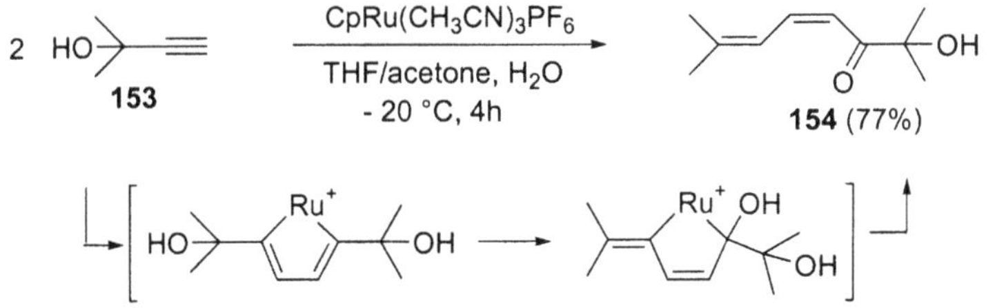

(70)

8.2 Intramolecular Coupling of Diynes

Ruthenium-catalyzed reactions involving diynes generally lead to the intramolecular oxidative coupling of the two C≡C bonds. Bicyclic compounds can be synthesized in the presence of another unsaturated molecule.

The reaction of 1,6-heptadiynes with alkenes led to a [2+2+2] cyclotrimerization in the case of cyclic or linear alkenes possessing heteroatoms at the allylic position. Bicyclic cyclohexadienes were thus produced in good yields with $RuCl(COD)C_5Me_5$ [92,93] (Eq. 71). A ruthenacyclopentadiene is invoked as an intermediate in the mechanism. Insertion of the alkene becomes possible by a heteroatom-assisted reaction.

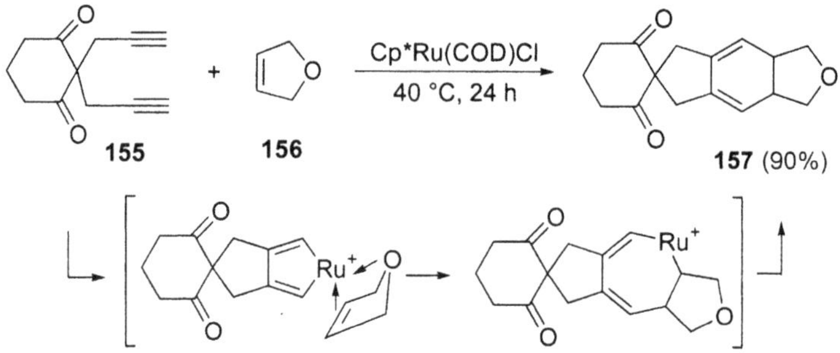

155 + 156 —Cp*Ru(COD)Cl, 40 °C, 24 h→ 157 (90%) (71)

With strained bicycloalkenes such as norbornene derivatives a ruthenium-catalyzed tandem cyclopropanation occurred together with common [2+2+2] cyclotrimerization, showing a biscarbenoid hybride structure for the ruthenacyclopentadiene intermediate [92] (Eq. 72).

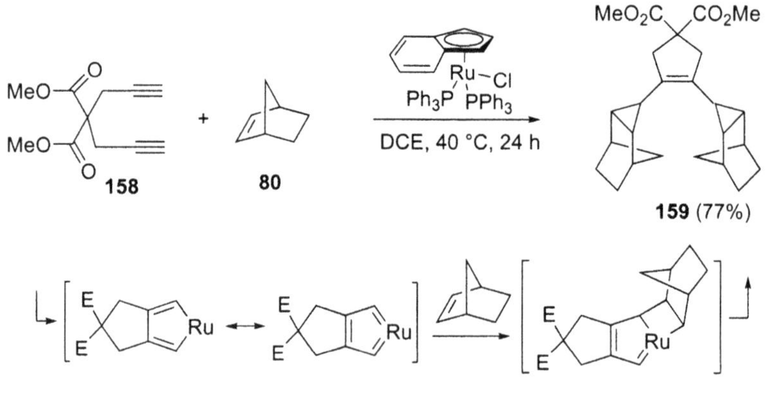

158 + 80 —Ru(indenyl)Cl(PPh_3)$_2$, DCE, 40 °C, 24 h→ 159 (77%) (72)

Recently, a formal ruthenium-catalyzed [4+2+2] cycloaddition of 1,6-diynes to 1,3-dienes gave conjugated 1,3,5-cyclooctatrienes and vinylcyclohexadienes [94] (Eq. 73). Insertion of a double bond in the ruthenacyclopentadiene can lead to the formation of tetraenes or vinyltrienes which undergo a thermal electrocyclization.

MeO, MeO, O, O **158** + **160** → $Cp^*Ru(CH_3CN)_3PF_6$, Et_4NCl, DMF, 80 °C, 4 h → MeO_2C, MeO_2C **161** + MeO_2C, MeO_2C **162**

161 + **162** (83% - 4.8/1)

[Ru intermediates: H, X, Ru, or] → [X, or, X] → △

$X = C(CO_2Me)_2$ (73)

When the [2+2+2] cycloaddition of 1,6-diynes occurred in the presence of C≡C bonds or C≡N bonds, substituted benzenes or pyridines were obtained in good yields [95–98]. For example, anthraquinones were produced by reaction of 1,2-bis(propiolyl)benzenes with a variety of monoalkynes [95] (Eq. 74).

163 + **144** (H, H) → Cp*Ru(COD)Cl, DCE, r. t., 0.5 h → **164** (92%) (74)

Bicyclic pyridines were regioselectively formed by reaction with electron-deficient nitriles [97] or dicyanides [98] (Eq. 75).

165 (MeO, MeO, O, O) + **166** ($H_2C(CN)_2$) → Cp*Ru(COD)Cl, DCE, 60 °C, 5 h → **167** (97%) (MeO_2C, MeO_2C, N, CN) (75)

The ruthenium complex Cp*RuCl(COD) catalyzed the [2+2+2] cycloaddition of 1,6-diynes with heterocumulenes such as isocyanates, isothiocyanates, or carbon disulfide [99, 100]. Bicyclic pyridones [99] and bicyclic thiopyrans [100] were thus obtained (Eq. 76).

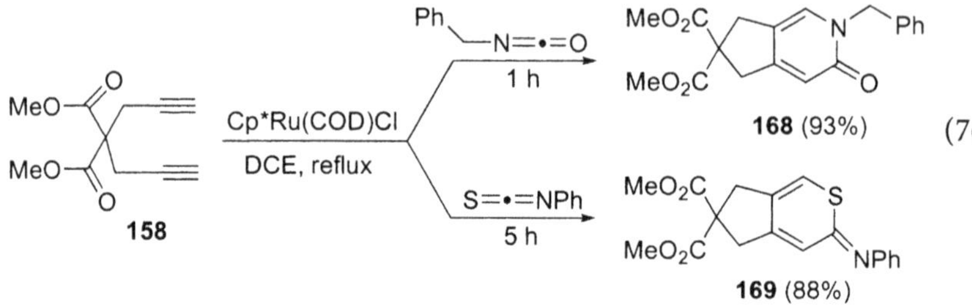

(76)

The formation of various heterocycles is developed in a specific chapter of this volume: Ruthenium-Catalyzed Synthesis of Heterocyclic Compounds.

The ruthenium-catalyzed [2+2+2] cycloaddition of 1,6-diynes was performed with an electron-deficient carbonyl double bond, activated with two electron-withdrawing groups, to produce conjugated dienones via electrocyclic ring opening of the expected cycloadduct [101] (Eq. 77).

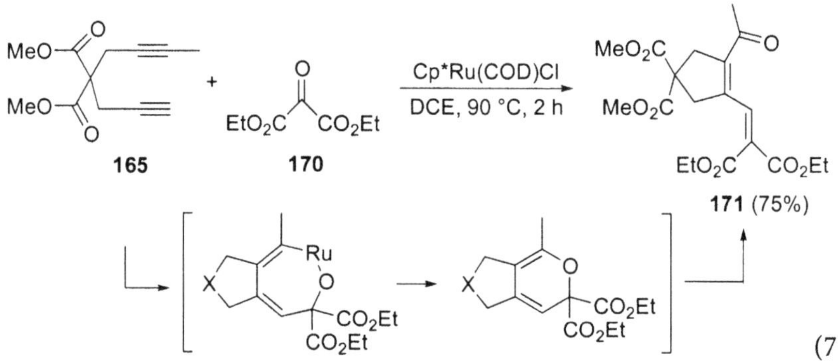

(77)

Conjugated dienones were also obtained by cycloisomerization of alkynes and propargyl alcohols by using the complex $CpRu(CH_3CN)_3PF_6$ as a catalyst in the presence of water [102, 103] (Eq. 78). The ruthenacyclopentadiene intermediate undergoes an elimination of the hydroxy group and adds water at the resulting carbene carbon.

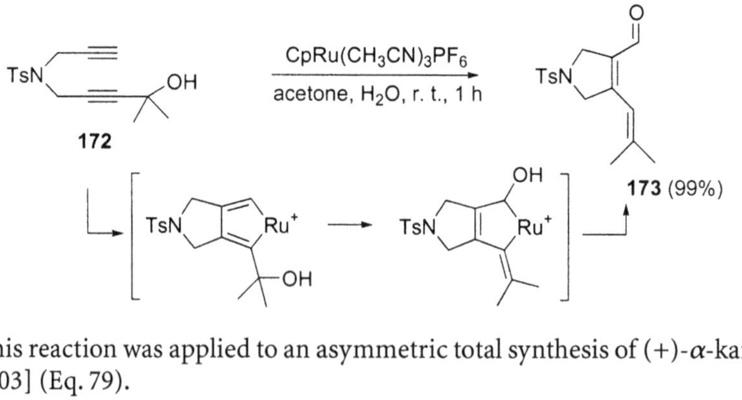

(78)

This reaction was applied to an asymmetric total synthesis of (+)-α-kainic acid [103] (Eq. 79).

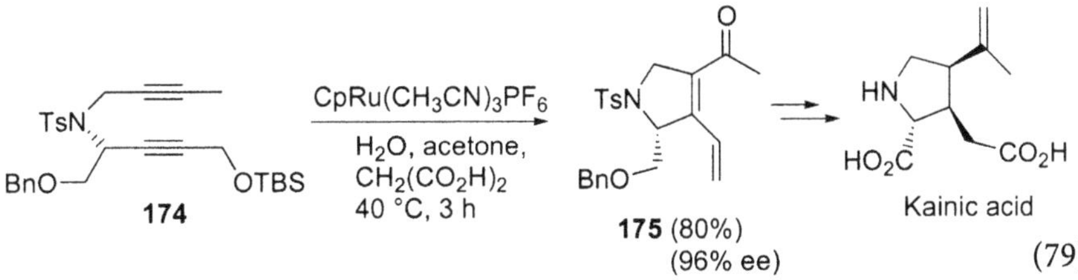

(79)

On the other hand, cyclization of diynes separated with a long chain catalyzed by thiolate-bridged diruthenium complexes led to endo-macrocyclic (*Z*)-conjugated enynes with 10–16-membered rings [104] (Eq. 80). A butenynyl intermediate is invoked in this mechanism, probably via a vinylidene intermediate.

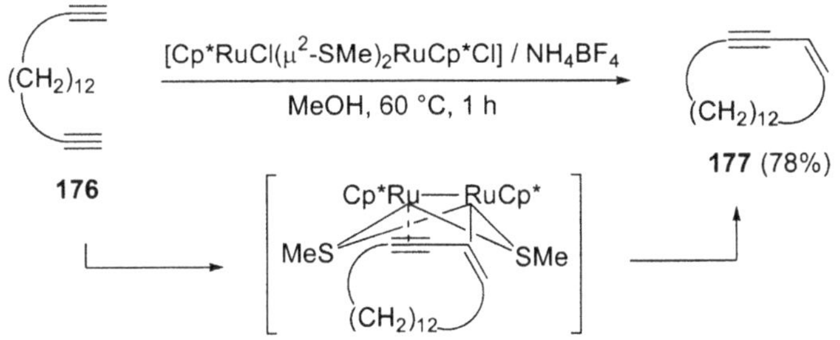

(80)

9 Addition of Diazo Compounds

Addition of diazo compounds to metallic complexes allows the formation of metal carbenoid species which can react with unsaturated molecules to form C–C or C=C bonds.

9.1 Addition to Alkenes

Reaction of a ruthenium carbenoid species, formed from a diazo compound, with an alkene produces cyclopropanes. A variety of ruthenium catalysts, notably chiral catalysts, have been developed to lead to efficient asymmetric cyclopropanation of alkenes as in the following example [105] (Eq. 81).

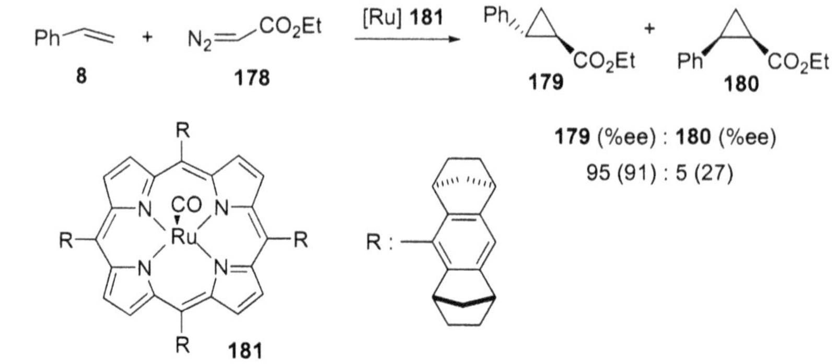

179 (%ee) : 180 (%ee)
95 (91) : 5 (27)

(81)

This reaction is developed in the chapter Cyclopropanation with Ruthenium Catalysts.

9.2 Addition to Alkynes

The ruthenium-catalyzed addition of diazo compounds to alkynes has led to the selective synthesis of functional 1,3-dienes by the combination of two molecules of diazoalkane and one of alkyne [106] (Eqs. 82, 83). The stereoselective formation of these conjugated dienes results from the selective creation of two C=C double bonds rather than leading to the cyclopropene derivative. This is expected to be due to the possibility for the C_5Me_5RuCl moiety to accomodate two cis carbene ligands.

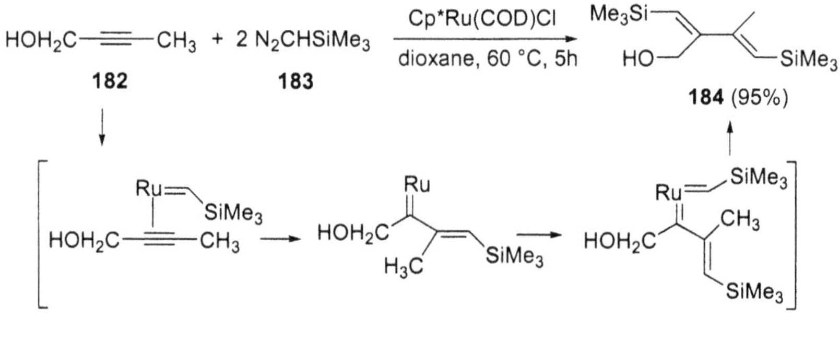

(82)

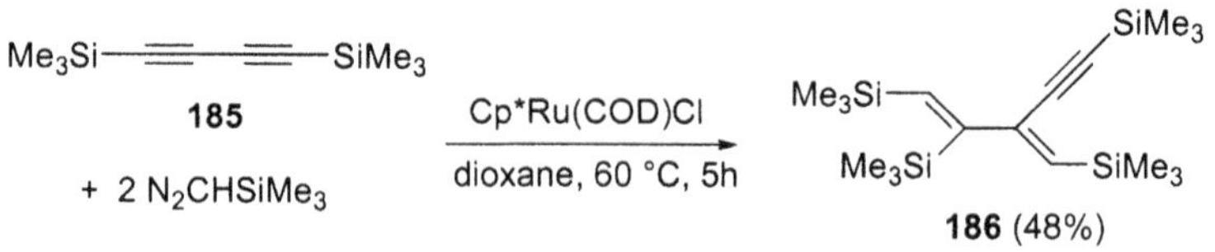

(83)

This reaction applied to enynes allowed the one-step selective synthesis of alkenylbicyclo[3.1.0]hexane derivatives by addition of one equivalent of diazoalkane to enyne. This novel reaction involves the stereoselective formation of three C–C bonds including a cyclopropanation step [107] (Eq. 84). It is noteworthy that the bulky Cp*Ru moiety favors reductive elimination and formation of a cyclopropane derivative with respect to the metathesis reaction.

TsN **187** + $N_2CHSiMe_3$ —[Cp*Ru(COD)Cl; dioxane, 60 °C, 1.5 h]→ **188** (85%) (84)

10
Allylic Alkylation Reaction

Several ruthenium catalysts have been tuned in order to perform catalytic allylation of nucleophiles, especially as an attempt to favor the nucleophilic addition on the substituted allylcarbon in order to create chiral molecules.

The catalytic activation of allylic carbonates for the alkylation of soft carbonucleophiles was first carried out with ruthenium hydride catalysts such as $RuH_2(PPh_3)_4$ [108] and Ru(COD)(COT) [109]. The efficiency of the cyclopentadienyl ruthenium complexes CpRu(COD)Cl [110] and Cp*Ru(amidinate) [111] was recently shown. An important catalyst, $[Ru(MeCN)_3Cp^*]PF_6$, was revealed to favor the nucleophilic substitution of optically active allycarbonates at the most substituted allyl carbon atom and the reaction took place with retention of configuration [112] (Eq. 85). The introduction of an optically pure chelating cyclopentadienylphosphine ligand with planar chirality leads to the creation of the new C–C bond with very high enantioselectivity from symmetrical carbonates and sodiomalonates [113].

189(*S*) 94% e. e. + **190** —[$Cp^*Ru(MeCN)_3PF_6$; NaH, DMF, r.t.]→ **191** (*R*) 94% e. e. (85)

The observation that the Ru(amidinate)C_5Me_5 complex could generate the first allyl ruthenium(IV) complex containing a nitrogen ligand led to the use of this complex as catalyst for simple allyl substitution of allylcarbonates [111]. Recently, it has been shown that Cp*Ru(bisimine)Cl also catalyzes the nucleophilic substitution of allylic carbonates [114] and that [Cp*Ru(bipyridine)(MeCN)]PF_6 catalysts provide a highly regioselective nucleophilic substitution with C, N, and O nucleophiles via a dicationic ruthenium(IV) intermediate without previous deprotonation of the carbonucleophile precursor [115].

It is noteworthy that (η^3-allyl)ruthenium species, which react with nucleophiles to give nucleophilic substitution of allylic substrates, are also active for the allylation of electrophiles. Thus, allyl acetate and carbonate react with aldehydes in the presence of catalytic amounts of $Ru_3(CO)_{12}$ to give homoallylic alcohols in good yields [116] (Eq. 86).

$$\mathbf{192} + \mathbf{193} \xrightarrow[\text{THF, 120 °C}]{Ru_3(CO)_{12}} \mathbf{194}\ (84\%) \quad (86)$$

The allylic activation of 1,3-dienes by Ru(COD)(COT) makes possible their hydroacylation to form β,γ-unsaturated ketones via C–H activation of aldehydes at the same metal center [117], and their selective coupling with acrylic compounds [18] (Eq. 87).

$$\mathbf{28} \xrightarrow[\text{Ru(cod)(cot)-PPh}_3\text{, 120 °C, 15 h}]{\text{PhCHO}} \mathbf{195}\ (54\%); \quad \mathbf{28} \xrightarrow[\text{Ru(cod)(cot), N-methylpiperidine, 80 °C, 2.5 h}]{\text{CH}_2\text{=CHCO}_2\text{Et}} \mathbf{196}\ (82\%) \quad (87)$$

11 Propargylic Substitution Reactions

Thiolate-bridged diruthenium complexes such as Cp*RuCl(μ_2-SR)$_2$RuCp*Cl catalyze the propargylic substitution reaction of propargylic alcohol derivatives with various carbon-centered nucleophiles [118–120]. Ketones [119] (Eq. 88), aromatic compounds [120] (Eq. 89), or alkenes thus selectively afford the corresponding propargylated products with C–C bond formation. An allenylidene intermediate is proposed in these reactions. They are detailed in the chapter Ruthenium Vinylidenes and Allenylidenes in Catalysis of this volume.

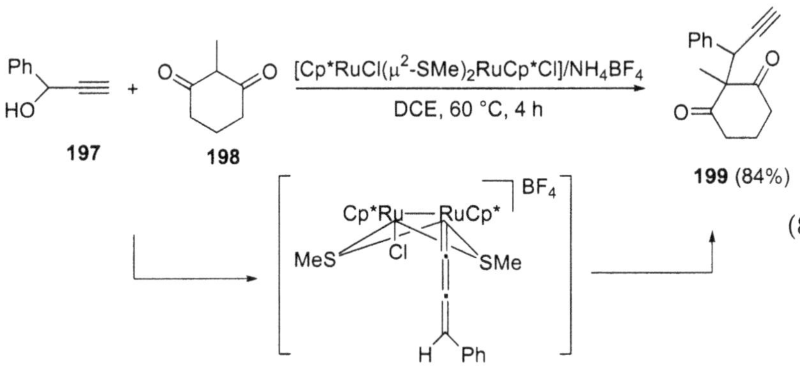

(88)

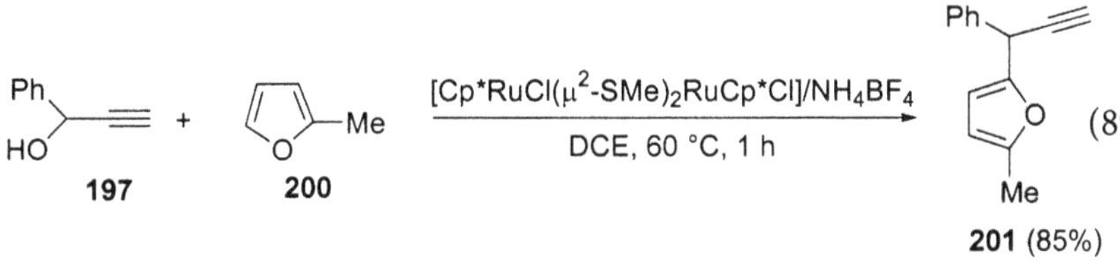

(89)

12 Reactions via C–H Bond Activation

Selective addition of alkenes and alkynes to aromatic compounds has also been performed by ruthenium-catalyzed aromatic C–H bond activation. Carbon–carbon bond formation occurs at the ortho positions of aromatic compounds, assisted by the neighboring functional group chelation. The reaction, catalyzed by $RuH_2(CO)(PPh_3)_3$, was efficient with aromatic and heteroaromatic compounds, with various functional groups, and a variety of alkenes and alkynes [121] (Eq. 90). Activation of vinylic C–H bonds can occur in a similar manner.

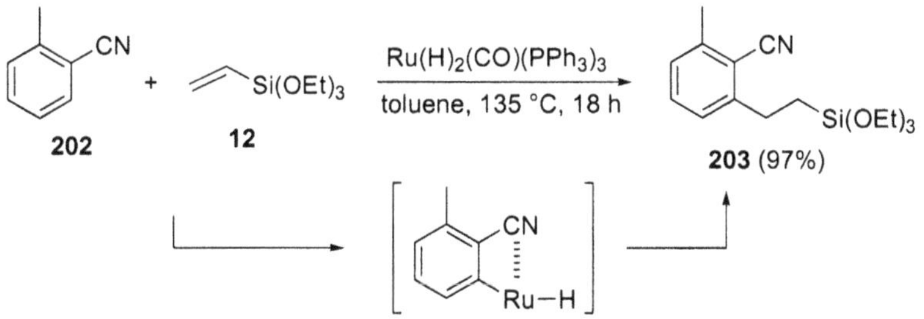

(90)

The sp^2 C–H bond of aldehydes, formamides, or formate esters undergoes oxidative addition to ruthenium complexes to generate acylruthenium hydride, which can insert alkenes leading to the overall H-COR addition to alkenes [122] (Eq. 91).

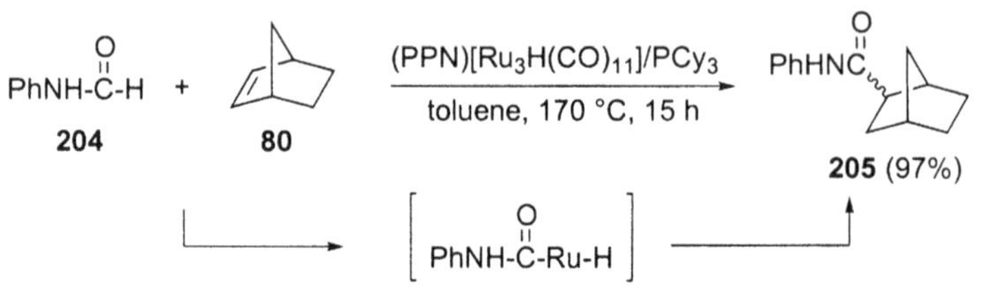

PhNH-C(O)-H + 80 → (PPN)[$Ru_3H(CO)_{11}$]/PCy_3, toluene, 170 °C, 15 h → PhHNC(O)-norbornyl **205** (97%) (91)

204 **80**

[PhNH-C(O)-Ru-H]

A ruthenium complex such as $Ru_3(CO)_{12}$ can activate the C–H bond of sp^3 carbons on the condition that a neighboring functional group can coordinate to the metal to favor intramolecular C–H bond activation [123] (Eq. 92).

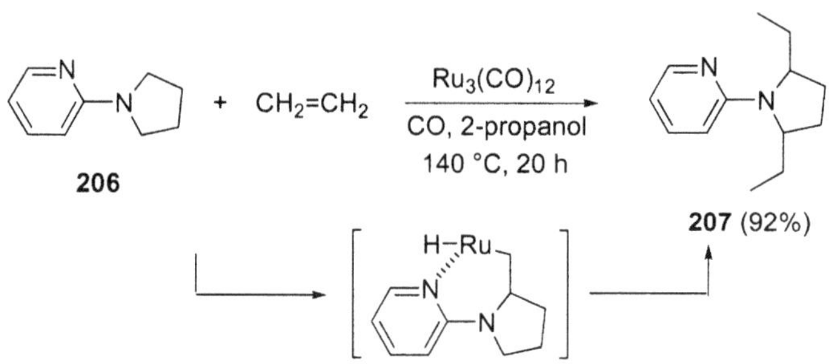

206 + CH_2=CH_2 → $Ru_3(CO)_{12}$, CO, 2-propanol, 140 °C, 20 h → **207** (92%) (92)

Emerging reactions involving C–H bond activation with ruthenium catalysts are detailed in the corresponding chapter Activation of Inert C–H Bonds of this volume.

13 Reactions Involving Carbonylations Promoted by Ruthenium Complexes

Ruthenium complexes are also suitable catalysts for carbonylation reactions of a variety of substrates. Indeed, when a reaction leads to C–Ru or heteroatom–Ru bond formation in the presence of carbon monoxide, CO insertion can take place at the coordinatively unsaturated ruthenium center, leading to linear ketones or lactones. Thus, ruthenium-catalyzed carbonylative cyclization was involved in the synthesis of cyclopentenones by reaction of allylic carbonates with alkenes in the presence of carbon monoxide [124] (Eq. 93).

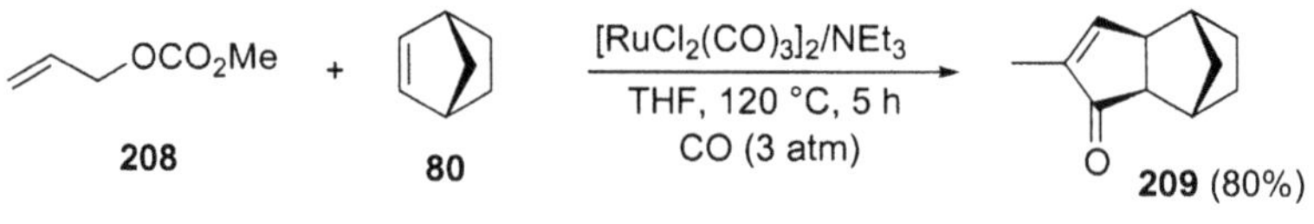

CH_2=CHCH$_2$OCO$_2$Me **208** + **80** → [$RuCl_2(CO)_3$]$_2$/NEt_3, THF, 120 °C, 5 h, CO (3 atm) → **209** (80%) (93)

Insertion of carbon monoxide can also allow the formation of the C–C bond. For example, γ-butyrolactones were produced by carbonylative cyclization of allenols [125] (Eq. 94).

138 → $Ru_3(CO)_{12}$, CO (10 atm), NEt_3, dioxane, 100 °C, 8 h → 210 (99%) (94)

[O–Ru–CO, H]

Reactions involving carbonylation are detailed in the chapter Selective Carbonylations with Ruthenium Catalysts of this volume.

14
Radical Reactions

Ruthenium catalysts can participate in electron-transfer processes. Thus, a variety of radical reactions of organic halides have been catalyzed by ruthenium complexes, as in the following example [126] (Eq. 95).

MeO_2C (211) + CCl_4 (212) → $Cp^*RuCl(PPh_3)_2$, toluene, 85 °C, 2 h → 213 (97%) (95)

[$^{\bullet}CCl_3$ + Ru-Cl] → [MeO_2C • CCl_3]

Reactions involving radical reactions are detailed in the chapter Ruthenium-Promoted Radical Processes Toward Fine Chemistry of this volume.

15
Concluding Remarks

The ruthenium-catalyzed reactions by their diversity, selectivity, and interest for the production of fine chemicals, especially during the last decade, show that molecular ruthenium catalysts are not only versatile but that they have now become unavoidable tools in organic synthesis [127]. They also appear to be complementary to organic, enzyme, or other metal catalysts as they have the power to generate original activation pathways for the combination of a variety of simple and complex molecules.

It seems that molecular ruthenium catalysts may play a crucial role, during the next decade, at the center of the current efforts to transform stoichiometric reactions into catalytic reactions and to perform innovative combinations of molecules with atom economy. Their ability to promote new methods in the selective formation of carbon–carbon single and double bonds should allow the development of new concepts for the building of multifunctional large molecules and polymers and for the profit of the emerging field of molecular materials.

References

1. Pertici P, Ballantini V, Salvadori P, Bennett MA (1995) Organometallics 14:2565
2. (a) Sustmann R, Hornung HJ, Schupp T, Patzke B (1993) J Mol Catal 85:149; (b) Ren CY, Cheng WC, Chan WC, Yeung CH, Lau CP (1990) J Mol Catal 59:L1; (c) McKinney RJ (1986) Organometallics 5:1752; (d) McKinney RJ, Colton MC (1986) Organometallics 5:1080; (e) Alderson T, Jenner EL, Lindsey RV (1965) J Am Chem Soc 87:5638
3. Ohgomori Y, Ichikawa S, Sumitani N (1994) Organometallics 13:3758
4. Fukuoka A, Nagano T, Furuta S, Yoshizawa M, Hirano M, Komiya S (1998) Bull Chem Soc Jpn 71:1409
5. Kashiwagi K, Sugise R, Shimakawa T, Matuura T, Shirai M, Kakiuchi F, Murai S (1997) Organometallics 16:2223
6. Yi CS, Liu N (1998) J Organomet Chem 553:157
7. (a) Mitsudo T-a, Suzuki T, Zhang S-W, Imai D, Fujita K-i, Manabe T, Shiotsuki M, Watanabe Y, Wada K, Kondo T (1999) J Am Chem Soc 121:1839; (b) Mitsudo T-A, Zhang S-W, Watanabe Y (1994) J Chem Soc Chem Commun 435
8. Yi CS, He Z, Lee DW (2001) Organometallics 20:802
9. (a) Trost BM, Imi K, Davies IW (1995) J Am Chem Soc 117:5371; (b) Kakiuchi F, Tanaka Y, Sato T, Chatani N, Murai S (1995) Chem Lett 679
10. Kakiuchi F, Murai S (1999) Top Organomet Chem 3:47
11. Mitsudo T-a, Takagi M, Zhang S-W, Watanabe Y (1992) J Organomet Chem 423:405
12. Murahashi S-I, Yamamura M, Yanagisawa K-i, Mita N, Kondo K (1979) J Org Chem 44:2408
13. (a) Yamamoto Y, Nakagai Y-i, Ohkoshi N, Itoh K (2001) J Am Chem Soc 123:6372; (b) Yamamoto Y, Ohkoshi N, Kameda M, Itoh K (1999) J Org Chem 64:2178
14. Michaut M, Santelli M, Parrain JL (2003) Tetrahedron Lett 44:2157
15. (a) Fujii N, Kakiuchi F, Yamada A, Chatani N, Murai S (1998) Bull Chem Soc Jpn 71:285; (b) Fujii N, Kakiuchi F, Chatani N, Murai S (1996) Chem Lett 939
16. Chatani N, Ishii Y, Ie Y, Kakiuchi F, Murai S (1998) J Org Chem 63:5129
17. (a) Berger D, Imhof W (2000) Tetrahedron 56:2015; (b) Imhof W, Berger D, Kötteritzsch M, Rost M, Schönecker B (2001) Adv Synth Catal 343:795; (c) Berger D, Imhof W (1999) Chem Commun 1457
18. Mitsudo T, Zhang S W, Kondo T, Watanabe Y (1992) Tetrahedron Lett. 33:341
19. He Z, Yi CS, Donaldson WA (2003) Org Lett 5:1567
20. Fujiwhara M, Nishikawa T, Hori Y (1999) Org Lett 1:1635
21. (a) Faller JW, Lavoie AR, Grimmond BJ (2002) Organometallics 21:1662; (b) Kündig EP, Saudan CM, Viton F (2001) Adv Synth Catal 343:51; (c) Faller JW, Lavoie A (2001) J Organomet Chem 630:17; (d) Davies DL, Fawcett J, Garratt SA, Russell DR (2001) Organometallics 20:3029; (e) Faller JW, Parr J (2000) Organometallics 19:1829; (f) Davenport AJ, Davies DL, Fawcett J, Garratt SA, Russell DR (2000) J Chem Soc Dalton Trans 4432; (g) Kündig EP, Saudan CM, Bernardinelli G (1999) Angew Chem Int Ed Engl 38:1220; (h) Davies DL, Fawcett J, Garratt SA, Russell DR (1997) Chem Commun 1351; (i) Carmona D, Cativiela C, Elipe S, Lahoz FJ, Lamata MP, Lopez-Ram de Viu MP, Oro

LA, Vega C, Viguri F (1997) Chem Commun 2351; (j) Hollis TK, Odenkirk W, Robinson NP, Whelan J, Bosnich B (1993) Tetrahedron 49:5415; (k) Odenkirk W, Rheingold AL, Bosnich B (1992) J Am Chem Soc 114:6392
22. Kündig EP, Saudan CM, Alezra V, Viton F, Bernardinelli G (2001) Angew Chem Int Ed Engl 40:4481
23. Faller JW, Grimmond BJ, D'Alliessi DG (2001) J Am Chem Soc 123:2525
24. Viton F, Bernardinelli G, Kündig EP (2002) J Am Chem Soc 124:4968
25. Ellis WW, Odenkirk W, Bosnich B (1998) Chem Commun 1311
26. Sato S, Matsuda I, Shibata M (1989) J Organomet Chem 377:347
27. (a) Itoh K, Masuda K, Fukahori T, Nakano K, Aoki K, Nagashima H (1994) Organometallics 13:1020; (b) Masuda K, Nakano K, Fukahori T, Nagashima H, Itoh K (1992) J Organomet Chem 428:C21
28. (a) Trost BM, Pinkerton AB, Seidel M (2001) J Am Chem Soc 123:12466; (b) Trost BM, Pinkerton AB (1999) J Am Chem Soc 121:4068
29. Trost BM, Pinkerton AB (1999) J Am Chem Soc 121:10842
30. Trost BM, Pinkerton AB, Kremzow D (2000) J Am Chem Soc 122:12007
31. Yoshida M, Sugimoto K, Ihara M (2001) Tetrahedron Lett 42:3877
32. Kang S-K, Ko B-S, Lee D-M (2002) Tetrahedron Lett 43:6693
33. Mitsudo T-a, Zhang S-W, Nagao M, Watanabe Y(1991) J Chem Soc Chem Commun 598
34. (a) Trost BM, Chisholm JD, Wrobleski ST, Jung M (2002) J Am Chem Soc 124:12420; (b) Lopez F, Castedo L, Mascarenas JL (2002) J Am Chem Soc 124:4218; (c) Trost BM, Shen HC, Pinkerton AB (2002) Chem Eur J 8:2341; (d) Trost BM, Toste FD (1999) Tetrahedron Lett 40:7739; (e) Trost BM, Roth GJ (1999) Org Lett 1:67; (f) Trost BM, Probst GD, Schoop A (1998) J Am Chem Soc 120:9228; (g) Trost BM, Indolese A (1993) J Am Chem Soc 115:4361
35. Trost BM, Indolese AF, Müller TJJ, Treptow B (1995) J Am Chem Soc 117:615
36. Trost BM, Pinkerton AB, Toste FD, Sperrle M (2001) J Am Chem Soc 123:12504
37. (a) Trost BM, Calkins TL, Bochet CG (1997) Angew Chem Int Ed Engl 36:2632; (b) Trost BM, Müller TJJ, Martinez J (1995) J Am Chem Soc 117:1888; (c) Trost BM, Calkins TL (1995) Tetrahedron Lett 36:6021; (d) Trost BM, Müller TJJ (1994) J Am Chem Soc 116:4985
38. (a) Trost BM, Machacek MR, Ball ZT (2003) Org Lett 5:1895; (b) Trost BM, Machacek MR (2002) Angew Chem Int Ed Engl 41:4693; (c) Trost BM, Machacek MR, Schnaderbeck MJ (2000) Org Lett 2:1761
39. Trost BM, Martinez JA, Kulawiec RJ, Indolese AF (1993) J Am Chem Soc 115:10402
40. (a) Trost BM, Surivet J-P (2001) Angew Chem Int Ed Engl 40:1468; (b) Trost BM, Surivet J-P, Toste FD (2001) J Am Chem Soc 123:2897
41. (a) Dérien S, Jan D, Dixneuf P-H (1996) Tetrahedron 52:5511; (b) Dérien S, Dixneuf P-H (1994) J Chem Soc Chem Commun 2551
42. (a) Dérien S, Ropartz L, Le Paih J, Dixneuf P-H (1999) J Org Chem 64:3524; (b) Dérien S, Gomez Vicente B, Dixneuf P-H (1997) Chem Commun 1405
43. Yi CS, Lee DW, Chen Y (1999) Organometallics 18:2043
44. Kakiuchi F, Uetsuhara T, Tanaka Y, Chatani N, Murai S (2002) J Mol Catal A 182–183: 511
45. Mitsudo T-a, Naruse H, Kondo T, Ozaki Y, Watanabe Y (1994) Angew Chem Int Ed Engl 33:580
46. Mitsudo T-a, Kokuryo K, Shinsugi T, Nakagawa Y, Watanabe Y, Takegami Y (1979) J Org Chem 44:4492
47. (a) Jordan RW, Tam W (2002) Tetrahedron Lett 43:6051; (b) Jordan RW, Tam W (2001) Org Lett 3:2367; (c) Jordan RW, Tam W (2000) Org Lett 2:3031

48. Warrener RN, Abbenante G, Kennard CHL (1994) J Am Chem Soc 116:3645
49. Trost BM, Imi K, Indolese AF (1993) J Am Chem Soc 115:8831
50. (a) Matsushima Y, Kikuchi H, Uno M, Takahashi S (1999) Bull Chem Soc Jpn 72:2475; (b) Kikuchi H, Uno M, Takahashi S (1997) Chem Lett 1273
51. Miki K, Ohe K, Uemura S (2003) Tetrahedron Lett 44:2019
52. Kondo T, Kaneko Y, Tsunawaki F, Okada T, Shiotsuki M, Morisaki Y, Mitsudo T-a (2002) Organometallics 21:4564
53. (a) Trost BM, Kulawiec RJ (1992) J Am Chem Soc 114:5579; (b) Trost BM, Vidal B, Thommen M (1999) Chem Eur J 5:1055; (c) Trost BM, Flygare JA (1994) J Org Chem 59:1078; (d) Trost BM, Kulawiec RJ, Hammes A (1993) Tetrahedron Lett 34:587; (e) Trost BM, Flygare JA (1994) Tetrahedron Lett 35:4059; (f) Trost BM, Flygare JA (1992) J Am Chem Soc 114:5476
54. Trost BM, Dyker G, Kulawiec RJ (1990) J Am Chem Soc 112:7809
55. Merlic CA, Pauly ME (1996) J Am Chem Soc 118:11319
56. Murakami M, Ubukata M, Ito Y (1998) Tetrahedron Lett 39:7361
57. Murakami M, Hori S (2003) J Am Chem Soc 125:4720
58. Trost BM, Portnoy M, Kurihara H (1997) J Am Chem Soc 119:836
59. Trost BM, Krause L, Portnoy M (1997) J Am Chem Soc 119:11319
60. Trost BM, Pinkerton AB (2002) J Am Chem Soc 124:7376
61. (a) Trost BM, Pinkerton AB (2000) Tetrahedron Lett 41:9627; (b) Trost BM, Pinkerton AB (2000) Angew Chem Int Ed Engl 39:360
62. Trost BM, Pinkerton AB (1999) J Am Chem Soc 121:1988
63. (a) Trost BM, Pinkerton AB (2001) J Org Chem 66:7714; (b) Trost BM, Pinkerton AB (2000) Org Lett 2:1601
64. Trost BM, Pinkerton AB (2000) J Am Chem Soc 122:8081
65. Chatani N, Morimoto T, Muto T, Murai S (1994) J Am Chem Soc 116:6049
66. Mori M (1998) Top Organomet Chem 1:133
67. Chatani N, Kataoka K, Murai S, Furukawa N, Seki Y (1998) J Am Chem Soc 120: 9104
68. Chatani N, Inoue H, Ikeda T, Murai S (2000) J Org Chem 65:4913
69. Mori M, Kozawa Y, Nishida M, Kanamaru M, Onozuka K, Takimoto M (2000) Org Lett 2:3245
70. Nishida M, Adachi N, Onozuka K, Matsumura H, Mori M (1998) J Org Chem 63: 9158
71. Le Paih J, Cuervo Rodriguez D, Dérien S, Dixneuf P-H (2000) Synlett 95
72. Trost BM, Toste FD (2002) J Am Chem Soc 124:5025
73. Trost BM, Toste FD (2000) J Am Chem Soc 122:714
74. Mendez M, Paz Munoz M, Nevado C, Cardenas DJ, Echavarren AM (2001) J Am Chem Soc 123:10511
75. Trost BM, Shen HC (2001) Angew Chem Int Ed Engl 40:2313
76. (a) Trost BM, Shen HC (2000) Org Lett 2:2523; (b) Trost BM, Toste FD, Shen H (2000) J Am Chem Soc 122:2379
77. Trost BM, Brown RE, Toste FD (2000) J Am Chem Soc 122:5877
78. (a) Kondo T, Suzuki N, Okada T, Mitsudo T-a (1997) J Am Chem Soc 119:6187; (b) Morimoto T, Chatani N, Fukumoto Y, Murai S (1997) J Org Chem 62:3762
79. Mori M, Saito N, Tanaka D, Takimoto M, Sato Y (2003) J Am Chem Soc 125:5606
80. Trost BM, Toste FD (1999) J Am Chem Soc 121:9728
81. (a) Fernandez-Rivas C, Mendez M, Nieto-Oberhuber C, Echavarren AM (2002) J Org Chem 67:5197; (b) Fernandez-Rivas C, Mendez M, Echavarren AM (2000) J Am Chem Soc 122:1221

82. (a) Mitsudo T-a, Hori Y, Watanabe Y (1987) J Organomet Chem 334:157; (b) Mitsudo T-a, Nakagawa Y, Watanabe K, Hori Y, Misawa H, Watanabe H, Watanabe Y (1985) J Org Chem 50:565
83. Yamaguchi M, Kido Y, Omata K, Hirama M (1995) Synlett 1181
84. Yi CS, Liu N (1999) Synlett 281
85. Slugovc C, Mereiter K, Zobetz E, Schmid R, Kirchner K (1996) Organometallics 15: 5275
86. Wakatsuki Y, Yamazaki H, Kumegawa N, Satoh T, Satoh JY (1991) J Am Chem Soc 113: 9604
87. Yi CS, Liu N (1998) Organometallics 17:3158
88. Yi CS, Torres-Lubian JR, Liu N (1998) Organometallics 17:1257
89. (a) Le Paih J, Monnier F, Dérien S, Dixneuf P-H, Clot E, Eisenstein O (2003) J Am Chem Soc 125:11964; (b) Le Paih J, Dérien S, Dixneuf P-H (1999) Chem Commun 1437
90. Le Paih J, Dérien S, Bruneau C, Demerseman B, Toupet L, Dixneuf PH (2001) Angew Chem Int Ed Engl 40:2912
91. Trost BM, Rudd MT (2001) J Am Chem Soc 123:8862
92. Yamamoto Y, Kitahara H, Ogawa R, Kawaguchi H, Tatsumi K, Itoh K (2000) J Am Chem Soc 122:4310
93. (a) Yamamoto Y, Kitahara H, Ogawa R, Itoh K (1998) J Org Chem 63:9610; (b) Yamamoto Y, Kitahara H, Hattori R, Itoh K (1998) Organometallics 17:1910
94. Varela JA, Castedo L, Saa C (2003) Org Lett 5:2841
95. Yamamoto Y, Hata K, Arakawa T, Itoh K (2003) Chem Commun 1290
96. Yamamoto Y, Ogawa R, Itoh K (2000) Chem Commun 549
97. Yamamoto Y, Okuda S, Itoh K (2001) Chem Commun 1102
98. Yamamoto Y, Ogawa R, Itoh K (2001) J Am Chem Soc 123:6189
99. Yamamoto Y, Takagishi H, Itoh K (2001) Org Lett 13:2117
100. Yamamoto Y, Takagishi H, Itoh K (2002) J Am Chem Soc 124:28
101. Yamamoto Y, Takagishi H, Itoh K (2002) J Am Chem Soc 124:6844
102. Trost BM, Rudd MT (2002) J Am Chem Soc 124:4178
103. Trost BM, Rudd MT (2003) Org Lett 9:1467
104. Nishibayashi Y, Yamanashi M, Wakiji I, Hidai M (2000) Angew Chem Int Ed Engl 39: 2909
105. Frauenkron M, Berkessel A (1997) Tetrahedron Lett 38:7175
106. Le Paih J, Dérien S, Ozdemir I, Dixneuf P-H (2000) J Am Chem Soc 122:7400
107. Monnier F, Castillo D, Dérien S, Toupet L, Dixneuf P-H (2003) Angew Chem Int Ed Engl 42:5474
108. (a) Hayashi Y, Komiya S, Yamamoto T, Yamamoto A (1984) Chem Lett 977; (b) Minami I, Shimizu I, Tsuji J (1985) J Organomet Chem 296:269
109. Zhang SW, Mitsudo T, Kondo T, Watanabe Y (1993) J Organomet Chem 450:197
110. Morisaki Y, Kondo T, Mitsudo T (1999) Organometallics 18:4742
111. Kondo H, Yamaguchi Y, Nagashima H (2000) Chem Commun 1075
112. Trost BM, Fraisse PL, Ball ZT (2002) Angew Chem Int Ed Engl 41:1059
113. Matsushima Y, Onitsuka K, Kondo T, Mitsudo T-a, Takahashi S (2001) J Am Chem Soc 123:10405
114. Renaud JL, Bruneau C, Demerseman B (2003) Synlett 408
115. Mbaye MD, Demerseman B, Renaud JL, Toupet L, Bruneau C (2003) Angew Chem Int Ed Engl 42:5066
116. (a) Tsuji Y, Mukai T, Kondo T, Watanabe Y (1989) J Organomet Chem 369:C51; (b) Kondo T, Ono H, Satake N, Mitsudo T-a, Watanabe Y (1995) Organometallics 14: 1945

117. Kondo T, Hiraishi N, Morisaki Y, Wada K, Watanabe Y, Mitsudo T-A (1998) Organometallics 17:2131
118. (a) Nishibayashi Y, Inada Y, Hidai M, Uemura S (2003) J Am Chem Soc 125:6060; (b) Nishibayashi Y, Onodera G, Inada Y, Hidai M, Uemura S (2003) Organometallics 22:873; (c) Nishibayashi Y, Inada Y, Hidai M, Uemura S (2002) J Am Chem Soc 124: 7900
119. Nishibayashi Y, Wakiji I, Ishii Y, Uemura S, Hidai M (2001) J Am Chem Soc 123:3393
120. Nishibayashi Y, Yoshikawa M, Inada Y, Hidai M, Uemura S (2002) J Am Chem Soc 124:11846
121. Kakiuchi F, Sonoda M, Tsujimoto T, Chatani N, Murai S (1999) Chem Lett 1083
122. Kondo T, Okada T, Mitsudo T-A (1999) Organometallics 18:4123
123. Chatani N, Asaumi T, Yorimitsu S, Ikeda T, Kakiuchi F, Murai S (2001) J Am Chem Soc 123:10935
124. Morisaki Y, Kondo T, Mitsudo T-A (2000) Org Lett 2:949
125. Yoneda E, Kaneko T, Zhang S-W, Onitsuka K, Takahashi S (2000) Org Lett 2:441
126. Simal F, Wlodarczak L, Demonceau A, Noels AF (2000) Tetrahedron Lett 41:6071
127. (a) Trost BM, Toste FD, Pinkerton AB (2001) Chem Rev 101:2067; (b) Naota T, Takaya H, Murahashi S-I (1998) Chem Rev 98:2599

Topics Organomet Chem (2004) 11: 45–79
DOI 10.1007/978-3-540-40027-1

Activation of Inert C–H Bonds

Fumitoshi Kakiuchi (✉) · Naoto Chatani

Department of Applied Chemistry, Faculty of Engineering, Osaka University, 565-0871 Suita, Osaka, Japan
kakiuchi@chem.eng.osaka-u.ac.jp

Abstract The development of catalytic reactions involving carbon–hydrogen bond cleavage is one of the most attractive research subjects in organic and organometallic chemistries. To date, a vast number of studies of C–H bonds cleavage using stoichiometric amounts of transition-metal complexes have appeared. In the last decade, a variety of catalytic reactions involving C–H bond cleavage have been reported. Among these reactions, ruthenium complexes have extensively been used as effective catalysts. In this review we briefly survey the results of these research activities with respect to the ruthenium-catalyzed reactions involving C–H bond cleavage.

Keywords C–H bond activation · Silylation · Arylation · Hydroacylation · Active methylene compounds

1 Introduction

When chemists open textbooks on organic synthesis, they find a large number of transition-metal-catalyzed reactions, such as selective carbon–carbon bond formation, carbon–heteroatom bond formation, enantioselective reactions, reductions, and oxidations. These transition-metal-catalyzed reactions are highly reliable because, in many cases, bond breaking and bond formation take place in a highly selective manner. To date, a variety of catalytic reactions and a large number of transition-metal complexes have been developed [1, 2]. It is no exaggeration to say palladium-catalyzed reactions represent the most extensively studied reactions in organic synthesis over the past 50 years [2]. Compared with these well-published subjects, catalytic reactions involving ruthenium complexes as a catalyst are immature research areas in organic synthesis [3]. Recently, organic syntheses involving ruthenium-catalyzed reactions have developed to a considerable extent. A number of impressive ruthenium-catalyzed reactions have been reported. For example, Noyori and coworkers [4] reported on various asymmetric reactions using a ruthenium 2,2′-bis(diphenylphosphino)-1,1′-binaphthyl catalyst system (Eq. 1). Murahashi reported on the ruthenium-catalyzed oxidation of C–H bonds adjacent to heteroatoms. A representative example of this type of oxidation reaction is the ruthenium-on-charcoal-catalyzed direct acetoxylation of a C–H bond adjacent to a nitrogen atom in β-lactames, giving acetoxylactames (Eq. 2) [5]. In addition, one of the flourishing research subjects in ruthenium-catalyzed organic synthesis involves ring-closing and ring-opening metathesis reactions, which have been extensively developed by Grubbs and coworkers [6] (Eq. 3).

$$R^2R^3C{=}C(R^1)COOH + H_2 \xrightarrow{Ru(OCOR)_2(binap)} R^2R^3C^{*}H{-}C^{*}H(R^1)COOH \tag{1}$$

$$\text{(R}^1\text{,R}^2\text{-azetidin-2-one, NH)} + AcOOH \xrightarrow{Ru/C} \text{(R}^1\text{,R}^2\text{-4-OAc-azetidin-2-one, NH)} \tag{2}$$

$$\text{diene (X)} \xrightarrow{RuCl_2(PCy_3)_2({=}CHPh)} \text{cycloalkene (X)} \tag{3}$$

Recently, an extensive research area with respect to the transition-metal-catalyzed manipulation of inert C–H bonds in organic synthesis has appeared [7, 8]. In 1965, the remarkable pioneering findings with respect to the oxidative addition of a C–H bond to a zero-valent ruthenium center were reported by Chatt and Davidson [9]. They revealed that the Ru(H)(2-naphthyl)(dmpe)$_2$ complex, where dmpe is 1,2-bis(dimethylphosphino)ethane, is in equilibrium with a π-coordinated naphthalene ruthenium complex, Ru(naphthalene)-(dmpe)$_2$ (Eq. 4) [9]. They also reported that the sp^3 C–H bond of a methyl group in the dmpe ligand can be cleaved by the ruthenium(0) complex.

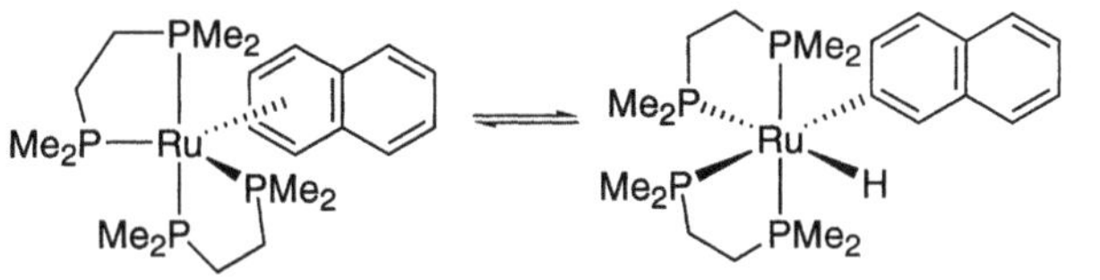

(4)

After these pioneering studies, a number of other research groups reported on the cleavage of C–H bonds via the use of a stoichiometric amount of transition-metal complexes [7]. To date, several types of catalytic reactions involving C–H bond cleavage, for example, alkyl, alkenyl, aryl, formyl, and active methylene C–H bonds have been developed [8]. In many cases, for these types of catalytic reactions, ruthenium, rhodium, iridium, platinum, and palladium complexes all show catalytic activity.

This review article will broadly survey the literature dealing with the ruthenium-catalyzed reactions involving otherwise unreactive C–H bond cleavage in organic synthesis up to early in the year 2003. Only limited numbers of examples which involve unusual significance, originality, or complexity will be presented in equation form. However, several areas, for example, reactions involving transition-metal carbenoids, reactions involving *sp* C–H bonds cleavage, and reactions in the presence of heterogeneous catalysts, will not be dealt with. The ruthenium-catalyzed carbonylation of C–H bonds is covered in the section on ruthenium-catalyzed carbonylation reactions.

2 Reaction of Aromatic Compounds

2.1 Addition to Olefins

One of the most important goals of catalytic methods involving C–H bond cleavage has been to achieve the one-step addition of a C–H bond across the double bond of an olefin (Eq. 5). If an unreactive C–H bond could participate in such a reaction without being converted into a reactive but sacrificing functional group such as a halogen, the overall transformation would be of great utility in organic synthesis. The transformation for synthetic purposes must be highly efficient, i.e., high yield, highly selective, and catalytic. Murai and coworkers [10–23] reported on a breakthrough discovery for such a process, i.e., a series of ruthenium-catalyzed reactions for carbon–carbon bond formation at unactivated

C–H bonds. In this section, among the transition-metal-catalyzed C–H/olefin coupling, ruthenium-catalyzed reactions will mainly be discussed.

C—H + >C=C< —catalyst→ C, H on —C—C— (5)

The pioneering study of ruthenium-catalyzed regioselective alkylation using olefins as an alkylating reagent was reported by Lewis and Smith [24]. The ortho-selective ethylation of phenols with ethylene can be attained with the aid of a ruthenium(II) phosphite complex as a catalyst. This alkylation takes place exclusively at the position ortho to the hydroxyl group, and the corresponding 1:2 addition product is the major product (Eq. 6). The use of potassium phenoxide is the key in this catalytic reaction. Unfortunately, however, the applicability of this reaction is narrow. Thus, phenol is the only applicable substrate in this reaction.

OH + = —cat. [Ru complex: PhO, OPh, O, P, Ru, O, P, OPh, PhO, OPh, OPhP(OPh)$_3$]; PhOK (cat.), THF, 177 °C, 3.5 h→ OH 12% + OH 13% + OH 75% (6)

In 1993, Murai et al. [10] reported on the first example of a highly efficient, selective C–H/olefin coupling reaction. The reaction of aromatic ketones with olefins in the presence of a $RuH_2(CO)(PPh_3)_3$ complex gave the corresponding ortho-alkylated compounds in high yields. A representative example of the C–H/olefin coupling reaction is given in Eq. (7). The reaction involves the cleavage and addition of an ortho C–H bond of acetophenone to an olefin.

O + Si(OEt)$_3$ —catalyst $RuH_2(CO)(PPh_3)_3$; toluene, 2 h, reflux (135 °C, bath temp.)→ O, Si(OEt)$_3$ 93%

Ru(0) → [O, Ru, H] → Si(OEt)$_3$ (7)

Table 1 The $RuH_2(CO)(PPh_3)_3$-catalyzed reaction of several aromatic ketones with triethoxyvinylsilane

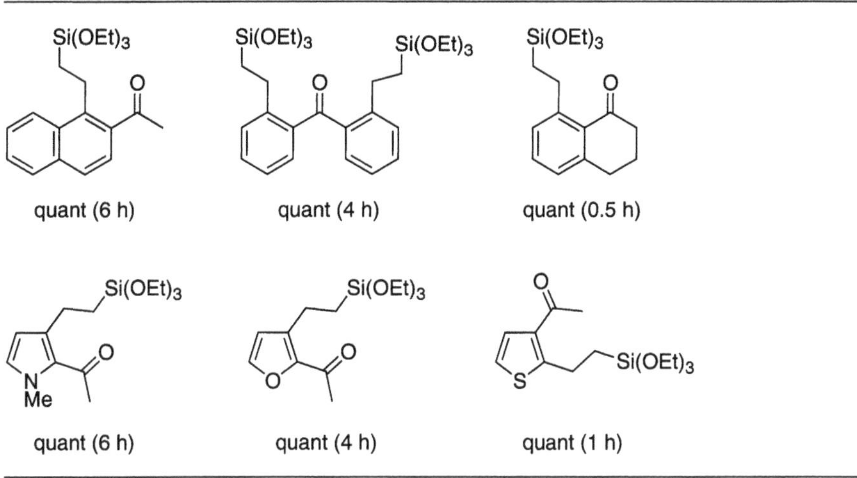

For this catalytic coupling reaction of aromatic ketones with olefins, among the transition-metal complexes screened by Murai, a ruthenium complex, $RuH_2(CO)(PPh_3)_3$, exhibited the highest catalytic activity. A variety of aromatic and heteroaromatic ketones can also be used in this coupling reaction (Table 1) [12, 13]. In many cases, the corresponding coupling products are obtained in quantitative yield. Terminal olefins such as vinylsilanes, *tert*-butylethylene, styrenes, and allylsilanes show high reactivity, but olefins having allylic hydrogens, such as 1-hexene, result in low yields owing to the isomerization of the double bond to the internal positions.

The $RuH_2(CO)(PPh_3)_3$-catalyzed coupling of aromatic ketones with olefins is tolerant of several functional groups [e.g., NMe_2, OMe, F, NEtC(O)Me, CF_3, CO_2Et, CN, acetals, $OC(O)CH_3$] [17]. Steric hindrance of a substituent on the aromatic ring is critical in determining the reaction site (Fig. 1). The regioselectivity of the C–C bond formation is largely affected by a steric factor. Thus, C–C bond formation, generally, takes place at the less congested position (6′-position). Interestingly, however, the reaction of *m*-methoxyacetophenone (**1**) with triethoxyvinylsilane takes place at the more congested ortho position, i.e.,

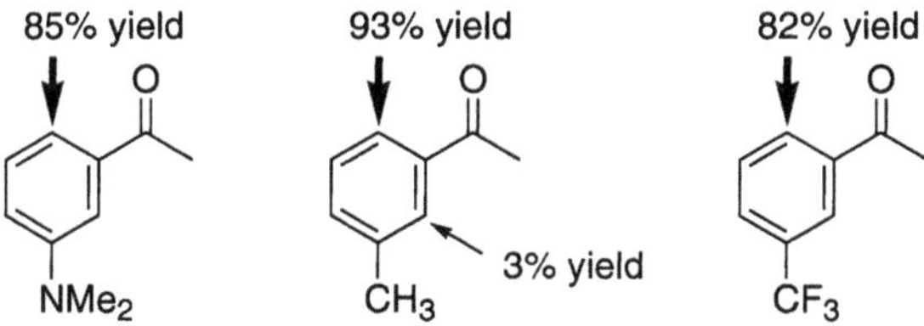

Fig. 1 Site-selective alkylation controlled by a steric congestion

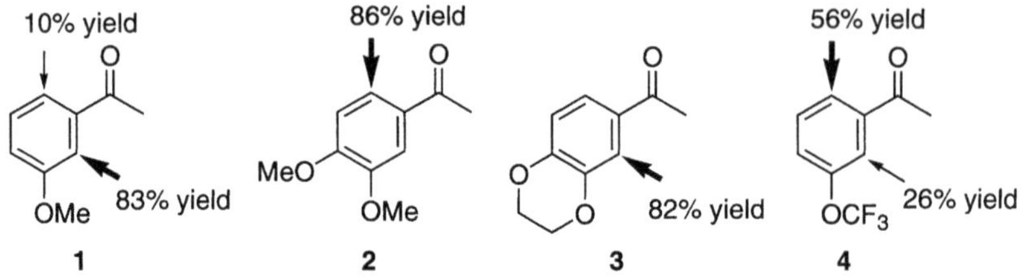

Fig. 2 Effect of the oxygen atom toward the site-selectivity

the 2′-position (Fig. 2) [17]. In the case of *m,p*-dimethoxyacetophenone (**2**), the reaction site moves to the opposite ortho position. This opposite site selectivity appears to be caused by a so-called buttressing effect between the methoxy groups [25]. Interestingly, however, in the case of ketone **3**, the reaction site is located at a more crowded position probably because free conformational rotation around C–O–C bonds is not possible owing to the ethylene bridge and, as a result, the lone pair of electrons point in the desired direction. The electron density of the oxygen atom is also important for the site selectivity. When a strong electron-withdrawing CF_3 group, which should decrease the electron density of the adjacent atom, is introduced on the ether oxygen (i.e., **4**), alkylation takes place preferentially at the less congested position. These results suggest that heteroatoms may additionally assist in the regioselectivity determination step.

Several related examples of the ruthenium-catalyzed addition of C–H bonds in ketones to olefins have been reported [26–30]. Application of C–H/olefin coupling to polymer chemistry has been reported by Weber's group [26]. They prepared a variety of polymers by reactions of aromatic ketones having two free ortho C–H bonds with α,ω-dienes such as 1,1,3,3-tetramethyl-1,3-divinyldisiloxane with the aid of the $RuH_2(CO)(PPh_3)_3$ complex as a catalyst (Eq. 8). Woodgate and coworkers [27] applied the $RuH_2(CO)(PPh_3)_3$-catalyzed coupling of aromatic ketones with olefins to the alkylation of aromatic diterpenoids (Eq. 9). The alkylation of phenyl 3-pyridyl ketone using

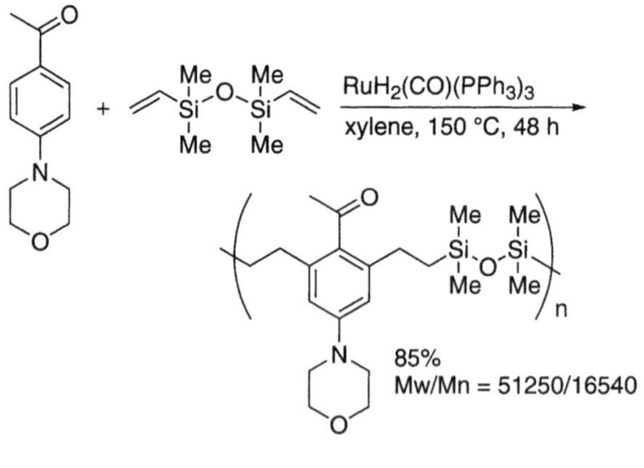

(8)

$RuH_2(CO)(PPh_3)_3$ as a catalyst proceeds exclusively at the pyridine ring (Eq. 10) [28]. This result indicates that C–C bond formation preferentially takes place at the electron-deficient aromatic ring. Chaudret et al. [29] prepared a reactive ruthenium complex, $RuH_2(H_2)(CO)(PCy_3)_2$, where Cy is cyclohexyl, and examined the catalytic activity of this complex in the reaction of benzophenone with ethylene (Eq. 11). The desired C–H/olefin coupling reaction giving the bis(alkylation) product proceeded at room temperature (Eq. 11). Busch and Leitner [30] subsequently reported a similar room-temperature C–H/olefin coupling reaction using the catalyst of Chaudret et al. (Eq. 12).

OMe, Me, Me, COOMe, O + Si(OEt)3 —Ru(CO)2(PPh3)3, toluene, reflux, 48 h→ OMe, Me, Si(OEt)3, Me, COOMe, O 100% (9)

O, N + Si(OEt)3 —RuH2(CO)(PPh3)3, toluene, reflux,→ O, Si(OEt)3, N (10)

O + = —RuH2(H2)2(PCy3)2, toluene, 48 h, 20 °C→ O 96% (11)

O + = —RuH2(H2)2(PCy3)2, toluene, 24 h, 30 °C→ O 99% (12)

In the case of the reaction of aromatic esters with olefins, the electronic effect of the substituent is critical for achieving a catalytic reaction. The reactions of methyl benzoate and methyl *o*-toluate with triethoxyvinylsilane in the presence of $RuH_2(CO)(PPh_3)_3$ as a catalyst result in no reaction [23]. Interestingly, however, the introduction of an electron-withdrawing group such as CF_3 and CN groups on the aromatic ring dramatically improved the reactivity of these substrates (Eq. 13) [14, 23]. The subsitutent on the silicon atom of vinylsilanes is important for improving the reactivity of vinylsilanes. The reaction of methyl benzoate, which is ineffective for the reaction with triethoxyvinylsilane, with trimethylvinylsilane gives the corresponding 1:1 and 1:2 coupling products in 51% and 21% yields, respectively. Trost et al. [31] concluded that the lack of reactivity of methyl benzoate resulted from the nature of the aromatic ester. However, the results of Murai and coworkers point out the inaccuracy of the result of Trost et al. with respect to the reactivity of aromatic esters.

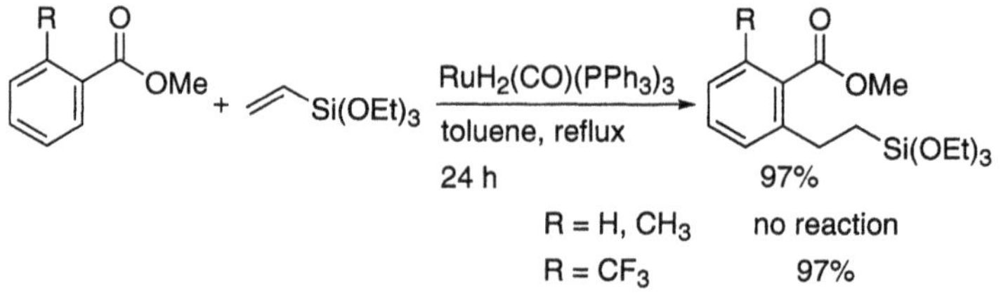

$RuH_2(CO)(PPh_3)_3$, toluene, reflux, 24 h; 97%

R = H, CH_3: no reaction

R = CF_3: 97% (13)

The use of a formyl group as a directing functionality is challenging because in the case of the low-valent transition-metal-catalyzed reaction of aldehydes with an olefin, aldehydes are prone to undergo decarbonylation or hydroacylation of olefins. Murai devised the following protocol, one being steric (Fig. 3) and the other electronic in nature (Fig. 4). A sterically bulky substituent on the ortho position or a heteroatom at the β-position of enals is believed to suppress undesired decarbonylation reactions (Eqs. 14, 15) [21]. The reaction of 2,4-di-*tert*-butylbenzaldehyde provides the corresponding coupling product in 69% yield. In the case of the reaction of 1-methylindole-3-carboxaldehyde with ethylene, the ethylation product is also obtained in quantitative yield.

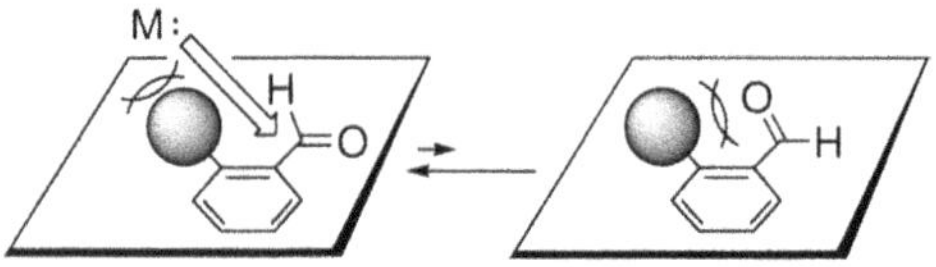

Fig. 3 The suppression of reactivity of the formyl group by steric effects

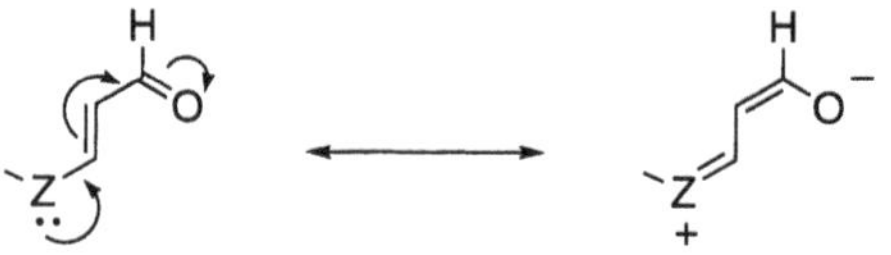

Fig. 4 The suppression of reactivity of the formyl group by electronic effects

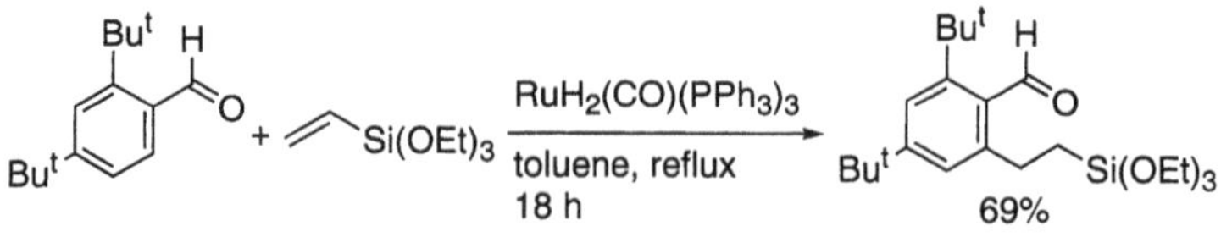

$RuH_2(CO)(PPh_3)_3$, toluene, reflux, 18 h; 69% (14)

$$\text{1-methylindole-3-carbaldehyde} + \text{ethylene} \xrightarrow[\text{toluene, 115 °C, 24 h}]{RuH_2(CO)(PPh_3)_3} \text{2-ethyl-1-methylindole-3-carbaldehyde (quant)} \quad (15)$$

An appropriate sp^2 nitrogen atom can also function as a directing group. In the case of the reaction of aromatic compounds having an sp^2 nitrogen directing group, $Ru_3(CO)_{12}$, which is an ineffective catalyst for the reaction of ketones, exhibits a higher activity (Eq. 16) [15]. The reaction of aldimines yields a mixture of the corresponding 1:1 coupling product and the dehydrogenation product. Interestingly, the reaction of aromatic ketimines derived from acetophenone affords the corresponding 1:1 coupling product as a single product (Eq. 17) [15].

$$\text{aldimine } (N\text{-}Bu^t) + CH_2{=}CHSi(OEt)_3 \xrightarrow[\text{toluene, reflux, 24 h}]{\text{catalyst}} \text{alkylation product (}CH_2CH_2Si(OEt)_3\text{)} + \text{alkenylation product (}CH{=}CHSi(OEt)_3\text{)} \quad (16)$$

catalyst		
$RuH_2(CO)(PPh_3)_3$	50%	22%
$Ru_3(CO)_{12}$	81%	10%

$$\text{ketimine } (N\text{-}Bu^t) + \text{ethylene} \xrightarrow[\text{toluene, reflux, 24 h}]{Ru_3(CO)_{12}} \text{ortho-ethyl ketimine, 85\% } (anti{:}syn = 57{:}43) \quad (17)$$

By taking advantage of these different catalytic activities of $RuH_2(CO)(PPh_3)_3$ and $Ru_3(CO)_{12}$ towards the ketones and imines, unique site-selective alkylation can be attained. When the reaction of 1-[3-(*tert*-butyliminomethyl)phenyl]-ethanone and triethoxyvinylsilane is conducted in the presence of the $RuH_2(CO)(PPh_3)_3$ catalyst, which shows a high catalytic activity for ketones, alkylation exclusively occurs at the position ortho to the acetyl group (6-position) (Eq. 18) [18]. On the other hand, in the case of the reaction using $Ru_3(CO)_{12}$, which is an effective catalyst for imines, the alkylation proceeds predominantly at the imino group side (Eq. 18). This protocol can also be applied to stepwise dialkylations using different olefins (Eq. 19) [18].

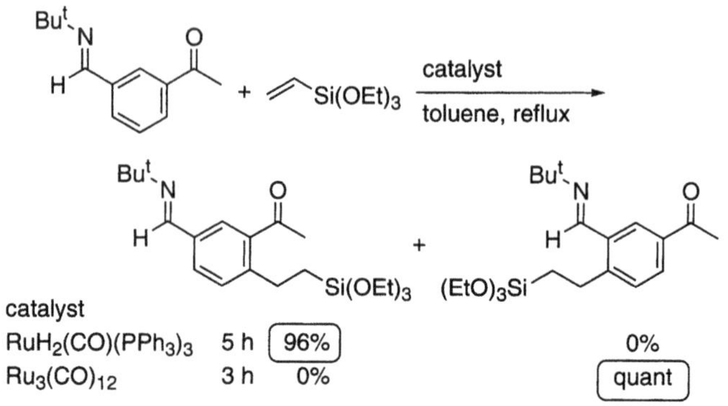

(18)

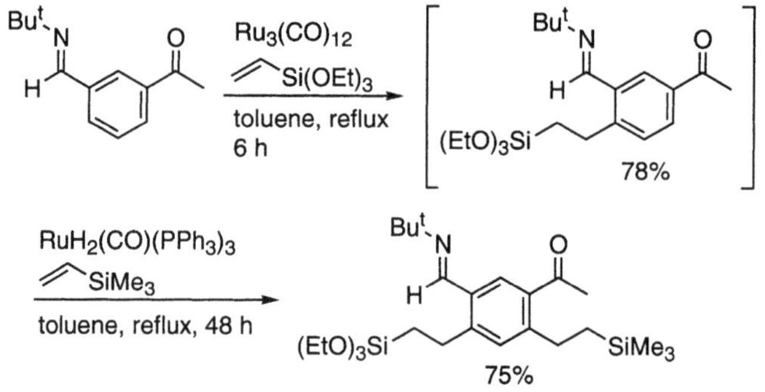

(19)

Hydrazone groups can also function as a directing group. In this case, both $RuH_2(CO)(PPh_3)_3$ and $Ru_3(CO)_{12}$ exhibit catalytic activity (Eq. 20) [22].

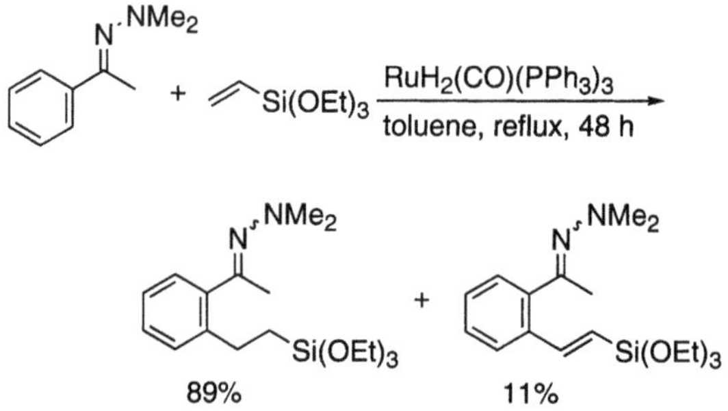

(20)

Chelation-assisted C–H/olefin coupling can be applied to the atroposelective alkylation of biaryl compounds. The reaction of 2-(1-naphthyl)-3-methylpyridine with ethylene using $[RhCl(coe)_2]_2$, where coe is cyclooctene, and PCy_3 results in the formation of an ethylation product in 92% yield (Eq. 21) [20]. In place of the PCy_3 ligand, the use of (*R*)-(1-[(*S*)-2-diphenylphosphino]ferrocenyl)ethyl methyl ether [(*R*),(*S*)-PPFOMe] leads to the atropselective alkylation product (37% yield, 49% ee). A catalytic system using Ru(1,5-cyclooctadiene)(1,3,5-cyclooctatriene) also shows activity for this atropselective alkylation albeit in low chemical and optical yields (15% yield and 15% ee, respectively) [20]. Although the chemical and optical yields are inadequate, these results suggest that the atropselective alkylation of biaryl compounds can be attained by means of chelation-assisted C–H/olefin coupling.

catalyst; toluene, 120 °C, 20 h (21)

catalyst		
$[RhCl(coe)_2]_2$/L	L = PCy_3	92%
$[RhCl(coe)_2]_2$/L	L = (*R*),(*S*)-PPFOMe	37%, 49% ee
Ru(cod)(cot)/L	L = (*R*),(*S*)-PPFOMe	15%, 15% ee

The C–H/olefin coupling of aryloxazolines proceeds with unusual product selectivity. In this case, alkylation products, i.e., formally dehydrogenation products, are obtained as a major product (Eq. 22) [11]. These types of dehydrogenation compounds are believed to be formed via a carbometalation pathway. The first example of this type of alkenylation of arenes with olefins using palladium(II) complexes via C–H bond cleavage was reported in 1967 [32]. Later, several efforts were made to perform this reaction in a catalytic manner [33]. In 2001, Milstein et al. [34] reported the oxidative alkenylation of arenes with olefins using a Ru/O_2/CO catalyst system (Eq. 23). Details of the reaction mechanism have not been elucidated.

+ $Si(OEt)_3$ — $Ru_3(CO)_{12}$; toluene, reflux, 24 h → $Si(OEt)_3$ 10% + $Si(OEt)_3$ 87% (22)

The dimerization of acrylonitrile is a cheaper route to the synthesis of highly valuable hexamethylenediamine, which is one component of the starting materials for nylon-6,6 [16, 35] In some cases of the dimerizations of acrylic acid

OMe + COOMe $\xrightarrow[\text{CO/O}_2\text{/HQ, 180 °C, 48 h}]{\text{RuCl}_3\ 3\text{H}_2\text{O}}$ OMe-C$_6$H$_4$-CH=CH-COOMe (23)

47% (118 TON)

o:*m*:*p* = 1.05:1.3:1.0

esters, acrylonitriles, and acroleins, a direct C–H bond cleavage step is believed to be involved in the catalytic reaction. At an early stage of the catalytic dimerization of acrylonitrile, *cis*-1,4-dicyanobut-1-ene is formed as the major product, and not the trans isomer [16, 35]. It has been proposed that this high cis selectivity indicates the selective cleavage of the C–H bond cis to CN by the metal coordinated to the nitrile group in a side-on fashion [36]. However, the participation of a π-bonded nitrile is still a matter of conjecture because several possible reaction pathways for the dimerization of olefins have been proposed. The ruthenium-catalyzed alkylation of benzonitriles with triethoxyvinylsilane takes place at the ortho position predominantly (Eq. 24) [19]. This regioselectivity indicates the possibility of π coordination of the CN group to the ruthenium in the catalytic cycle.

CN + Si(OEt)$_3$ $\xrightarrow[\text{toluene, reflux, 24 h}]{\text{RuH}_2\text{(CO)(PPh}_3\text{)}_3}$ Si(OEt)$_3$ / CN / Si(OEt)$_3$ (24)

97%

[C≡N–Ru, H]

Several studies on the transition-metal-catalyzed coupling of aromatic ketones with olefins have appeared. Almost all of these studies have focused on investigating the scope and limitations of this type of coupling reaction. A limited number of mechanistic studies have been reported. On the basis of product analysis, Murai et al. [13] proposed two possible reaction mechanisms (Scheme 1). One is the usual oxidative addition of the C–H bond to ruthenium which proceeds through transition state **TS-A** and a reductive elimination of the C–C bond from the ruthenium center through **TS-B** (path a, concerted pathway). In the other case, C–H bond cleavage through intermediate **C** occurs in two steps and C–C bond formation through intermediate **D** also proceeds in two steps (path b, stepwise pathway). Both mechanisms satisfactorily explain the ortho selectivity observed. The highly important difference between these paths is that C–H bond breaking and C–C bond formation in path a take place simultaneously, but in path b Ru–C and C–C bond formations occur before C–H and C–Ru bond breakings, respectively. The ab initio theoretical calculation by Morokuma et al. [37] suggested that in both the C–H bond cleavage and

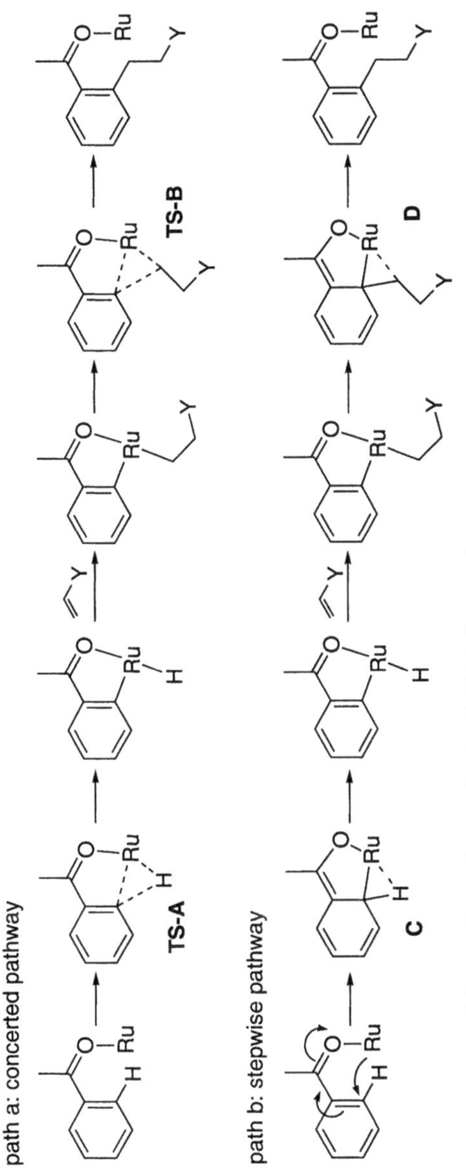

path a: concerted pathway

TS-A

TS-B

path b: stepwise pathway

C

D

Scheme 1 Possible reaction pathways for the C–H/olefin coupling

the C–C bond formation steps the stepwise reaction pathway (path b) would be highly preferable compared with the concerted pathway (path a).

Murai et al. [23] have elucidated the rate-determining step for the $RuH_2(CO)(PPh_3)_3$-catalyzed reactions of aromatic esters and aromatic ketones to olefins by means of deuterium-labeling experiments and ^{13}C kinetic isotope effects (KIE) at natural abundance. When the reaction of methyl benzoate-d_5 with triethoxyvinylsilane was carried out in the presence of $RuH_2(CO)(PPh_3)_3$ as a catalyst, no coupling product was observed even after refluxing for 24 h (Eq. 25). The ^1H NMR spectra of the recovered starting materials indicate that the hydrogen intensity of the two ortho positions of the benzoate and the three vinylic positions of the vinylsilane was around 0.6 H. This observation suggests that extensive H/D scrambling among these five positions took place. Therefore, the C–H (or C–D) bond cleavage is not rate-determining and a rapid equilibrium occurs prior to the reductive elimination. Thus, the experimental results of Murai et al. are consistent with the theoretical calculation of Morokuma et al. [37]. The rate-determining step in the reaction of methyl *o*-toluate with trimethylvinylsilane was determined by means of ^{13}C KIE (Table 2) [23]. The ^{13}C KIE was observed only at the C6 carbon. This result indicates that the ortho carbon participates in the rate-determining step; thus, C–C bond formation, i.e., the reductive elimination step, is rate-determining.

D O
D OMe
+ Si(OEt)3
RuH2(CO)(PPH3)3
toluene, reflux, 24 h
(0% conversion)
no coupling product
D D
D

(25)

[observed] → [1.21 H]
(theoritical) → (1.20 H)
O
OMe
[0.60 H]
(0.60 H)
[0.59 H]
(0.60 H)
Si(OEt)3
[0.59 H]
(0.60 H)

Table 2 Experimental ^{13}C kinetic isotope effect

2 1 O
3 OMe
4 6
5

KIEs (average)			
C1	0.997	C5	1.000 (assumed)
C2	0.998	C6	1.033
C3+C4	2.005		

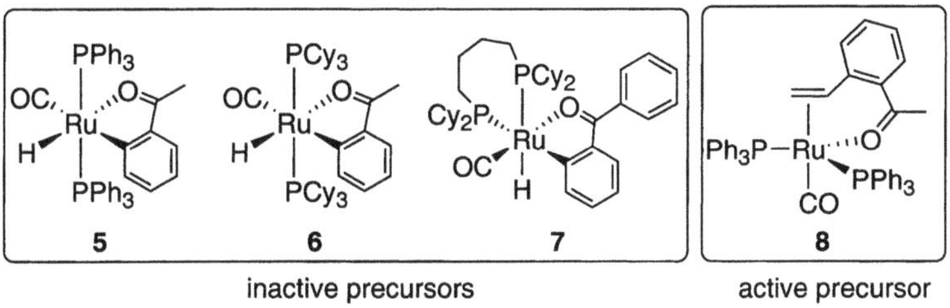

Fig. 5 Plausible intermediates of C–H/olefin coupling

Several attempts have been made to understand the reaction mechanism and the intermediates involved in the catalytic reaction. For the reaction of aromatic ketones with olefins $RuH_2(CO)(PPh_3)_3$, $RuH_2(PPh_3)_4$, $Ru(CO)_2(PPh_3)_3$, and $Ru(CO)_3(PPh_3)_2$ show catalytic activity, but $Ru_3(CO)_{12}$ does not [10, 13]. These results suggest that neither H nor CO is a necessary ligand and a zero-valent ruthenium having at least two PPh_3 constitutes the essential part of the catalyst [10]. Whittlesey et al. [38] synthesized $RuH(o\text{-}C_6H_4C(O)CH_3)(CO)(PPh_3)_2$ (**5**) and examined the catalytic activity of this ortho-metalated complex. Chaudret et al. [29] prepared a similar ortho-metalated $RuH(o\text{-}C_6H_4C(O)CH_3)(CO)(PCy_3)_2$ (**6**), which is a PCy_3 analogue of **5**. Fogg et al. [39] reported another type of an ortho-metalated ruthenium complex, $RuH(o\text{-}C_6H_4C(O)\ Ph(CO)[Cy_2P(CH_2)_4PCy_2]$ (**7**), prepared by the reaction of $RuH_2[Cy_2P(CH_2)_4PCy_2](CO)$ with benzophenone (Fig. 5). The hydride ligand is located at the apical position. This stereochemistry around the ruthenium center is different from the stereochemistry in the complexes of Whittlesey et al. and Chaudret et al. The catalytic activities of **5**–**7** for the reaction of aromatic ketones with olefin were examined. However, these three complexes are ineffective for this coupling reaction. It was proposed that the CO ligand suppresses the catalytic activity of the ruthenium complexes since Trost et al. [31] reported that a CO atmosphere completely inhibited the $RuH_2(CO)(PPh_3)_3$-catalyzed C–H/olefin coupling. In some cases, ruthenium complexes having a CO ligand show catalytic activity for the reaction of aromatic ketones with olefins. For example, Hiraki et al. [40] carried out an NMR study of the $RuH_2(CO)(PPh_3)_3$-catalyzed reaction of aromatic ketones with olefins and found that several ruthenium hydride species were present during the catalytic reaction on the basis of ^1H and ^{31}P NMR spectroscopy. They concluded that the carbonyl ligand is bound to the ruthenium throughout the catalytic reaction. Weber et al. [41] synthesized a zero-valent ruthenium complex, Ru(*o*-vinylacetophenone)$(CO)(PPh_3)_2$ (**8**) (Fig. 5). Interestingly, this complex had catalytic activity for the copolymerization of acetophenone with 1,3-divinyltetramethyldisiloxane, although it contains a CO ligand. These results suggest that the relation between the structures of the catalyst precursor and the catalyst activity is currently poorly understood and it is premature to conclude that the presence of a CO ligand on the ruthenium center retards catalytic activity. Further studies to elucidate the structure of the actual active species are awaited.

2.2 Addition to Acetylenes

Substituted styrenes and vinylic compounds are versatile intermediates in organic synthesis, and various methods for achieving them have been published in the literature [42–44]. The addition of C–H bonds to olefins gives alkylation products. In the case of a reaction using acetylenes as an acceptor of the C–H bond, alkenylation can be accomplished. The first example of aromatic C–H/acetylene coupling was reported by Yamazaki et al. in 1979 [45]. Later some catalytic reactions concerning aromatic C–H/acetylene coupling using rhodium [46], iridium [47], and palladium [48] were developed. As regards the ruthenium-catalyzed C–H/acetylene coupling, only three studies have been reported [49–51].

The successful result of Murai et al. [49] is shown in (Eq. 26). When 1-trimethylsilylpropyne is used, the desired coupling product is obtained in excellent yield and the regiochemical and stereochemical outcome is perfect [49]. The E isomer is the predominant product. This result indicates that the addition of C–H bonds to the C–C triple bond proceeds with syn selectivity. In the case of the reaction with 1-trimethylsilyl-1-octyne, stereoselectivity is slightly decreased. This suggests that the small difference in steric bulkiness between methyl and hexyl groups affects the stereoselectivity.

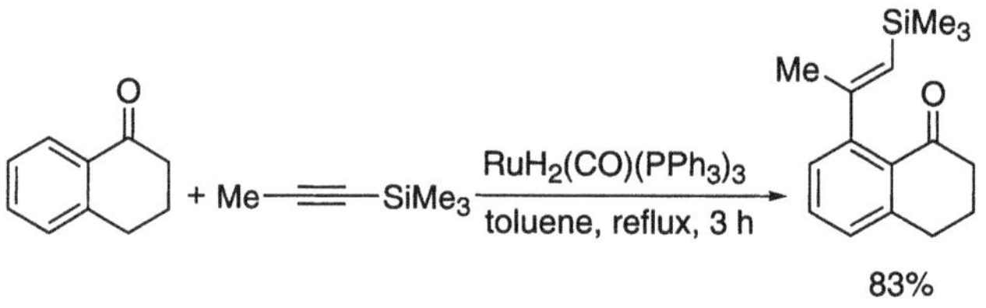

(26)

Woodgate et al. [51] applied the C–H/acetylene coupling to the ortho-selective alkenylation of terpene derivatives (Eq. 27). The basic feature of this reaction is the same as the alkenylation reaction of Murai et al. The combination of acetophenone and diynes provides a new entry for the copolymerization of aromatic ketones with acetylenes. Weber et al. [50] studied extensive reactions of ruthenium-catalyzed C–H/acetylene coupling with respect to the step-growth copolymerization of aromatic ketones and acetylenes (Eq. 28). These coupling reactions provide a new route to the preparation of trisubstituted styrene derivatives.

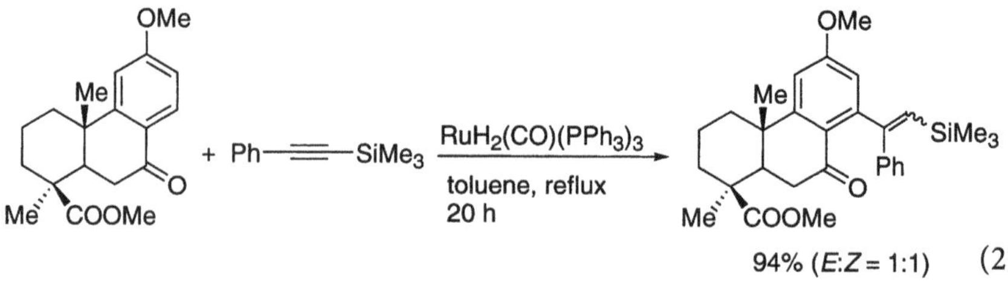

(27)

$RuH_2(CO)(PPh_3)_3$, toluene, 135 °C, 24 h; 79% (28)

2.3 Arylation

Additions of aromatic C–H bond to olefins and acetylenes result in the formation of aryl–alkyl and aryl–alkenyl bonds. This type of addition reaction is not applicable to aryl–aryl bond formation. Catellani and Chiusoli [52] reported the first example of this type of arylation in 1985. To date, several arylation reactions of aromatic rings have been developed. In almost all cases, C–H bond cleavage proceeds through electrophilic substitution with transition-metal complexes [53].

In 2001, Oi et al. [54] reported on the ruthenium(II) phosphine catalyzed regioselective arylation of 2-arylpyridines using aryl halides (Eq. 29). C–C bond formation occurs predominantly at the position ortho to the pyridyl group. The same catalyst system is also effective for the arylation of aromatic imines (Eq. 30) [55]. Although the reaction mechanism has not been elucidated, it was proposed that a tetravalent arylruthenium complex, for example, $Ru(Ph)(Br)(Cl)_2(L)_n$, reacts electrophilically with the arylimines. Therefore, C–H bond cleavage is believed to proceed via an electrophilic substitution pathway.

+ PhBr; $[RuCl_2(\eta^6\text{-}C_6H_6)]_2$, PPh_3; NMP, K_2CO_3, 120 °C, 20 h; 95% (29)

+ PhBr; $[RuCl_2(\eta^6\text{-}C_6H_6)]_2$, PPh_3; NMP, K_2CO_3, 120 °C, 20 h; 90%; Ar = 4-$MeOC_6H_4$ (30)

Very recently, Kakiuchi et al. [56] reported that the ruthenium-catalyzed coupling reaction of aromatic ketones with arylboronates resulted in ortho-arylated aromatic ketones (Eq. 31). This arylation reaction using arylboronates can be applied to a variety of aromatic ketones. They concluded that this reaction involves an oxidative addition of a C–H bond to the Ru(0) species and transmetalation from the arylboron compounds to the ruthenium complex.

+ Ph–B ; $RuH_2(CO)(PPh_3)_3$; toluene, reflux, 1 h ; Ph ; 95% (31)

2.4 Silylation

The direct silylation of C–H bonds with hydrosilanes or disilanes is one of the simplest procedures for obtaining arylsilanes. Curtis et al. [57] reported, to the best of our knowledge, the first example of the dehydrogenative silylation of benzene with pentamethyldisiloxane using an $IrCl(CO)(PPh_3)_2$ catalyst under thermal reaction conditions in the absence of a hydrogen acceptor. Unfortunately, however, the efficiency and the selectivity of this reaction were low. After this discovery, several attempts to achieve a high efficiency and selectivity were made.

In 1994, Berry et al. [58] reported the first example of the ruthenium-catalyzed silylation of arene C–H bonds. In this study, they reported that (η^6-arene)$Ru(H)_2(SiEt_3)_2$ and (η^5-C_5Me_5)$Rh(H)_2(SiEt_3)_2$ catalyze the transfer dehydrogenative coupling of triethylsilane in the presence of a hydrogen scavenger to give the dimer of the hydrosilane (Eq. 32). They later applied this catalytic system to the silylation of arenes having an electron-withdrawing substituent (Eq. 33) [59]. The relative reactivity ratios of the arylsilanes to phenylsilane are CF_3 (2.8)>F (1.4)>H (1.0)>CH_3 (0.32). This indicates that an electron-withdrawing group enhances the C–H functionalization. The silylation procedure of Berry et al. is promising, but the low regioselectivity poses an inevitable drawback.

Et_3SiH + Bu^t ; (η^5-C_5Me_5)$Rh(H)_2(SiEt_3)_2$; c-C_6H_{12}, 150 °C, 12 h

HEt_2Si–C(Me)(H)–$SiEt_3$ + Et_3Si Bu^t + Et_3Si Bu^t (32)

83% 9% 3%

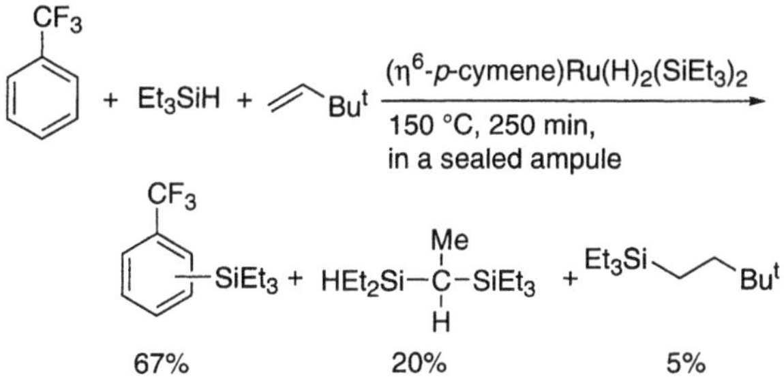

$$\text{PhCF}_3 + Et_3SiH + \text{CH}_2{=}\text{CHBu}^t \xrightarrow[\text{150 °C, 250 min, in a sealed ampule}]{(\eta^6\text{-}p\text{-cymene})Ru(H)_2(SiEt_3)_2} CF_3C_6H_4SiEt_3\ (67\%) + HEt_2Si{-}C(Me)(H){-}SiEt_3\ (20\%) + Et_3SiCH_2CH_2Bu^t\ (5\%) \quad (33)$$

Murai and coworkers [60–62] applied the chelation-assisted C–H bond cleavage protocol, one of the most reliable methods for attaining high regioselectivity, to the silylation reaction. The $Ru_3(CO)_{12}$-catalyzed silylation of aryloxazolines with hydrosilanes gives ortho-selective silylation products in good-to-excellent yields (Eq. 34) [60]. For the dehydrogenative silylation of C–H bonds, the use of an olefin as a hydrogen scavenger is required for the reaction to proceed in a catalytic manner. Triorganosilane, especially triethylsilane, is highly reactive. The functional group compatibility of this reaction is high. This reaction is tolerant to both electron-donating (Me, OMe, and NMe_2) and electron-withdrawing (CF_3 and F) groups.

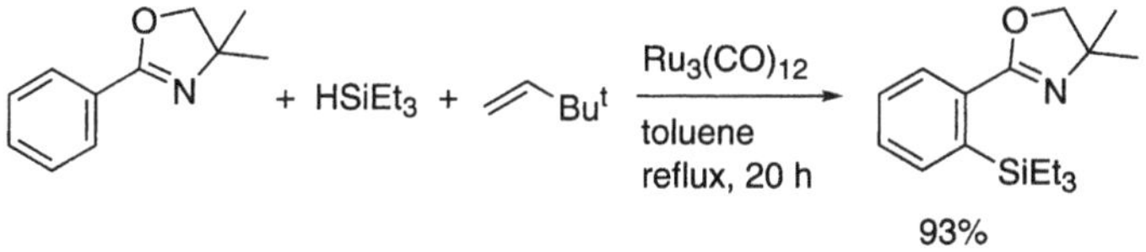

$$\text{2-phenyl-4,4-dimethyloxazoline} + HSiEt_3 + \text{CH}_2{=}\text{CHBu}^t \xrightarrow[\text{toluene, reflux, 20 h}]{Ru_3(CO)_{12}} \text{2-(2-SiEt}_3\text{-phenyl)-4,4-dimethyloxazoline}\ (93\%) \quad (34)$$

Aromatic imines are also effective in this silylation reaction (Scheme 2) [62]. The silylation products are obtained in high yields. A nitrogen atom in a heteroaromatic ring can also function as a directing group. Several azoles, such as phenyltetrazoles and phenylimidazoles, are also effective. In the case of the reaction of 2-(1-naphthyl)-3-methylpyridine, the silylation product is obtained in quantitative yield. This result indicates an important feature of this silylation reaction. In the C–H bond cleavage step for the C–H/SiR_3 coupling, the conjugation between the directing group and the aromatic ring is not so important because 2-(1-naphthyl)-3-methylpyridine attains a coplanar geometry with great difficulty, owing to steric repulsion between the methyl group of the pi-

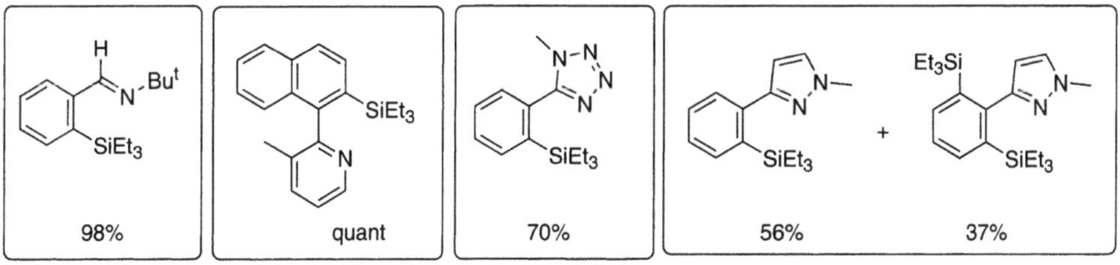

Product	Yield
H, N-Bu^t imine, SiEt$_3$	98%
1-naphthyl-3-methylpyridine, SiEt$_3$	quant
N-methyltetrazolyl, SiEt$_3$	70%
N-methylpyrazolyl, SiEt$_3$ + Et$_3$Si, N-methylpyrazolyl, SiEt$_3$	56% + 37%

reaction time = 20 h

Scheme 2 Silylation of aromatic compounds

coline moiety and the peri-hydrogen of the naphthalene moiety. Much more promising results were observed when *N,N*-dimethylbenzylamine, 2-benzylpyridine, and 2-pyridyl(phenyl)ether were used in the silylation reaction; the silylation products are obtained in high yields in an ortho-selective manner (Eqs. 35, 36) [61]. These results suggest that predicting the relationship between the structures of substrates and reactivity is difficult.

$Ru_3(CO)_{12}$, norbornene; toluene, 20 h, reflux (bath. temp. 115 °C) (35)

$Ru_3(CO)_{12}$, norbornene; toluene, 20 h, reflux (bath. temp. 115 °C)

R	X	mono-$SiEt_3$	bis-$SiEt_3$
R = H	X = CH_2	77%	13%
R = H	X = O	30%	55%
R = Cl	X = O	30%	51%

(36)

Murai et al. [63] reported on a unique system for the dehydrogenative silylation of heteroaromatic compounds in which triorganovinylsilane was used as a silylating reagent. In this reaction, the vinyl moiety functions as a hydrogen acceptor. Thus, ethylene should be generated after the reaction. When the reaction of 3-acetylthiophene with trimethylvinylsilane is conducted using $Ru_3(CO)_{12}$ as a catalyst, silylation occurs at the 2-position of 3-acetylthiophene (Eq. 37). The important step of this reaction is a β-silyl elimination, yielding a metal silyl species [64]. This silylation protocol using vinylsilanes can be applied only to heteroaromatic compounds.

$Ru_3(CO)_{12}$; toluene, 115 °C, 20 h; 64% (37)

3
Addition of Olefinic and Aliphatic C–H Bonds to C–C Multiple Bonds

3.1
Addition to Olefins

In this section, the catalytic functionalization of olefinic and aliphatic C–H bonds is discussed. Murai et al. [65, 66] reported that olefinic C–H bonds in conjugated enones are able to add across C–C double bonds with the aid of the $RuH_2(CO)(PPh_3)_3$ catalyst (Eq. 38). Among the cyclohexenes investigated, 1-pivaloyl-1-cyclohexene exhibits a high reactivity and the presence of an oxygen atom at the allylic position in the six-membered ring increases the reactivity of the enones. The reactions of conjugated ketones, esters, and amides with olefins provide the corresponding β-alkylation products in good-to-excellent yields (Table 3) [65]

$$\xrightarrow[\text{toluene, 15 h, reflux (135 °C, bath temp.)}]{RuH_2(CO)(PPh_3)_3} \quad 96\% \tag{38}$$

These reactions can be applied to an acyclic system [66]. When reactions of *trans*-4,4-dimethyl-1-phenyl-1-peneten-3-one with styrene and triethoxyvinylsilane are conducted using $RuH_2(CO)(PPh_3)_3$ as the catalyst, the expected olefinic C–H/olefin coupling products are obtained in good yields (Eq. 39) [66].

The reaction of conjugate enals having a heteroatom at the β-position with olefins proceeds with the aid of $RuH_2(CO)(PPh_3)_3$. The reaction of 5,6-dihydro-4*H*-pyran-3-carboxaldehyde with styrene gives the alkylation product in quan-

Table 3 Catalytic reaction of olefinic compounds with olefins

$$\xrightarrow[\text{toluene, reflux}]{\text{catalyst } RuH_2(CO)(PPh_3)_3}$$

run	X	R	Y	time	yield
1	CH_2	CH_3	$Si(OEt)_3$	24 h	50%
2	CH_2	^tBu	$Si(OEt)_3$	0.5 h	quant
3	O	CH_3	$Si(OEt)_3$	10 h	quant
4	O	^tBu	c-C_6H_{11}	12 h	98%
5	O	^tBu	2-pyridyl	12 h	81% (n:i = 78:22)
6	O	OMe	$Si(OEt)_3$	24 h	73%
7	O	N^iPr_2	$Si(OEt)_3$	24 h	73%

$$\text{PhCH=CHC(O)Bu}^t + \text{CH}_2\text{=CHY} \xrightarrow[\text{toluene, reflux}]{\text{RuH}_2\text{(CO)(PPh}_3)_3} \text{product} \quad (39)$$

Y = Ph 3 h 64%
Y = $Si(OEt)_3$ 1 h 79% (E:Z = 96:4)

titative yield (Eq. 40) [21]. For this reaction, the heteroatom of the enals is essential in achieving a catalytic reaction. This substituent is believed to suppress undesired decarbonylation reactions by electronic effects (Fig. 4) [21].

$$+ \text{CH}_2\text{=CHPh} \xrightarrow[\text{toluene, reflux, 24 h}]{\text{RuH}_2\text{(CO)(PPh}_3)_3} \quad (40)$$

quant. (n/i = 75/25)

Trost et al. [31] reported on a similar coupling reaction of a conjugated ester with olefins using $RuH_2(CO)(PPh_3)_3$ (Eq. 41). Both cyclic and acyclic conjugated esters can be applied to the coupling reaction. This coupling reaction tolerates various functional groups on the ester moiety.

$$+ \text{CH}_2\text{=CHSi(OEt)}_3 \xrightarrow[\text{toluene, reflux, 18 h}]{\text{RuH}_2\text{(CO)(PPh}_3)_3} \quad (41)$$

97%

Intramolecular olefinic C–H/olefin coupling with the aid of $Ru(CO)_3(PPh_3)_2$, which is also effective for the reaction of aromatic ketones with olefins, yields the carbocyclic compounds in excellent yield (Eq. 42) [67]. This type of cyclization reaction can be extended to an asymmetric version when the $[RhCl(coe)_2]_2$/PPFOMe catalyst system is employed [68].

$$\xrightarrow[\text{THF, reflux, 40 h}]{\text{Ru(CO)}_3\text{(PPh}_3)_2} \quad (42)$$

quant

Catalytic C–C bond formation via sp^3 C–H bond cleavage represents the ultimate reaction in organic synthesis. A relatively ideal catalytic reaction system involves the use of sp^3 C–H bonds adjacent to a heteroatom such as nitrogen and oxygen atoms. Recently, Jun et al. [69] succeeded in the $Ru_3(CO)_{12}$-catalyzed alkylation of an sp^3 C–H bond α to the nitrogen atom in benzyl-(3-methyl-2-pyridinyl)amine by means of chelation assistance (Eq. 43). In this case, the coordination of the pyridine nitrogen to the ruthenium complex followed by C–H bond cleavage, which allows the formation of a five-membered ruthenacycle, was proposed to be important in this catalytic reaction. Murai et al. [70] also reported on the ruthenium-catalyzed coupling of 1-(2-pyridiny)-1,2,3,4-tetrahydroquinoline (Eq. 44). The use of 2-propanol as a solvent dramatically improves the yield of the product.

$Ru_3(CO)_{12}$, toluene, 130 °C, 6 h; + C_4H_9; NH; Ph; N; 95% (43)

$Ru_3(CO)_{12}$, CO, 2-PrOH, 140 °C, 20 h; N; 73% (44)

Transition-metal-catalyzed annulation reactions of anilines with tertiary amines is another protocol for the catalytic functionalization of sp^3 C–H bonds (Eq. 45). Several reaction systems resulting in the formation of heteroaromatic compounds which are modifications of the preceding annulation reaction using aniline and ethylene with the aid of rhodium catalyst [71] have recently been developed [72].

NH_2 + Bu_3N; $RuCl_3$-nH_2O, dppm, $SnCl_2$-nH_2O, 1-hexene, dioxane, 180 °C, 20 h; N; 51% (45)

3.2 Addition to Acetylenes

Reactions of aromatic compounds with acetylenes give styrene derivatives. In the case of the reaction of α,β-enones, conjugated dienones would be expected. The olefinic C–H/acetylene coupling using several conjugate enones was examined (Eq. 46) [73]. Cyclohexene derivatives and dihydropyran derivatives are also applicable to this coupling reaction. This reaction gives highly congested conjugate dienones. The reaction using phenyl(trimethylsilyl)acetylene results in regioselective alkenylation. This regioselectivity is the same as in the reaction of aromatic ketones (Eq. 46, run 3]. Trost et al. [31] also reported on the alkenylation of α,β-unsaturated esters with acetylenes using $RuH_2(CO)(PPh_3)_3$ catalyst (Eq. 47).

Bu^t + Ph—≡—R $\xrightarrow[\text{toluene, reflux}]{RuH_2(CO)(PPh_3)_3}$ Bu^t R Ph (46)

run				
1	X = CH_2	R = Ph	2 h	85% (60:40)
2	X = O	R = Ph	1 h	71% (*E*:*Z* = 76:24)
3	X = O	R = $SiMe_3$	1 h	96% (*E*:*Z* = 94:6)

OMe + C_4H_9—≡—$SiMe_3$ $\xrightarrow[\text{THF, reflux}]{RuH_2(CO)(PPh_3)_3}$ OMe $SiMe_3$ C_4H_9 82% (47)

4
Reaction of Aldehydes and Related Compounds

The addition of a C–H bond of a formyl group to a C–C multiple bond is a highly useful method for synthesizing various types of ketones. The transition-metal-catalyzed intramolecular cyclization of enals to the corresponding ketones has been extensively studied, since this methodology provides a new route to the construction of a cyclopentanone framework from readily obtainable 1,4-pentanals [74]. The asymmetric version of this type of cyclization is of current interest. For these reactions, rhodium complexes often show high activity. To the best of our knowledge, there is one example of the ruthenium-catalyzed intramolecular hydroacylation of olefins. Eilbracht et al. [75] reported ruthenium-catalyzed one-pot synthesis of cyclopentanone from allyl vinyl ether via tandem Claisen rearrangement and hydroacylation (Eq. 48). This procedure requires a high temperature (140–220 °C) and also requires alkyl or aryl substituents at the terminal position of the allylic double bond to prevent undesirable double-bond migration in the intermediary formed, an unsaturated aldehyde.

$\xrightarrow[\text{octane, CO 50 bar; 200 °C, 16 h}]{RuCl_2(PPh_3)_3}$ 73% (48)

Research on intermolecular hydroacylation has also attracted considerable attention. The transition-metal-catalyzed addition of a formyl C–H bond to C–C multiple bonds gives the corresponding unsymmetrically substituted ketones. For the intermolecular hydroacylation of C–C multiple bonds, ruthenium complexes, as well as rhodium complexes, are effective [76–84]. In this section, intermolecular hydroacylation reactions of alkenes and alkynes using ruthenium catalysts are described.

In 1980, Miller et al. [76] reported the first example of an intermolecular hydroacylation of an aldehyde with an olefin to give a ketone, during their studies of the mechanism of the rhodium-catalyzed intramolecular cyclization of 4-pentenal using ethylene-saturated chloroform as the solvent. Later James and Young [77] reported that the reaction of propionaldehyde with ethylene can be conducted in the presence of $RuCl_2(PPh_3)_3$ as the catalyst without any solvent at 210 °C, resulting in the formation of 3-pentanone in 2–4% yield (turnover number of 230) (Eq. 49).

$$\text{EtCHO} + \text{=} \xrightarrow[\text{210°C, 18 h}]{RuCl_2(PPh_3)_3} \text{3-pentanone} \quad \text{2-4\% yield (230 TON)} \tag{49}$$

Watanabe et al. [78] reported that the addition of C–H bonds in aldehydes to olefins took place efficiently with the aid of $Ru_3(CO)_{12}$ under a CO atmosphere at 200 °C (Eq. 50). In the case of the reaction with 1-hexene, a mixture of linear and branched ketones was obtained in 35% and 12% yields, respectively. To accomplish this reaction in a catalytic manner, the presence of carbon monoxide appears to be essential for suppressing the decarbonylation of aldehydes and for stabilizing the active catalyst species on the basis of the following observations:

1. When the reaction of $Ar^{13}CHO$ with cyclohexene was carried our under a ^{12}CO atmosphere, the ^{12}CO-incorporated ketone was obtained in addition to the corresponding ^{13}CO-enriched ketone (Eq. 50).
2. When the reaction of the aldehyde was carried out under an argon atmosphere, a variety of products were produced.

$$PhC^{13}HO + \text{cyclohexene} \xrightarrow[\text{CO 20 kg cm}^{-2}\text{, 200 °C, 48 h}]{Ru_3(CO)_{12}} Ph^{13}C(O)C_6H_{11} + Ph^{12}C(O)C_6H_{11} \quad 44\%\ (1:2) \tag{50}$$

Later, they also reported an intermolecular hydroacylations of 1,3-dienes with aromatic aldehydes yielding the corresponding β,γ-unsaturated ketones (Eq. 51) [79]. This reaction does not require a CO atmosphere. The addition of formyl C–H bond in formic acid esters and amides to olefins and conjugate

+ PhCHO $\xrightarrow[\text{120 °C, 15 h}]{\text{Ru(cod)(cot)/PPh}_3}$ Ph (51)

dienes proceeds with the aid of a ruthenium catalyst [80–82]. For the reaction of alkylformates, a $Ru_3(CO)_{12}$–$(CH_3)_3NO(2H_2O)$ catalyst system showed a high catalyst activity (Eq. 52) [80]. This reaction requires the use of trimethylamine oxide as an additive. They proposed that this amine oxide was necessary to offer a coordinatively unsaturated position. The hydroamidation of cyclopentene takes place in the presence of $Ru_3(CO)_{12}$ as a catalyst (Eq. 53) [81]. Internal olefins such as cyclohexene and cyclopentene exhibit a high reactivity compared with terminal olefins. They later reported that the [bis(triphenylphosphoranylidene)ammonium] $[HRu_3(CO)_{11}]/PCy_3$ catalyst system showed a high catalytic activity [82]. The reaction of *N*-phenylformamide with norbornene in the presence of a [bis(triphenylphosphoranylidene)-ammonium]$[Ru_3H(CO)_{11}]$ catalyst gave the corresponding hydroamidation product in high yield.

$Ru_3(CO)_{12}$, $(CH_3)_3NO(2H_2O)$, CO 20 kg cm^{-2}, 200 °C, 6 h; 68% (52)

$Ru_3(CO)_{12}$, CO 20 kg cm^{-2}, 200 °C, 24 h; 90% (53)

Very recently, a new strategy for the hydroesterification and hydroamidation of olefins was reported by Chang and coworkers [83]. They used a chelation-assisted protocol for the hydroesterification of olefins. The reaction of 2-pyridylmethyl formate with 1-hexene in the presence of a $Ru_3(CO)_{12}$ catalyst gave the hydroesterification product in 98% yield as a mixture of linear and branched isomers (Eq. 54). The chain length of the methylene tether is important for a successful reaction. Thus, the reaction of 2-pyridyl formate (n=0) afforded 2-hydroxypyridine, a decarbonylation product, and the reaction of 2-pyridiylethyl formate (n=2) resulted in a low conversion (7% conversion) of the starting formate. From these results, the formation of a six-membered ruthenacycle intermediate is crucial for this chelation-assisted hydroesterification.

(54)

run

run	n			
1	n = 1	98% (linear:branched = 74:26)	<1%
2	n = 0		95%
3	n = 2		5%

Interestingly, however, in the case of the reaction of formamide, *N*-(2-pyridyl)formamide showed a high reactivity [84]. This result indicates that the reaction proceeds through a five-membered ruthenacycle intermediate. The olefins having a bulky substituent, such as *tert*-butyl and trimethylsilyl groups, exhibited a high regioselectivity.

(55)

69%

The unique transformation of formamides to ureas was reported by Watanabe and coworkers [85]. In place of carbon monoxide, formamide derivatives are used as a carbonyl source. The reaction of formanilide with aniline was conducted in the presence of a catalytic amount of $RuCl_2(PPh_3)_3$ in refluxing mesitylene, leading to *N,N′*-diphenylurea in 92% yield (Eq. 56) [85]. They proposed that the catalysis starts with the oxidative addition of the formyl C–H bond to the active ruthenium center. In the case of the reaction of formamide, $HCONH_2$, with amines, two molecules of the amine react with the amide to afford the symmetrically substituted ureas in good yields. This reaction evolves one molecule of NH_3 and one molecule of H_2.

$$\text{R-NH-C(=O)-H} + \text{H}_2\text{N-Ph} \xrightarrow[\text{mesitylene, reflux, 12 h}]{\text{RuCl}_2(\text{PPh}_3)_3} \text{Ph-NH-C(=O)-NH-Ph} \tag{56}$$

R = Ph 92%
R = H 77%

Chang et al. [84] reported on the unprecedented aminocarbonylation of aryl iodide. In this case, the $Ru_3(CO)_{12}$–$PdCl_2$ cooperative catalyst system is effective. The reaction of 4-acetyliodobenzene with *N*-(2-pyridyl)formamide with the aid of $Ru_3(CO)_{12}$ and $PdCl_2$ gives 4-acetyl-*N*-(2-pyridyl)benzamide in 83% yield (Eq. 57).

$$\text{4-IC}_6\text{H}_4\text{COCH}_3 + \text{HCONH(2-pyridyl)} \xrightarrow[\text{CH}_3\text{CN, NaHCO}_3,\ 135\,^\circ\text{C, 10 h}]{\text{Ru}_3(\text{CO})_{12},\ \text{PdCl}_2} \text{4-CH}_3\text{COC}_6\text{H}_4\text{CONH(2-pyridyl)}\ \ 83\% \tag{57}$$

In place of formamides, the use of alkyl formates resulted in the alkylation of arenes [86]. When the reaction of alkyl formates with arenes is conducted with the aid of the $Ru_3(CO)_{12}$ catalyst, decarboxylation of alkyl formate proceeds selectively and the subsequent alkylation of the arenes occurs with the evolution of molecular hydrogen (Eq. 58). This alkylation procedure is unique even though the site selectivity is low.

$$\text{C}_6\text{H}_5\text{CH}_3 + \text{H-C(=O)-OCH}_2\text{Ph} \xrightarrow[\text{mesitylene, reflux, 12 h}]{\text{RuCl}_2(\text{PPh}_3)_3} \text{CH}_3\text{C}_6\text{H}_4\text{CH}_2\text{Ph}\ \ 77\%\ (o{:}m{:}p = 40{:}8{:}52) \tag{58}$$

Another type of unique coupling reaction was reported by Jones and coworkers [87]. The low-valent ruthenium phosphine complex $RuH_2(dmpe)_2$ catalyzed intramolecular insertion of isocyanide into the benzyl C–H bond of 2,6-xylylisonitrile under thermal conditions (Eq. 59). Their finding provided a new route to the synthesis of indoles.

$$\text{2,6-(CH}_3)_2\text{C}_6\text{H}_3\text{NC} \xrightarrow[\text{C}_6\text{D}_6,\ 140\,^\circ\text{C, 25 h}]{\text{RuH}_2(\text{dmpe})_2} \text{7-methylindole}\ \ 98\% \tag{59}$$

5 Reaction of Active Methylene and Related Compounds

Recently, the transition-metal-catalyzed addition of active methylene C–H bonds to electron-deficient olefins having a carbonyl, a nitrile, or a sulfonyl group has been extensively studied by several research groups. In particular, the asymmetric version of this type of catalytic reaction provides a new route to the enantioselective construction of quaternary carbon centers [88]. Another topic of recent interest is the catalytic addition of active methylene C–H bonds to acetylenes, allenes, conjugate ene-ynes, and nitrile C–N triple bonds. In this section, the ruthenium-catalyzed addition of C–H bonds in active methylene compounds to carbonyl groups and C–C multiple bonds is described.

The addition of active methylene compounds to aldehydes and α,β-unsaturated carbonyl compounds is catalyzed by several transition-metal complexes. Murahashi and coworkers [89–92] reported on the $RuH_2(PPh_3)_4$-catalyzed addition of activated nitriles to aldehydes and ketones (Eq. 60). In the case of the reaction of the nitriles with aldehydes and ketones, condensation products corresponding to a Knoevenagel reaction are obtained in high yields. The quite similar Knoevenagel reaction of aldehydes with cyanoacetate using $RuH_2(PPh_3)_4$ was reported by Lin et al. [93] in 1993.

H, HO, O, + CO₂Et, CN, RuH₂(PPh₃)₄, THF, r.t., over night, H, HO, CO₂Et, CN (60)

Interestingly, the reaction of active methylene compounds having a nitrile group with α,β-unsaturated carbonyl compounds give Michael adducts without contamination by the corresponding aldol products (Eq. 61) [89–92]. Murahashi and coworkers [89–91] proposed that the addition of the C–H bond to a low-valent ruthenium constitutes the initial step. Recently, Takaya and Murahashi [94] applied their aldol and Michael addition reactions to solid-phase synthesis using polymer-supported nitriles.

EtO₂C, CO₂Et, NC, CO₂Et, Me, +, O, RuH₂(PPh₃)₄, THF, -78 °C, 6 h, O, EtO₂C, CO₂Et, NC, CO₂Et, Me, +, O, EtO₂C, CO₂Et, NC, CO₂Et, Me, 72% (97/3) (61)

Details of the mechanism of ruthenium-catalyzed aldol and Michael reactions of active methylene compounds having a nitrile group have been obtained by means of kinetic studies, X-ray analyses, and NMR studies [90, 91]. The stoichiometric reaction of $RuH(C_2H_4)(PPh_3)_2(PPh_2C_6H_4)$ with ethyl cyanoacetates gives *mer*-$RuH(NCCHCO_2Et)(NCCH_2CO_2Et)(PPh_3)_3$, which has been characterized by spectroscopic and analytical methods, with the liberation of a quantitative amount of ethylene (Eq. 62) [90, 95]. The IR spectrum of the complex showed ν(Ru–H) at around 1,960 cm^{-1} and the NMR spectrum also indicates the presence of a Ru–H bond. X-ray analysis indicates that both cyanoacetate molecules are bonded to the ruthenium center with a nitrogen atom of the cyano group. One cyanoacetate ligand is coordinated in the enolate form. Kinetic studies of the reaction of ethyl cyanoacetate with benzaldehyde were made using the hydrido(enolate)ruthenium(II) catalyst. The results suggest

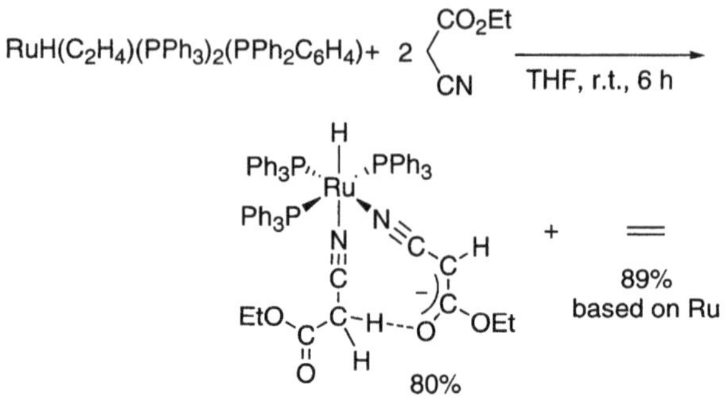

$$RuH(C_2H_4)(PPh_3)_2(PPh_2C_6H_4) + 2\ NCCH_2CO_2Et \xrightarrow{\text{THF, r.t., 6 h}} RuH(PPh_3)_3[NCCH_2CO_2Et][NC\text{-}CH\text{-}C(O)OEt] \ (80\%) + CH_2{=}CH_2 \ (89\% \text{ based on Ru}) \qquad (62)$$

that the rate is first order with respect to benzaldehyde and the ruthenium catalyst, and zero order with respect to ethyl cyanoacetate. The Michael reaction of nitriles with olefins having electron-withdrawing groups can be rationalized by the pathway shown in Scheme 3.

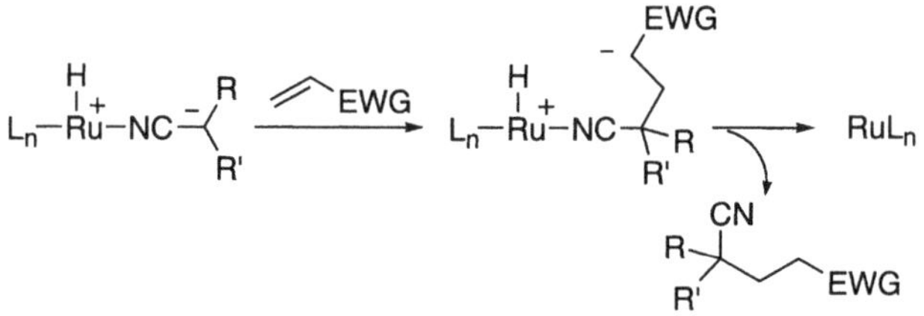

$$L_nRu^{+}(H){-}NC{-}C^{-}RR' \xrightarrow{CH_2{=}CH\text{-}EWG} L_nRu^{+}(H){-}NC{-}CRR'{-}CH_2CH^{-}{-}EWG \longrightarrow RuL_n + NC{-}CRR'{-}CH_2CH_2{-}EWG$$

Scheme 3 Plausible reaction pathway of the ruthenium-catalyzed Michael addition

In place of active methylene compounds having a nitrile group, malonates, β-ketoesters, 1,3-diketones, 1,1-disulfones, nitro compounds, Meldrum acid, and anthrone can also be used as the Michael donors for these ruthenium-catalyzed aldol and Michael reactions. The reaction proceeds well in acetonitrile under mild and neutral conditions (Eq. 63) [96]. The role of the phosphine ligand in the Michael addition reaction was investigated. When the reaction of dimethyl malonate with 3-butene-2-one was carried out in the presence of PPh_3 as the catalyst, a Michael reaction took place to some extent (Eq. 64). The use of trialkylphosphine improves the reaction rate as well as the yield of the addition product. Although triphenylphosphine exhibits catalytic activity for some of the reactions examined, very significant rate differences were found in Michael reactions catalyzed by $RuH_2(PPh_3)_4$ or triphenylphosphine (Eq. 64). Interestingly, the observation of identical chemoselectivity, regioselectivity, or stereoselectivity in reactions catalyzed by ruthenium complexes or triphenylphosphine suggests that similar pathways are followed in both processes. Dixneuf et al. [97] reported that a similar reaction of active methylene compounds with but-3-en-2-one using a non-hydride ruthenium complex, $[Ru(O_2CH)(CO)_2(PPh_3)]_2$, as a catalyst occurs (Eq. 65). In this case, a higher reaction temperature is required than for the ruthenium hydride complexes such as $RuH_2(PPh_3)_4$ [89–92].

PhO_2S, PhO_2S + (methyl vinyl ketone) $\xrightarrow[CH_3CN,\ r.t.,\ 24\ h]{RuH_2(PPh_3)_4}$ PhO_2S, PhO_2S … 94% (63)

MeO_2C, MeO_2C + (methyl acrylate, OMe) $\xrightarrow[CH_3CN,\ r.t.,\ 24\ h]{catalyst}$ MeO_2C, MeO_2C, CO_2Me, CO_2Me (64)

catalyst = $RuH_2(PPh_3)_4$ 99%
PPh_3 13%

(2-acetyl-1-tetralone) + (methyl vinyl ketone) $\xrightarrow[CH_3CN,\ 100\ °C,\ 20\ h]{[Ru(O_2CH)(CO)_2(PPh_3)]_2}$ … 98% (65)

Moreno-Mañas et al. [98] reported on a similar effect of triphenylphosphine for the Michael addition of active methylene compounds to π-acceptor olefins such as methyl vinyl ketone, acrylonitrile, and 2-vinylpyridine and dialkyl azodicarboxylates. They compared the reactivity of $RuH_2(PPh_3)_4$, $RuCl_2(PPh_3)_3$, and PPh_3 and concluded that for β-diketones, ketoesters, and ketoamides, triphenylphosphine released from the ruthenium complexes contributes totally or partially to the catalysis.

Dixneuf et al. [97] reported on a unique example of a Michael reaction using terminal acetylenes. Addition of the C–H bond in terminal alkynes to the C–C double bond in α,β-enones took place with the aid of $[Ru(O_2CH)(CO)_2PPh_3]_2$ or $[Ru(O_2CH)(CO)_2PMe_3]_2$ complexes as catalysts, giving γ,δ-ynones. The reaction of phenylacetylene with but-3-en-2-one afforded the corresponding ynone in 74% yield (Eq. 66). In the case of the reaction of alkylacetylenes, the use of $[Ru(O_2CH)(CO)_2PMe_3]_2$ as a catalyst is essential for attaining improved yield. The reaction with cyclohexenone was unsuccessful. This suggests that this reaction is sensitive to steric hindrance at the β-carbon. A similar conjugate addition of terminal acetylenes to α,β-enones was reported by Chang et al. [99]. The reaction of 1-decyne with phenyl vinyl ketone in the presence of $[RuCl_2(p\text{-cymene})]_2$ and pyrrolidine as catalysts gave the γ,δ-ynones in 98% yield (Eq. 67). This reaction is also sensitive to the steric factor. Amines appeared to be essential for the generation of catalytic active ruthenium acetylide species. A variety of alkynes, for example, trimethylsilylacetylene, 5-hexyn-1-ol, 5-chloropent-1-yne, and hex-5-ynenitrile, can be used for this addition reaction.

$$\mathrm{Ph{-}C{\equiv}C{-}H} + \mathrm{CH_2{=}CHC(O)CH_3} \xrightarrow[\mathrm{CH_3CN,\ 100\ ^\circ C,\ 20\ h}]{[\mathrm{Ru(O_2CH)(CO)_2(PPh_3)}]_2} \mathrm{Ph{-}C{\equiv}C{-}CH_2CH_2C(O)CH_3}\ \ 74\% \tag{66}$$

$$\mathrm{Ph{-}C{\equiv}C{-}H} + \mathrm{CH_2{=}CHC(O)Ph} \xrightarrow[\mathrm{benzene,\ 60\ ^\circ C,\ 12\ h}]{[\mathrm{RuCl_2(p\text{-}cymene)}]_2,\ \mathrm{pyrrolidine}} \mathrm{Ph{-}C{\equiv}C{-}CH_2CH_2C(O)Ph}\ \ 98\% \tag{67}$$

6 Conclusion

The use of C–H bonds is obviously one of the simplest and most straightforward methods in organic synthesis. From the synthetic point of view, the alkylation, alkenylation, arylation, and silylation of C–H bonds are regarded as practical tools since these reactions exhibit high selectivity, high efficiency, and are widely applicable, all of which are essential for practical organic synthesis. The hydroacylation of olefins provides unsymmetrical ketones, which are highly versatile synthetic intermediates. Transition-metal-catalyzed aldol and Michael addition reactions of active methylene compounds are now widely used for enantioselective and diastereoselective C–C bond formation reactions under neutral conditions.

In the past few years, the chemistry of the catalytic use of the C–H bond in organic synthesis has been rapidly expanding to various other fields, such as polymer chemistry. In the coming decade, it is likely that fascinating developments will be made for the direct use of C–H bonds in organic synthesis.

References

1. Larock RC (1989) Comprehensive organic transformations. VCH, New York
2. Tsuji J (1995) Palladium reagents and catalysts. Wiley, Chichester, UK
3. (a) Wakatsuki Y, Yamazaki H (1995) J Organomet Chem 500:349; (b) Naota T, Takaya H, Murahashi S-I (1998) Chem Rev 98:2599; (c) Trost BM, Toste FD, Pinkerton AB (2001) Chem Rev 101:2067
4. (a) Noyori R, Takaya H (1990) Acc Chem Res 23:345; (b) Noyori R, Okuma T (2001) Angew Chem Int Ed Engl 40:40; (c) Noyori R (2002) Angew Chem Int Ed Engl 41:2008
5. Murahashi S-I, Naota T, Kuwabara T, Saito T, Kumobayashi H, Akutagawa S (1990) J Am Chem Soc 112:7820
6. (a) Grubbs RH, Miller SJ, Fu GC (1995) Acc Chem Res 28:446; (b) Trnka TN, Grubbs RH (2001) Acc Chem Res 34:18
7. (a) Crabtree RH (1985) Chem Rev 85:245; (b) Shilov AE, Shul'pin GG (1997) Chem Rev 97:2879; (c) Crabtree RH (2001) J Chem Soc Dalton Trans 2437; (d) Labinger JA, Bercaw JE (2002) Nature 417:507
8. (a) Kakiuchi F, Murai S (1999) Top Organomet Chem 3:47; (b) Guari Y, Sabo-Etienne S, Chaudret S (1999) Eur J Inorg Chem 1047; (c) Ritleng V, Sirlin C, Pfeffer M (2002) Chem Rev 102:1731; (d) Kakiuchi F, Chatani N (2003) Adv Synth Catal 345:1077

9. Chatt J, Davidson JM (1965) J Chem Soc 843
10. Murai S, Kakiuchi F, Sekine S, Tanaka Y, Kamatani A, Sonoda M, Chatani N (1993) Nature 366:529
11. Kakiuchi F, Sato T, Yamauchi M, Chatani N, Murai S (1999) Chem Lett 19
12. (a) Murai S, Kakiuchi F, Sekine S, Tanaka Y, Kamatani A, Sonoda M, Chatani N (1994) Pure Appl Chem 66:1527; (b) Murai S, Chatani N, Kakiuchi F (1997) Pure Appl Chem 69:589; (c) Murai S, Chatani N, Kakiuchi F (1997) Catal SurvJpn 1:35; (d) Kakiuchi F, Murai S (2002) Acc Chem Res 35:826; (e) Kakiuchi F, Murai S (2003) Org Synth 80:104
13. Kakiuchi F, Sekine S, Tanaka Y, Kamatani A, Sonoda M, Chatani N, Murai S (1995) Bull Chem Soc Jpn 68:62
14. Sonoda M, Kakiuchi F, Kamatani A, Chatani N, Murai S (1996) Chem Lett 109
15. Kakiuchi F, Yamauchi M, Chatani N, Murai S (1996) Chem Lett 111
16. Kashiwagi K, Sugise R, Shimakawa T, Matuura T, Shirai M, Kakiuchi F, Murai S (1997) Organometallics 16:2233
17. (a) Sonoda M, Kakiuchi F, Chatani N, Murai S (1995) J Organomet Chem 504:151; (c) Sonoda M, Kakiuchi F, Chatani N, Murai S (1997) Bull Chem Soc Jpn 70:3117
18. Kakiuchi F, Sato T, Tsujimoto T, Yamauchi M, Chatani N, Murai S (1998) Chem Lett 1053
19. Kakiuchi F, Sonoda M, Tsujimoto T, Chatani N, Murai S (1999) Chem Lett 1083
20. Kakiuchi F, Le Gendre P, Yamada A, Ohtaki H, Murai S (2000) Tetrahedron: Asymmetry 11:2647
21. Kakiuchi F, Sato T, Igi K, Chatani N, Murai S (2001) Chem Lett 386
22. Kakiuchi F, Tsujimoto T, Sonoda M, Chatani N, Murai S (2001) Synlett 948
23. Kakiuchi F, Ohtaki H, Sonoda M, Chatani N, Murai S (2001) Chem Lett 918
24. Lewis LN, Smith JF (1986) J Am Chem Soc 108:2728
25. Westheimer FH (1956) In: Newman MS (ed) Steric effects in organic chemistry. Wiley, New York, pp 523–555
26. (a) H. Guo, Weber WP (1994) Polym Bull 32:525; (b) Lu P, Paulasaari JK, Bau R, Weber WP (1998) Organometallics 17:584; (c) Gupta SK; Weber WP (2002) Macromolecules 35:3369 and references therein
27. (a) Harris PWR, Woodgate PD (1996) J Organomet Chem 506:339; (b) Harris PWR, Woodgate PD (1997) J Organomet Chem 530:211; (c) Harris PWR, Rickard CEF, Woodgate PD (2000) J Organomet Chem 601:172
28. Grigg R, Savic V (1997) Tetrahedron Lett 38:5737
29. Guari Y, Sabo-Etienne S, Chaudret B (1998) J Am Chem Soc 120:4228
30. (a) Busch S, Leitner W (1999) Chem Commun 2305; (b) Busch S, Leitner W, (2001) Adv Synth Catal 343:192
31. Trost BM, Imi K, Davies IW (1995) J Am Chem Soc 117:5371
32. (a) Moritani I, Fujiwara Y (1967) Tetrahedron Lett 8:1119; (b) Moritani I, Fujiwara Y (1973) Synthesis 524
33. (a) Fujiwara Y, Moritani I, Danno S, Asano R, Teranishi S (1969) J Am Chem Soc 91:7166; (b) Tsuji J, Nagashima H (1984) Tetrahedron 40:2699
34. Weissman H, Song X, Milstein D (2001) J Am Chem Soc 123:337
35. (a) Misono A, Uchida Y, Hidai M, Kanai H (1967) Chem Commun 357; (b) Fukuoka A, Nagano T, Furuta S, Yoshizawa M, Hirano M, Komiya S (1998) Bull Chem Soc Jpn 71:1409 and references therein
36. (a) Chetcuti PA, Knobler CB, Hawthorne MF (1988) Organometallics 7:650; (b) Barrera J, Sabat M, Harman WD (1993) Organometallics 12:4381
37. (a) Matsubara T, Koga N, Musaev DG, Morokuma K (1998) J Am Chem Soc 120:12692; (b) Matsubara T, Koga N, Musaev DG, Morokuma K (2000) Organometallics 19:2318
38. Jazzar RFR, Mahon MF, Whittlesey MK (2001) Organometallics 20:3745

39. Drouin SD, Amoroso D, Yap GP, Fogg DE (2002) Organometallics 21:1042
40. Hiraki K, Ishimoto T, Kawano H (2000) Bull Chem Soc Jpn 73:2099
41. Lu P, Paulasaari J, Jin K Bau R, Weber WP (1998) Organometallics 17:584
42. (a) Corey EJ, Cheng X-M (1989) The logic of chemical synthesis. Wiley New York; (b) Fuhrhop J, Penzlin G (1994) Organic synthesis, 2nd edn. VCH, Weinheim
43. (a) Larock RC (1989) Comprehensive organic transformations. VCH, New York, pp 105–282; (b) Farina V, Krishnamurthy V, Scott WJ (1998) The Stille reaction. Wiley, New York
44. Mizoroki T, Mori K, Ozaki A (1971) Bull Chem Soc Jpn 44:581
45. (a) Hong P, Cho B-R, Yamazaki H (1979) Chem Lett 339; (b) Hong P, Cho B-R, Yamazaki H (1980) Chem Lett 507
46. (a) Halbritter G, Knoch F, Wolski A, Kisch H (1994) Angew Chem Int Ed Engl 33:1603; (b) Aulwurm UR, Melchinger JU, Kisch H (1995) Organometallics 14:3385; (c) Dürr U, Kisch H (1997) Synlett 1335; (d) Lim Y-G, Lee KH, Koo BT, Kang J-B (2001) Tetrahedron Lett 42:7609; (e) Hayashi T, Inoue K, Taniguchi N, Ogasawara M (2001) J Am Chem Soc 123:9918; (f) Campo MA, Larock RC (2002) J Am Chem Soc 124:14326
47. Sakaguchi S, Kubo T, Ishii Y (2001) Angew Chem Int Ed Engl 40:2534
48. (a) Satoh T, Nishinaka Y, Miura M, Nomura M (1999) Chem Lett 615; (b) Jia C, Piao D, Oyamada J, Lu W, Kitamura T, FujiwaraY (2000) Science 287:1992; (c) Jia C, Lu W, Oyamada J, Kitamura T, Matsuda K, Irie M, Fujiwara Y (2000) J Am Chem Soc 122:7252; (d) Jia C, Piao D, Kitamura T, Fujiwara Y (2000) J Org Chem (2000) 65:7516; (e) Jia C, Kitamura T, Fujiwara Y (2001) Acc Chem Res 34:633; (f) Oyamada J, Jia C, Fujiwara Y, Kitamura T (2002) Chem Lett 380
49. Kakiuchi F, Yamamoto Y, Chatani N, Murai S (1995) Chem Lett 681
50. Londergan TM, You Y, Thompson ME, Weber WP (1998) Macromolecules 31:2784
51. Harris PWR, Rickard CEF, Woodgate PD (1999) J Organomet Chem 589:168
52. (a) Catellani M, Chiusoli GP (1985) J Organomet Chem 286:13; (b) Catellani M, Chiusoli GP, Castagnoli C (1991) J Organomet Chem 407:C30
53. Dyker G (1999) Angew Chem Int Ed Engl 38:1698
54. Oi S, Fukita S, Hirata N, Watanuki N, Miyano S, Inoue Y (2001) Org Lett 3:2579
55. Oi S, Ogino Y, Fukita S, Inoue Y (2002) Org Lett 4:1783
56. Kakiuchi F, Kan S, Igi K, Chatani N, Murai S (2003) J Am Chem Soc 125:1698
57. Gustavson WA, Epstein PA, Curtis MD (1992) Organometallics 1:884
58. Djurovich PI, Dolich AR, Berry DH (1994) J Chem Soc Chem Commun 1897
59. Ezbiansky K, Djurovich PI, LaForest M, Sinning DJ, Zayes R, Berry DH (1998) Organometallics 17:1455
60. Kakiuchi F, Igi K, Matsumoto M, Chatani N, Murai S (2001) Chem Lett 422
61. Kakiuchi F, Igi K, Matsumoto M, Hayamizu T, Chatani N, Murai S (2002) Chem Lett 396
62. Kakiuchi F, Matsumoto M, Tsuchiya K, Igi K, Hayamizu T, Chatani N, Murai S J Organomet Chem 686:134
63. Kakiuchi F, Matsumoto M, Sonoda M, Fukuyama T, Chatani N, Murai S, Furukawa N, Seki Y (2000) Chem Lett 750
64. (a) Seki Y, Takeshita K, Kawamoto K (1989) J Organomet Chem 369:117; (b) Wakatsuki Y, Yamazaki H, Nakano M, Yamamoto Y (1991) J Chem Soc Chem Commun 703; (c) Kakiuchi F, Yamada A, Chatani N, Murai S (1998) Organometallics 18:2033
65. Kakiuchi F, Tanaka Y, Sato T, Chatani N, Murai S (1995) Chem Lett 679
66. Sato T, Kakiuchi F, Chatani N, Murai S (1998) Chem Lett 893
67. (a) Fujii N, Kakiuchi F, Chatani N, Murai S (1996) Chem Lett 939; (b) Fujii N, Kakiuchi F, Yamada A, Chatani N, Murai S (1998) Bull Chem Soc Jpn 71:285
68. Fujii N, Kakiuchi F, Yamada A, Chatani N, Murai S (1997) Chem Lett 425
69. Jun C-H, Hwang D-C, Na S-J (1998) Chem Commun 1405

70. Chatani N, Asaumi T, Yorimitsu S, Ikeda T, Kakiuchi F, Murai S (2001) J Am Chem Soc 123:10935
71. Diamond SE, Szalkiewicz A, Mares F (1979) J Am Chem Soc 101:490
72. Cho CS, Kim TK, Kim BT, Kim T-J, Shim SC (2002) J Organomet Chem 650:65
73. Kakiuchi F, Uetsuhara T, Tanaka Y, Chatani N, Murai S (2002) J Mol Catal A 182–183:511
74. Bosnich B (1998) Acc Chem Res 31:667 and references therein
75. Eilbracht P, Gersmeier A, Lennartz D, Huber T (1995) Synthesis 330
76. (a) Lochow CF, Miller RG (1976) J Am Chem Soc 98:1281; (b) Vora KP, Lochow CF, Miller RG (1980) J Organomet Chem 192:257
77. (a) James BR, Young CG (1983) J Chem Soc Chem Commun1215; (b) James BR, Young CG (1985) J Organomet Chem 285:321
78. (a) Kondo T, Tsuji Y, WatanabeY (1987) Tetrahedron Lett 28:6229; (b) Kondo T, Akazome M, Tsuji Y, Watanabe Y (1990) J Org Chem 55:1286
79. Kondo T, Hiraishi N, Morisaki Y, Wada K, Watanabe Y, Mitsudo T (1998) Organometallics 17:2131
80. (a) Tsuji Y, Yoshii S, Ohsumi T, Kondo T, Watanabe Y (1987) J Organomet Chem 331:379; (b) Kondo T, Yoshii S, Tsuji Y, Watanabe Y (1989) J Mol Catal 50:31
81. Kondo T, Tantayanon S, Tsuji Y, Watanabe Y (1989) Tetrahedron Lett 30:4137
82. Kondo T, Okada T, Mitsudo T (1999) Organometallics 18:4123
83. (a) Ko S, Na Y, Chang S (2002) J Am Chem Soc 124:750; (b) Na Y, Ko S, Hwang LK, Chang S (2003) Tetrahedron Lett 44:4475
84. Ko S, Han H, Chang S (2003) Org Lett 5:2687
85. (a) Kotachi S, Tsuji Y, Kondo T, Watanabe Y (1990) J Chem Soc Chem Commun 594; (b) Kondo T, Kotachi S, Tsuji Y, Watanabe Y, Mitsudo T (1997) Organometallics 16: 2562
86. Kondo T, Kajiya S, Tantayanon S, Watanabe Y (1995) J Organomet Chem 489:83
87. (a) Jones WD, Kosar WP (1986) J Am Chem Soc 108:5640; (b) Hsu GC, Kosar WP, Jones WD (1994) Organomet Chem 13:385
88. Corey EJ, Guzman-Perez A (1998) Angew Chem Int Ed Engl 37:388
89. Naota T, Taki H, Mizuno M, Murahashi S-I (1989) J Am Chem Soc 111:5954
90. Murahashi S-I, Naota T, Taki H, Mizuno M, Takaya H, Komiya S, Mizuho Y, Oyasato N, Hiraoka M, Hirano M, Fukuoka A (1995) J Am Chem Soc 117:12436
91. Murahashi S-I, Naota T (1996) Bull Chem Soc Jpn 69:1805
92. (a) Murahasahi S-I, Take K, Naota T, Takaya H (2000) Synlett 1016; (b) Murahashi S-I, Takaya H (2000) Acc Chem Res 33:225; (c) Naota T, Tannna A, Murahashi S-I (2001) Chem Commun 63; (d) Naota T, Tannna A, Kamuro S, Murahashi S-I (2002) J Am Chem Soc 124:6842
93. Lin Y, Zhu X, Xian M (1993) J Organomet Chem 448:215
94. Takaya H, Murahashi S-I (2001) Synlett 991
95. (a) Komiya S, Mizuho Y, Kasuga N (1991) Chem Lett 2127; (b) Hirano M, Hirai M, Ito Y, Tsurumaki T, Baba A, Fukuoka A, Komiya S (1998) J Organomet Chem 569:3; (c) Komiya S, Hirano M (2003) Dalton Trans 1439
96. Gómez-Bengoa E, Cuerva JM, Mateo C, Echavarren AM (1996) J Am Chem Soc 118:8553
97. Picquet M, Bruneau C, Dixneuf PH (1999) Tetrahedron 55:3937
98. Lumbierres M, Marchi C, Moreno-Mañas M, Sebastián RM, Vallribera A, Lago E, Molins E (2001) Eur J Org Chem 2321
99. Chang S, Na Y, Choi E, Kim S (2001) Org Lett 3:2089

Topics Organomet Chem (2004) 11: 81–92
DOI 10.1007/978-3-540-40027-1

Cyclopropanation with Ruthenium Catalysts

Hisao Nishiyama (✉)

Department of Applied Chemistry, Graduate School of Engineering, Nagoya University, 464-8603 Chikusa, Nagoya, Japan
hnishi@apchem.nagoya-u.ac.jp

Abstract In this decade, a variety of ruthenium complexes have been intensively investigated that exhibit catalytic activity for cyclopropanation of olefins and diazoester derivatives producing cyclopropanecarboxylates with high stereoselectivity (*trans*>*cis* or *cis*>*trans*) and enantioselectivity (over 90% ee). In order to attain their high performance, chiral or achiral nitrogen-based compounds have been synthesized and applied as ligands to control the activity of the metal center and to construct appropriate setereochemical environments. Some of the related carbene complexes of ruthenium, thought to be intermediates for the corresponding cyclopropanation reactions, have been isolated and characterized.

Keywords Cyclopropanation · Diazoesters · Carbene transfer · Chiral ligand · Enantioselective

1 Introduction

Ruthenium complexes have been well investigated as molecular catalysts in synthetic chemistry to clarify their great potential for a wide range of organic reactions, such as reduction, oxidation, and carbon–carbon bond formations [1]. As their reactivities in various catalysis strongly depend on their auxiliaries, for example, phosphines, nitrogen ligands, carbonyl, or halogens, design and selection of the auxiliaries suitable for an objective reaction become very important for the discovery of a new catalyst.

Incidentally, cyclopropanation, one of the carbon–carbon bond formation reactions, is a useful reaction by using diazo compounds to give a variety of keto or alkoxycarbonyl cyclopropane derivatives, for example, crythantimates and their analogues as popular insectsides [2]. The cyclopropanation reactions have often been carried out with the use of copper or rhodium catalysts; however, this results in stereochemical problems of *trans–cis* selectivity or asymmetric induction [3].

Here, I focus on application of ruthenium complexes as catalysts for the cyclopropanation of olefins with diazoesters to describe their catalytic activity, stereoselectivity, and enantioselectivity together with structural analysis of intermediary carbene complexes, especially with nitrogen-based ligands including porphyrin derivatives [4, 5].

2 Ruthenium-Catalyzed Cyclopropanation

Catalytic activity of a ruthenium complex for the cyclopropanation of olefins with diazoacetate was first found in 1980 by Hubert and Noels [6]; however, the activity of $Ru_2(OAc)_4Cl$ was lower than that of palladium, copper, and rhodium complexes (Scheme 1).

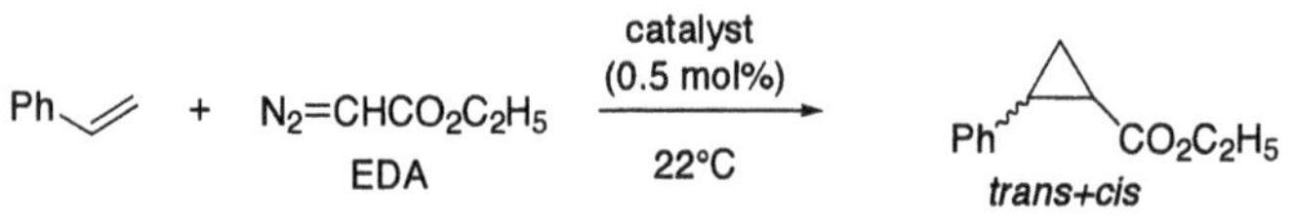

Ph + N_2=$CHCO_2C_2H_5$ (EDA) —catalyst (0.5 mol%), 22°C→ Ph–cyclopropane–$CO_2C_2H_5$ *trans+cis*

catalyst	$Pd(OAc)_2$	$Pd(PPh_3)_4$	$Cu(acac)_2$	$Cu(OTf)_2$	$Rh_2(OAc)_4$	$RhCl(PPh_3)_3$	$Ru_2(OAc)_4Cl$
yield(%)	98	57	65	80	92	12	38
trans:cis	67:33	69:31	68:32	-	60:40	-	64:36

Scheme 1

Similarly, $Ru_3(CO)_{12}$ exhibits a catalytic potential comparable to or a little lower than that of copper and rhodium catalysts for cyclopropanation of ethyl diazoacetate (EDA) with *n*-butoxyethylene: catalyst (0.5 mol %) (yield of cyclopropane product, reaction temperature) $Ru_3(CO)_{12}$ (65%, 60 °C), Cu(acetylacetonate)$_2$ (71%, 60 °C), $Rh_6(CO)_{16}$ (86%, 25 °C), $Rh_2(OAc)_4$ (86%, 25 °C), $PdCl_2(PhCN)_2$ (34%, 25 °C) [7]. Until early the 1990s, cyclopropanation was reported with several ruthenium catalysts, $Ru_2(OAc)_4$ [8], polyethylene carboxylates of ruthenium(II) [9], ruthenacarborane clusters [10], and polymeric $Ru_2(CO)_4(\mu\text{-OAc})_2/_n$ [11]. These reactions were preferably carried out at 60–100 °C to give good-to-moderate up to high yields and a *trans*-to-*cis* ratio of 60:40–70:30. As shown in Scheme 1, the triphenylphosphine ligand decreased the catalytic activity in the case of the corresponding palladium and rhodium complexes [7]. In comparison, $RuCl_2(Ph_3P)_3$ exhibited high efficiency at 60 °C for the cyclopropanation of substituted styrenes with EDA in 89–94% yields (67:33 of *trans* to *cis*), but it gave low yields, 4–7%, for 1-hexene, 1-octene, and cyclohexene [12]. Although the yields and the reaction temperatures strongly depend on the substrate olefins and diazoesters, it may be thought that the order of catalytic activity is almost Rh>Pd≈Cu>Ru.

From 1995 to 2000, catalyst profiles of several ruthenium catalysts bearing pyridine-diimide **1** [13], diiminocarbene **2** [14], diamine-arene **3** [15], phosphino-arene **4** [16], and substituted cyclopentadienyl **5** and **6** [17, 18] were shown to have good activity for the cyclopropanation (Fig. 1). At the relatively high reaction temperature of 60–100 °C, they also gave moderate-to-high yields over 90%. It is interesting in that the dipyridine-diimide complex **1** and the *p*-cymene–carbene complex **2** show high *trans* selectivity, 86:14 and 82:18, respectively.

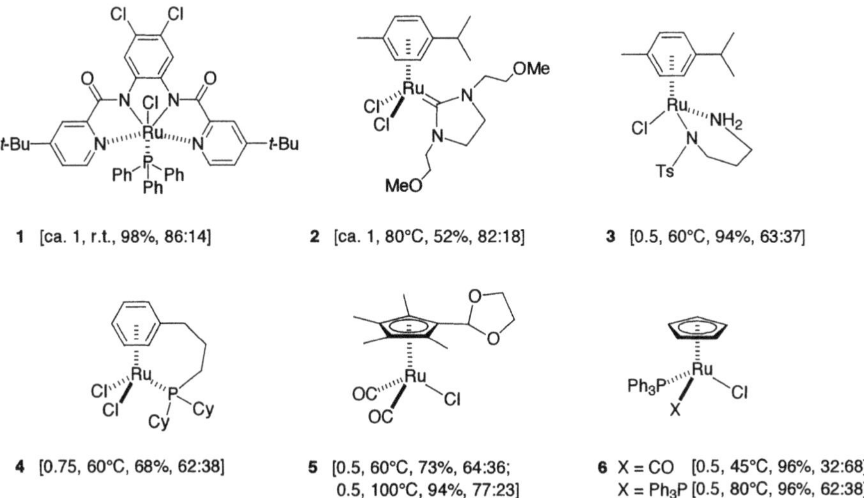

Fig. 1 Catalytic cyclopropanation of ethyl diazoacetate and styrene with ruthenium complexes **1–6**: [catalyst roading (mol%), reaction temperature, yield of the cyclopropanacarboxylate, ratio of *trans:cis*]

3
Asymmetric Catalytic Cyclopropanation

In 1994, asymmetric cyclopropanation (ACP) with ruthenium catalysts was first reported by Nishiyama and coworkers [19, 20] by adoption of their chiral bis(oxazolinyl)pyridine (Pybox) ligands. The reaction profiles of Ru Pybox catalysts reveal extremely high trans selectivity with high enantioselectivity (or diastereoselectivity) of cyclopropane products at the relatively low reaction temperatures (around 20–50 °C) so far reported for ruthenium catalysts. After 1997, different types of chiral nitrogen-based ligands in combination with the ruthenium atom have shown good-to-excellent catalytic activity comparable to chiral Ru Pybox and other copper or rhodium catalysts.

3.1 Ru Pybox Catalysts

In 1989, Pybox was first reported as a chiral nitrogen-based ligand for asymmetric hydrosilylation of ketones [21]. Pybox has a central pyridine bearing two homochiral oxazoline rings derived from optically active β–aminoalcohols giving a C_2-symmetric concave environment around the active metal site (Fig. 2). In asymmetric reactions, the C_2-symmetric design of the chiral auxiliaries has preferably been adopted to give high enantioselectivity [22].

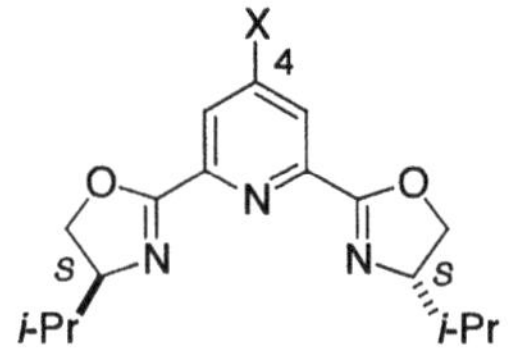

Fig. 2 Pybox

[*S*,*S*]-Pybox-*ip* X = H

As for ACP, from 1986 to 1991, several nitrogen ligands of bidentate corrin mimics and bis(oxazoline) were successfully developed as chiral ligands with copper catalysts [23, 24]. Pybox families reported by Singh et al. [25] resulted in good-to-excellent activity for ACP in combination with copper.

We came up with the idea of the combination of Pybox and a ruthenium atom, like a bolt from the blue, after screening several metals. The new catalytic system was eventually reported in 1994 [19]. A combination of Pybox ligand with a ruthenium(II) cymene complex exhibits high stereochemical efficiency as an *in situ* catalyst (Scheme 2). The Ru Pybox-*ip in situ* catalyst (catalyst A)

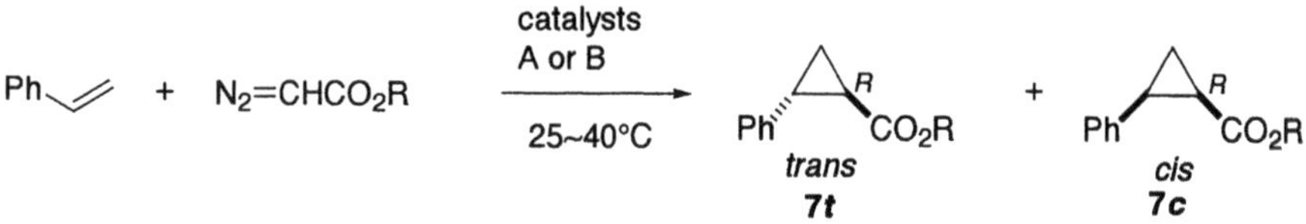

Ph + N_2=$CHCO_2R$ —(catalysts A or B, 25~40°C)→ *trans* **7t** + *cis* **7c**

catalysts	**A**: [Ru(*p*-cymen)Cl_2/2 + Pybox-*ip*]			**B**: $RuCl_2(C_2H_4)$(Pybox-*ip*) **8**		
R =	yield (%)	*trans:cis*	%ee (*trans,cis*)	yield (%)	*trans:cis*	%ee (*trans,cis*)
Et	69	92:8	(89,75)	73	91:9	(89,79)
t-Bu	81	97:3	(94,85)	65	97:3	(94,87)
l-menthyl	87	95:5	(95,76)	83	97:3	(96,80)

Scheme 2

could produce a mixture of *trans*- and *cis*-cyclopropanecarboxylates (**7t** and **7c**) in 69% yield, 92:8 *trans–cis* ratio, and 89% ee for the *trans* and 78% ee for the *cis* forms by the reaction of EDA and styrene. An increase of the bulkiness of the ester groups from ethyl to *tert*-butyl and *d*-menthyl or *l*-menthyl groups gave high *trans* selectivity up to 97:3 with 95% ee for the *trans* form. It was found that $RuCl_2$(Pybox-*ip*)(C_2H_4) (catalyst B), obtained by treatment of the Ru Pybox *in situ* mixture derived from the ruthenium cymene complex under ethylene gas, also exhibited the same catalytic activity and efficiency: 96% ee for the *trans* form with *l*-menthyl diazoacetate; 97% ee for the *cis* form with *d*-menthyl diazoacetate.

Other terminal olefins were transformed to the corresponding cyclopropane esters with *l*-menthyl and *d*-menthyl diazoacetates with high stereoselectivity up to 98% ee (Scheme 3). Intramolecular reaction of the phenylallyl ester **9** was carried out to give the bicyclic compound **10** with 86% ee and 93% yield. The enantioselectivity for intramolecular cyclopropanation of the 3-methylbutenyl ester **11** was compared with chiral Cu(I), Rh(II), and Ru Pybox catalysts: Rh>Ru>Cu [26].

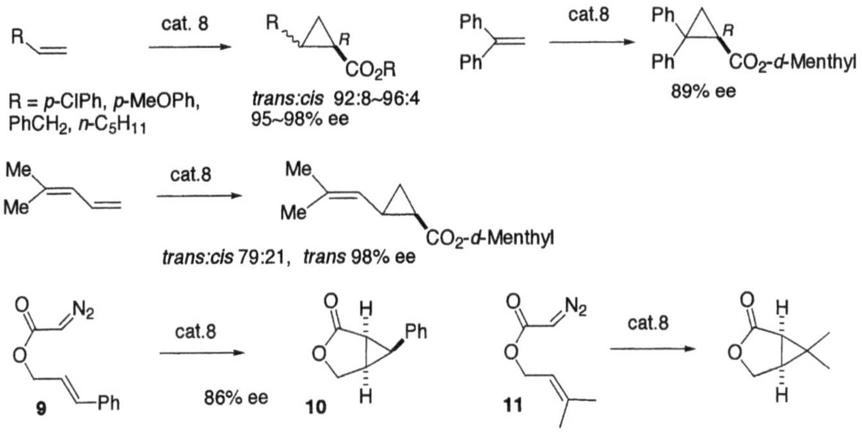

Scheme 3

By use of Pybox substituted at the 4-position of the pyridine skeleton, the remote electronic substituent effect for asymmetric induction was found to show that enantiomeric excesses became higher with electron-withdrawing groups and lower with electron-donating groups, but the ratios of the *trans*-to-*cis* products were not influenced by the substituents [27].

On the basis of analysis of the stereochemical course of the cyclopropanation, we came up with the idea that a single chiral Pybox could work well to obtain higher enantioselectivity (Fig. 3) [28]. Because we expected that the intermediate carbene complex **I** can open the *re*-face for the approaching olefin as the double homochiral intermediate **II** does. The catalyst of the single chiral Pybox **12** gave 94% ee for the *trans* form.

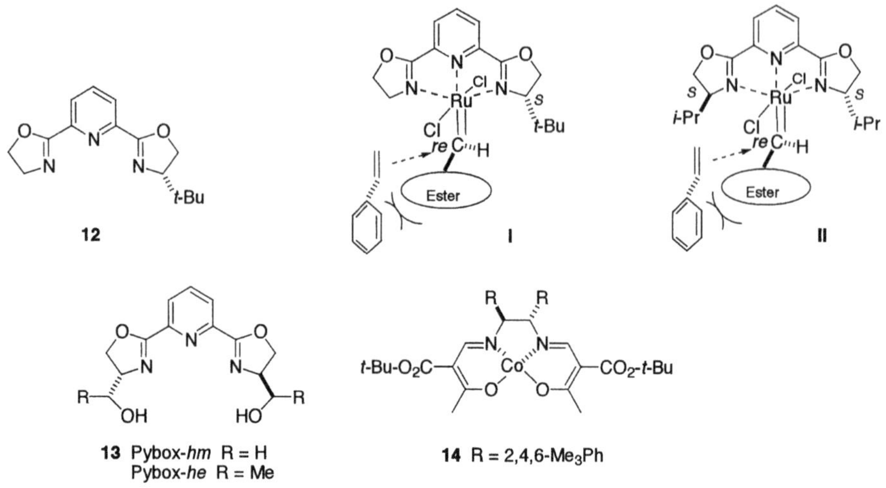

Fig. 3 Pybox derivatives and Yamada's Co catalyst

In 2000, we synthesized a new type of Pybox, Pybox-*hm* **13**, having a hydroxymethyl group on the oxazoline rings [29]. It was found that the ruthenium chloride complex of **13** was very soluble in water and alcohols; therefore, the phenomenon reminded us of applications of the *in situ* ruthenium complex to ACP in protic media, on the basis of environmental concerns, as nonhalogenated solvent systems [30, 31].

There had been no reports of catalytic cyclopropanation systems effective in aqueous or protic solvents until our report and the cobalt catalysts of Yamada and coworkers in 2001 [32]. Some of the Rh catalysts decrease their catalytic activity or decompose diazo compounds in the presence of water or alcohols giving alcohols or ethers [33]. In the case of copper catalysts, the free hydroxy groups on ligands do not interfere with the cyclopropanations [23, 34].

The ACP with Ru Pybox-*hm* revealed that the use of single organic solvents, such as toluene and tetrahydrofuran (THF), resulted in lower yields and lower enantioselectivity. However, when water was added to THF or toluene solutions, the reaction proceeded smoothly, improving the enantioselectivity and the yields slightly. This phenomenon accounted for the increase of the solubility of the Ru(Pybox-*hm*)Cl_2(vacant or solvent) species. The ACP carried out in toluene/water biphasic media attained 94% ee for the *trans* form (Scheme 4). As the active Ru Pybox-*hm* species still remained in the aqueous phase after the reaction, the second run(*) could be carried out by addition of diazoacetate and styrene to give a similar result. Thus, the water-soluble catalyst can be recycled.

In place of water, alcohols such as ethanol, isopropyl alcohol, and *tert*-butyl alcohol were used to prepare homogeneous protic media. Isopropyl alcohol resulted in the best enantioselectivity, up to 96% ee for the *trans* form and 88% ee for the *cis* form. Use of single solvent, only isopropyl alcohol, resulted in moderate enantiomeric excess. At present, these stereochemical outcomes depend-

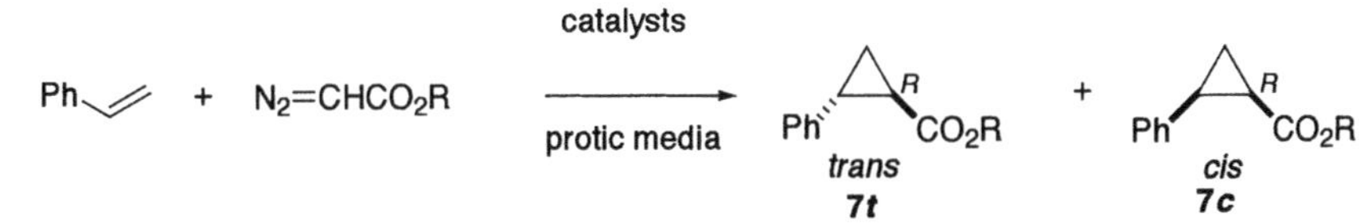

Ph-CH=CH$_2$ + N_2=CHCO$_2$R —catalysts, protic media→ *trans* **7t** + *cis* **7c**

catalyst	R =	solvent	temp.	yield (%)	*trans:cis*	%ee (*trans,cis*)
Ru cat.	*t*-Bu	$C_6H_5CH_3/H_2O$	40°C	30	92:8	(57,26)
Ru-**13**	*d*-Menthyl	$C_6H_5CH_3/H_2O$		57(62)*	97:3(98:2) *	(94,76)(97:3)*
(5 mol%)		$C_6H_5CH_3$/*i*-PrOH		52	97:3	(96,88)
Co cat.**14** (5 mol%)	*t*-Bu	MeOH/H_2O(5%)	50°C	89	83:17	(92,-)

* second run

Scheme 4

ing on the solvents cannot be clearly defined. In the protic system, hydroxyethyl and bulky siloxymethyl substituents on the oxazoline of Pybox were not effective. Vinyl ethers as substrates were also readily cyclopropanated in this toluene/water medium.

In an exciting new challenge the Bristol-Myers-Squibb group carried out an ACP on a 100-kg scale with a chiral Ru Pybox catalyst, especially in two-phase media of water and *tert*-butyl methyl ether (Scheme 5) [35]. The operations produced good yields and enantioselectivity, but separation was difficult. Similarly, Wurz and Charette [36] demonstrated ACP in aqueous media by using Ru, Rh, and Co catalysts including an O–H insertion reaction of carbenes.

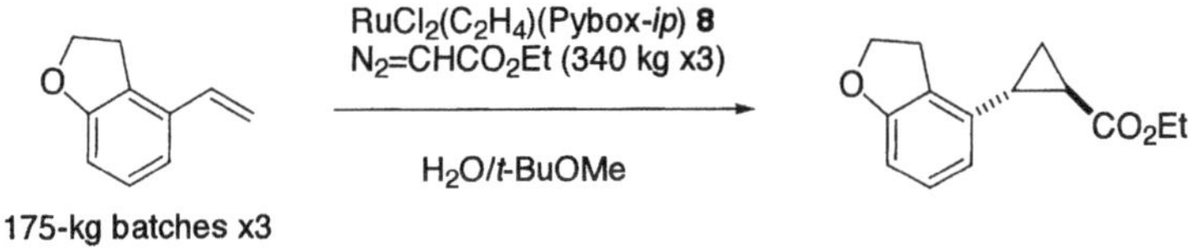

175-kg batches x3 —$RuCl_2(C_2H_4)$(Pybox-*ip*) **8**, N_2=CHCO$_2$Et (340 kg x3), H_2O/*t*-BuOMe→ CO$_2$Et

Scheme 5

Analogues of chiral Pybox have been reported by other chemists and have been applied to ACP with ruthenium catalysts [37,38]. For example, Pybox substituted by a vinyl group at the 4-position of the pyridine skeleton was polymerized with styrene and divinylbenzene to give immobilized ligands, the ruthenium complexes of which were used to give 85% ee for ACP with EDA and styrene [38].

3.2 Ruthenium Salen, Ruthenium Porphyrin, and Related Catalysts

In 1997 after the introduction of Ru Pybox catalysts, chiral ruthenium porphyrin derivatives were found by three groups to be effective catalysts for ACP

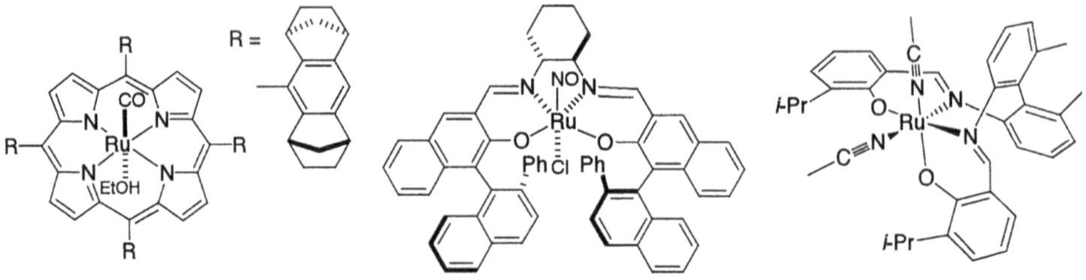

Fig. 4

[39–41]. Complex **15** exhibited higher efficiency of lower catalyst loading (0.15–0.33 mol %), a high *trans*-to-*cis* ratio (96:4) and high enantioselectivity (91% ee) at 0 °C [39, 40]. Dendritic ruthenium porphyrins were synthesized by Zhang et al. [42] and were examined as a catalyst for cyclopropanation to show high *trans* selectivity.

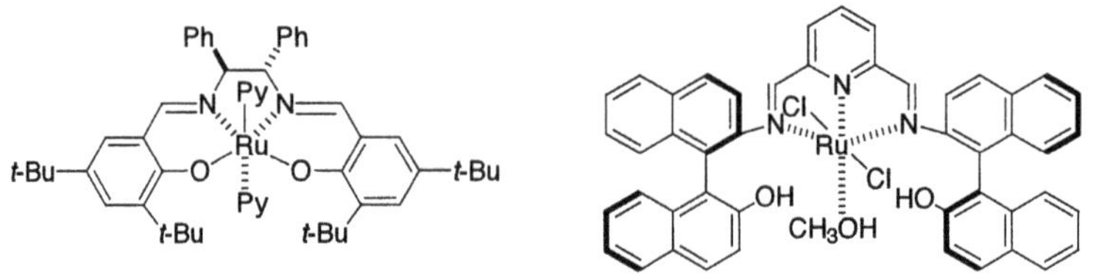

Ru(ON$^+$)(salen) complexes were also applied to ACP [43]. Complex **16** afforded mainly the *cis* product (9:91 *trans–cis* ratio) with 91% ee for the *cis* form under irradiation and it was also applied to intramolecular cyclopropanation giving 94% ee. Several Ru salen derivatives were investigated for ACP [44].

From 1999 to 2001, ACP giving moderate selectivity was reported with other ruthenium complexes containing phosphorous ligands, chiral tridentate phosphine ligands by Lee et al. [45], chiral diphenylphosphine(oxazolinyl)quinoline ligands by Park et al. [46], chiral diphosphines by Stoop et al. and Zheng et al. [47], NPN ligands by Braunstein et al. [48], and the PNNP ligand by Bachmann and Mezzetti [49]. High *cis* selectivity was attained with the PNNP ligand in a nonasymmetric system [50].

The Schiff-base complex **17** of ruthenium was developed by Scott et al. [51] and shows substantially high efficiency for ACP. The cyclopropanes derived from 4-nitrostyrene and EDA were obtained in 92% yield with a 99:1 *trans*-to-*cis* ratio and 98% ee for the *trans* form [51]. In most cases, in order to attain high *trans–cis* stereoselectivity, bulky ester groups of diazoesters were effective. Nevertheless, Nguyen et al. [52] reported in 2002 that the reaction of the smaller and common EDA with styrene assisted by Ru–salen–pyridine complexes **18** (1 mol %) at room temperature produced the cyclopropane products in high yield, 90–96%, and 98–99% ee for the *trans* form and 95–96% ee for the *cis* form. Zhang et al. [53] reported that a N,O-mixed polydentate ligand produced the corresponding ruthenium complex, **19**, which (1 mol %) showed high efficiency at room temperature in 88% yield, 90:10 *trans*-to-*cis* ratio, and 96% ee for the *trans* form.

4 Related Carbene Complexes

It is normally accepted that related metal carbene complexes are very important for the clarification of the mechanism of metal-catalyzed cyclopropanation. In the case of the Ru Pybox system, trimethylsilylcarbene and vinylcarbene complexes **20** and **21** derived from trimethylsilyldiazomethane and diphenylcyclopropene, respectively, were isolated and characterized by NMR (Fig. 5) [54, 55]. Characteristic signals corresponding to Ru=*CHR* were observed: for **20** δ_H=26.5 ppm (s), δ_C=388.9 ppm; for **21** δ_H=20.7 ppm (d, J=13.2 Hz), δ_C=314.0 ppm ($^1J_{C\text{-}H}$=136.1 Hz). Diazomalonate similarly gave a stable dicarbonylcarbene complex, **22**, which was completely analyzed by X-ray and NMR; $\delta_{\text{carbene carbon}}$=296.1 ppm [56]. The geometry around the carbene carbon atom of Ru=C(CO_2Me)$_2$ is almost planar with sp^2 configuration. The length of the Ru–C bond is 1.88 Å. At 110 °C, the carbene moiety of complex **22** was transferred to styrene to give the phenylcyclopropanedicarboxylate in 11% yield with 36% ee.

In 1996, chiral intermediate Ru–Pybox–carbene complexes **23** and **24** were isolated by the reaction with Ru–Pybox–ethylene complex **8** and diazoesters having bulky ester groups, 2,6-di-*tert*-butyltolyl or 1,3,5-trimethylphenyl; for **24**

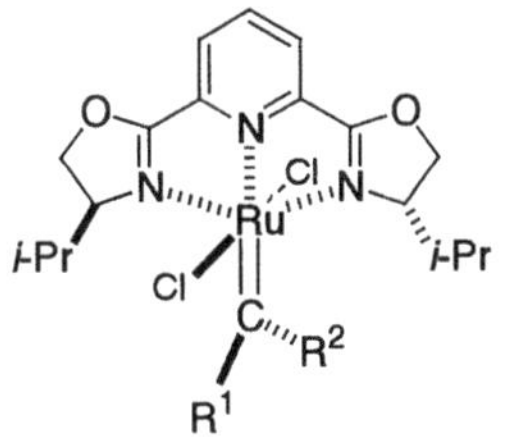

	R^1	R^2
20	H	$SiMe_3$
21	H	CH=CHPh
22	CO_2Me	CO_2Me
23	H	C_6H_2(2,6-di-*t*-Bu)(4-Me)
24	H	C_6H_2(2,4,6-tri-Me)

Fig. 5

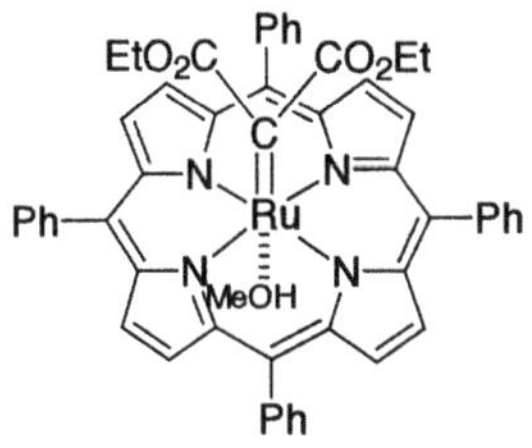

Fig. 6 **25** **26**

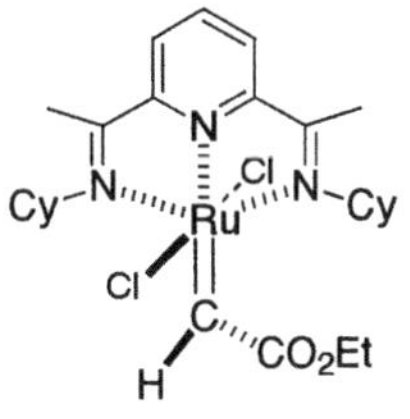

Ru=*CH*-, δ_H=21.7 ppm (s), δ_C=305.7 ppm ($^1J_{C\text{-}H}$=142.4 Hz) [57]. Complex **24** released only *trans*-phenylcyclopropanecarboxylate at 60 °C by the reaction with styrene in 82% yield and 97% ee. Moreover, complex **24** acted as a catalyst in the same way as the ethylene complex **8**; at 50 °C, 95% yield, 98:2 *trans*-to-*cis* ratio with 93% ee for the *trans* form. Thus, the mechanism of ACP catalyzed by Ru Pybox was explained by isolation of the corresponding carbene complexes and realization of the asymmetric carbene transfer reaction.

In 1998, Galardon et al. [58] reported the crystal structure of tetraphenylporphyrinate–ruthenium–(diethoxycarbonyl)carbene complex **25**, which exhibited catalytic activity for cyclopropanation of EDA and styrene, giving 85% of the product with 93:7 *trans*-to-*cis* ratio. The Ru–C distance is 1.829 Å and the carbon resonance is at δ_C=271.3 ppm. In 2000, Bianchini and Lee [59] isolated the similar ruthenium carbene complexes **26** with tridentate imine ligands and EDA; Ru=*CH*–, δ_H=20.44 ppm (s), δ_C=299.9 ppm [59]. Simonneaux et al. [60] isolated the phosphonate carbene complexes of ruthenium porphyrins.

5 Conclusion

We have demonstrated the ACP reaction catalyzed by Ru Pybox complexes. The catalytic activity of ruthenium complexes is commonly not strong. Nevertheless, ruthenium catalysts activated by newly designed ligands have recently received much attention not only for ACP but also for the nonasymmetric version in terms of coordination chemistry and also industrial curiosity because of high stereoselectivity. We believe that further improvement of the ruthenium catalysts will be in environmental interest to realize industrially applicable process.

Acknowledgements I thank all the members of Toyohashi University of Technology and Aoki Katsuyuki for his assistance with X-ray analysis.

References

1. Naota T, Takaya H, Murahashi S (1998) Chem Rev 98:2599
2. (a) Doyle MP (1986) Chem Rev 86:919; (b) Brookhart M, Studebaker WB (1987) Chem Rev 89:411; (c) Noyori R (1994) Asymmetric catalysis in organic synthesis. Wiley, New York, chap 3
3. (a) Doyle MP, Protopopova MN (1998) Tetrahedron 54:7919; (b) Doyle MP, Forbes DC (1998) Chem Rev 98:911
4. (a) Doyle MP (2000) In: Ojima I (ed) Catalytic asymmetric synthesis, 2nd edn. VCH, New York, p 191; (b) Pfaltz A, Lyndon KM, Mckervey MA, Charette AB, Lebel H (1999) In: Jacobsen EN, Pfaltz A, Yamamoto H (eds) Comprehensive asymmetric catalysis II. Springer, Berlin Heidelberg New York, pp 513–603
5. (a) Simonneaux G, Le Maux P (2002) Coord Chem Rev 228:43; (b) Che C-M, Huang J-S (2002) Coord Chem Rev 231:151

6. Anciaux AJ, Hubert AJ, Noels AF, Petiniot N, Teyssie P (1980) J Org Chem 45:695
7. Tamblyn, WH, Hoffmann SR, Doyle MP (1981) J Organomet Chem 216:C64
8. Noels AF, Demonceau A, Carlier E, Hubert AJ, Marquez-Silva RL, Sanchez-Delgado RA (1988) J Chem Soc Chem Commun 783
9. Bergbreiter DE, Morvant M, Chen B (1991) Tetrahedron Lett 32:2731
10. (a) Demonceau A, Saive E, de Froidmont Y, Noels AF, Hubert AJ (1992) Tetrahedron Lett 33:2009; (b) Teixidor F, Nunez R, Flores MA, Demonceau A, Vinas C (2000) J Organomet Chem 614–615:48
11. Maas G, Werle T, Alt M, Mayer D (1993) Tetrahedron 49:881
12. (a) Demonceau A, Abreu DE, Lemoine CA, Stumpf AW, Noels AF, Pietraszuk C, Gulinski J, Marciniec B (1995) Tetrahedron Lett 36:3519; (b) Demonceau A, Lemoine CA, Noels AF, Chizhevsky IT, Sorokin PV (1995) Tetrahedron Lett 36:8419
13. Ko P-H, Chen T-Y, Zhu J, Cheng K-F, Peng, S-M, Che C-M (1995) J Chem Soc Dalton Trans 2215
14. Cetinkaya B, Özdemir I, Dixneuf PH (1997) J Organomet Chem 534:153
15. Simal F, Demonceau A, Noels AF (1998) Tetrahedron Lett 39:3493
16. Simal F, Jan D, Demonceau A, Noels AF (1999) Tetrahedron Lett 40:1653
17. Simal F, Demonceau A, Noels AF, Knowles DRT, O'Leary S, Maitlis PM, Gusev O (1998) J Organomet Chem 558:163
18. (a) Baratta W, Herrmann WA, Kratzer RM, Rigo P (2000) Organometallics 19:3664; (b) Tutusaus O, Delfosse S, Demonceau A, Noels AF (2002) Tetrahedron Lett 43:983
19. Nishiyama H, Itoh Y, Matsumoto H, Park SB, Itoh K (1994) J Am Chem Soc 116:2223
20. Nishiyama H, Itoh Y, Sugawara Y, Matsumoto H, Aoki K, Itoh K (1995) Bull Chem Soc Jpn 68:1247
21. (a) Nishiyama H, Sakaguchi H, Nakamura T, Horihata M, Kondo N, Itoh K (1989) Organometallics 8:846; (b) Nishiyama H, Kondo M, Nakamura T, Itoh K (1991) Organometallics 10:500
22. Whitesell JK (1989) Chem Rev 89:1581
23. (a) Fritschi H, Leutenegger U, Pfaltz A (1986) Angew Chem Int Ed Engl 25:1005; (b) Fritschi H, Leutenegger U, Pfaltz A (1988) Helv Chim Acta 71:1553
24. (a) Lowenthal RE, Abiko A, Masamune S (1990) Tetrahedron Lett 31:6005; (b) Evans DA, Woerpel KA, Hinman MM, Faul MM (1991) J Am Chem Soc 113:726
25. Gupta AD, Bhuniya D, Singh VK (1994) Tetrahedron 50:13725
26. Doyle MP, Peterson CS, Zhou Q-L, Nishiyama H (1997) Chem Commum 211
27. Park SB, Murata K, Matsumoto H, Nishiyama H (1995) Tetrahedron Asymmetry 6:2487
28. Nishiyama H, Soeda N, Naito T, Motoyama Y (1998) Tetrahedron Asymmetry 9:2865
29. Iwasa S, Nakamura H, Nishiyama H (2000) Heterocycles 52:939
30. Iwasa S, Takezawa F, Tuchiya Y, Nishiyama H (2001) Chem Commum 59
31. (a) Iwasa S, Tsushima S, Nishiyama K, Tsuchiya Y, Takezawa F, Nishiyama H (2003) Tetrahedron Asymmetry 14:855; (b) Wurz RP, Charette AB (2002) Org Lett 4:4531
32. (a) Ikeno T, Nishizuka A, Sato M, Yamada T (2001) Synlett 406; (b) Ikeno T, Sato M, Sekino H, Nishizuka A, Yamada T (2001) Bull Chem Soc Jpn 74:2139
33. Wynne DCM, Olmstead M, Jessop PG (2000) J Am Chem Soc 122:7638
34. Hoarau O, Ait-Haddou H, Daran JC, Cramailere D, Balavoine GGA (1999) Organometallics 18:4718
35. (a) Stinson SC (2001) Chem Eng News December 35; (b) Totleben MJ, Prasad JS, Sompson JH, Chan SH, Vanyo DJ, Kuehner DE, Deshpande R, Kodersha GA (2001) J Org Chem 66:1057
36. Wurz RP, Charette AB (2002) Org Lett 4:4531
37. Davies IW, Gerena L, Cai D, Larsen RD, Verhoeven TR, Reider PJ (1997) Tetrahedron Lett 38:1145

38. Cornejo A, Fraile JM, Garcia JI, Garcia-Verdugo E, Gil MJ, Legarreta G, Luis SV, Martinez-Merino V, Mayoral JA (2002) Org Lett 4:3927
39. (a) Lo, W-C, Che C-M, Cheng K-F, Mak TCW (1997) Chem Commun 1205; (b) Che C-M, Huang J-S, Lee F-W, Li Y, Lai T-S, Kwong H-L, Teng P-F, Lee W-S, Lo W-C, Peng S-M, Zhou Z-Y (2001) J Am Chem Soc 123:4119
40. Frauenkron M, Berkessel A (1997) Tetrahedron Lett 38:7175
41. (a) Galardon E, Le Maux P, Simonneaux G (1997) Chem Commun 927; (b) Galardon E, Roue S, Le Maux P, Simonneaux G (1998) 39:2333
42. Zhang J-L, Zhou H-B, Huang J-S, Che C-M (2002) Chem Eur J 8:1554
43. (a) Uchida T, Irie R, Katsuki T (2000) Tetrahedron 56:3501; Uchida T, Irie R, Katsuki T (1999) Synlett 1163; (b) Uchida T, Irie R, Katsuki T (2000) Tetrahedron 56:3501; Uchida T, Irie R, Katsuki T (1999) Synlett 1793; (c) Saha B, Uchida T, Katsuki T (2000) Synlett 114; (d) Saha B, Uchida T, Katsuki T (2002) Chem Lett 846
44. Yao X, Qiu M, Lü W, Chen H, Zheng Z (2001) Tetrahedron Asymmetry 12:197
45. Lee HM, Bianchini C, Jia G, Barbaro P (1999) Organometallics 18:1961
46. Park S-W, Son J-H, Kim S-G, Ahn KH (1999) Tetrahedron Asymmetry 10:1903
47. (a) Stoop RM, Bauer C, Setz P, Wörle M, Wong TYH, Mezzetti A (1999) Organometallics 18:5691; (b) Zheng Z, Yao Z, Li C, Chen H, Hu X (2001) Tetrahedron Lett 42:2847
48. Braunstein P, Naud F, Pfaltz A, Rettig SJ (2000) Organometallics 19:2676
49. Bachmann S, Mezzetti A (2001) Helv Chim Acta 84:3063
50. Bachmann S, Furler M, Mezzetti A (2001) Organometallics 20:2102
51. Munslow IJ, Gillespie KM, Deeth RJ, Scott P (2001) Chem Commun 1638
52. Miller JA, Jin W, Nguyen SBT (2002) Angew Chem Int Ed Engl 41:2953
53. Tang W, Hu X, Zhang X (2002) Tetrahedron Lett 43:3075
54. Park SB, Nishiyama H, Itoh Y, Itoh K (1994) J Chem Soc Chem Commun 1315
55. Nishiyama H, Park SB, Itoh K (1995) Chem Lett 599
56. Nishiyama H, Aoki K, Itoh H, Iwamura T, Sakata N, Kurihara O, Motoyama Y (1996) Chem Lett 1071
57. Park SB, Sakata N, Nishiyama H (1996) Chem Eur J 2:303
58. Galardon E, Le Maux P, Toupet L, Simonneaux G (1998) Organometallics 17:565
59. Bianchini C, Lee HM (2000) Organometallics 19:1833
60. Simonneaux G, De Montigny F, Paul-Roth C, Gulea M, Masson S (2002) Tetrahedron Lett 43:3685

Topics Organomet Chem (2004) 11: 93–124
DOI 10.1007/978-3-540-40027-1

Recent Advances in Alkene Metathesis

Stephen J. Connon[1] · Siegfried Blechert[2] (✉)

[1] Chemistry Department, University of Dublin, Trinity College, 2 Dublin, Ireland
[2] Institut für Chemie, Technische Universität Berlin, Strasse des 17. Juni 135, 10623 Berlin, Germany
blechert@chem.tu-berlin.de

Abstract The last half decade has been a period of unprecedented development for the range of transition-metal-catalysed alkylidene exchange reactions collectively known as alkene metathesis. These carbon–carbon bond forming processes have, in a relatively short time, evolved from relative obscurity into a major research area at the forefront of both modern organometallic and synthetic organic chemistry, driven by the rational design of ever more robust and powerful catalytic systems. The advent of modern well-defined catalysts has allowed practitioners to develop alkene metathesis beyond traditional intramolecular processes into diverse areas such as chemoselective cross-metathesis, intramolecular and intramolecular enyne metathesis, domino metathesis reactions and enantioselective alkene metathesis, together with applications in solid-phase organic synthesis. The major recent developments in these areas are discussed.

Keywords Alkenes · Ruthenium · Catalysis · Metathesis · Asymmetric synthesis

Abbreviations

CM	Cross-metathesis
RCM	Ring-closing metathesis
ROM-CM	Ring-opening cross-metathesis
RRM	Ring-rearrangement metathesis
NHC	*N*-Heterocyclic carbene
KR	Kinetic resolution
PS	Polystyrene

1 Introduction

1.1 Scope

This work consists of an overview of the major developments in the alkene metathesis reaction since 1997. In view of the breadth of the subject area and the rapid pace of advancement in the field in recent years, this review is not intended to serve as a comprehensive survey, but rather as an account of how the development of novel catalyst systems has made a dramatic impact on the reaction in terms of scope and efficiency/selectivity.

1.2 General

Alkene metathesis can be considered as the formal exchange of carbene (alkylidene) units between olefins catalysed by a transition-metal alkylidene catalyst. From humble beginnings it has evolved into a powerful methodology for the formation and cleavage of carbon–carbon bonds under mild conditions [1–3]. The currently accepted mechanism involves a series of [2+2] cycloaddition/cycloreversion reactions through transient metal carbene and metallacyclobutane intermediates [4]. Several synthetically useful intramolecular and intermolecular variants of the process are known (Fig. 1), including ring-closing metathesis (RCM, reaction A) cross-metathesis (CM, reaction B), ring-opening cross-metathesis (ROM-CM, reaction C), enyne metathesis (reaction D) and ring-rearrangement metathesis (RRM, reaction E). Two synthetically useful polymer-forming metathesis reactions outside the scope of this survey are also

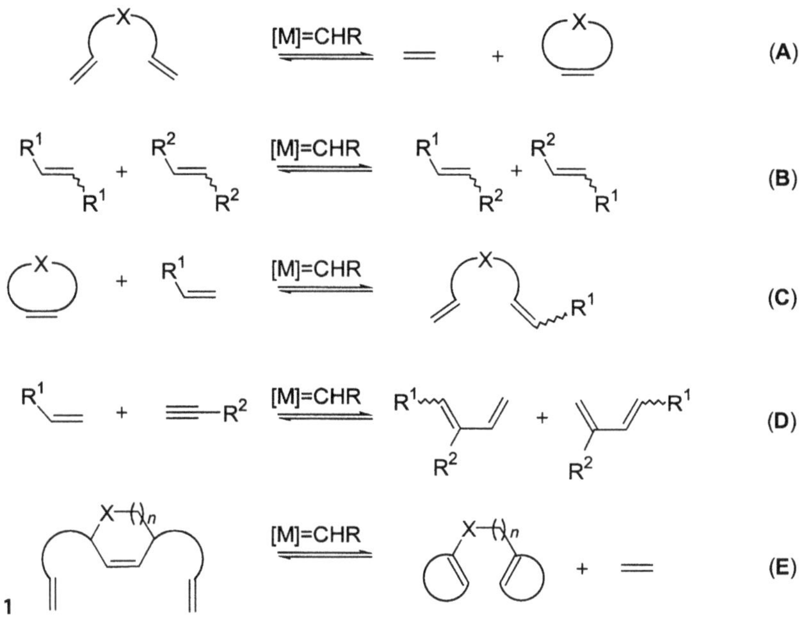

Fig. 1

known: ring-opening metathesis polymerisation (ROMP) and acyclic diene metathesis.

Metathesis reactions are essentially reversible, which under normal circumstances ensures the preferential formation of the thermodynamic product. The employment of terminal olefins (particularly in RCM and CM processes) results in the liberation of ethylene gas from the equilibrium, a tactic often used to drive metathesis reactions. Over the last decade, the advent of active and functional-group-insensitive catalysts have broadened the scope of the reaction to the extent that currently only minimal (if any) protection of Lewis-basic functionality is required.

1.3 Catalysts

The first well-defined catalysts to enjoy widespread use were Schrock's [5] air/moisture sensitive yet highly active molybdenum carbene **1** and the less active yet more practical and robust ruthenium-based catalyst **2** of Schwab and coworkers [6]. The more recent introduction of catalysts bearing relatively non-labile sterically hindered *N*-heterocyclic carbene (NHC) ligands such as **3** [7–9], **4** [10] and **6** [11, 12] has further galvanised research in this field (Fig. 2). The NHC ligands possess strong *σ-donor and poor π*-acceptor properties, which are thought to help stabilise the 14-electron ruthenium intermediates during metathesis. In particular, catalysts **3** and **4** represent a significant improvement on **2**, as they possesses a similar (often superior) functional-group tolerance

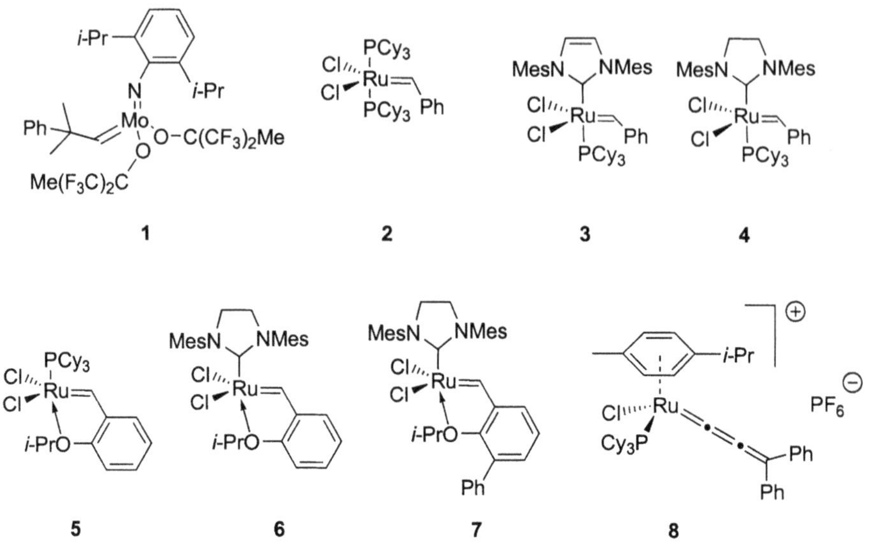

Mes = 2,4,6 Trimethylphenyl
Cy = Cyclohexyl

Fig. 2

while exhibiting a reactivity closer to the highly active yet oxyphilic molybdenum catalyst **1** [13]. The exchange of a phosphine for an isopropoxybenzylidene ligand results in exceptional air and moisture stability: **5** [14] and **6** can be recycled after reaction by column chromatography and can outperform phosphine-based systems in cases where catalyst decomposition is a serious limiting factor. Catalyst **6** is also amenable to structural modification, leading to hindered analogues (such as **7**) capable of unprecedented initiation rates in metathesis reactions [15, 16].

An alternative generation of a ruthenium catalyst has also emerged which is not based on the benzylidene structural motif. Easily accessible catalyst **8** is typical of a class of cationic catalyst from the groups of Fürstner and Dixneuf [17]. This species can promote highly efficient RCM reactions and has the flexibility associated with both thermal and photochemical modes of activation [18].

1.4 Ring-Closing Metathesis

The search for improved methods for the formation of ring structures is and always has been at the core of organic chemistry research [19]. This drive has led to the extensive investigation of the RCM reaction and its acceptance as one of the most facile, flexible and selective cyclisation methodologies in the contemporary chemist's repertoire. (A scifinder 2002 search found over 1,400 papers concerning RCM since 1998; 113 were found from 1992 to 1997.) This section highlights selected recent achievements in this area.

1.5 Total Synthesis

Nowhere has RCM proved more valuable than in the total synthesis of natural products [20]. The preference for convergent over linear synthetic strategies often necessitates cyclisation at a late stage in a synthetic route, in the presence of a variety of functional groups. The ability of modern robust catalysts to ring-close relatively inert and readily prepared diene substrates (as opposed to boranes, stannanes, halides, triflates, epoxides, etc.) to form small, medium or macrocyclic rings in the presence of mild Lewis-basic and Lewis-acidic functionality therefore often offers a distinct advantage over alternative coupling protocols. In an outstanding example, Nicolaou et al. [21] utilised a late-stage RCM reaction to prepare the 11-membered heterocycle of coleophomone B. Treatment of precursor **9** with catalytic **4** results in the formation of adduct **10** in high yield. The 100:0 Z/E ratio of the newly formed olefin and the complete absence of spirocyclopentene adducts (derived from RCM between the prenyl groups) attest to the high levels of regioselectivity and stereoselectivity attainable (Scheme 1).

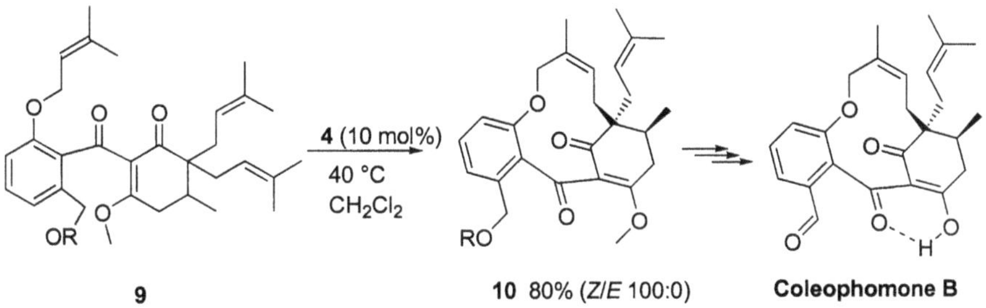

4 (10 mol%)
40 °C
CH_2Cl_2

9 **10** 80% (*Z*/*E* 100:0) **Coleophomone B**

Scheme 1

The formation of the A,B ring fragment had been traditionally regarded as one of the most challenging tasks in the total synthesis of Taxol [22]. Among the most problematic of issues is the presence of the *gem*-dimethyl groups, between which eight-membered ring formation reactions must often proceed. Wenz et al. [23] have reported a RCM-based solution to this problem: cyclisation of (–)-β-pinene-derived **11** in the presence of **2** gave **12** (Scheme 2), which was isolated as a single diastereomer at C-9 (Taxol numbering). For a RCM approach to the B,C Taxol ring system see Ref. [24]. It is interesting that the β-acetate diastereomer underwent exclusive CM dimerisation under these conditions.

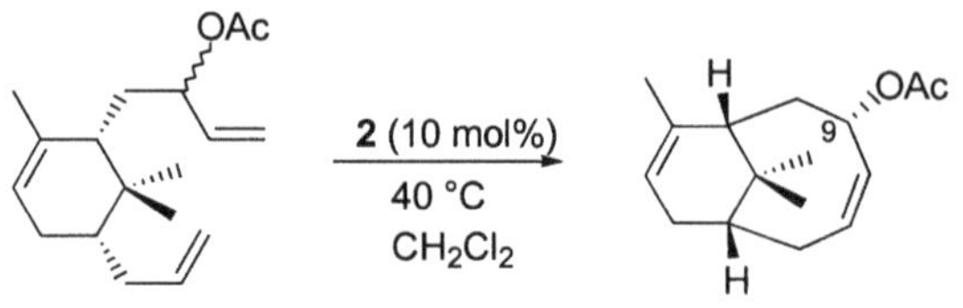

Scheme 2 **11** **12** 59%

Andreana et al. [25] have recently invoked RCM to prepare *β,γ*-unsaturated *δ*-lactones (Scheme 3). Exposure of dienes of general type **13** to either **2** or **4** (which could be used at lower loadings) readily furnished lactones **14.** For other examples of *α,β*-unsaturated *δ*- and *γ*-lactone synthesis by RCM see Ref. [26]. Variation of the configuration at the chiral carbons and the ligand for the asymmetric dihydroxylation reaction allows access to an array of biologically important dideoxy-sugar derivatives.

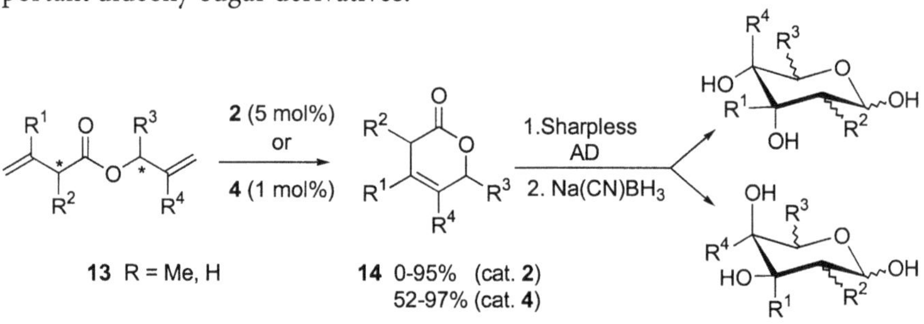

Scheme 3

1.6 Diene-Ene Ring-Closing Metathesis

The increased catalytic activity of **4** relative to **2** has sparked a recent surge of interest in diene-ene RCM. Danishefsky and coworkers [27] have used this strategy to prepare the natural products epothilone 490 and radicicol [28] (Fig. 3), neither of which could be efficiently synthesised using **2.** For the use of RCM in the synthesis of various epothilones and analogues see Ref. [29].

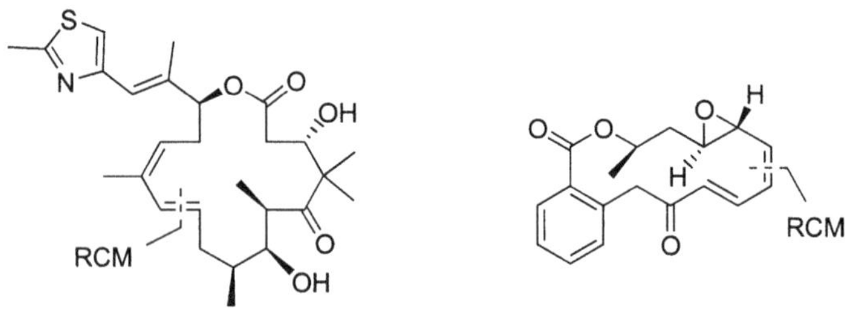

Fig. 3 **Epothilone 490** **Radicicol**

Basu and coworkers [30, 31] have reported that the chemoselectivity of diene-ene RCM reactions can be catalyst-dependent. In a systematic study it was found that in the RCM of **15**, increasing the chain length (and hence product ring size) led to a divergence of catalyst behaviour, with **2** favouring diene products **17**, and **4** favouring the formation of monoene heterocycles **16** (Scheme 4). This was explained in terms of the reversibility of RCM; the stabler and more active catalyst **4** promotes the formation of the kinetic product **17** initially, which then equilibrates to **16** over time. Similar catalyst-dependent selectivity had previously been observed [32]. Using **4**, diene versus monoene product temperature dependence has also been reported [33].

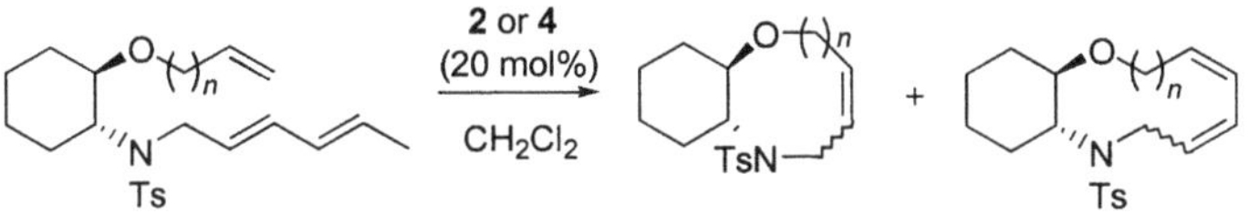

2 or 4 (20 mol%), CH_2Cl_2

15 n = 1, 5, 8 → **16** + **17**

Cat.	n	Ratio **16**:**17**
2	1	1:0
4	1	1:0
2	5	1:0.9
4	5	1:4.3
2	8	1:1.9
4	8	1:0.4

Scheme 4

1.7 Ring-Closing Metathesis in Tandem with Hydrogenation

In a landmark communication, Grubbs' group [34] demonstrated that after metathesis reactions **2** and **4** are both capable of promoting hydrogenation or hydrogen-transfer reactions under appropriate conditions. Cossy et al. [35] have shown that this sequence is also possible using **6**. Treatment of **2** with hydrogen gas was found to quantitatively afford the hydride complex $RuHCl(H_2)(PCy_3)_2$, where Cy is cyclohexyl. This species was capable of the efficient catalytic hydrogenation of a variety of olefins. The one-pot synthesis of the natural product (–)-muscone (Scheme 5) is illustrative of the potential utility of this discovery.

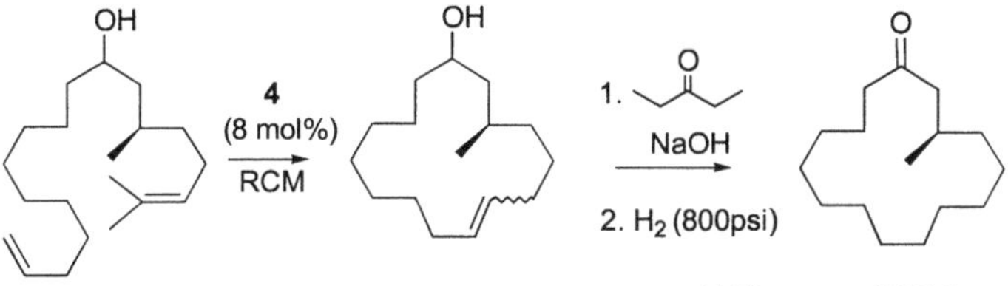

18 → **4** (8 mol%), RCM → **19** → 1. (butanone), NaOH; 2. H_2 (800psi) → **(-)-Muscone** (56% from **18**)

Scheme 5

RCM of **18** with **4** gives **19**, to which sodium hydroxide and 3-pentanone are added. Catalyst **4** (or possibly decomposition products thereof) can then act as a transfer (de)hydrogenation catalyst at this point, oxidising **19** and reducing the pentanone. On completion, the ruthenium residue can promote the quantitative hydrogenation of the alkene moiety under high pressure. Thus, olefin coupling/macrocycle formation, oxidation of the alcohol and hydrogenation of the olefin are possible in reasonable overall yield in one pot. It is clear from this elegant work that the potential of the Grubbs catalysts **2** and **4** for functional group interconversion extends far beyond olefin alkylidene exchange. Sutton et al. [36] have recently reported a tandem RCM–isomerisation sequence for the preparation of enol ethers. For a short treatise on nonmetathetical uses for **2** and **4** see Ref. [37].

2 Cross-Metathesis

2.1 Selective Cross-Metathesis Using Electron-Deficient Alkenes

The key obstacle to the acceptance of CM as a mainstream carbon–carbon coupling tool has traditionally been the issue of chemo- and stereoselectivity [1] For a previous review in this series see Ref. [38]. For example, the CM of two terminal olefins can give rise to six distinct alkenes (cross-product, homodimers and E/Z isomers, Scheme 6).

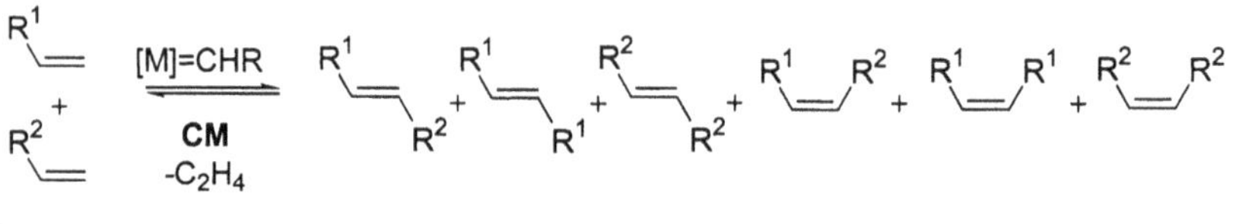

R^1 + R^2 — [M]=CHR, CM, $-C_2H_4$ →

Scheme 6

To achieve selectivity in these reactions, a steric or electronic bias is required to favour one particular product or (more importantly given the reversible nature of CM) one metal–alkylidene precursor in the catalytic cycle.[1] In particular, it has been known for some time that metathesis reactions involving one highly electron deficient olefin partner can be selective (for the first example using acrylonitrile or styrene and **1** see Ref. [40]); however, readily available potential substrates such as enones, acrylates and acrylamides are generally incompatible with either **1** or **2** (for two reported exceptions see Ref. [41]). This was partially overcome by the use of acrolein acetals as α,β-unsaturated car-

[1] Chelation of Lewis-basic groups to ruthenium during catalysis can also strongly influence selectivity [39].

bonyl equivalents [42] and the employment of CM dimers as coupling partners [43], although these reactions still generate significant amounts of dimeric products with overall moderate E/Z selectivity.

A significant breakthrough came with the discovery that Grubbs' second-generation catalyst **4** could promote efficacious and highly selective CM reactions (Scheme 7) involving α,β-unsaturated olefins [44,45] (for a similar report using in situ prepared **4** see Ref. [46]). Enones, acrylates, acrylamides [47] and vinyl/allyl phosphonates [48, 49] were all reactive, while vinylic halides, phthalimides, sulfones (Grela and Bieniek later reported good yields with phenyl vinyl sulfone, however in our hands this was not a suitable CM substrate [50]), silanes, acetates, ethers, alkylstannanes and acrylonitriles gave either poor results or no reaction at all. The source of the CM selectivity observed with electron-deficient alkenes is purported to stem from the inherent instability of the β-acceptor-substituted 14-electron intermediate **20** (Scheme 7). (It is accepted that olefin metathesis proceeds through 14-electron alkylidene intermediates; see Ref [13].) Since any dimerisation of the electron-deficient component must pass through **20**, the relatively slow formation of this intermediate naturally increases the selectivity of the CM process. We [12, 51, 52] and others [53] have shown that catalyst **6** can complement benchmark catalyst **4**, in that it is capable of the CM functionalisation of challenging substrates such as acrylonitrile (Love et al. [53] have recently developed a bis-pyridine based analogue of **4** which catalyses efficient CM reactions involving acrylonitrile; see also Refs. [54, 55]), phenyl vinyl sulfone and polyfluorinated olefins in good-to-excellent yields and high stereoselectivity.

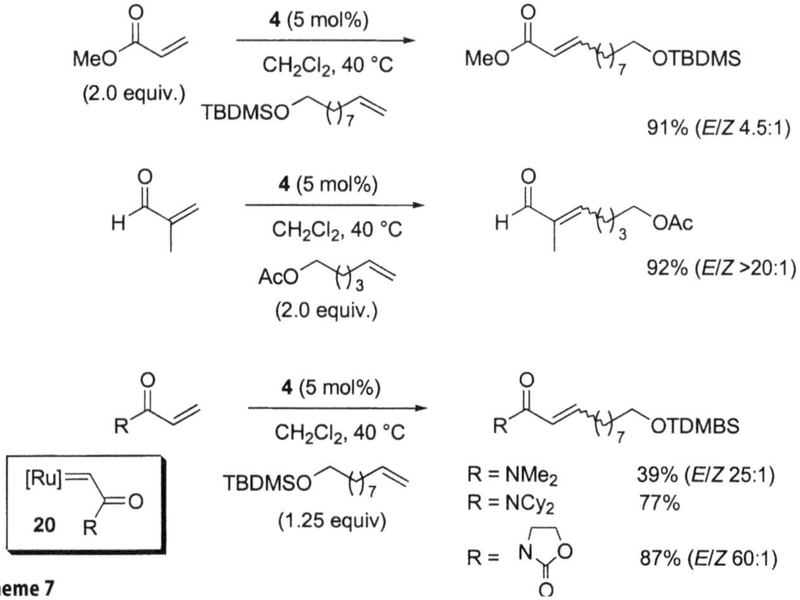

Scheme 7

2.2 Selective Cross-Metathesis with Vinylsiloxanes

Like α,β-unsaturated carbonyl compounds, vinyl silanes and particularly vinylsiloxanes do not dimerise readily and can thus be used in excess to drive selective CM reactions. Pietraszuk et al. [56] have exploited this to couple vinyltriethoxysilane with a variety of olefin partners in good yield and with high stereoselectivity (Scheme 8) [57]. Given the importance of vinylsiloxanes as nucleophilic components in Pd-catalysed coupling reactions [58], their facile and stereoselective modification by CM is a significant development.

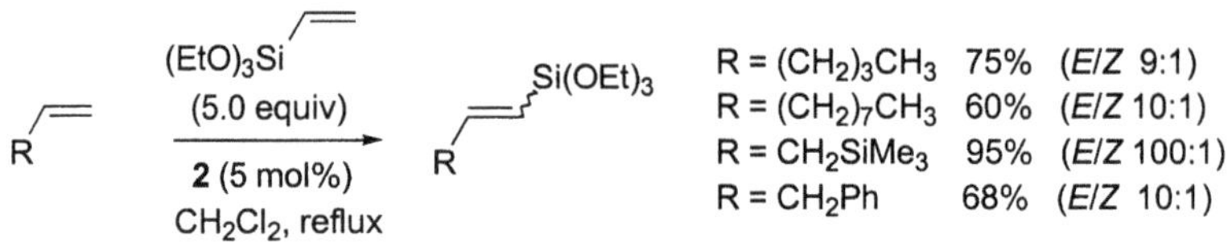

(EtO)$_3$Si (5.0 equiv); **2** (5 mol%), CH$_2$Cl$_2$, reflux

R = (CH$_2$)$_3$CH$_3$ 75% (*E*/*Z* 9:1)
R = (CH$_2$)$_7$CH$_3$ 60% (*E*/*Z* 10:1)
R = CH$_2$SiMe$_3$ 95% (*E*/*Z* 100:1)
R = CH$_2$Ph 68% (*E*/*Z* 10:1)

Scheme 8

2.3 Cross-Metathesis in Biomolecule and Natural Product Synthesis

The ready availability of *O*-vinylglycosides, *C*-vinylglycosides and allylglycosides has facilitated the extensive use of CM methodologies for carbohydrate coupling/modification [3, 59]. Gan and Roy [60] have used CM to prepare sialoside derivatives (Scheme 9): dimers **24** and **25** were isolated in high yield from the metathetical homocoupling of monomers **21** and **22** with moderate E/Z selectivity. This paper [60] also detailed the first example of ruthenium-catalysed CM involving a sulfide (known transition-metal-based catalyst poisons), affording thiosialoside dimer **26**, albeit in poor yield. For more recent examples of CM in carbohydrate synthesis see Ref. [61].

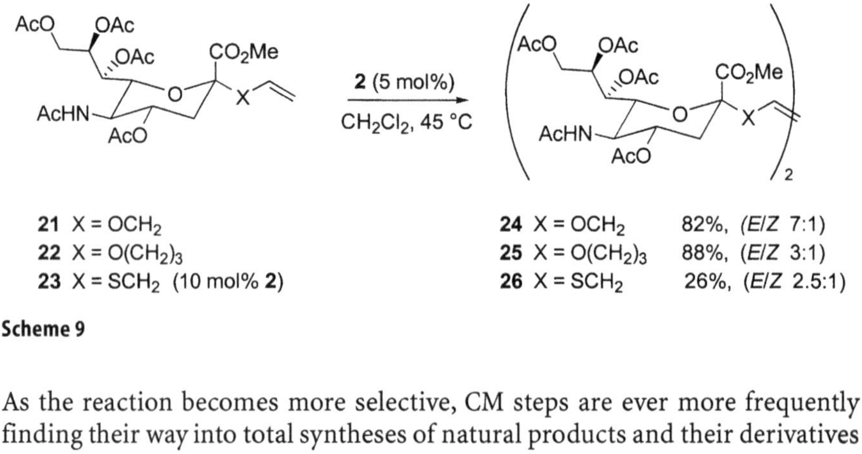

2 (5 mol%), CH$_2$Cl$_2$, 45 °C

21 X = OCH$_2$
22 X = O(CH$_2$)$_3$
23 X = SCH$_2$ (10 mol% **2**)

24 X = OCH$_2$ 82%, (*E*/*Z* 7:1)
25 X = O(CH$_2$)$_3$ 88%, (*E*/*Z* 3:1)
26 X = SCH$_2$ 26%, (*E*/*Z* 2.5:1)

Scheme 9

As the reaction becomes more selective, CM steps are ever more frequently finding their way into total syntheses of natural products and their derivatives

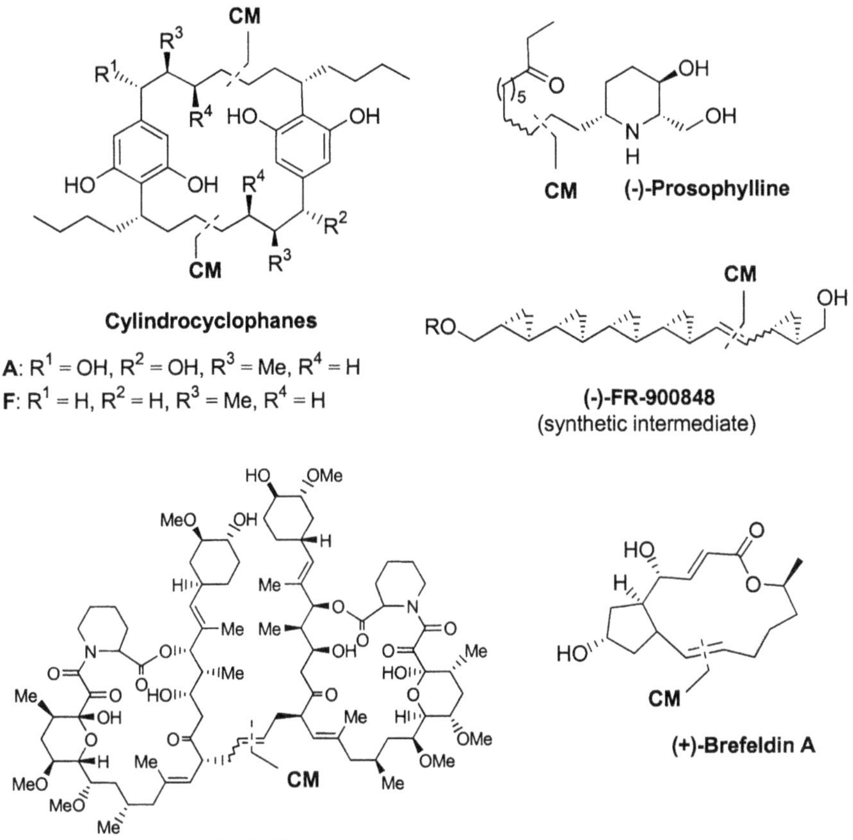

Fig. 4 FK1012

(Fig. 4). Some noteworthy examples are the construction of (or the synthesis of fragments/derivatives thereof) (–)-prosophylline [62], (–)-FR-900848 [63], ciguatoxin [64], thyrsiferol/venustratriol [65], garsubelin A [66, 67], vancomycin dimers [68], cyclindrocyclophanes A and F [69], amphidinol III [70], cyclosporin A [71], (+)-brefeldin A [72] and FK1012 (a dimer of FK506) [73].

In an interesting novel CM application, Ratnayake and Heimscheidt [74] unambiguously assigned the previously unknown stereochemistry of (+)-falcarindiol at C-3 using CM degradation. CM cleavage of the internal alkene in natural product **27** by ethylene promoted by **4** gave rise to *meso*-**28a.** The absence of product optical activity clearly demonstrated that the natural material possesses a (3*R*,8*S*) configuration, as a (3*S*,8*S*) starting material would have given rise to optically active **28b** (Scheme 10).

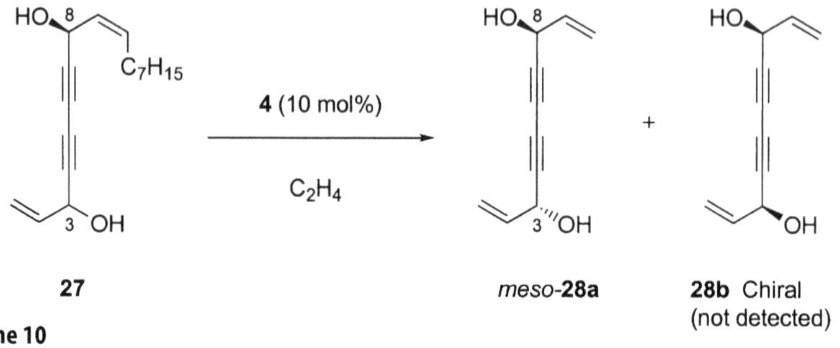

Scheme 10

2.4 Ring-Opening Cross-Metathesis

The ROM-CM of norbornenes, oxanorbornenes and cyclobutenes are among the most efficient and atom-economic of the metathesis reactions [75, 76]. The proposed catalytic cycle for this transformation is shown in Fig. 5.

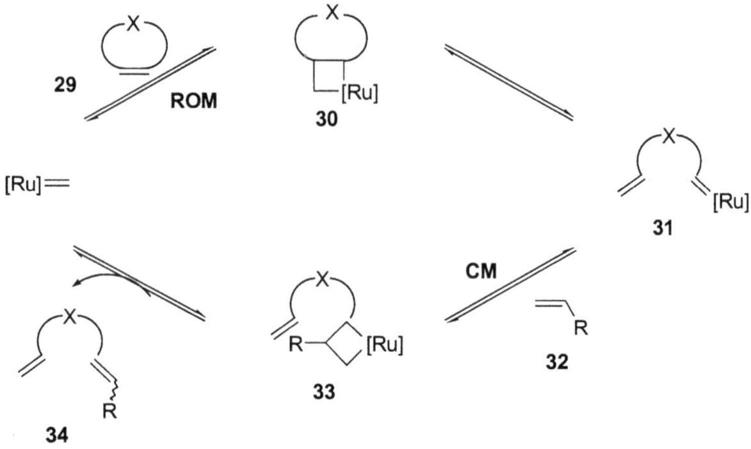

Fig. 5

Reaction between ruthenium methylidene and olefin substrate **29** gives a ring-opened alkylidene **31** via metallacyclobutane **30**. This step is most efficient using highly strained cyclic olefin substrates, where relief from ring strain provides an energetic counterweight to the entropically favoured reverse RCM reaction (**31** → **29**). CM between **31** and added terminal olefin **32** (internal olefins may also serve as CM partners) then affords ROM-CM product **34**. For ROM-CM to be efficient, CM between **32** and **31** must be faster than the reaction between **31** and **29** (a competing ROMP pathway); a factor which very much depends on the nature of the cyclic olefin and the CM partner used. For highly efficient and functional-group-tolerant ROM-CM with catalyst **7** see Ref. [77].

Insights into the origins of regioselectivty [78] in these processes have been obtained from a ROM-CM study (Scheme 11) of various 2-substituted 7-oxanorbornenes of general type **35** [79]. Selectivity was found to depend strongly on the nature of the substituents at C-2. Small *Y* substituents had no effect; however even exchanging the hydroxy (*Y*) group for an acetate led to an impressive increase in selectivity. Also interesting is that the introduction of a substituent (*X*) other than hydrogen does not improve the selectivity further, and that the major products all have the alkyl side chain on the same side of the ring as the *Y* substiuent (i.e. cis products with respect to *Y*), providing evidence that interaction between the *Y* substituent and the metal moiety in the putative intermediates is critical.

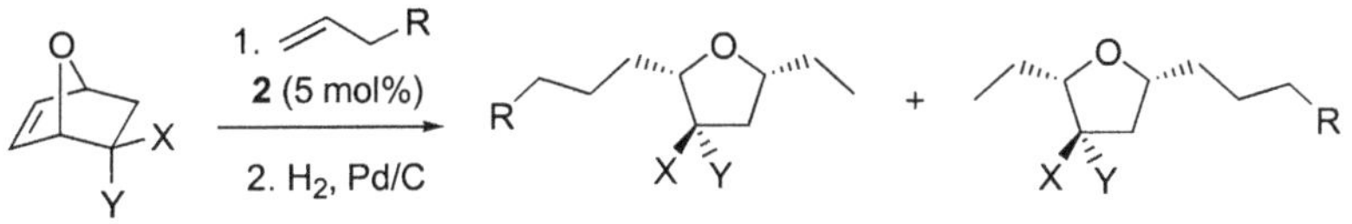

1. (alkene)-R, **2** (5 mol%); 2. H_2, Pd/C

35 → **A** + **B**

X, Y = C=O, R = OAc 75% (**A**:**B** = 50:50)
X = H, Y = OH, R = OAc 80% (**A**:**B** = 50:50)
X = H, Y = OAc, R = OAc 75% (**A**:**B** = 19:81)
X = H, Y = $OCOCH=CH_2$, R = OAc 75% (**A**:**B** = 23:77)
X, Y = OCH_2CH_2O, R = OAc 80% (**A**:**B** = 21:79)
X, Y = OCH_2CH_2O, R = OBn 70% (**A**:**B** = 23:77)

Scheme 11

Limanto and Snapper [80] have reported the clever use of ROM-CM in conjunction with a sigmatropic rearrangement in the total synthesis of asteriscanolide. ROM-CM of tetracycle **36** using **4** with excess ethylene as the CM partner gave intermediate diene **37**, which smoothly underwent a Cope rearrangement under the reaction conditions to give **38** (which could be converted to the natural product in three steps) in good yield (Scheme 12).

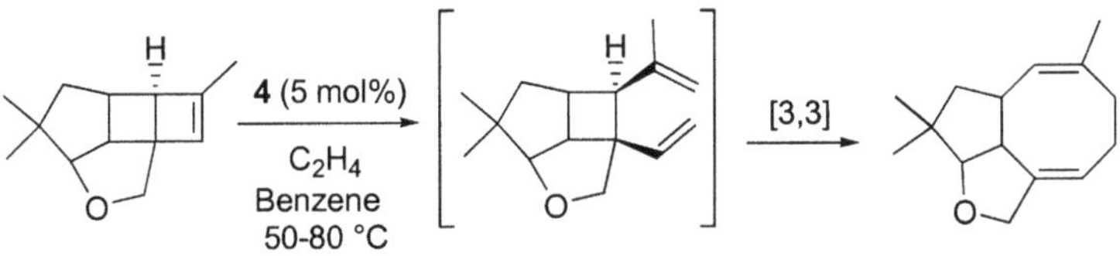

36 → **4** (5 mol%), C_2H_4, Benzene, 50-80 °C → [**37**] → [3,3] → **38** 74%

Scheme 12

As stated earlier, the ROM-CM reaction generally relies on relief from ring strain to drive the reaction. For an example of reversible ROM-CM using a relatively unstrained substrate see Ref. [81]. As such, unstrained olefins such as cyclopentene and cyclohexene have (until recently) been considered inert to ruthenium alkylidene catalysts under norbornene-opening conditions. The extension of the scope of the CM reaction to include α,β-unsaturated carbonyl compounds has resulted in a re-examination of these substrates in ROM-CM reactions. In 2000, Ulman et al. [82] reported that highly reactive ruthenium alkylidenes of general type **20** could react stoichiometrically with cyclohexene to afford ring-opened metal carbene complexes. Both our group [83] and that of Grubbs [84] subsequently developed catalytic variants of this reaction. We found it possible to bis-functionalise cyclopentene, cyclohexene, cycloheptene (Scheme 13) and certain heterocycles using catalysts **4** or **6** in the presence of highly electron deficient alkenes. Acrylonitrile, acrylamides and vinyl sulfones were unfortunately found to be unreactive. Cycloalkene reactivity unsurprisingly increased with ring strain and catalyst **6** proved superior to **4** in all cases tested. It is assumed that the initial reaction is between the catalyst and the acyclic olefin to give the highly reactive intermediate **20**, which proceeds to ring-open the carbocycle/heterocycle via alkylidene **39**.

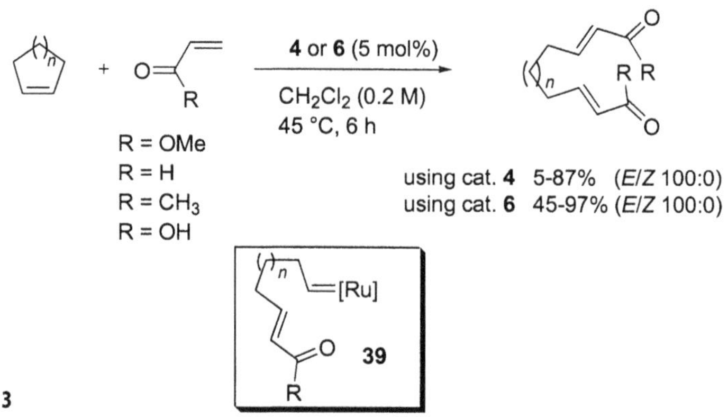

Scheme 13

If these reactions can be developed further they would have considerable potential as an attractive alternative to ozonolysis, whereby the carbocycle-derived chain is extended and variable functionality is installed directly by the metathesis reaction.

3 Enantioselective Alkene Metathesis

One does not immediately associate a reaction which generates sp^2 carbon centres with asymmetric inductive capability, however the development of non-racemic catalysts such as **40**, **41** and **42** (Fig. 6) has allowed the efficient synthesis of optically active alkenes via the kinetic resolution (KR) of dienes and the desymmetrisation of *meso*-alkenes via either RCM or ROM-CM. For a short review of asymmetric metathesis see Ref. [85].

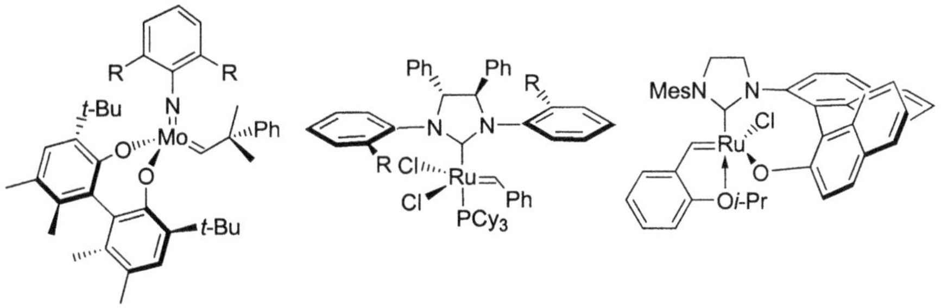
t-Bu
Mo
Ph
t-Bu
Ph
Ph
R
R Cl
Ru
Cl
Ph
PCy3
MesN
Cl
Ru
O
Oi-Pr

40a R = Me
40b R = *i*-Pr

41a R = Me
41b R = *i*-Pr

42

Fig. 6

One of the main challenges associated with the design of a chiral alkene metathesis catalyst is ensuring that the chiral information-carrying ligand remains attached to the metal centre during bond-breaking/bond-forming stages of the catalytic cycle. Chiral catalysts based on the highly active molybdenum species **1** enjoy a significant advantage over Grubbs-catalyst analogues for two main reasons: firstly, their alkoxy and imido ligands are nondissociative, therefore chirality installed at these moieties stands a reasonable chance of influencing the product stereochemistry, and, secondly, enantiopure bis-alkoxy or phenoxy ligands (e.g. BINOL) are readily available, and do not significantly reduce the catalyst efficiency. Ruthenium-based catalysts, on the other hand, operate via a dissociative mechanism, while replacement of chloride for alkoxy ligands results in a marked reduction in activity [86].

Catalysts **40a** and **40b** have proven to be remarkably successful promoters of asymmetric variants of ROM-CM [87], KR-RCM [88], RCM [89] and domino processes [90] For further reports concerning molybdenum- and tungsten-based catalysts see Ref. [91]. Hoveyda and coworkers have demonstrated the high utility of **40a** in the asymmetric ROM-CM of *meso*-**43** with vinyltrimethoxysilane, affording **44** in quantitative conversion and greater than 98% ee. Cyclopentane **44** could then undergo palladium-catalysed coupling with 1-iodonaphtalene to give **45** in 51% overall yield from **43** (Scheme 14).

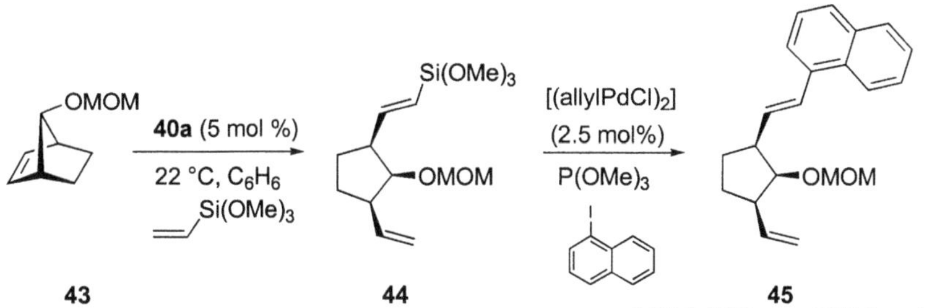
OMOM
40a (5 mol %)
22 °C, C6H6
Si(OMe)3
Si(OMe)3
OMOM
[(allylPdCl)2]
(2.5 mol%)
P(OMe)3
I
OMOM

43 **44** **45**

51% (>98% *ee*, >98% *trans*)

Scheme 14

Seiders et al. [92] have prepared enantiopure ruthenium-based catalysts **41a** and **41b** for use in asymmetric RCM reactions. Using catalyst **41b**, the desymmetrisation of achiral triene **46** was not as selective as was reported by Hoveyda and Schrock [85] using **40a**. Nonetheless, under optimised conditions (involving the formation of an iodo derivative of **41b** in situ) a product (**47**) enantiomeric excess of 90% was possible (Scheme 15), thus showing that there is some potential for analogues of the more robust **4** to serve as efficient chiral catalysts.

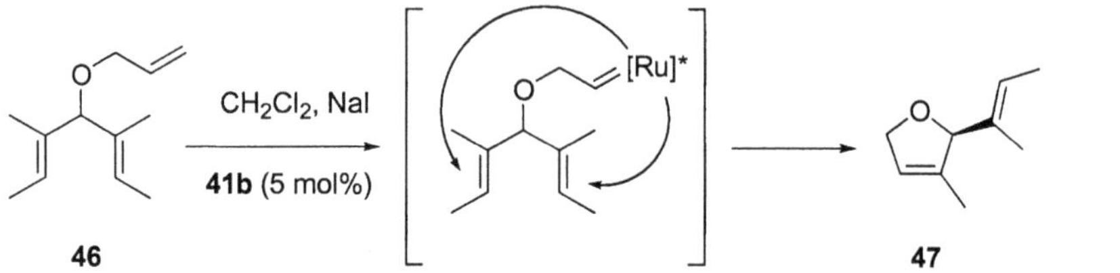

CH_2Cl_2, NaI

41b (5 mol%)

[Ru]*

46

47

(82% conversion, 90% ee)

Scheme 15

This concept was expanded upon by Van Veldhuizen et al. [93] through the preparation of **42**: a highly stable recyclable analogue of **6** capable of enantioselective asymmetric ROM-CM under an air atmosphere in undistilled solvent.

4 Enyne Metathesis

As the name suggests, enyne metathesis [94, 95] is the metathetical reaction between an alkene and an alkyne to generate an often synthetically useful butadiene product.[2] Two distinct reaction types can be classified: enyne RCM and enyne CM.

4.1 Enyne Ring-Closing Metathesis

Although enyne RCM has been known for some time (for the first report of enyne metathesis see Ref. [97]), it has not yet established itself as a mainstream ring-forming method, nor has it received the considerable level of synthetic attention focused on the analogous diene RCM reaction (see section Ring-Closing Metathesis). This is largely a selectivity issue: while high yields of enyne RCM product are usually attainable, two product ring sizes are possible (and sometimes obtained) depending on the substrate. Nevertheless, as more and

[2] It should be noted that nonmetathetical enyne reactions to afford 1,3-dienes are known to be catalysed by low-valent transition metals; for a recent review see Ref. [96].

more is learnt about this atom-economic transformation its high potential for the synthesis of functionalised cyclic building blocks is becoming apparent. Mechanistically speaking, three possibilities arise (Fig. 7). Mechanism A: initial attack of a ruthenium alkylidene on the alkene moiety followed by cyclisation, leading to **49** via carbene **48**. Mechanism B: the first reaction is between the metal carbene and the alkyne, giving rise to **49** via vinyl carbene **50**. Mechanism C: primary formation of intermediate **51**, furnishing *exo*-methylene product **52** after RCM.

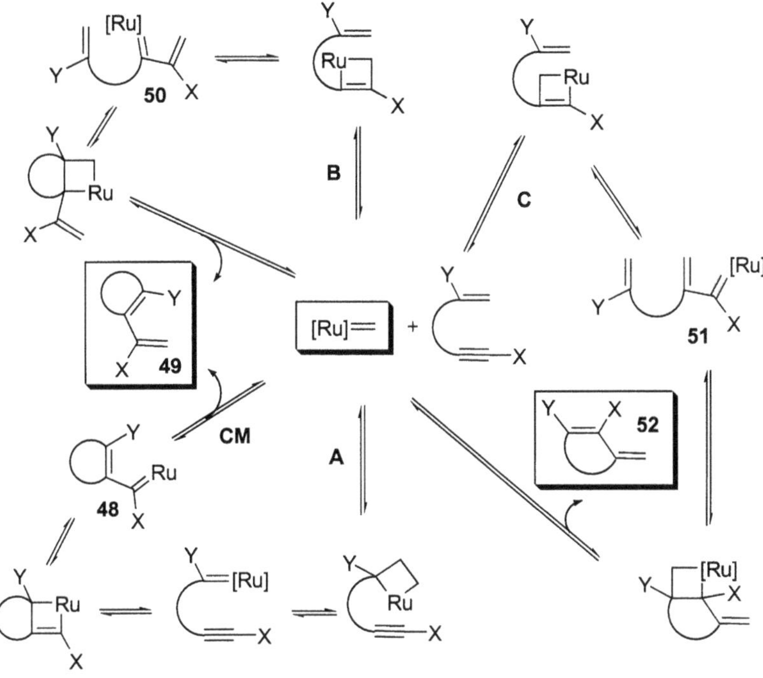

Fig. 7

Evidence suggests that both mechanisms A and B are the major pathways involved in the enyne RCM of relatively unsubstituted alkenes. For evidence supporting preferential initial attack at the alkyne (path B) see Ref. [98]. For example, monosubstituted enyne **53a** undergoes high-yielding enyne RCM metathesis catalysed by **3** to form cyclopentene **54a** exclusively (Scheme 16) [99]. On the other hand, the less reactive gem-disubstituted olefin **53b** gives significant amounts of *exo*-methylene **55b** and even some cyclopropane derivative **56b** [100]. These results point towards a shift in mechanism away from pathway A/B towards pathway C as the alkene moiety becomes less susceptible to initial attack by the catalyst. Internal olefin enyne RCM substrates have also been found to give significant amounts of *exo*-methylene products [95].

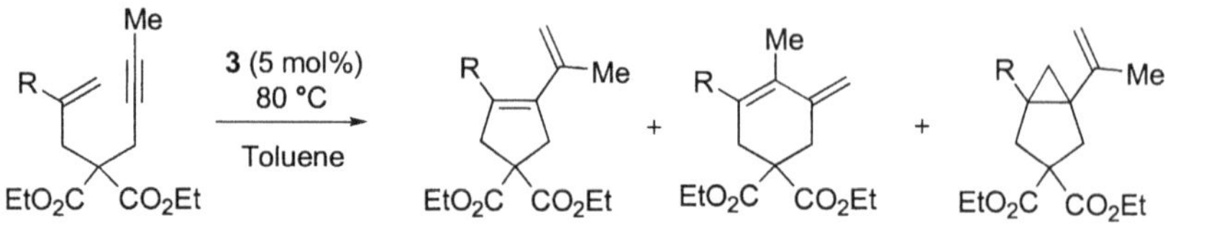

53a R = H
53b R = Me

3 (5 mol%)
80 °C
Toluene

54a R = H 89%
54b R = Me 43%

55a R = H 0%
55b R = Me 42%

56a R = H 0%
56b R = Me 5%

Scheme 16

In terms of reactivity, substrates with terminal monosubstituted olefins and internal alkynes generally react smoothly with either **2**, **3** or **4** to form five- and six-membered rings [101]. Medium ring-forming substrates are somewhat more recalcitrant and require a structural bias towards cyclisation (quaternary centres, ring junctions or heteroatoms) for efficient ring closure to occur [102]. The low reactivity of terminal alkynes can be overcome in some cases by performing the reaction under an ethylene atmosphere, which is thought to aid the formation of the ruthenium methylidene from the stable vinyl carbene **48** via CM (Fig. 7, mechanism A) [101].

It should be noted that while enyne metathesis is considered incompatible with molybdenum catalyst **1**, ruthenium catalysts other than the Grubbs type also promote the reaction. Sémeril et al. [103] reported efficient enyne RCM with a catalyst conveniently generated in situ from $[RuCl_2(p\text{-cymene})]_2$, 1,3-bis(mesityl)imidazolium chloride and caesium carbonate. Interestingly the authors found that the in situ derived system gave better results than the isolated catalyst. One of the most impressive examples of the use of enyne RCM is the total synthesis of (–)-longithorone by Layton et al. [104]. Inspired by a pro-

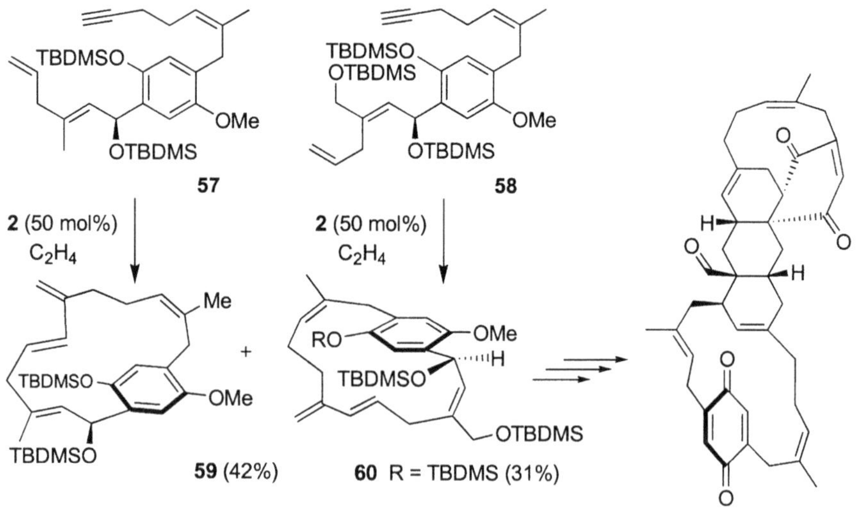

57

58

2 (50 mol%)
C_2H_4

59 (42%)

60 R = TBDMS (31%)

(-)-Longithorone

Scheme 17

posed biosynthetic pathway, the Harvard group prepared both **59** and **60** from **57** and **58**, respectively, via unprecendented enyne RCM macrocyclisations (Scheme 17) to give 1,3-substituted diene products (indicating mechanism B is the predominant pathway). These dienes contained the critical functionality required to execute (after further functional group interconversion) two sequential Diels–Alder cycloadditions to afford the natural product in a biomimetic fashion. Other noteworthy examples of enyne RCM in total synthesis include those in Ref. [105].

4.2 Enyne Cross-Metathesis

Until recently, intermolecular enyne metathesis received scant attention. Competing CM homodimerisation of the alkene, alkyne metathesis and polymerisation were issues of concern which hampered the development of the enyne CM reaction. The first report of a selective ruthenium-catalysed enyne CM reaction came from our laboratories [106]. Reaction of various terminal alkynes **61** with terminal olefins **62** gave 1,3-substituted diene products **63** in good-to-excellent yields (Scheme 18). It is interesting that in these and all enyne CM reactions subsequently reported, terminal alkynes are more reactive than internal analogues, and 1,2-substituted diene products are never formed; thus, in terms of reactivity and selectivity enyne CM is the antithesis of enyne RCM. The mechanism of enyne CM is not well understood. It would appear that initial attack is at the alkyne; however, one report has demonstrated initial attack at the alkene (substrate-dependent) is also possible, see Ref. [107].

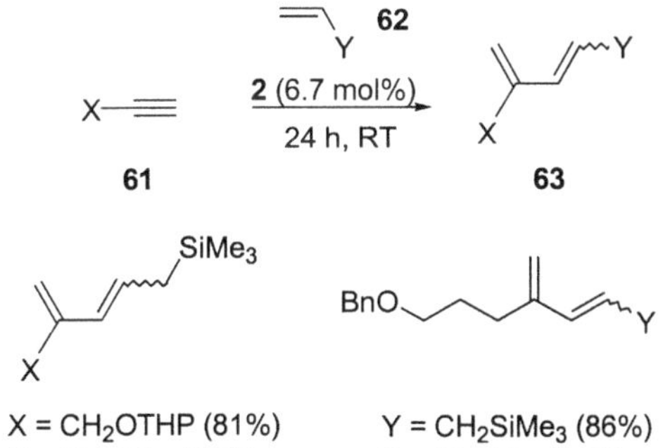

62
Y
2 (6.7 mol%)
24 h, RT
X
61
63
Y
X
SiMe3
BnO
Y

X = CH_2OTHP (81%)
X = CH_2OAc (90%)
X = $CH(CH_3)OAc$ (63%)
X = CH_2OCO_2Me (89%)

Y = CH_2SiMe_3 (86%)
Y = CH_2t-Bu (82%)
Y = $(CH_2)_2CH(CO_2Me)_2$ (76%, *E/Z* 3:1)
Y = $CH_2OTBDMS$ (83%)

Scheme 18

Enyne CM of either terminal or internal alkynes is possible if the ethylene is employed as the CM partner. Initial attempts using bis-phosphine catalyst **2** (3–10%) were only successful if the substrate contained heteroatomic ester or amide functionality at the propargylic position [108]. Later endeavours from several groups with both **2** and the more active promoter **4** (and in some cases high ethylene pressures) broadened the scope of the reaction to include a wide range of functionalised and unfunctionalised alkynes [109].

The ability of NHC-based catalysts to promote efficient enyne CM has been recently exploited in the concise synthesis of HIV reverse transcriptase inhibitors anolignan A and B (Scheme 19) [110]. Reaction between internal alkyne **64** and ethylene promoted by **4** gave the corresponding diene product in excellent yield. Subsequent palladium-catalysed removal of the acetoxy groups and deprotection afforded the natural product. Over the last 5 years enyne CM has evolved into a useful synthetic tool for the formation of conjugated 1,3-dienes. For another example of enyne CM in natural product synthesis see Ref. [111]. Given the high selectivity attainable, the broad range of easily prepared substrates compatible with the process and the amenability of the products to further transformations, interest in this reaction seems set to grow further in the future.[3]

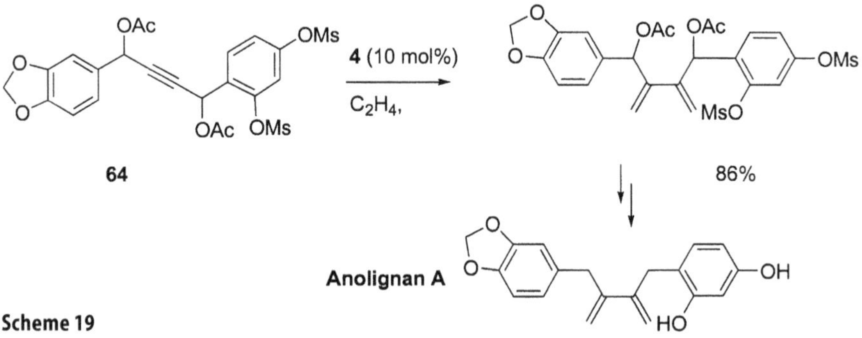

OAc OMs **64** OAc OMs — **4** (10 mol%), C_2H_4, → OAc OAc OMs MsO 86% → **Anolignan A** OH HO

Scheme 19

4.3 Domino Metathesis

Domino reactions are processes in which the product from one synthetic operation undergoes subsequent transformation(s) in situ without added reagents or intermediate isolation. These reactions are highly prized by organic chemists, as they often allow the rapid construction of structurally intricate compounds from relatively simple starting materials. Alkene metathesis is particularly suited for use in domino processes, as it is highly chemoselective (only acting on alkenes and alkynes). It is an equilibrium process (the formation of the most thermodynamically stable product is usual) and a number of diverse metathesis reactions are available for use in sequence or in concert (RCM, CM, ROM-CM, enyne metathesis, etc.), all of which can be promoted by a single metal carbene catalyst.

[3] For use in subsequent cycloaddition reactons see Refs. [107, 112].

4.4 Ring-Rearrangement Metathesis

The combination of RCM and ROM to produce rearranged ring structures is an excellent general method for the formation of heterocycles. For the seminal work in this area see Ref. [113]. For RRM involving electron-deficient alkenes catalysed by **4** see Ref. [114]. A generic example of such a reaction involving a tandem RCM–ROM–RCM sequence is shown in Fig. 1. The driving force for these reactions is the release of ethylene gas (for terminal olefin substrates) and also often the formation of heterocyclic as opposed to carbocyclic ring systems. It is also possible to replace the second RCM step with a CM operation if required [115]. We [116] and others [117] have utilised RRM reactions extensively in the synthesis of natural products. One such report which exemplifies the power of RRM reactions is the synthesis of (+)-astrophylline, an alkaloid which incorporates an unusual and synthetically challenging 2,3-bipiperidinyl ring system [118] (Scheme 20).

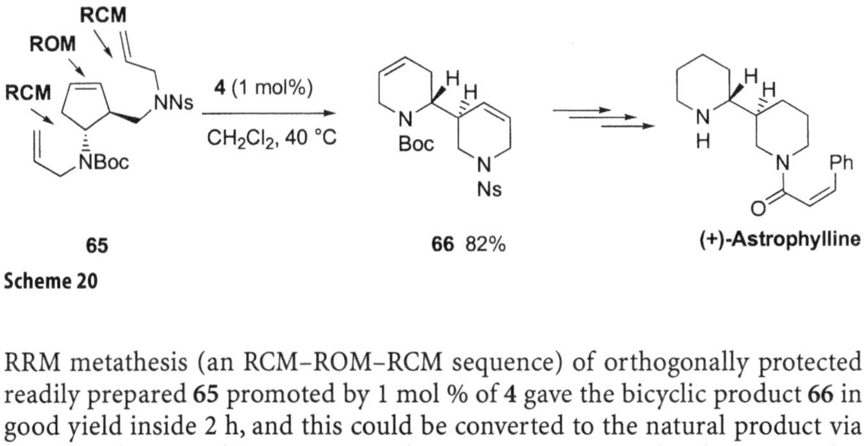

Scheme 20

RRM metathesis (an RCM–ROM–RCM sequence) of orthogonally protected readily prepared **65** promoted by 1 mol % of **4** gave the bicyclic product **66** in good yield inside 2 h, and this could be converted to the natural product via standard functional group manipulations. It is noteworthy that using this methodology, the chiral information associated with triene **65** was quantitatively transferred to the product **66**.

4.5 Domino Enyne Metathesis Reactions

Enyne systems are also capable of impressive multiple RRM transformations. For the first such example see Ref. [119]. The reaction of a ruthenium alkylidene with an alkyne produces a new vinyl alkylidene, which can participate in further intramolecular or intermolecular metathesis reactions to form fused ring systems. This has led Grubbs to designate alkynes in such systems as “relays”. In a noteworthy example, Zuercher et al. [120] constructed the four fused rings of the steroid backbone **68** in one efficient step using tandem enyne reactions in a "zipper"-type process (Scheme 21). For further examples of enynes in domino metathesis reactions see Ref. [121].

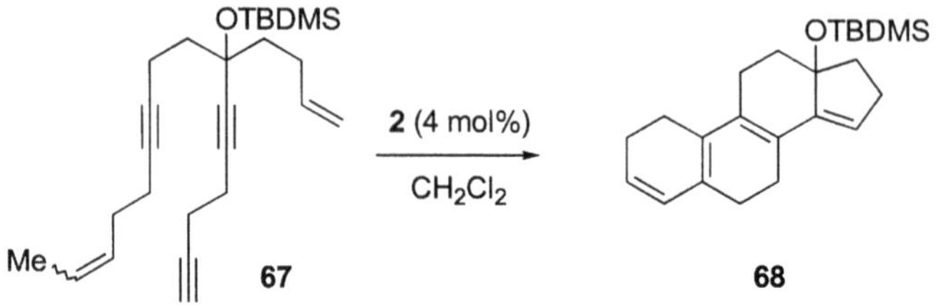

Scheme 21

We have found that a combination of intermolecular and intramolecular domino enyne metathesis reactions is also feasible [122]. Reaction between 1,6-heptadiynes **69–72** and allyltrimethylsilane promoted by **2** gave triene cycloadducts **73–76** in moderate-to-good yields (Scheme 22).

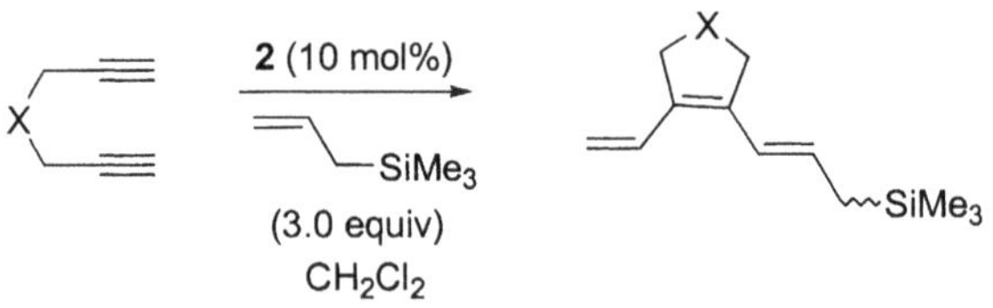

69 X = $C(CO_2Et)_2$
70 X = NTs
71 X = NAc
72 X = O

73 X = $C(CO_2Et)_2$ (65%)
74 X = NTs (64%)
75 X = NAc (73%)
76 X = O (49%)

Scheme 22

As is evident from the selected examples discussed here, domino metathesis reactions are fast becoming a method of choice for the rapid synthesis of complex ring structures from simple building blocks. The recent advances in catalyst technology and associated widening of the scope of alkene metathesis seems certain to further augment the importance of these relatively unexplored reactions in years to come.

5 Alkene Metathesis on Solid Supports

5.1 Immobilised Catalysts

The expense of ruthenium and the current rise of importance attached to environmentally friendly and cost-effective synthetic protocols gives the search for recyclable alkene metathesis catalysts ever-increasing impetus. Since the

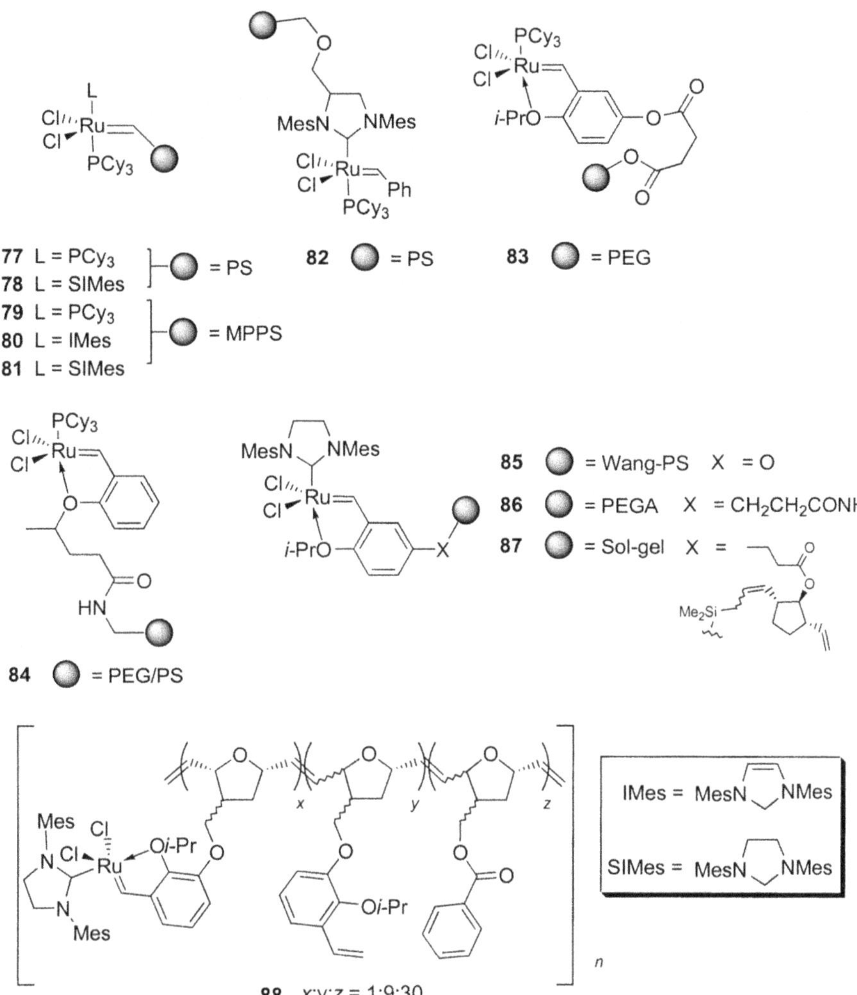

Fig. 8

first report of a well-defined ruthenium catalyst immobilised on polystyrene (PS) [123], considerable progress has been made; selected examples of recently designed alkylidenes (**77–88**) are shown in Fig. 8.

Ahmed and coworkers have developed analogues of **2** and **4** immobilised on PS via the alkylidene moiety. Catalysts **77** [124] and **78** [125] promoted efficient RCM; they could also be recovered (by simple filtration) and recycled several times. Jafarpour and coworkers [126] later prepared less-polymer-swelling-dependent analogues **79–81** bound to macroporous PS. We have developed the permanently immobilised catalyst **82**, which was the first polymer-supported catalyst reported to be active across a range of metathesis reactions such as RCM, enyne CM and RRM [127]. Catalysts which incorporate readily modified isopropoxybenzylidene ligands are ideal for adaptation to solid-phase catalysis; with variable catalyst physical properties obtainable depending on the particular solid support employed. Dichloromethane-soluble **83** [128] and highly air stable **84** [129] are examples of supported analogues of **5**, while impressive recyclability in RCM reactions has been reported using variants of **6** attached to the Wang resin (**85** [130]) or a sol–gel glass (**87** [131]). For a report concerning the immobilisation of **4** on monolithic material by ROM-CM see Ref. [132]. Use of a hydrophilic yet water-insoluble polyacrylamide–poly(ethylene glycol) resin has been shown to lend catalyst **86** high reactivity in RCM and CM reactions in water, methanol and dichloromethane [133].

We have reported a novel self-generating method of catalyst immobilisation with the synthesis of **88**; an analogue of **6** which catalyses the formation of its polymer support by ROMP and subsequently loads itself onto the polymer by CM in one pot. The catalyst is soluble in dichloromethane and most organic solvents with the exception of ether and hexane, from which it can be selectively precipitated after metathesis reactions, allowing for homogeneous reactivity profiles (one of the major disadvantages of polymer-supported catalysts is their slow reaction times relative to homogeneous analogues) together with a simple catalyst-recovery methodology. ROMP catalyst **88** was active in RCM, RRM and ROM-CM reactions and exhibited unprecedented recyclability in the RCM of *N*-tosyldiallylamine [134].

Another intriguing and potentially convenient and useful immobilisation technique is microencapsulation [135]. Kobayashi and Akiyama [136] have developed a microencapsulated analogue of Dixneuf's catalyst by simply heating a derivative of **8** in a solution of PS in cyclohexane. The resulting alkylidene was active and recyclable in a variety of RCM reactions. This methodology has the advantage that binding of the ligand/metal alkylidene to the resin does not require a potentially unselective or inefficient nucleophilic substitution reaction, although admittedly further work is required before the recyclability of microencapsulated systems is in the range obtainable using **77–88**. For a microencapsulated version of **4** see Ref. [137]. For recyclable alkene metathesis catalysis in ionic liquids see Ref. [138].

5.2 Metathesis of Resin-Bound Alkenes

Since our initial reports [139, 140] on the immobilisation and CM reactions of olefins on solid supports significant progress has been made. As is the case with most reactions on solid supports, resin-bound metathesis reactions (particularly CM reactions and variations thereof) are more selective as dimerisation pathways are suppressed (although intraresin metathesis reactions are by no means excluded [141, 142]) and the reaction can be driven to completion by using an excess of the other olefin substrates (the dimers of which can be removed by filtration). In many cases the required products can be cleaved from the resin after the reaction, and as such are available in relatively pure form for further use. An illustrative example is the enyne CM between PS-bound allyltrimethylsilane and alkynated acetals, malonates, esters, acrylates, protected amino acids and carbohydrates (Scheme 23) [143]. Treatment of the immobilised butadiene products resulted in protodesilylative cleavage from the resin to afford pure products. Enyne CM involving a resin-bound alkyne is also possible [144].

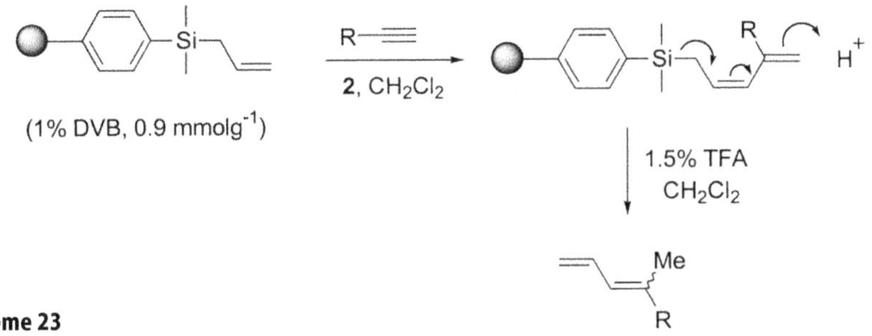

(1% DVB, 0.9 mmolg^{-1})

R—≡ ; 2, CH_2Cl_2 ; H^+ ; 1.5% TFA CH_2Cl_2 ; Me ; R

Scheme 23

Knerr and Schmidt [145] have utilised RCM as a method for efficiently cleaving glycosides from PS. For example, treatment of resin-bound disaccharide **89** with 16 mol% of **2** in two batches gave allyl ether **90** in good yield. It was found that this method was convenient and generally applicable for the formation of di-, tri- and tetrasaccharides (Scheme 24) For other recent examples see Ref. [146]. For early examples of RCM cleavage see Ref. [147].

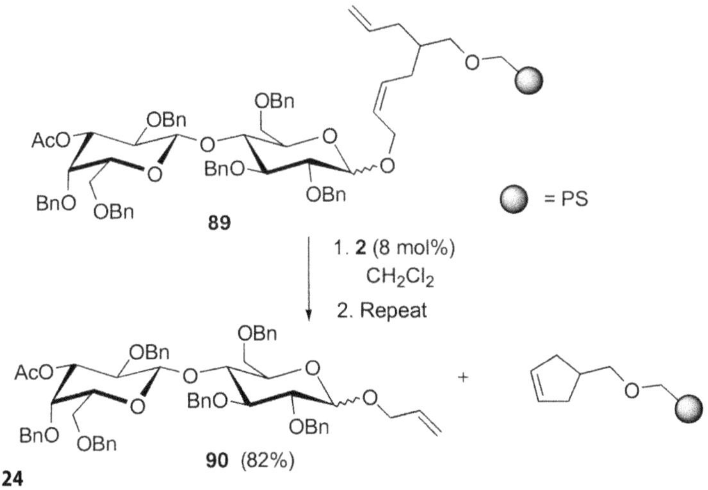

89 ; = PS ; 1. **2** (8 mol%) CH_2Cl_2 2. Repeat ; **90** (82%)

Scheme 24

In a similar fashion, Seeberger's group has developed an octenediol-derived linker for polysaccharide synthesis which is cleavable by CM with ethylene gas [148, 149].

Barrett et al. [150] have used ROM-CM to attach ROMP norbornene polymers to vinyl-PS via resin-bound catalyst **77** (Scheme 25). For further examples of ROM-CM of norbornene derivatives on solid supports see Ref. [151]. The products have been called “ROMP spheres” and have swelling properties differing from those of the PS starting material, with polymer functional diversity and properties being tuneable simply by variation of the norbornene moiety used in the ROM-CM reaction [152]. This is a high-potential area for further alkene metathesis research and has been the subject of a recent review [153].

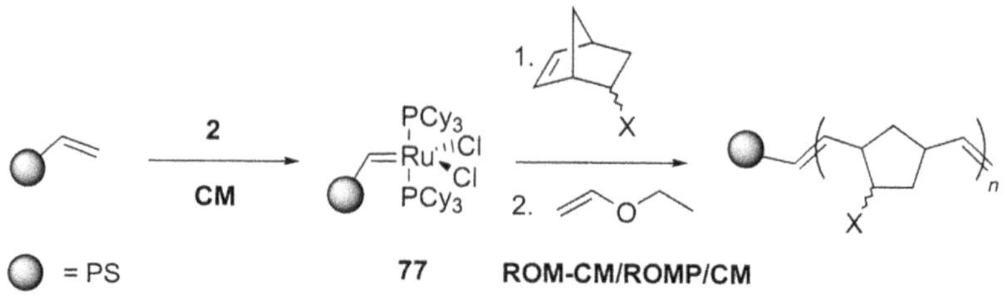

Scheme 25

6 Conclusions

The past 5 years have witnessed a period of unprecedented research activity in the alkene metathesis field. This is due in no small part to the development of novel NHC-based catalysts, which have undoubtedly ushered in a new era in carbon–carbon bond forming organic chemistry. In particular, the ability of modern robust catalysts to tolerate previously incompatible highly electron deficient alkenes and unfunctionalised alkynes has revolutionised intermolecular metathetical processes to the extent that they are beginning to gain recognition as viable, mild and selective alternatives to established coupling methods. The development of asymmetric variants of RCM and ROM-CM is also highly significant, further widening the utility of alkene metathesis reactions in total synthesis and medicinal/pharmaceutical chemistry. With research interest in the field increasing year on year, we can fully expect the continual evolution of this versatile reaction to provide the organic chemist with ever more powerful tools to meet the synthetic challenges of the twenty first century.

Acknowledgements We are grateful to the Fonds der Chemischen Industrie for generous financial support. S.J.C. thanks the Alexander von Humboldt foundation for a postdoctoral fellowship.

References

1. Connon SJ, Blechert S (2003) Angew Chem Int Ed Engl 42:1900
2. (a) Trnka TM, Grubbs RH (2001) Acc Chem Res 34:18; (b) Fürstner A (2000) Angew Chem Int Ed Engl 39:3012; (c) Philips AJ, Abell AD (1999) Aldrichima Acta 32:75; (d) Schrock RR (1999) Tetrahedron 55:8141; (e) Pandit UK, Overkleeft HS, Borer BC, Bieräugel H (1999) Eur J Org Chem 959; (f) Armstrong SK (1998) J Chem Soc Perkin Trans 1 371; (g) Grubbs RH, Chang S (1998) Tetrahedron 54:4413; (h) Ivin KJ (1998) J Mol Catal A 133:1; (i) Fürstner A (1998) (ed) Alkene metathesis in organic synthesis. Springer, Berlin Heidelberg New York; (l) Randall ML, Snapper ML (1998) J Mol Cat A 133:29
3. Roy R, Das SK (2000) Chem Commun 519
4. Hérisson JL, Chauvin Y (1970) Makromol Chem 141:161
5. Schrock RR, Murdzek JS, Bazan GC, Robbins J, DiMare M, O'Regan M (1990) J Am Chem Soc 112:3875
6. (a) Schwab P, France MB, Ziller JW, Grubbs RH (1995) Angew Chem Int Ed Engl 34:2039; (b) Schwab P, Grubbs RH, Ziller JW (1996) J Am Chem Soc 118:100
7. Ackermann L, Fürstner A, Weskamp T, Kohl FJ, Herrmann WA (1999) Tetrahedron Lett 40:4787
8. Huang J, Stevens ED, Nolan SP, Petersen JL (1999) J Am Chem Soc 121:2674
9. Scholl M, Trnka TM, Morgan JP, Grubbs RH (1999) Tetrahedron Lett 40:2247
10. Scholl M, Ding S, Lee CW, Grubbs RH (1999) Org Lett 1:953
11. Garber SB, Kingsbury JS, Gray BL, Hoveyda AH (2000) J Am Chem Soc 122:8168
12. Gessler S, Randl S, Blechert S (2000) Tetrahedron Lett 41:9973
13. (a) Sanford MS, Love JA, Grubbs RH (2001) J Am Chem Soc 123:6543; (b) Sanford MS, Ulman MJ, Grubbs RH (2001) J Am Chem Soc 123:749; (c) Trnka TM, Morgan JP, Sanford MS, Wilhelm TE, Scholl M, Choi, T-L, Ding S, Day MW, Grubbs RH (2003) J Am Chem Soc 125:2546
14. (a) Harrity JPA, La DS, Visser MS, Hoveyda AH (1998) J Am Chem Soc 120:2343; (b) Kingsbury JS, Harrity JPA, Bonitatebus PJ, Hoveya AH (1999) J Am Chem Soc 121:791
15. Wakamatsu H, Blechert S (2002) Angew Chem Int Ed Engl 41:2059
16. Wakamatsu H, Blechert S (2002) Angew Chem Int Ed Engl 41:794
17. (a) Fürstner A, Picquet M, Bruneau C, Dixneuf PH (1998) Chem Commun 1315; (b) Picquet M, Bruneau C, Dixneuf PH (1998) Chem Commun 2249
18. (a) Picquet M, Touchard D, Bruneau C, Dixneuf PH (1999) New J Chem 141; (b) Fürstner A, Liebl M, Lehmann CW, Picquet M, Kunz R, Bruneau C, Touchard D, Dixneuf PH (2000) Chem Eur J 6:1849
19. Nicolaou KC, Vourloumis D Winssinger N, Baran P (2000) Angew Chem Int Ed Engl 39:44
20. Maier ME (2000) Angew Chem Int Ed Engl 39:2073
21. Nicolaou KC, Vassilikogiannakis G, Montagnon T (2002) Angew Chem Int Ed Engl 41:3276
22. (a) Illuminati G, Mandolini L (1981) Acc Chem Res 14:95; (b) Petasis NA, Patane MA (1992) Tetrahedron 48:6953
23. Wenz M, Grossbach D, Beitzel M, Blechert S (1999) Synthesis 607
24. Bourgeois D, Pancrazi A, Ricard L, Prunet J (2000) Angew Chem Int Ed Engl 39:726
25. Andreana PR, McLellan JS, Yongchen C, Wang PG (2002) Org Lett 4:3875
26. (a) Fürstner A, Dierkes T (2000) Org Lett 2:2463; (b) Cossy J, Bauer D, Bellosta V (1999) Tetrahedron Lett 40:4187; (c) Ghosh AK, Liu C (1999) Chem Commun 17:1743; (d) Nicolaou KC, Rodriguez RM, Mitchell HJ, van Delft FL (1998) Angew Chem Int Ed Engl 37:1874

27. Biswas K, Lin H, Njardarson JT, Chappell MD, Chou T-C, Guan Y, Tong WP, He L, Horwitz SB, Danishefsky SJ (2002) J Am Chem Soc 124:9825
28. (a) Garbaccio RM, Danishefsky SJ (2000) Org Lett 2:3127; (b) Garbaccio RM, Stachel SJ, Baeschlin DK, Danishefsky SJ (2001) J Am Chem Soc 123:10903
29. (a) Nicolaou KC, He D, Vourloumis D, Vallberg H, Roschangar F, Sarabia F, Ninkovic S, Yang Z, Trujillo JI (1997) J Am Chem Soc 119:7960; (b) Meng D, Su D-S, Balog A, Bertinato P, Sorensen EJ, Danishefsky SJ, Zheng Y-H, Chou T-C, He L, Horwitz SB (1997) J Am Chem Soc 119:2733; (c) Rivkin A, Njardarson JT, Biswas K, Chou T-C, Danishefsky SJ (2002) J Org Chem 67:7737; (d) Rivkin A, Biswas K, Chou T-C, Danishefsky SJ (2002) Org Lett 4:4081
30. Basu K, Eppich JC, Paquette LA (2002) Adv Synth Catal 344:615
31. Paquette LA, Basu K, Eppich JC, Hofferberth JE (2002) Helv Chim Acta 85:3033
32. Wagner J, Cabrejas LMM, Grossmith CE, Papagerogiou C, Senia F, Wagner D, France J, Nolan SP (2000) J Org Chem 65:9255
33. Yamamoto K, Biswas K, Gaul C, Danishefsky SJ (2003) Tetrahedron Lett 44:3297
34. Louie J, Bielawski CW, Grubbs RH (2001) J Am Chem Soc 123:11312
35. Cossy J, Bargiggia F, BouzBouz S (2003) Org Lett 5:459
36. Sutton AE, Seigal BA, Finnegan DF, Snapper ML (2002) J Am Chem Soc 124:13391
37. Alcaide B, Almendros P (2003) Chem Eur J 9:1258
38. Gibson SE, Keen SP (1998) Top Organomet Chem 1:155
39. (a) BouzBouz S, Cossy J (2001) Org Lett 3:1451; (b) Engelhardt FC, Schmitt MJ, Taylor RE (2001) Org Lett 3:2209; (c) Engelhardt FC, Schmitt MJ, Yuan H, Taylor RE (2001) J Am Chem Soc 123:2964
40. (a) Crowe WE, Zhang ZJ (1993) J Am Chem Soc 115:10998; (b) Crowe WE, Goldberg DR (1995) J Am Chem Soc 117:5162; (c) Crowe WE, Goldberg DR, Zhang ZJ (1996) Tetrahedron Lett 37:2117
41. (a) Blanco OM, Castedo L (1999) Synlett 557; (b) Seshadri H, Lovely CJ (2000) Org Lett 3:327
42. Grubbs RH, Blackwell HE, O'Leary DJ, Washenfelder RA, Miura K (1999) Tetrahedron Lett 40:1091
43. (a) Grubbs RH, O'Leary DJ, Blackwell HE, Washenfelder RA (1998) Tetrahedron Lett 39:7427; (b) Grubbs RH, Blackwell HE, Chatterjee AK, O'Leary DJ, Washenfelder RA, Bussmann DA (2000) J Am Chem Soc 122:58
44. Chatterjee AK, Morgan JP, Scholl M, Grubbs RH (2000) J Am Chem Soc 122:3783
45. Chatterjee AK, Grubbs RH (2002) Angew Chem Int Ed Engl 41:3171
46. Morgan JP, Grubbs RH (2000) Org Lett 2:3153
47. Choi T-L, Chatterjee AK, Grubbs RH (2001) Angew Chem Int Ed Engl 40:1277
48. Chatterjee AK, Choi T-L, Grubbs RH (2001) Synlett 1034
49. Lera M, Hayes CJ (2001) Org Lett 3:2765
50. Grela K, Bieniek M (2001) Tetrahedron Lett 42:6425
51. Gessler S, Randl S, Wakamatsu H, Blechert S (2001) Synlett 430
52. Imhof S, Randl S, Blechert S (2001) Chem Commun 1692
53. Cossy J, BouzBouz S, Hoveyda AH (2001) J Organomet Chem 624:327
54. Love JA, Morgan JP, Trnka TM, Grubbs RH (2002) Angew Chem Int Ed Engl 41:4035
55. Rivard M, Blechert S (2003) Eur J Org Chem 2225
56. Pietraszuk C, Fischer H, Kujawa M, Marciniec B (2001) Tetrahedron Lett 42:1175
57. (a) C Pietraszuk, Marciniec B, Fischer H (2000) Organometallics 19:913; (b) Pietraszuk C, Fischer H (2000) Chem Commun 2463; (c) Pietraszuk C, Fischer H (2001) Organometallics 20:4641; (d) Marciniec B, Kujawa M, Pietraszuk C (2000) New J Chem 24:671; (e) Kujawa-Welten M, Pietraszuk C, Marciniec B (2002) Organometallics 21:840

58. Denmark SE, Sweis RF (2002) Acc Chem Res 35:835
59. Jorgensen M, Hadwiger P, Madsen R, Stutz AE, Wrodnigg TM (2000) Curr Org Chem 4:565
60. Gan Z, Roy R (2000) Tetrahedron 56:1423
61. (a) Dominique R, Liu B, Das SK, Roy R (2000) Synthesis 862; (b) Liu B, Roy R (2002) Chem Commun 594; (c) Dondoni A, Giovannini PP, Marra A (2001) J Chem Soc Perkin Trans 1 2380
62. (a) Cossy J, Willis C, Bellosta V (2001) Synlett 1578; (b) Cossy J, Willis C, Bellosta V, BouzBouz S (2002) J Org Chem 67:1982
63. Verbicky CA, Zercher CK (2000) Tetrahedron Lett 41:8723
64. Oguri H, Sasaki SY, Oishi T, Hirama M (1999) Tetrahedron Lett 40:5405
65. McDonald FE, Wei X (2002) Org Lett 4:593
66. Spessard SJ, Stoltz BM (2002) Org Lett 4:1943
67. (a) Chatterjee AK, Grubbs RH (1999) Org Lett 1:1751; (b) Chatterjee AK, Sanders DP, Grubbs RH (2002) Org Lett 4:1939
68. Nicolaou KC, Hughes R, Cho SY, Winssinger N, Labischinski H, Endermann R (2001) Chem Eur J 7:3824
69. (a) Smith AB III, Kozmin SA, Adams CM, Paone DV (2000) J Am Chem Soc 122:4984; (b) Smith AB III, Adams CM, Kozmin SA (2001) J Am Chem Soc 123:990; (c) Smith AB III, Adams CM, Kozmin SA, Paone DV (2001) J Am Chem Soc 123:5925
70. BouzBouz S, Cossy J (2001) Org Lett 3:1451
71. Smulik JA, Diver ST, Pan F, Liu JO (2002) Org Lett 4:2051
72. Wang Y, Romo D (2002) Org Lett 4:3231
73. Diver ST, Schreiber SL (1997) J Am Chem Soc 119:5106
74. Ratnayake AS, Heimscheidt T (2002) Org Lett 4:4667
75. (a) Schneider MF, Blechert S (1996) Angew Chem Int Ed Engl 35:411; (b) Schneider MF, Lucas N, Velder J, Blechert S (1997) Angew Chem Int Ed Engl 36:257
76. (a) Randall ML, Tallarico JA, Snapper ML (1995) J Am Chem Soc 117:9610; (b) Snapper ML, Tallarico JA, Randall ML (1997) J Am Chem Soc 119:1478; (c) Tallarico JA, Bonitatebus PJ, Snapper ML (1997) J Am Chem Soc 119:7157
77. Dunne AM, Mix S, Blechert S (2003) Tetrahedron Lett 44:2733
78. Arjona O, Csàkÿ AG, Plumet J (2000) Synthesis 857
79. (a) Arjona O, Csàkÿ AG, Murcia MC, Plumet J (1999) J Org Chem 64:9739; (b) Arjona O, Csàkÿ AG, Murcia MC, Plumet J, Mula MB (2001) J Organomet Chem 627:105
80. Limanto J, Snapper ML (2000) J Am Chem Soc 122:8071
81. Wright DL, Usher LC, Estrella-Jimenez M (2001) Org Lett 3:4275
82. Ulman M, Belderrain TR, Grubbs RH (2000) Tetrahedron Lett 41:4689
83. Randl S, Connon SJ, Blechert S (2001) Chem Commun 1796
84. (a) Lee CW, Choi T-L, Grubbs RH (2002) J Am Chem Soc 124:3224; (b) Morgan JP, Morrill C, Grubbs RH (2002) Org Lett 4:67
85. Hoveyda AH, Schrock RR (2001) Chem Eur J 7:945
86. Dias EL, Nguyen ST, Grubbs RH (1997) J Am Chem Soc 119:3887
87. (a) La DS, Ford JG, Sattely ES, Bonitatebus PJ, Schrock RR, Hoveyda AH (1999) J Am Chem Soc 121:11603; (b) La DS, Sattely ES, Ford JG, Schrock RR, Hoveyda AH (2001) J Am Chem Soc 123:7767
88. (a) Alexander JB, La DS, Cefalo DR, Hoveyda AH, Schrock RR (1998) J Am Chem Soc 120:4041; (b) La DS, Alexander JB, Cefalo DR, Graf DD, Hoveyda AH, Schrock RR (1998) J Am Chem Soc 120:9720
89. (a) Weatherhead GS, Houser JH, Ford GJ, Jamieson JY, Schrock RR, Hoveyda AH (2000) Tetrahedron Lett 41:9553; (b) Kiely AF, Jernelius JA, Schrock RR, Hoveyda AH (2002) J

Am Chem Soc 124:2868; (c) Dolman SJ, Sattely ES, Hoveyda AH, Schrock RR (2002) J Am Chem Soc 124:6991
90. Weatherhead GS, Ford JG, Alexanian EJ, Schrock RR, Hoveyda AH (2000) J Am Chem Soc 122:1828
91. (a) Zhu S, Cefalo DR, La DS, Jamieson JY, Davis WM, Hoveyda AH, Schrock RR (1999) J Am Chem Soc 121:8251; (b) Cefalo DR, Kiely AF, Wucher M, Jamieson JY, Schrock RR, Hoveyda AH (2001) J Am Chem Soc 123:3139; (c) Tsang WCP, Schrock RR, Hoveyda AH (2001) Organometallics 20:5658; (d) Hultzsch KC, Jernelius JA, Hoveyda AH, Schrock RR (2002) Angew Chem Int Ed Engl 41:589; (e) Teng X, Cefalo DR, Schrock RR, Hoveyda AH (2002) J Am Chem Soc 124:10779; (f) Schrock RR, Jamieson JY, Dolman SJ, Miller SA, Bonitatebus PJ, Hoveyda AH (2002) Organometallics 21:409; (g) Tsang WCP, Hultzsch KC, Alexander JB, Bonitatebus PJ, Schrock RR, Hoveyda AH (2003) J Am Chem Soc 125:2652; (h) Tsang WCP, Jernelius JA, Cortez GA, Weatherhead GS, Schrock RR, Hoveyda AH (2003) J Am Chem Soc 125:2591
92. Seiders TJ, Ward DW, Grubbs RH (2001) Org Lett 3:3225
93. Van Veldhuizen JJ, Garber SB, Kingsbury JS, Hoveyda AH (2002) J Am Chem Soc 124:4954
94. (a) Mori M (1998) Top Organomet Chem 1:133
95. Poulsen CS, Madsen R (2003) Synthesis 1
96. Aubert C, Buisine O, Malacria M (2002) Chem Rev 102:813
97. Katz TJ, Sivavec TM (1985) J Am Chem Soc 107:737
98. Mori M (1998) Top Organomet Chem 1:146
99. Fürstner A, Ackermann L, Gabor B, Goddard R, Lehmann CW, Mynott R, Stelzer F, Thiel OR (2001) Chem Eur J 7:3236
100. (a) Kitamura T, Sato Y, Mori M (2001) Chem Commun 1258; (b) Kitamura T, Sato Y, Mori M (2002) Adv Synth Catal 344:678
101. (a) Schramm MP, Reddy DS, Kozmin SA (2001) Angew Chem Int Ed Engl 40:4274; (b) Mori M, Sakakibara N, Kinoshita A (1998) J Org Chem 63:6082; (c) Renaud J, Graf CD, Oberer L (2000) Angew Chem Int Ed Engl 39:3101; (d) Clark JS, Elustondo F, Trevitt GP, Boyall D, Robertson J, Blake AJ, Wilson C, Stammen B (2002) Tetrahedron 58:1973
102. Mori M, Kitamura T, Sato Y (2001) Synthesis 654
103. Sémeril D, Bruneau C, Dixneuf PH (2002) Adv Synth Catal 344:585
104. Layton ME, Morales CA, Shair MD (2002) J Am Chem Soc 124:773
105. (a) Kinoshita A, Mori M (1996) J Org Chem 61:8356; (b) Kinoshita A, Mori M (1997) Heterocycles 46:287; (c) Hoye TR, Donaldson SM, Vos TJ (1999) Org Lett 1:277; (d) Clark JS, Townsend RJ, Blake AJ, Teat SJ, Johns A (2001) Tetrahedron Lett 42:3235
106. Stragies R, Schuster M, Blechert S (1997) Angew Chem Int Ed Engl 36:2518
107. Randl S, Lucas N, Connon SJ, Blechert S (2002) Adv Synth Catal 344:631
108. (a) Kinoshita A, Sakakibara N, Mori M (1997) J Am Chem Soc 119:12388; (b) Kinoshita A, Sakakibara N, Mori M (1999) Tetrahedron 55:8155; (c) Smulik JA, Diver ST (2000) J Org Chem 65:1788
109. (a) Stragies R, Voigtmann U, Blechert S (2001) Tetrahedron Lett 41:5465; (b) Tonogaki K, Mori M (2002) Tetrahedron Lett 43:2235; (c) Smulik JA, Diver ST Tetrahedron Lett (2001) 42:171; (d) Smulik JA, Giessert AJ, Diver ST (2002) Tetrahedron Lett 43:209; (e) Smulik JA, Diver ST (2000) Org Lett 2:2271
110. Mori M, Tonogaki K, Nishiguchi N (2002) J Org Chem 67:224
111. Rodriguez-Conesa S, Candal P, Jiménez C, Rodriguez J (2001) Tetrahedron Lett 42:6699
112. (a) Schürer SC, Blechert S (1999) Chem Commun 1203; (b) Schürer SC, Blechert S (1999) Tetrahedron Lett 40:1877; (c) Zheng G, Graham A, Shibata M, Missert JR, Oseroff AR,

Dougherty TJ, Pandey RK (2001) J Org Chem 66:8709; (d) Zheng G, Dougherty TJ, Pandey RK (1999) Chem Commun 2469; (e) Kotha S, Halder S, Brahmachary E, Ganesh T (2000) Synlett 853; (f) Schürer SC, Blechert S (1999) Synlett 1879
113. Zuercher WJ, Hashimoto M, Grubbs RH (1996) J Am Chem Soc 118:6634
114. Choi T-L, Grubbs RH (2001) Chem Commun 2648
115. (a) Stragies R, Blechert S (1998) Synlett 169; (b) Arjona O, Csàkÿ AG, Murcia MC, Plumet J (2000) Tetrahedron Lett 41:9777; (c) Arjona O, Csàkÿ AG, Medel R, Plumet J (2002) J Org Chem 67:1380
116. (a) Stragies R, Blechert S (1999) Tetrahedron 55:8179; (b) Stragies R, Blechert S (2000) J Am Chem Soc 122:9584; (c) Voigtmann U, Blechert S (2000) Org Lett 2:3971; (d) Voigtmann U, Blechert S (2000) Synthesis 893; (e) Buschmann N, Rückert A, Blechert S (2002) J Org Chem 67:4325; (f) Zaminer J, Stapper C, Blechert S (2002) Tetrahedron Lett 43:6739; (g) Stapper C, Blechert S (2002) J Org Chem 67:6456. (h) Stapper C, Blechert S (2002) Eur J Org Chem 2855
117. (a) Nicolaou KC, Vega JA, Vassilikogiannakis G (2001) Angew Chem Int Ed Engl 40:4441; (b) Johannes CW, Visser MS, Weatherhead GS, Hoveyda AH (1998) 120:8340; (c) Burke SD, Quinn KJ, Chen VJ (1998) J Org Chem 63:8626; (d) Wrobleski A, Sahasrabudhe K, Aubé (2002) J Am Chem Soc 124:9974
118. Schaudt M, Blechert S (2003) J Org Chem 68:2913
119. Kim S-H, Bowden N, Grubbs RH (1994) J Am Chem Soc 116:10801
120. Zuercher WJ, Scholl M, Grubbs RH (1998) J Org Chem 63:4291
121. Boyer FD, Hanna I, Ricard L (2001) Org Lett 3:3095; (b) Codesido EM, Castedo L, Granja JR (2001) Org Lett 3:1483; (c) Huang J, Xiong H, Hsung RP, Rameshkumar C, Mulder JA, Grebe TP (2002) Org Lett 4:2417; (d) Efskind J, Römming C, Undheim K (2001) J Chem Soc Perkin Trans 1 2697; (e) Undheim K, Efskind J (2000) Tetrahedron 56:4847; (f) Banti D, North M (2002) Adv Synth Catal 344:694
122. Stragies R, Schuster M, Blechert S (1998) Chem Commun 237
123. Nguyen ST, Grubbs RH (1995) J Organomet Chem 497:195
124. Ahmed M, Barrett AGM, Braddock DC, Cramp SM, Procopiou PA (1999) Tetrahedron Lett 40:8657
125. Ahmed M, Arnauld T, Barrett AGM, Braddock DC, Procopiou PA (2000) Synlett 1007
126. (a) Jafarpour L, Nolan SP (2000) Org Lett 2:4075; (b) Jafarpour L, Heck M-P, Baylon C, Lee HM, Mioskowski C, Nolan SP (2002) Organometallics 21:671
127. Schürer SC, Gessler S, Buschmann N, Blechert S (2000) Angew Chem Int Ed Engl 39:3898
128. Yao Q (2000) Angew Chem Int Ed Engl 39:3896
129. Dowden J, Savović J (2001) Chem Commun 37
130. Randl S, Buschmann N, Connon SJ, Blechert S (2001) Synlett 1547
131. Kingsbury JS, Garber SB, Giftos JM, Gray BL, Okamoto MM, Farrer RA, Fourkas JT, Hoveyda AH (2001) Angew Chem Int Ed Engl 40:4251
132. Mayr M, Mayr B Buchmeiser MR (2001) Angew Chem Int Ed Engl 40:3839
133. Connon SJ, S Blechert (2002) Bioorg Med Chem Lett 12:1873
134. Connon SJ, Dunne AM, Blechert S (2002) Angew Chem Int Ed Engl 41:3835
135. Kobayashi S, Akiyama R (2003) Chem Commun 449
136. Kobayashi S, Akiyama R (2002) Angew Chem Int Ed Engl 41:2602
137. Gibson SE, Swamy VM (2002) Adv Synth Catal 344:619
138. Csihony I, Fischmeister C, Bruneau C, Horvath IT, Dixneuf PH (2002) New J Chem 1667
139. Blechert S, Schuster M, Pernerstorfer J (1996) Angew Chem Int Ed Engl 35:1979
140. Blechert S, Schuster M, Lucas N (1997) Chem Commun 823
141. Tang Q, Wareing JR (2001) Tetrahedron Lett 42:1399
142. Blackwell HE, Clemons PA, Schreiber SA (2001) Org Lett 3:1185

143. Schuster M, Blechert S (1998) Tetrahedron Lett 39:2295
144. Schürer SC, Blechert S (1999) Synlett 1879
145. Knerr L, Schmidt RR (1999) Synlett 1802
146. (a) Veerman JJN, van Maarseveen JH, Visser GM, Kruse CG, Schoemaker HE, Hiemstra H, Rutjes FPJT (1998) Eur J Org Chem 2853; (b) Piscopio AD, Miller JF, Kock K (1999) Tetrahedron 55:8189
147. (a) Peters J-U, Blechert S (1997) Synlett 348; (b) Nicolaou KC, Winssinger, N, Pastor J, Ninkovic S, Sarabia F, He Y, Vourloumis D, Yang Z, Li T, Giannakakou P, Hamel E (1997) Nature 387:268
148. Andrade RB, Plante OJ, Melean LG, Seeberger PH (1999) Org Lett 1:1811
149. Melean LG, Haase W-C, Seeberger PH (2000) Tetrahedron Lett 41:4329
150. Barrett AGM, Cramp SM, Roberts RS (1999) Org Lett 1:1083
151. (a) Cao J, Cuny GD, Hauske JR (1998) Molecular Diversity 3:173; (b) Cuny GD, Cao J, Hauske JR (1997) Tetrahedron Lett 38:5237
152. (a) Barrett AGM, Cramp SM, Roberts RS, Zecri FJ (2000) Org Lett 2:261; (b) Barrett AGM, Cramp S, Hennessy AJ, Procopiou PA, Roberts RS (2001) Org Lett 3:271; (c) Buchmeiser MR (2001) Angew Chem Int Ed Engl 40:3795
153. Barrett AGM, Hopkins BT, Kobberling J (2002) Chem Rev 102:3301

Topics Organomet Chem (2004) 11: 125–153
DOI 10.1007/978-3-540-40027-1

Ruthenium Vinylidenes and Allenylidenes in Catalysis

Christian Bruneau (✉)

Institut de Chimie de Rennes, UMR6509, Université de Rennes 1-CNRS, Organométalliques et Catalyse, Campus de Beaulieu, 35042 Rennes Cedex, France
christian.bruneau@univ-rennes1.fr

Abstract During the last decade, ruthenium catalysis, including asymmetric catalysis, has attracted increasing interest owing to its wide range of applications in organic synthesis [T. Naota, H. Takaya, S.I. Murahashi (1998) *Chem. Rev.* 98:2599; B.M. Trost, F.D. Toste, A.B. Pinkerton (2001) *Chem. Rev.* 101:2067; B.M. Trost BM (2002) *Acc. Chem. Res.* 35:695]. Ruthenium complexes bearing a carbenic ligand, such as a carbene, a vinylidene or an allenylidene group have found useful applications for selective transformations of unsaturated substrates. After the discovery of the first ruthenium vinylidene complex by Bruce at the end of the 1970s [M.I. Bruce, R.C. Wallis (1978) *J. Organomet. Chem.* 161:C1; M.I. Bruce, A.G. Swincer, R.C. Wallis (1979) *J. Organomet. Chem.* 171:C5], the impact of these reactive species, easily generated from terminal alkynes, became obvious in catalysis, and led to the development of new selective transformations of alkynes, most of them with atom economy [C. Bruneau, P.H. Dixneuf (1999) *Acc. Chem. Res.* 32:311]. Their advantages include the unusual regioselectivity of additions to alkynes. A few years later, the adventure of ruthenium allenylidene derivatives started when their straightforward preparation via activation/dehydration of propargylic alcohols at selected ruthenium centres was demonstrated [J.P. Selegue (1982) *Organometallics* 1:217; J.P. Selegue (1983) *J. Am. Chem. Soc.* 105:5921]. Their involvement in catalysis is currently leading to a variety of unprecedented reactions from propargylic compounds. Recent results on transformations of alkynes with ruthenium vinylidenes and allenylidenes as postulated active species in catalytic reactions will be presented. The use of these unsaturated species as catalyst precursors for other types of reactions will also be reported.

Keywords Ruthenium vinylidene · Ruthenium allenylidene · Ruthenium catalysis

1 Introduction

Stabilization of organic vinylidene and allenylidene species via coordination to a ruthenium centre is now well established, and the stoichiometric reactivity of these highly unsaturated ligands is still under intense investigation [1–4], and theoretical studies are being carried out [5,6]. Most of the chemical properties of cumulenylidene structures arise from the alternate electronic distribution along the carbon chain (Fig. 1).

From ruthenium vinylidene complexes, three main chemical processes leading to catalytic reactions are involved:

- Addition of nucleophiles at Cα.
- Carbometallation followed by migration of an alkynyl group to Cα.
- [2+2] cycloaddition and formation of ruthenacyclobutane intermediates.

From ruthenium allenylidene complexes, nucleophilic addition at the less hindered Cγ represents the most classical initial step leading to catalytic transformations.

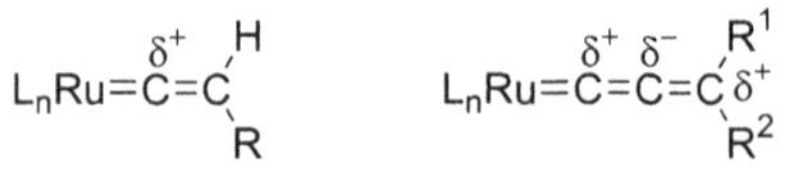

$$L_nRu{=}\overset{\delta^+}{C}{=}C(H)(R) \qquad L_nRu{=}\overset{\delta^+}{C}{=}\overset{\delta^-}{C}{=}\overset{\delta^+}{C}(R^1)(R^2)$$

Fig. 1

2 Ruthenium Vinylidenes as Active Catalytic Intermediates

2.1 Nucleophilic Addition

Several ruthenium complexes are able to promote the classical Markovnikov addition of O nucleophiles to alkynes via Lewis-acid-type activation of triple bonds. Starting from terminal alkynes, the anti-Markovnikov addition to form vinyl derivatives of type 1 (Scheme 1) is less common and requires selected catalysts. This regioselectivity corresponding to the addition of the nucleophile at the less substituted carbon of the C≡C triple bond is expected to result from the formation of a ruthenium vinylidene intermediate featuring a highly reactive electrophilic Cα atom.

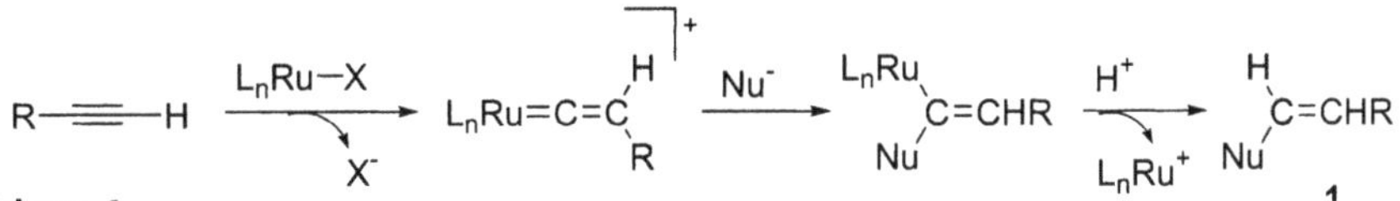

$R{-}C{\equiv}C{-}H \xrightarrow[-X^-]{L_nRu{-}X} [L_nRu{=}C{=}C(H)R]^+ \xrightarrow{Nu^-} L_nRu(Nu)C{=}CHR \xrightarrow[-L_nRu^+]{H^+} H(Nu)C{=}CHR$ (1)

Scheme 1

2.1.1 Addition of Carbamates: Synthesis of Vinylic Carbamates and Ureas

The first example of anti-Markovnikov addition of O nucleophiles to terminal alkynes was the catalytic addition of ammonium carbamates generated in situ from secondary amines and carbon dioxide to terminal alkynes, which selectively produced vinylcarbamates (Scheme 2) [7].

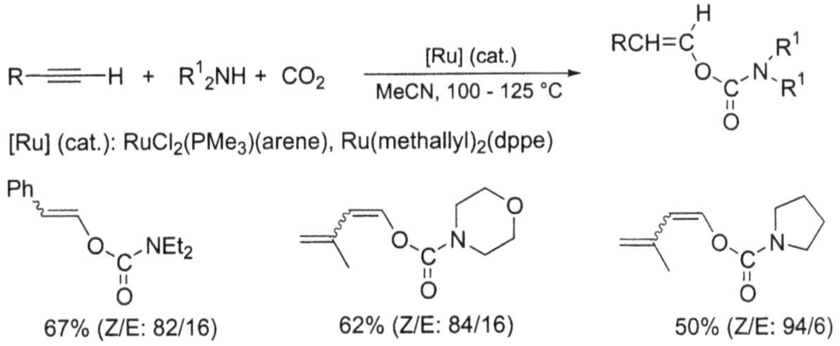

$R{-}C{\equiv}C{-}H + R^1{}_2NH + CO_2 \xrightarrow[\text{MeCN, 100 - 125 °C}]{\text{[Ru] (cat.)}} RCH{=}C(H){-}O{-}C(O){-}NR^1{}_2$

[Ru] (cat.): $RuCl_2(PMe_3)$(arene), $Ru(methallyl)_2(dppe)$

67% (Z/E: 82/16) 62% (Z/E: 84/16) 50% (Z/E: 94/6)

Scheme 2

The most efficient catalyst precursors were found in the $RuCl_2$(arene)(phosphine) series. These complexes are known to produce ruthenium vinylidene species upon reaction with terminal alkynes under stoichiometric conditions, and thus are able to generate potential catalysts active for anti-Markovnikov addition [8]. Dienylcarbamates could also be selectively prepared from conjugated enynes and secondary aliphatic amines but in this case, the best catalyst precursor was Ru(methallyl)$_2$(diphenylphosphinoethane) [9].

The formation of vinylcarbamates is restricted to terminal alkynes, which is in line with the formation of a metal vinylidene intermediate, and also to secondary amines. Indeed, a catalytic reaction also takes place under similar conditions with primary aliphatic amines but it leads to the formation of symmetrical ureas (Scheme 3) [10]. The catalytic system generated in this case is also thought to proceed via a ruthenium vinylidene active species and is very efficient for the formal elimination of water by formation of an organic adduct. The proposed general catalytic cycle, which applies for the formation of vinylcarbamates and ureas, is shown in Scheme 4 [11].

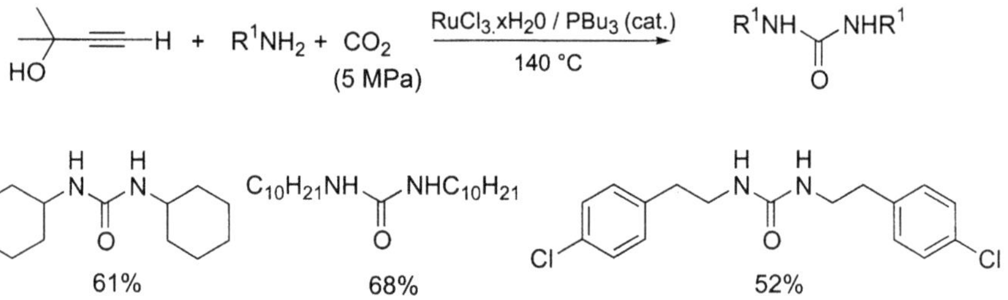

Scheme 3

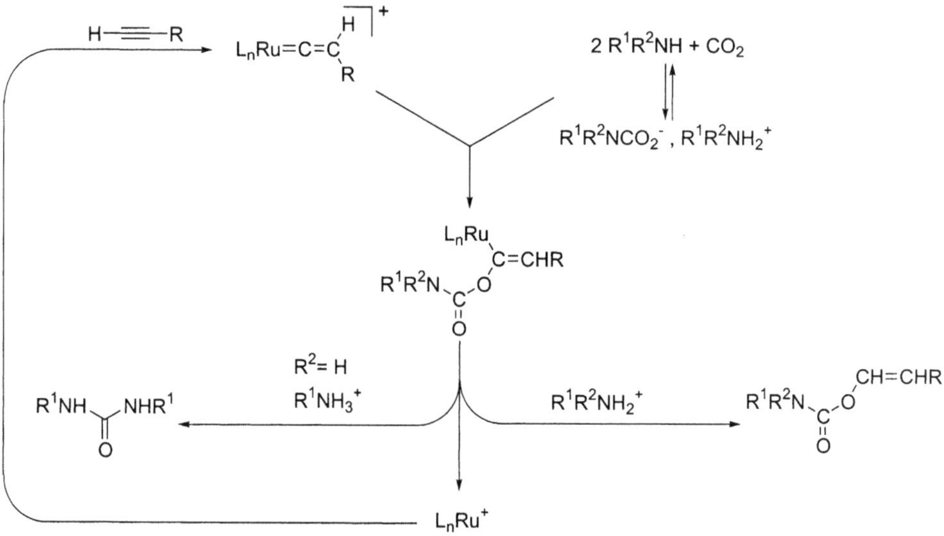

Scheme 4

2.1.2 Addition of Carboxylates: Synthesis of Enol Esters

Carboxylic acids also add to terminal alkynes to produce enol esters. A variety of ruthenium precursors such as $Ru_3(CO)_{12}$ [12], $Ru(cod)_2/PR_3$, where cod is cyclooctadiene, [13] $RuCl_2(PR_3)$(arene) [14, 15] or $[Ru(O_2CH)(CO)_2(PR_3)]_2$ [16] are good catalysts to perform the selective addition of the carboxylate to the C2 position of alkynes to afford geminal enol esters bearing a methylene group.

In contrast, some π-allyl ruthenium complexes containing a chelating diphosphine ligand were the first metal complexes which favoured the anti-Markovnikov addition of carboxylic acids to terminal alkynes to form (*Z*)-enol and (*E*)-enol esters with high regioselectivity and stereoselectivity [17–19] according to Eq. (1).

R—≡—H + HO(C=O)R¹ —[Ru] (cat.)→ R–CH=CH–O(C=O)R¹

[Ru] : Ru (methallyl)$_2$(dppb) ; Ru (methallyl)$_2$(dppe)

(1)

The best catalyst precursors are Ru(methallyl)$_2$(1,4-bis(diphenylphosphino)butane) (**A**) and Ru(methallyl)$_2$(dppe) (**B**), where dppe is 1,2-bis(diphenylphosphino)ethane, the choice of the appropriate complex depending on the steric demand of both the alkyne and the carboxylic acid. A large variety of carboxylic acids and alkynes have been used, including *N*-protected amino acids, α-hydroxy acids and functionalized alkynes such as enynes and propargylic ethers (Table 1) [20, 21].

The regioselective anti-Markovnikov addition of benzoic acid to phenylacetylene has also been carried out with success at 111 °C in the presence of ruthenium complexes containing a tris(pyrazolyl)borate (Tp) ligand [RuCl(Tp)(cod), RuCl(Tp)(pyridine), RuCl(Tp)(*N*,*N*,*N'*,*N'*-tetramethylethylenediamine)] with a stereoselectivity in favour of the (*E*)-enol ester isomer [22]. The σ-enynyl complex Ru(Tp)[PhC=C(Ph)C≡CPh](PMe*i*-Pr$_2$) (**C**) efficiently catalyses the regioselective cyclization of α,ω-alkynoic acids to give endocyclic enol lactones [23] (Eq. 2).

HC≡C–(CH$_2$)$_n$–COOH —Ru(Tp)[PhC=C(Ph)CCPh](PMe^{i}Pr$_2$) (**C**) (2 mol%), toluene, 100 °C→ endocyclic enol lactone; n= 1 (97%), 2 (95%), 3 (45%), 7 (84%)

(2)

Very recently, new catalysts precursors derived from $[RuCl_2(p\text{-cymene})]_2$ such as $RuCl_2$(triazol-5-ylidene)(*p*-cymene) (**D**) and RuCl(*p*-cymene)(*o*-Ph-(triazol-5-ylidene) (**E**) [24], or the in situ generated catalytic system based on [RuCl$_2$(*p*-

Table 1 Synthesis of *Z*-enol esters

Catalyst	Acid	Alkyne	*T* (°C)	Yield (%)	Z selectivity	Reference
A	$PhCO_2H$	$C_4H_9C{\equiv}CH$	65	95	98	[18]
A	$PhCO_2H$	$PhC{\equiv}CH$	100	97	96	[18]
A	CH_3CO_2H	$PhC{\equiv}CH$	45	90	99	[18]
A	$CHCl_2CO_2H$	$PhC{\equiv}CH$	20	78	100	[18]
A	CF_3CO_2H	$PhC{\equiv}CH$	0	61	100	[18]
A	Ph_2CHCO_2H	$C_4H_9C{\equiv}CH$	65	97	100	[18]
A	(*L*)-Boc–PheOH	$C_4H_9C{\equiv}CH$	65	97	100	[18]
A	(*L*)-*Z*-AlaOH	$PhC{\equiv}CH$	65	98	100	[18]
A	$CH_2{=}CHCO_2H$	$PhC{\equiv}CH$	45	65	99	[18]
A	$MeOCH_2CO_2H$	(*Z*)-MeOCH=CHC≡CH	65	69	99	[20]
A	$PhCO_2H$	(*Z*)-MeOCH=CHC≡CH	65	81	98	[20]
A	$PhCO_2H$	$CH_2{=}C(Me)C{\equiv}CH$	65	92	99	[20]
B	$PhCO_2H$	$Me_3SiC{\equiv}CH$	60	88	100	[18]
B	$CH_2{=}C(Me)CO_2H$	$Me_3SiC{\equiv}CH$	50	76	100	[18]
B	(*L*)-Boc–AlaOH	$Me_3SiC{\equiv}CH$	50	75	100	[18]
B	$PhCO_2H$	$MeOC(Me)_2C{\equiv}CH$	80	86	94	[21]
B	$PhCO_2H$	$MeOC(Me)(Ph)C{\equiv}CH$	80	95	98	[21]

cymene)]$_2$/P(p-C_6H_4Cl)$_3$/4-dimethylaminopyridine [25] have revealed their potential to perform the anti-Markonikov addition of a variety of carboxylic acids to phenylacetylene and terminal aliphatic alkynes. In contrast, ruthenium vinylidene complexes such as **F** have been reported as active catalysts for the addition of carboxylic acids to alkynes, but in most cases they favour the Markovnikov addition [26] (Fig. 2).

It is postulated that the catalytic cycle accounting for this regioselectivity involves a ruthenium vinylidene intermediate and is quite similar to Scheme 4 with the carboxylate nucleophile instead of the carbamate.

The addition to propargylic alcohols in the presence of Ru(methallyl)$_2$-(dppe) (**B**) at 65 °C leads to hydroxylated alk-1-en-1-yl esters via formation of a hydroxy vinylidene intermediate [27, 28]. These esters can easily be cleaved

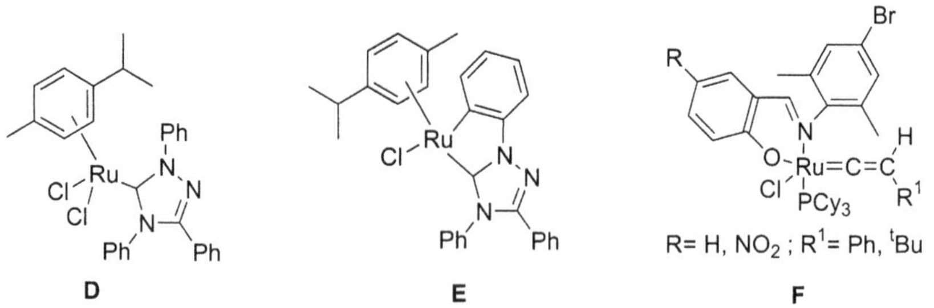

Ph N N Cl Ru Cl N Ph Ph

D

Cl Ru N N N Ph Ph

E

Br R N H O Ru=C=C Cl PCy3 R1

R= H, NO$_2$; R^1= Ph, ^{t}Bu

F

Fig. 2

under thermal or acidic conditions to give conjugated enals, corresponding to the formal isomerization products of the starting alcohols (Scheme 5).

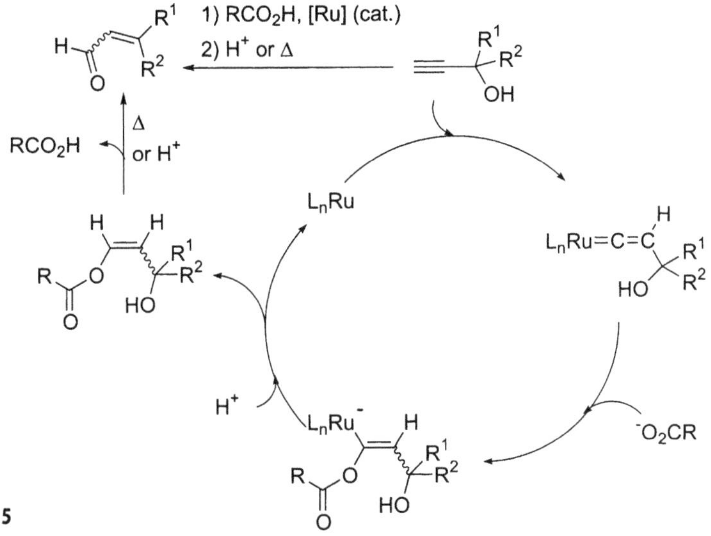

Scheme 5

2.1.3
Addition of Water: Synthesis of Aldehydes

The metal-catalysed addition of water to terminal alkynes usually leads to ketones following Markovnikov's rule. The first selective catalytic formation of aldehydes was reported by Tokunaga and Wakatsuki [29], who used $RuCl_2(C_6H_6)(PPh_2(C_6F_5))$+3 $PPh_2(C_6F_5)$ or $[RuCl_2(C_6H_6)]_2$ associated to 8 Eq of the water-soluble ligand $P(3\text{-}C_6H_5SO_3Na)_3$ in alcohol at 65–100 °C (Eq. 3).

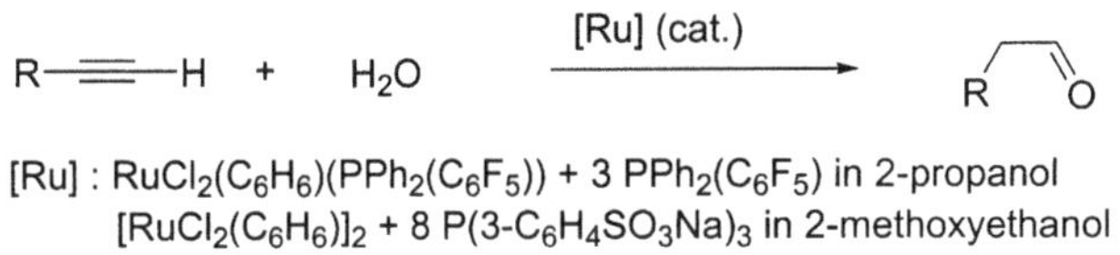

$$R{-}{\equiv}{-}H \;+\; H_2O \xrightarrow{[Ru]\ (cat.)} RCH_2CHO \tag{3}$$

[Ru] : $RuCl_2(C_6H_6)(PPh_2(C_6F_5))$ + 3 $PPh_2(C_6F_5)$ in 2-propanol
$[RuCl_2(C_6H_6)]_2$ + 8 $P(3\text{-}C_6H_4SO_3Na)_3$ in 2-methoxyethanol

Under these conditions, a variety of linear aliphatic terminal alkynes were transformed into aldehydes with good selectivity. The efficiency, regioselectivity of the addition and substituent tolerance were improved by using $RuCl(Cp)(phosphine)_2$, where Cp is cyclopentadienyl, or RuCl(Cp)(diphosphine) as catalyst precursors [30]. The best results were obtained with diphenylphosphinomethane as a ligand, which made possible the preparation of aldehydes from bulky aliphatic alkynes (*tert*-BuC≡CH), aromatic alkynes (PhC≡CH), diynes $[HC{\equiv}C(CH_2)_6C{\equiv}CH]$ and functional terminal alkynes $[NC(CH_2)_3C{\equiv}CH$, $PhCH_2O(CH_2)_2C{\equiv}CH$,...]. The mechanism of this reaction was investigated in details by isolation of intermediates, deuterium-labelling experiments and density functional theory calculations [31]. The most probable catalytic cycle involves first protonation of a ruthenium(II) π-alkyne species to give a Ru(IV) vinylidene intermediate via a Ru(IV) vinyl species. The nucleophilic addition of water to the α-carbon of the vinylidene ligand followed by reductive elimination affords the aldehyde (Scheme 6).

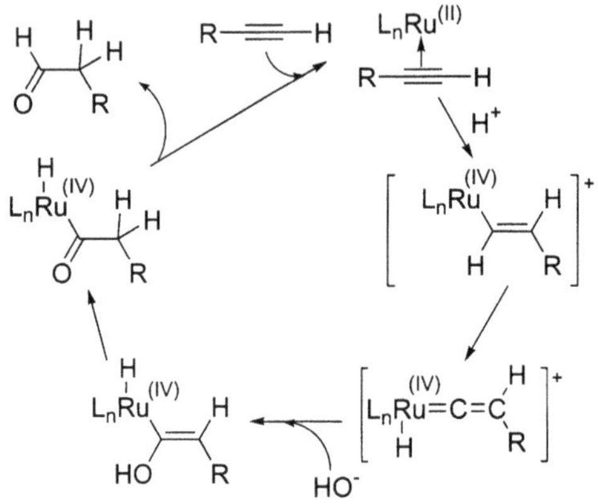

Scheme 6

It is noteworthy that the indenyl complex $RuCl(\eta^5\text{-}C_9H_7)(PPh_3)_2$ provides an efficient catalyst precursor for the anti-Markovnikov hydration of terminal alkynes in aqueous media and micellar solutions with either anionic (sodium dodecyl sulfate) or cationic (hexadecyltrimethylammonium bromide) surfactants [32]. This system can be applied to the hydration of propargylic alcohols to selectively produce β-hydroxyaldehydes (Eq. 4).

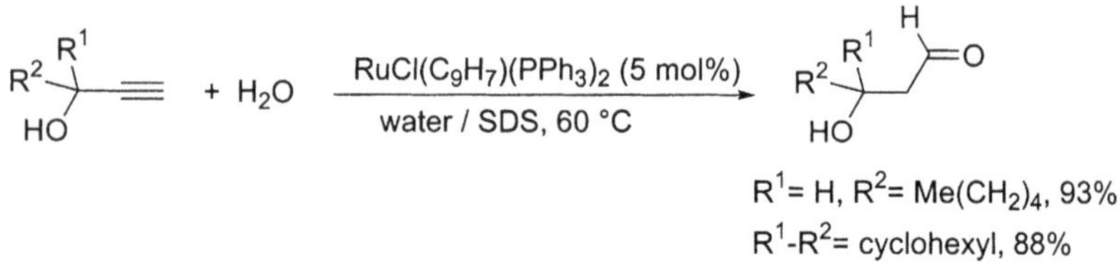

$$R^2R^1C(OH)C\equiv CH + H_2O \xrightarrow[\text{water / SDS, 60 °C}]{RuCl(C_9H_7)(PPh_3)_2\ (5\ mol\%)} R^2R^1C(OH)CH_2CHO \quad (4)$$

R^1= H, R^2= $Me(CH_2)_4$, 93%
R^1-R^2= cyclohexyl, 88%

2.1.4 Addition of Alcohols: Synthesis of Ethers and Ketones

2.1.4.1 Formation of Unsaturated Ethers and Furans

The ruthenium-catalysed direct addition of saturated aliphatic alcohols to non-activated alkynes remains a challenge. Only allyl alcohol has been successfully involved in the intermolecular addition to phenylacetylene to produce an ether and the enal resulting from Claisen rearrangement (Scheme 7) [22]. Thus, in refluxing toluene, in the presence of a catalytic amount of RuCl(tris(pyrazolyl)borate)(pyridine)$_2$, a 1:1 mixture of allyl β-styryl ether and 2-phenylpent-4-enal was obtained in 72% overall yield.

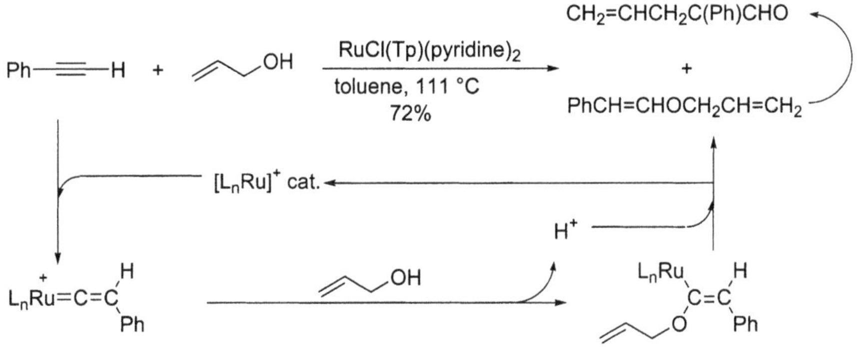

Scheme 7

The recent synthesis of furans via isomerization of terminal epoxyalkynes catalysed by $RuCl(Tp)(MeCN)_2$ in the presence of a base at 80 °C in 1,2-dichloroethane is explained by a related intramolecular nucleophilic addition of the oxygen atom of the epoxide onto the α-carbon atom of a ruthenium vinylidene intermediate (Scheme 8) [33]. This reaction is specific of terminal alkynes and tolerates a variety of functional groups (ether, ester, acetal, tosylamide, nitrile).

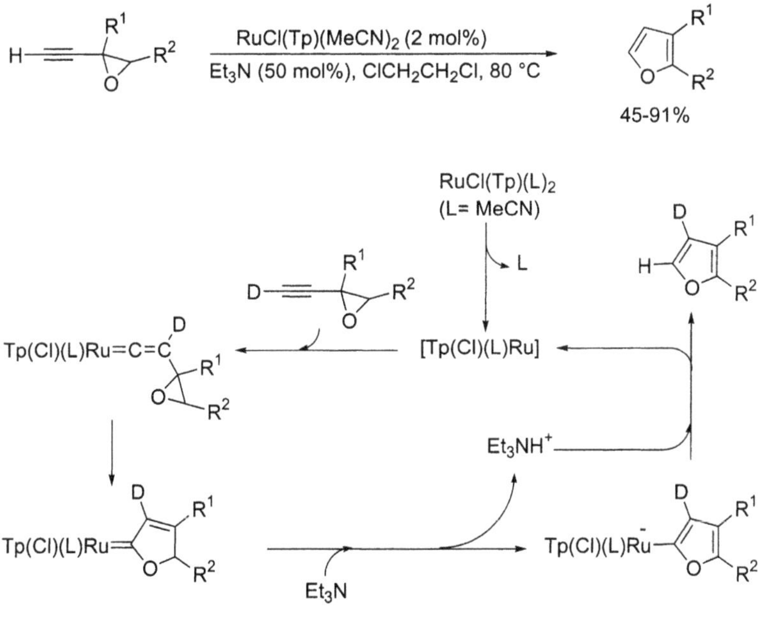

Scheme 8 R^1= H ; R^2= C_7H_{15} (84%), CH_2OBn (71%)
R^1= Me ; R^2= CH_2OH (86%), CH_2OCOPh (81%), CH_2OBn (91%)

2.1.4.2 Formation of Unsaturated Ketones

Another selective reaction involving C–O bond formation followed by rearrangement and C–C bond formation occurs when Cp-containing ruthenium complexes are used as catalytic precursors. With $RuCl(Cp)(PPh_3)_2$ in the presence of NH_4PF_6, AgOTf, where Tf is CF_3SO_2, or $In(OTf)_2$, additives which are known to facilitate chloride abstraction from the metal centre, the addition of allylic alcohols to terminal alkynes affords unsaturated ketones [34, 35]. It has been shown that the key steps are the nucleophilic addition of the allylic alcohol to a ruthenium vinylidene species followed by formation of an allyl metal intermediate via sigmatropic rearrangement (Scheme 9) [35]. This transformation of terminal alkynes via coupling with allylic alcohol and formation of a C–C bond with atom economy has been applied to the synthesis and modification of natural compounds such as rosefuran and steroids [36, 37].

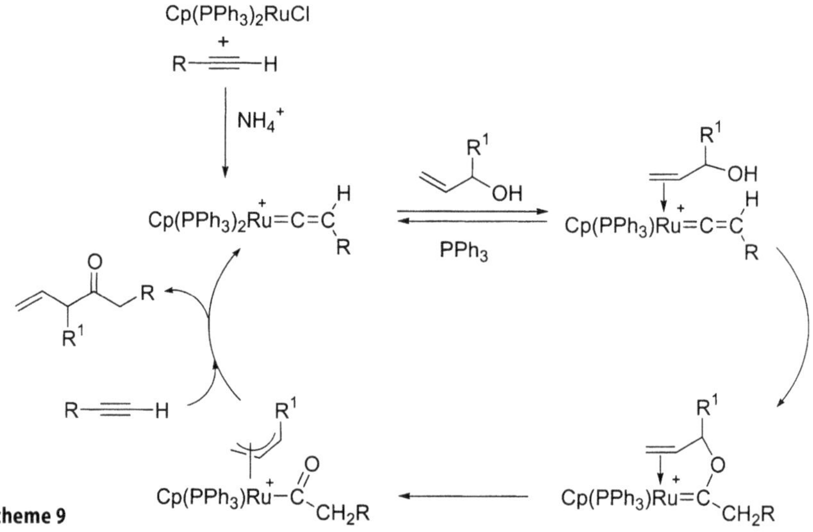

Cp(PPh3)2RuCl + R—≡—H → (NH4+) Cp(PPh3)2Ru=C=C(H)R ⇌ (allylic alcohol / PPh3) Cp(PPh3)Ru=C=C(H)R (coordinated allylic alcohol) → Cp(PPh3)Ru=C(O-allyl)CH2R → Cp(PPh3)Ru(allyl)–C(=O)CH2R → unsaturated ketone + (R—≡—H)

Scheme 9

The preparation of optically active ruthenium vinylidene complexes to perform asymmetric reconstitutive condensation has been carried out with moderate enantioselectivities [38, 39].

2.1.4.3 Formation of Lactones from but-3-yn-1-ols and pent-4-yn-1-ols

Homopropargylic alcohols (but-3-ynols) as well as propargylic epoxides are suitable products to form cyclic ruthenium alcoxycarbenes upon intramolecular nucleophilic addition of the OH group to the electrophilic α-carbon of ruthenium vinylidene species. The recovery of the organic ligand as a lactone

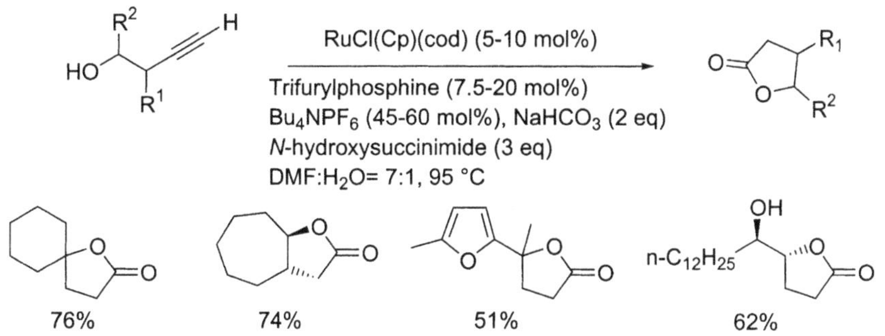

Fig. 3

is made possible by mild oxidation with an oxidant which does not destroy the catalyst. The best catalytic system reported up to now is based on $RuCl(C_5H_5)(cod)$, tris(2-furyl)phosphine as an ancillary ligand and $NaHCO_3$ as a base, in the presence of n-Bu_4NBr or n-Bu_4PF_6, and N-hydroxysuccinimide as the oxidant in dimethylformamide–water at 95 °C (Fig. 3, Scheme 10) [40].

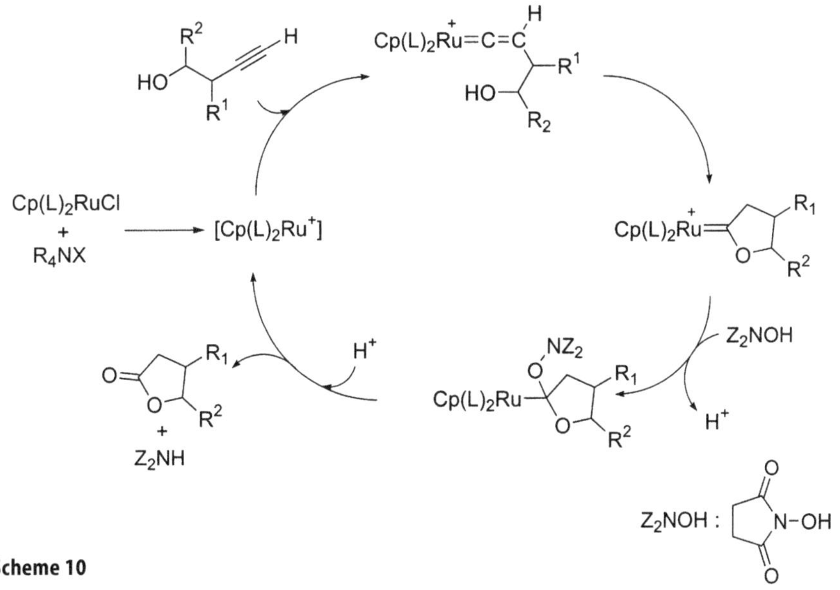

Scheme 10

Starting from pent-4-yn-1-ols, the previously described catalytic system led to a mixture of lactone and cyclic enol ether [41]. However, in the presence of (Cp)ruthenium complexes bearing an electron-rich ligand such as tris(p-methoxyphenyl)phosphine in the presence of a large excess of the same ligand [system A], the selective formation of lactones was obtained. A simple modification of the catalyst precursor such as the switch to the electron-deficient tris(p-fluorophenyl)phosphine [system B] completely reversed the chemoselectivity of the reaction towards the formation of dihydropyrans resulting from cycloisomerization of the starting alkynol (Scheme 11).

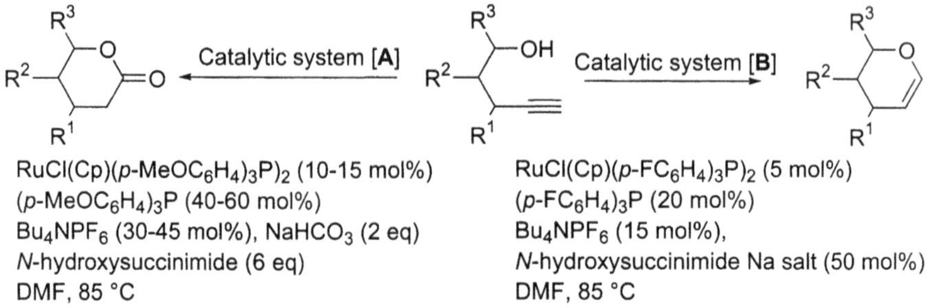

RuCl(Cp)(p-MeOC$_6$H$_4$)$_3$P)$_2$ (10-15 mol%)
(p-MeOC$_6$H$_4$)$_3$P (40-60 mol%)
Bu$_4$NPF$_6$ (30-45 mol%), NaHCO$_3$ (2 eq)
N-hydroxysuccinimide (6 eq)
DMF, 85 °C

RuCl(Cp)(p-FC$_6$H$_4$)$_3$P)$_2$ (5 mol%)
(p-FC$_6$H$_4$)$_3$P (20 mol%)
Bu$_4$NPF$_6$ (15 mol%),
N-hydroxysuccinimide Na salt (50 mol%)
DMF, 85 °C

Scheme 11

Both oxidative cyclization and cycloisomerization were applied to a variety of substrates including sugar derivatives; the only restriction to the formation of lactones was the presence of a tertiary alcohol functionality (Fig. 4).

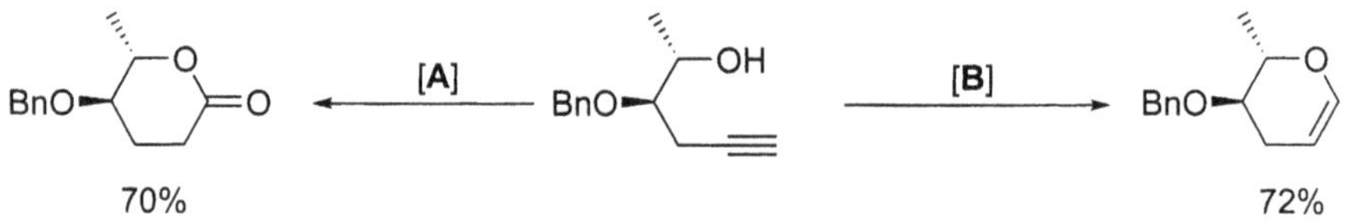

70% 72%

Fig. 4

2.1.5 Formation of Nitriles via Addition of Hydrazines

The formation of a ruthenium vinylidene is proposed as the key intermediate in the regioselective addition of hydrazine to terminal alkynes [42]. This new reaction, which proceeds via addition of the primary amino group of a 1,1-disubstituted hydrazine followed by deamination, provides access to a variety of aromatic and aliphatic nitriles. The tris(pyrazolyl)borate complex RuCl(Tp)(PPh$_3$)$_2$ gave the best catalytic activities in the absence of any chloride abstractor (Fig. 5, Scheme 12).

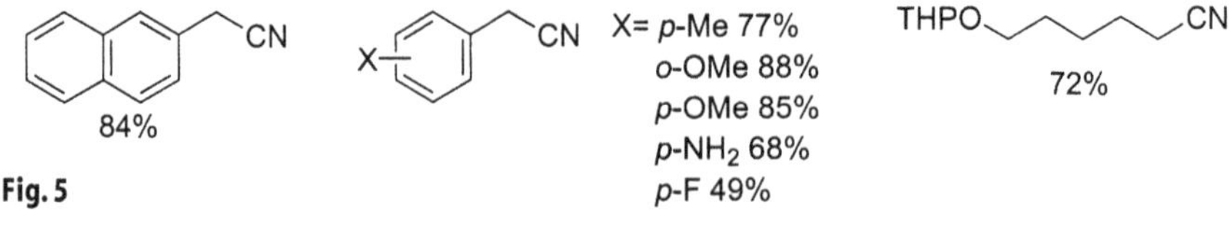

84%

X= p-Me 77%
o-OMe 88%
p-OMe 85%
p-NH$_2$ 68%
p-F 49%

72%

Fig. 5

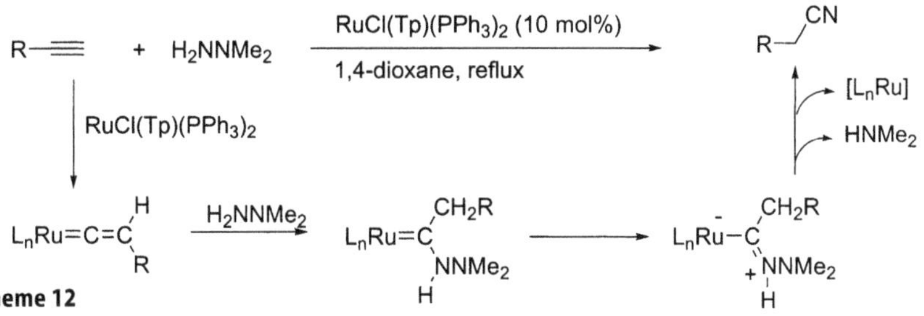

Scheme 12

2.1.6 Hydrophosphination: Synthesis of Vinylic Phosphine

The addition of secondary phosphines HPR_2 to prop-2-ynols in the presence of $RuCl(C_5Me_5)(cod)$ or $RuCl(C_5Me_5)(PPh_3)_2$ provides the first regioselective and stereoselective direct hydrophosphination of propargylic alcohols and leads to bifunctional (*Z*)-olefins [43]. Indeed, the reaction of tertiary propargylic alcohols with diphenylphosphine in the presence of $NaCO_3$ in refluxing $CHCl_3$ led to 3-diphenylphosphinoprop-2-enols in good yields and high stereoselectivity in favour of the Z isomer (from 75/25 to 95/5) (Fig. 6, Scheme 13).

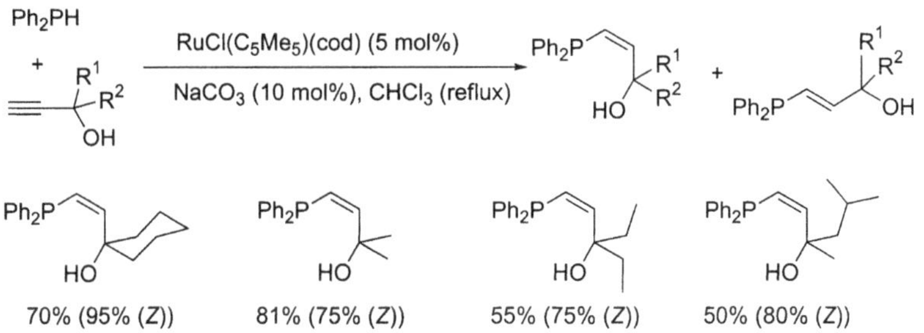

Fig. 6 70% (95% (Z)) 81% (75% (Z)) 55% (75% (Z)) 50% (80% (Z))

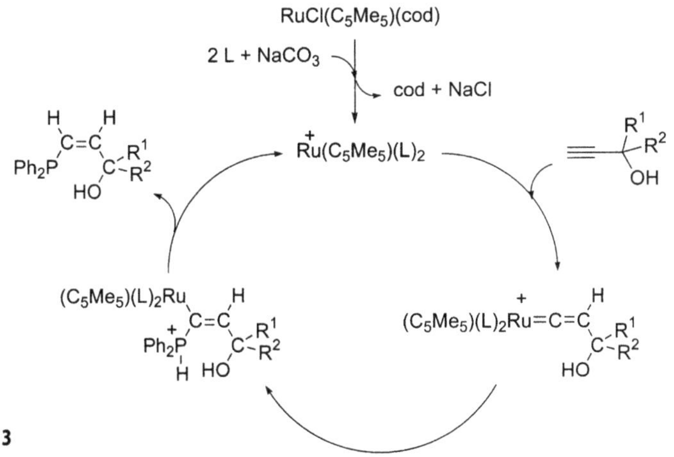

Scheme 13

2.1.7
Addition of Carbonucleophiles: C–C Bond Formation

Addition of carbonucleophiles to ruthenium vinylidenes is still rare and the only examples concern intramolecular additions. Cycloaromatization of conjugated dienynes has been performed under mild conditions (refluxing CH_2Cl_2) with $RuCl_2(PPh_3)(p\text{-cymene})/NH_4PF_6$ as a catalyst precursor (Eq. 5) [44]. Fused aromatic heterocycles have thus been obtained in good yields but when the terminal double bond is functionalized by an ether, an allylic alcohol or an ester group, the yields are very low (Scheme 14).

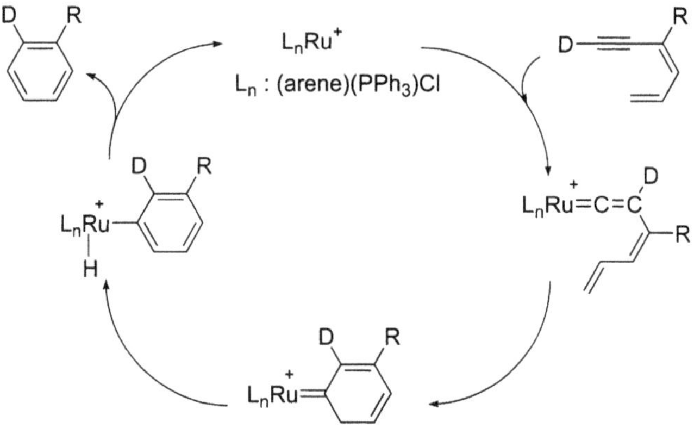

L_nRu^+

L_n : (arene)(PPh_3)Cl

Scheme 14

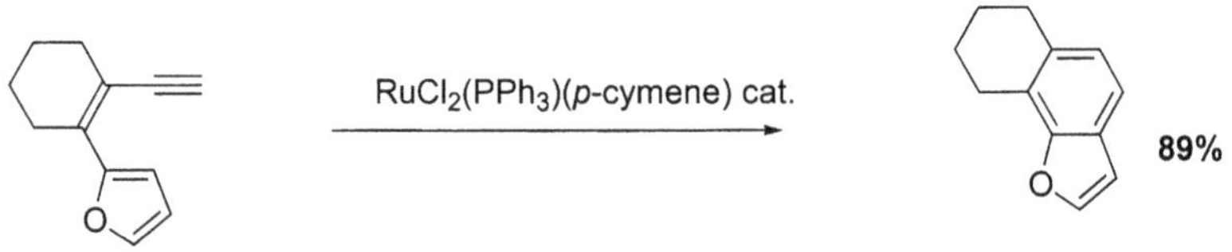

$RuCl_2(PPh_3)(p\text{-cymene})$ cat. → **89%** (5)

A related stoichiometric cycloaromatization of enediyne involving a ruthenium vinylidene intermediate takes place in the presence of $RuCl(Cp)(PMe_3)_2$/ NH_4PF_6 but a radical process has been proposed [45].

2.2
Dimerization of Terminal Alkynes

Terminal alkynes can undergo several types of interaction with ruthenium centres. In addition to the formation of ruthenium vinylidene species, a second type of activation provides alkynyl ruthenium complexes via oxidative addition. When these two types of coordination take place at the same metal centre, the migration of the alkynyl ligand onto the Cα atom of the vinylidene can occur to form enynyl intermediates, which upon protonation by the terminal alkyne lead to the formation of enynes corresponding to alkyne dimerization

(Eq. 6, Scheme 15, cycle A). In special cases, the rearrangement of the enynyl ligand to an allenylidenyl ligand can occur and the formation of the butatriene dimer is observed (Scheme 15, cycle B)

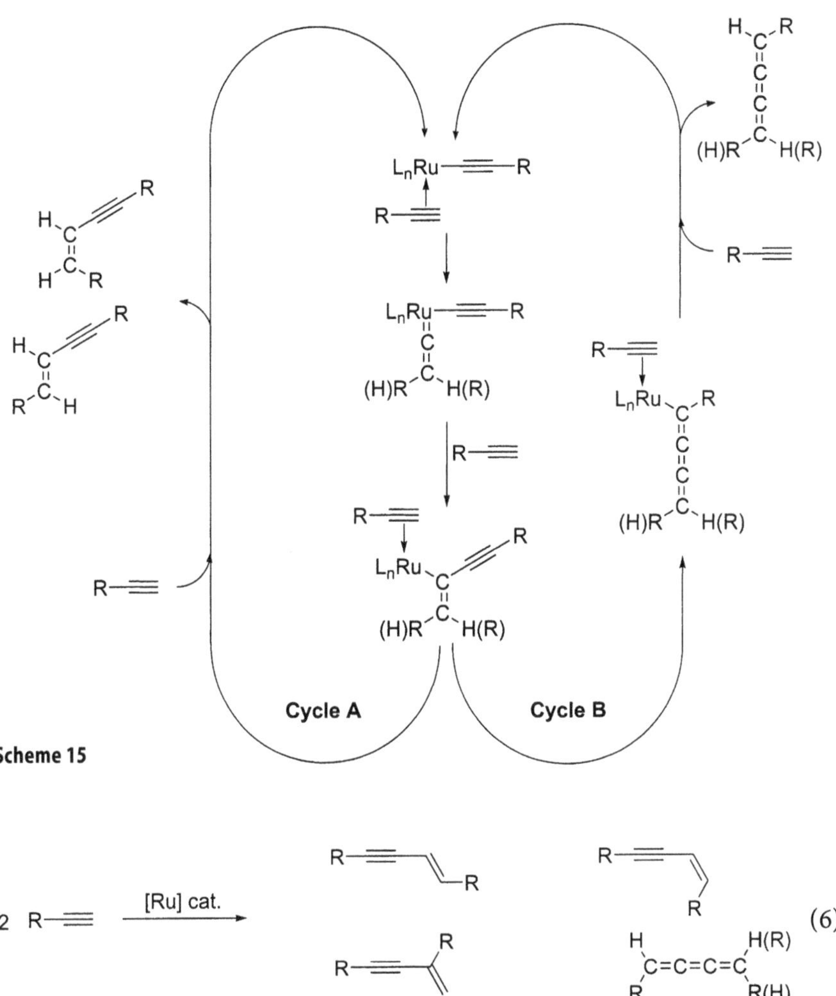

Scheme 15

(6)

Thus, ruthenium complexes containing a bulky electron-donating polydentate nitrogen ligand, such as a Tp (**G**) [46–48], a polypodal phosphorus ligand like $P(CH_2CH_2PPh_2)_3$ (**H, I**) [49, 50], a pentamethylcyclopentaniedyl (Cp*) (**J**) [51] or an indenyl (**K**) [52] ligand, are efficient catalysts for the selective head-to-head dimerization of terminal alkynes to enynes (Fig. 7). Most of these catalytic systems are able to dimerize either aromatic alkynes such as phenylacetylene derivatives or aliphatic alkynes such as trimethylsilylacetylene, *tert*-butylacetylene and benzylacetylene. The stereochemistry of the resulting enynes

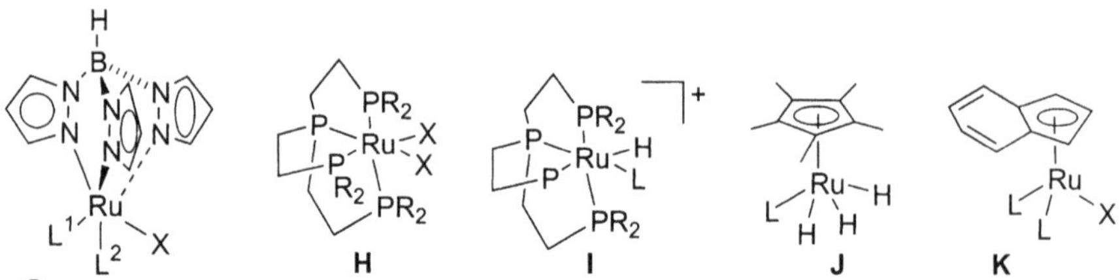

G

X: Cl ; L^1: PPh_3
L^2: PPh_3
pyridine
=C=CHPh

X: C≡CPh
L^1-L^2: $Ph_2PCH_2CH_2OMe$

H

X: H
C≡CPh

I

L: H_2, N_2

J

L: PPh_3
PCy_3
PMe_3

K

L: PPh_3
L-L : dppm
dppe
X: H
C≡CPh

Fig. 7

strongly depends of both the alkyne and the catalyst precursor. It is noteworthy that the ruthenium vinylidene complex RuCl(Cp*)(PPh_3)(=C=CHPh) catalyses the dimerization of phenylacetylene and methylpropiolate with high stereoselectivity towards the (*E*)-enyne [53, 54] and that head-to-tail dimerization is scarcely favoured with these catalysts. A dinuclear bis(pentamethylcyclopentadienyl)diruthenium complex bearing bridging thiolates has also shown very good catalytic efficiency to produce the (*Z*)-1,4-diferrocenylbutenyne from ferrocenylacetylene as the major compound at 60 °C in 1,2-dichloroethane [55].

Recently, it was shown that the metathesis catalyst $RuCl_2(PCy_3)_2$(=CHPh), where Cy is cyclohexyl, reacted in refluxing toluene with phenylacetylene to produce a ruthenium vinylidene species which promoted the regioselective dimerization of phenylacetylene into (*E*)-1,4-diphenylbutenyne [56]. The addition of 1 Eq acetic acid did not lead to enol esters but to a faster reaction and the stereoselective dimerization of phenylacetylene into the Z dimer.

The head-to-head dimerization with formation of a butatriene derivative was scarcely observed as the main catalytic route (Scheme 15, catalytic cycle B). Nevertheless, this was the case from benzylacetylene in the presence of RuH_3Cp*(PCy_3) as a catalyst precursor in tetrahydrofuran at 80 °C, which gave more than 95% of (*Z*)-1,4-dibenzylbutatriene [54], and from *tert*-butylacetylene with two efficient catalytic systems capable of generating zero-valent ruthenium species, $RuH_2(PPh_3)_3$(CO) and Ru(cod)(cyclooctatetraene) in the presence of an excess of triisopropylphosphine, which led to (*Z*)-1,4-di-*tert*-butylbutatriene as the major compound [57–59].

The formation of butadiynes from bis(alkynyl)mercury compounds in the presence of a catalytic amount of RuHCl(CO)$(PPh_3)_3$ is closely related to the ruthenium-catalysed dimerization reaction in terms of the mechanism [60]. The proposed catalytic cycle involves the formation of a (alkynyl)(vinylidene)ruthenium intermediate, which rearranges into an isolated enynyl ruthenium complex which liberates the diyne via β-elimination and regenerates the ruthenium hydride precursor.

2.3 C–C Bond Formation via Cycloaddition

The system $RuCl(Cp)(PPh_3)_2$ (5 mol %)/$NaPF_6$ (6 mol %) in pyridine at 100 °C catalyses the coupling of unactivated olefins with terminal alkynes to form conjugated dienes with a favoured head-to-head coupling (Eq. 7) [61].

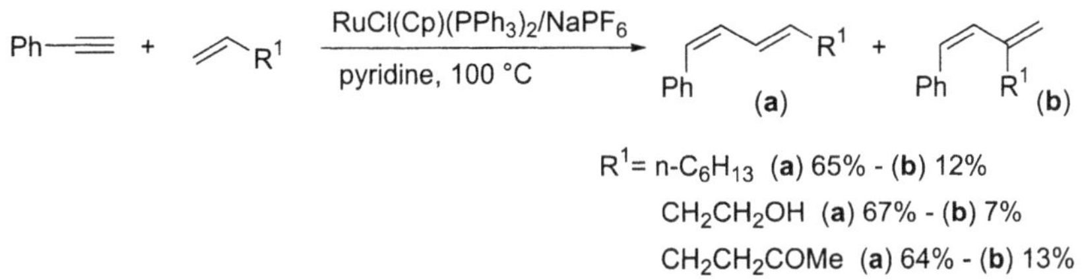

$RuCl(Cp)(PPh_3)_2/NaPF_6$, pyridine, 100 °C

R^1 = n-C_6H_{13} (**a**) 65% - (**b**) 12%
CH_2CH_2OH (**a**) 67% - (**b**) 7%
CH_2CH_2COMe (**a**) 64% - (**b**) 13%

(7)

The cycloaddition of the alkene to the ruthenium vinylidene species leads to a ruthenacyclobutane which rearranges into an allylic ruthenium species resulting from β-elimination or deprotonation assisted by pyridine and produces the diene after reductive elimination (Scheme 16). This mechanism is supported by the stoichiometric C–C bond formation between a terminal alkyne and an olefin, leading to η^3-butatrienyl and η^2-butadienyl complexes via a ruthenacyclobutane resulting from [2+2] cycloaddition [62].

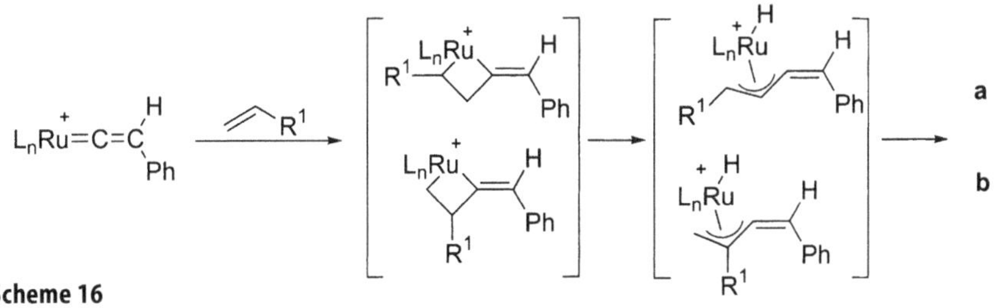

Scheme 16

Another example is the ruthenium-catalysed alkenylation of pyridine which is performed in the presence of the same catalyst precursor $RuCl(Cp)(PPh_3)_2$ (20 mol %)/$NaPF_6$ (20 mol %) at 150 °C [63]. The use of trimethylsilylalkynes, which are also known to produce vinylidene complexes rather than terminal alkynes, avoids the dimerization of the alkyne and favours the formation of the (*E*)-vinylpyridine (Scheme 17). The reaction has been applied to a variety of silylated alkynes and substituted pyridines (Fig. 8).

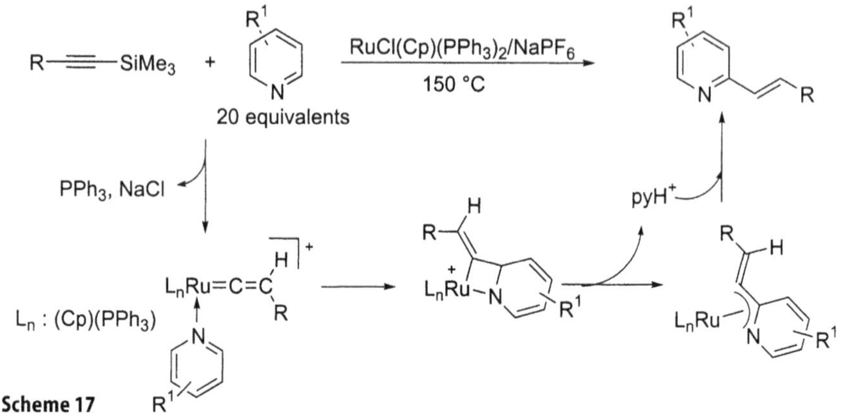

Scheme 17

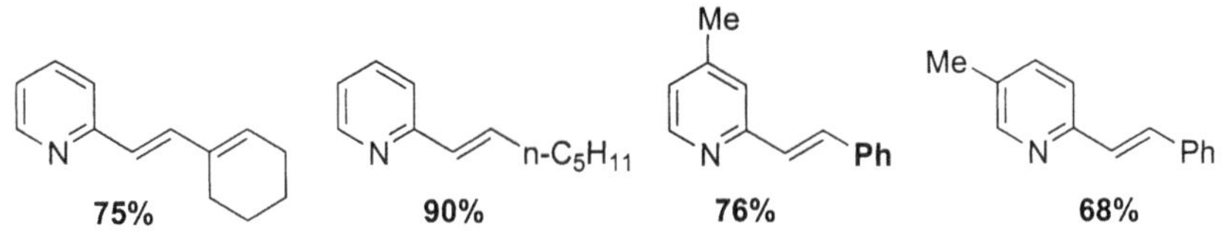

Fig. 8

A related mechanism involving insertion of a triple bond into a vinylidene ruthenium Ru=C bond has been postulated for the cyclocarbonylation of 1,1′-bis(trimethylsilyl)ethynylferrocene [64] according to Eq. (8).

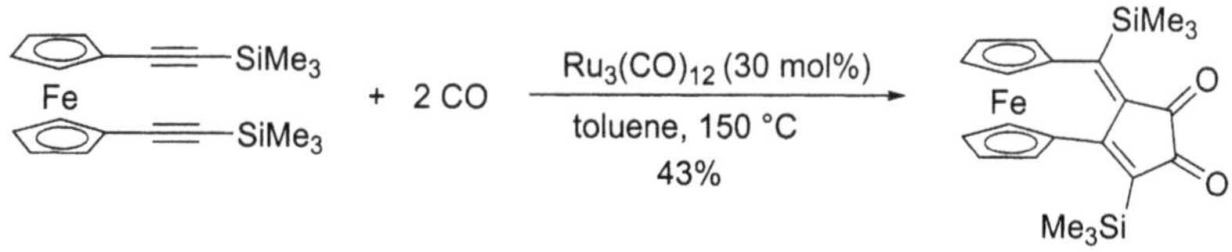

(8)

3 Ruthenium Vinylidenes as Catalyst Precursors

3.1 Olefin Metathesis: Ring-Opening Metathesis Polymerisation and Ring-Closing Metathesis

Well-defined ruthenium vinylidene complexes are efficient catalyst precursors for the ring-opening metathesis polymerization (ROMP) of cyclic olefins (Fig. 9). Most of them are neutral 16-electron complexes of the type $RuCl_2(L)_2(=C=CHR)$ (**L1**) [65–68] and the more active precursors contain

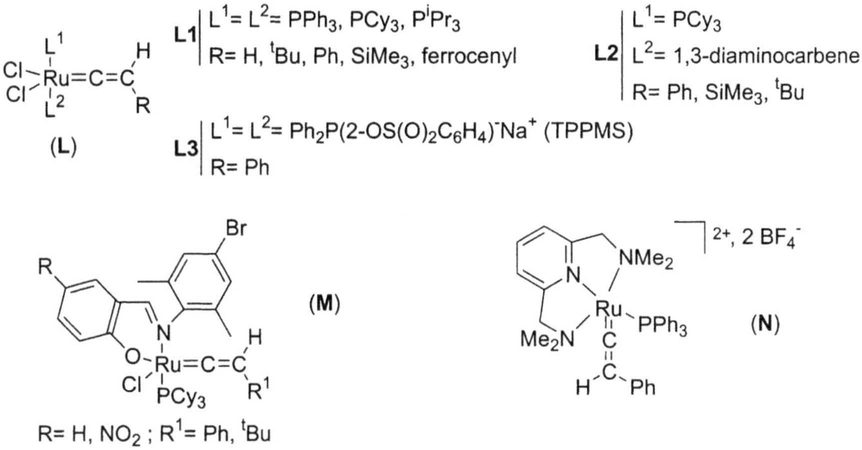

Fig. 9

bulky basic phosphines such as PCy_3 and $Pi\text{-}Pr_3$. A variety of cyclic olefins, including cyclopentene, cyclohexene, norbornene and oxanorbornene, and 5,6-difunctional norbornenes, such as diesters, anhydrides and dicarboximides, have been polymerized to produce polymeric materials with specific physical properties [69,70]. Some of these ruthenium vinylidene complexes are also active for the ring-closing metathesis (RCM) of dienes and enynes. The replacement of one phosphine by a diaminocarbene ligand to form $RuCl_2(PCy_3)$(1,3-bis(mesityl)imidazolinylidene)(=C=CHSiMe$_3$) (**L2**), for instance, leads to a new family of efficient catalysts for the ROMP of low-strained and high-strained cyclic olefins [71].

The water-soluble complex [RuCl(triphenylphosphine monosulfonate)(=C=CHPh)][Na]$_2$ (**L3**) made possible the following cascade ring-opening/cross-metathesis reaction (Eq. 9) [72].

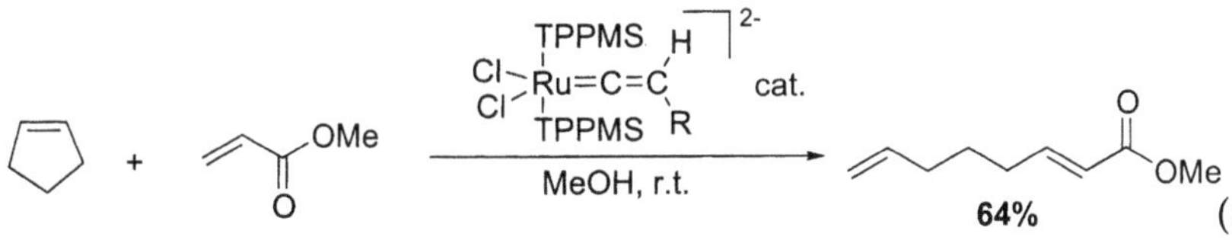

(9)

The neutral 18-electron $RuCl(Tp)(PPh_3)$(=C=CHPh) and RuCl(Cp*)(PPh$_3$)-(=C=CHPh) complexes also catalyse the ROMP of norbornene and their catalytic activity is enhanced by addition of a Lewis acid [66]. Other ruthenium vinylidene complexes featuring a bidentate (N,O) (**M**) [71] or a tridentate (N,N,N) (**N**) ligand [73] are also precatalysts of metathesis polymerisation.

Besides these stable single-site catalyst precursors, some in situ generated ruthenium vinylidene species have been postulated as initiators of diene RCM. Ru(1,3-bis(mesityl)imidazolyl-2-ylidene)(vinylidene) [74] and Ru(1,3-

bis(mesityl)imidazolylin-2-ylidene)(vinylidene) [75,76] intermediates generated upon reaction of [$RuCl_2$(arene)]$_2$ with the corresponding in situ prepared diaminocarbene in the presence of a terminal alkyne are key intermediates in the formation of [Ru=CH_2] active species.

3.2 Alkyne Cross-Coupling

In the presence of a base (Et_3N) the ruthenium vinylidene complex RuCl(Cp*)(PPh_3)(=C=CHPh) promotes the selective cross-coupling of a bulky terminal alkyne with internal alkynes at room temperature to yield functionalized enynes (Eq. 10) [77].

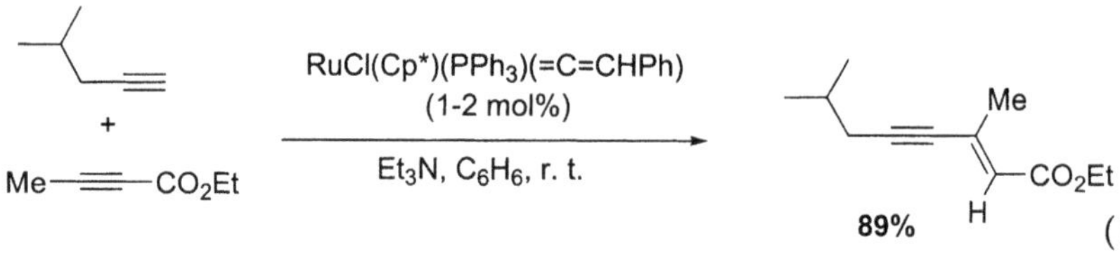

RuCl(Cp*)(PPh_3)(=C=CHPh) (1-2 mol%); Et_3N, C_6H_6, r. t.; Me—≡—CO_2Et; Me; CO_2Et; H; 89%

(10)

Complexes $RuCl_2(PCy_3)$(L)(=C=CH*t*-Bu) (**L**) are also good precursors for the Karasch reaction: they catalyse the atom-transfer radical addition of CCl_4 and $CHCl_3$ to various olefins such as acrylates, styrene and 1-octene [78].

4 Ruthenium Allenylidenes as Active Catalytic Intermediates

4.1 Nucleophilic Addition at Cγ

The selective intramolecular nucleophilic addition of a hydroxy group at Cγ of a ruthenium allenylidene generated by activation of propargylic alcohol by RuCl(Cp)(PPh_3)$_2$/NH_4PF_6 provides a ruthenium vinylidene species, which reacts with allylic alcohols as previously described in the section Formation of Unsaturated Ketones (Eq. 11, Scheme 18) [79]. This unprecedented tandem reaction makes possible the construction of tetrahydrofuran derivatives in good yields and has been used as a key step in the synthesis of (–)calyculin A [80].

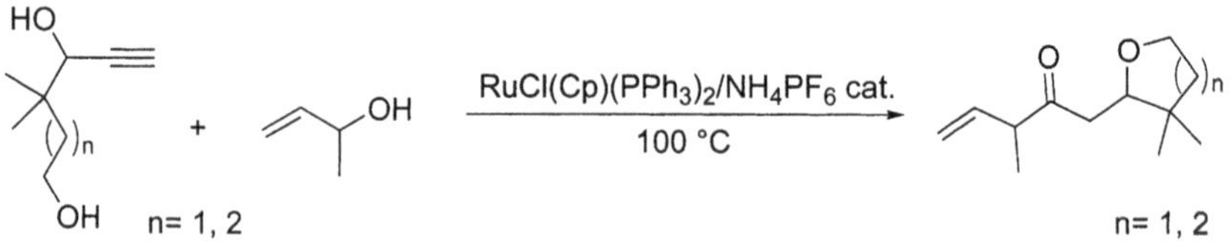

HO; OH; n= 1, 2; OH; RuCl(Cp)(PPh_3)$_2$/NH_4PF_6 cat.; 100 °C; O; O; n= 1, 2

(11)

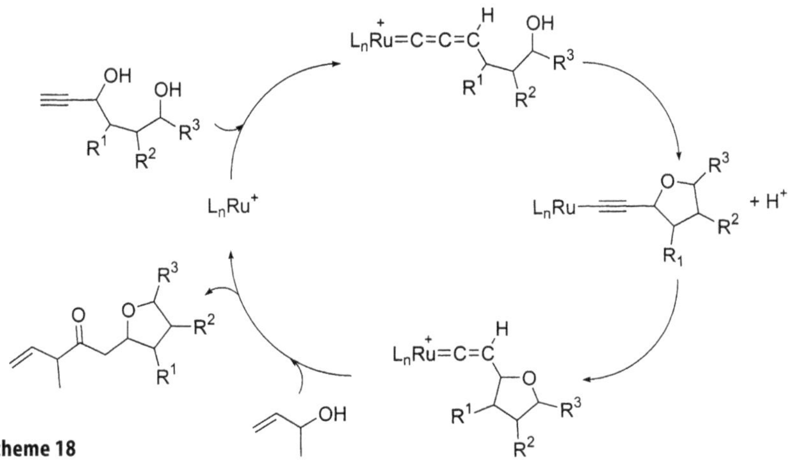

Scheme 18

The ability of the binuclear complex [Cp*RuCl(μ^2-S*R*)$_2$RuCl(Cp*)] to generate cationic allenylidene complexes by activation of terminal prop-2-ynols in the presence of NH_4BF_4 as a chloride abstractor opens the way to a variety of catalytic transformations of propargylic alcohols involving nucleophilic addition at the Cγ atom of the ruthenium allenylidene intermediate (Scheme 19). This leads to the formation of a functional ruthenium vinylidene species which tautomerizes into an η^2-coordinated alkyne that is removed from the ruthenium centre in the presence of the substrate.

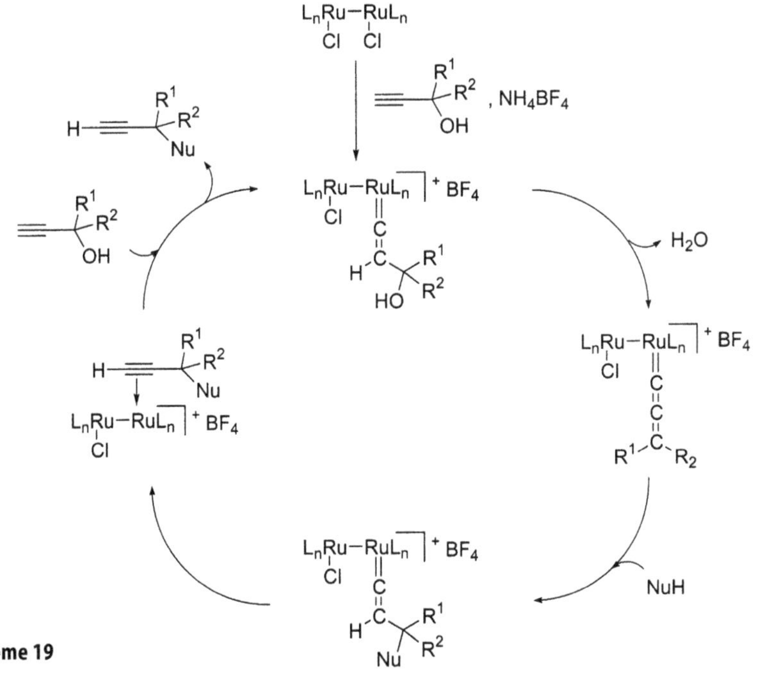

Scheme 19

The reaction, which is formally a nucleophilic substitution of the OH group, has been successfully performed with a variety of O, N, P, S and C nucleophiles, including aliphatic alcohols, amides, phosphine oxide, aromatic amines [81], phenols [82], ketones [83, 84], thiols [81, 85], aromatic heterocycles [86] and olefins [87] (Scheme 20). Under typical conditions, the reactions were carried out at 60 °C in $ClCH_2CH_2Cl$ in the presence of 5 mol % of catalyst.

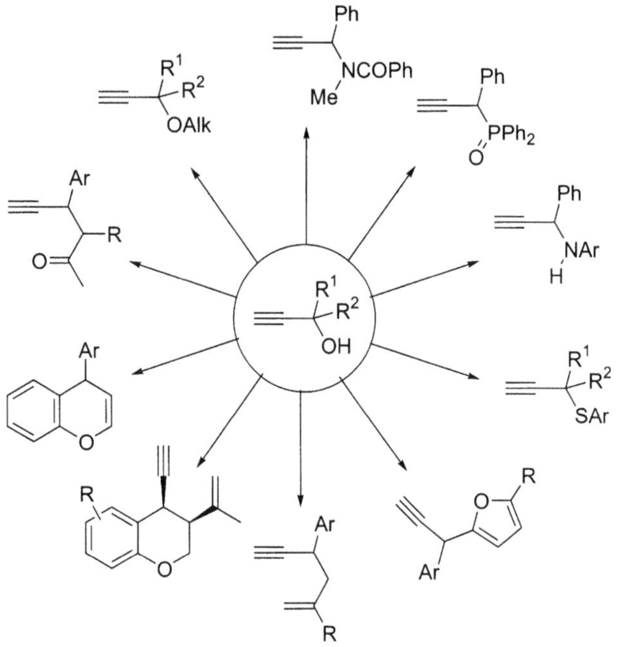

Scheme 20

It is noteworthy that the addition of thiols to form propargylic sulfides is not catalysed by the neutral complex [Cp*RuCl(μ^2-S*R*)$_2$RuCl(Cp*)], but requires the utilization of a cationic precursor such as [Cp*RuCl(μ^2-SMe)$_2$Ru(Cp*)(H_2O)]OTf [85]. With this catalytic system, propargylic alcohols bearing an internal triple bond are also transformed into propargylic sulfides, which indicates that in this special case, the reaction does not involve a ruthenium allenylidene as an active species.

4.2 Nucleophilic Addition at Cα

In contrast with the catalytic system based on RuCl(η^5-C_9H_7)(PPh_3)$_2$ in micellar solutions [32], the reaction of secondary propargylic alcohols in 2-propanol/H_2O at 100 °C in the presence of 5 mol % of RuCl(Cp)(PMe_3)$_2$ leads to conjugated enals with E stereoselectivity (Eq. 12) [88].

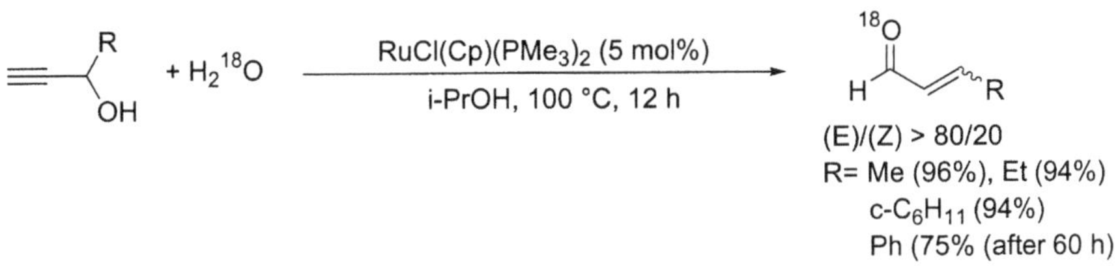

$+ H_2{}^{18}O$ — $RuCl(Cp)(PMe_3)_2$ (5 mol%), i-PrOH, 100 °C, 12 h →

(E)/(Z) > 80/20
R= Me (96%), Et (94%)
$c\text{-}C_6H_{11}$ (94%)
Ph (75% (after 60 h)

(12)

Labelling experiments clearly showed that the oxygen atom of the aldehyde came from external water, which confirmed the anti-Markovnikov addition of H_2O with concomitant dehydration of the alkynol. A Ru(IV) hydride bearing an allenylidene $[RuH(Cp)(PMe_3)_2(=C=C=CHR)]^+$ (**O**) or hydroxyvinylidene $[RuH(Cp)(PMe_3)_2(=C=CH=C(OH)HR)]^+$ (**P**) ligand is claimed as the key intermediate (see also Scheme 6).

The activation of propargylic ethers also provides the generation of ruthenium allenylidene species with elimination of alcohols (Eq. 13). This reaction has been used in the catalytic transformation of benzyl propargyl ethers into 1,3-dienes via dealkoxylation, addition of benzyl alcohol to the α-carbon atom of the allenylidene intermediate and hydrogen-transfer reactions according to Scheme 21 [89].

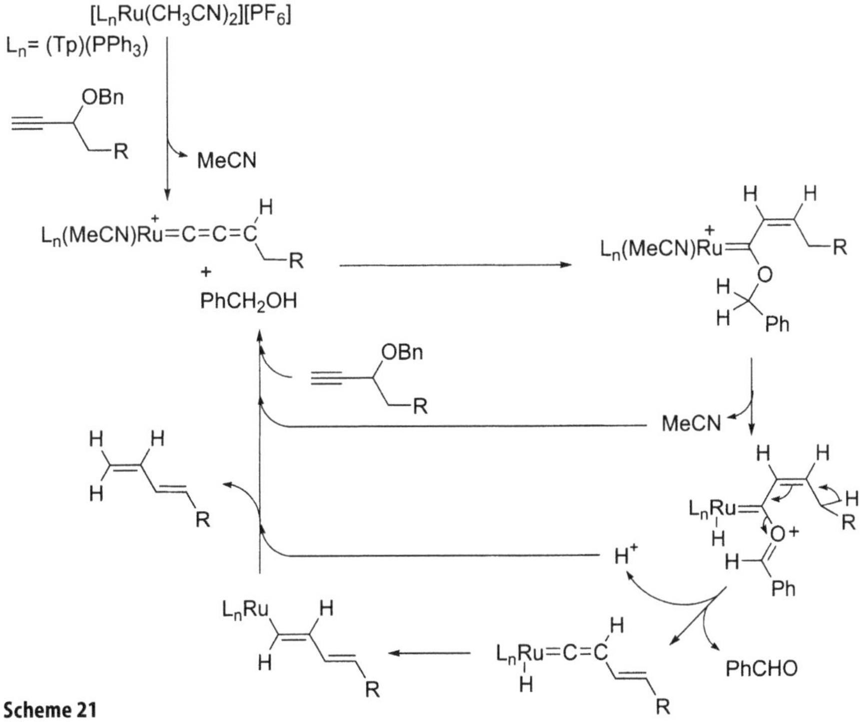

$[L_nRu(CH_3CN)_2][PF_6]$
L_n= (Tp)(PPh_3)

Scheme 21

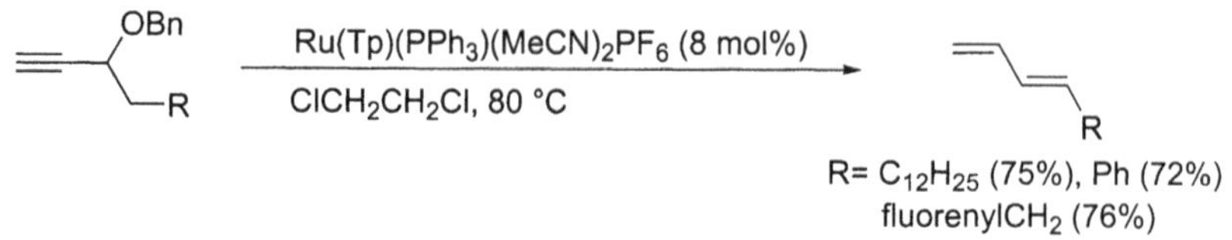

$Ru(Tp)(PPh_3)(MeCN)_2PF_6$ (8 mol%), $ClCH_2CH_2Cl$, 80 °C

R= $C_{12}H_{25}$ (75%), Ph (72%)
fluorenylCH_2 (76%)

(13)

The same authors have recently shown that in the presence of LiOTf as an additive (20 mol %), the same precatalyst $[RuTp(PPh_3)(MeCN)_2]PF_6$ (10 mol %) provided the selective cleavage of the carbon–carbon triple bond of terminal propargylic alcohols to form an olefin and CO (Eq. 14) [90].

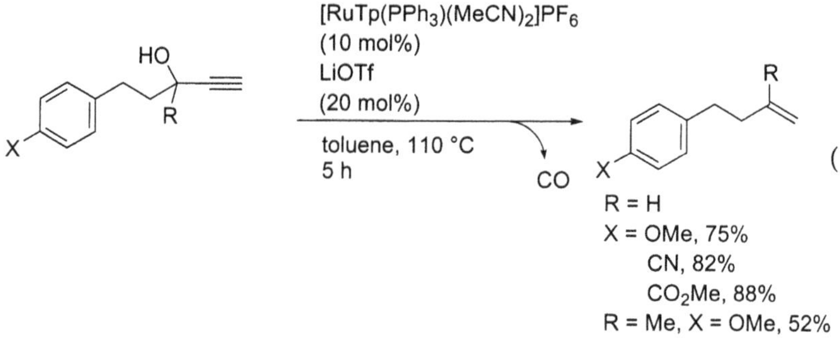

$[RuTp(PPh_3)(MeCN)_2]PF_6$ (10 mol%), LiOTf (20 mol%), toluene, 110 °C, 5 h; − CO

R = H
X = OMe, 75%
CN, 82%
CO_2Me, 88%
R = Me, X = OMe, 52%

(14)

The reaction proceeds via formation of an allenylidene ruthenium intermediate and addition of OH^- to the α-carbon to generate an acyl ligand, which produces an olefin after decarbonylation.

5 Ruthenium Allenylidenes as Catalyst Precursors

5.1 Olefin and Enyne Metathesis

The efficiency of cationic ruthenium allenylidene complexes **Q** (Fig. 10) containing a bulky basic electron-releasing ligand (PCy_3, Pi-Pr_3) in RCM of dienes was first reported in 1998 [91]. These catalyst precursors, which are usually activated in toluene at 80 °C, tolerate a variety of functional groups, such as amides, alcohols, esters and acetals, and they have been used for the preparation of small, medium and large rings and natural compound analogues [92–96]. Some variations have been performed to the initial allenylidene complexes [97] and it has been shown that complexes **S**, **T** [98, 99] and **R** [100], which contains a diaminocarbene ligand instead of a phosphine, are also efficient metathesis catalyst initiators. These ruthenium complexes also constitute a new family of enyne metathesis precatalysts [101, 102].

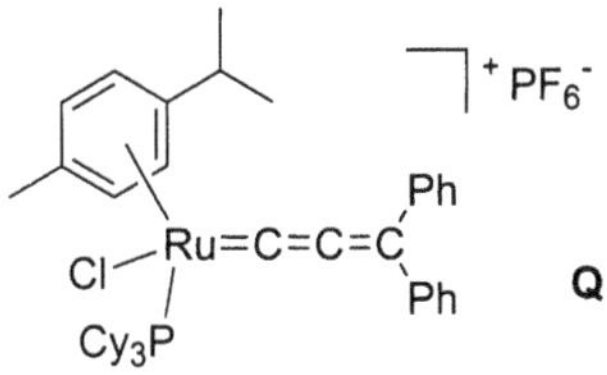

Q, 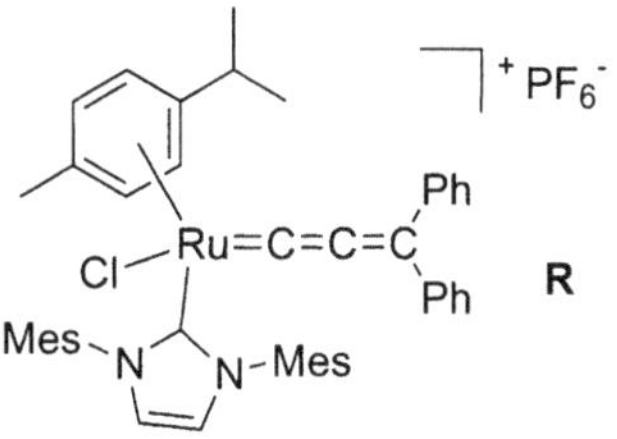 R, 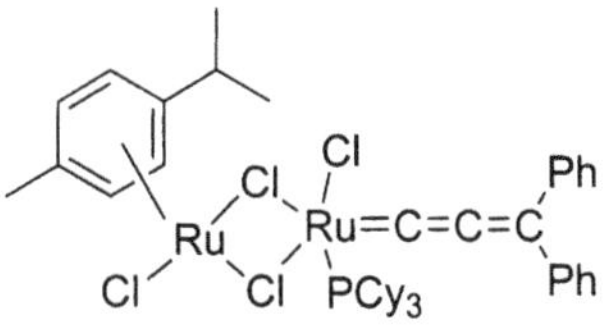 S, 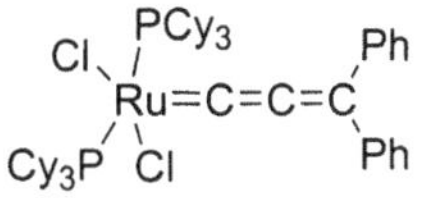 T

$RuCl_2(=C=C=CPh_2)(PCy_3)_n(DMSO)_{3-n}$
n= 1, 2
U

$[RuCl(=C=C=CPh_2)(PCy_3)_n(DMSO)_{3-n}][OTf]$
n= 1, 2
V

Fig. 10

Recently, complex **Q** [103] and other neutral and cationic ruthenium allenylidene complexes (**U**, **V**) [104] have been reported as efficient catalyst precursors for the ROMP of cyclic olefins such as cyclooctene and norbornene.

In an effort to recover and recycle the ionic precursor **Q**, two strategies have been tested in RCM of diallyltosylamide and ROMP of norbornene: (1) the catalytic precursor has been supported on a polystyrene polymer [105, 106] and (2) the metathesis reaction has been carried out in ionic liquids [107, 108].

5.2 Dimerization of Tin Hydrides

The ruthenium allenylidene complexes **W** are excellent precursors for the catalytic dimerization of tributyltin hydride under mild conditions [109] (Eq. 15). In the presence of Bu_3SnH, the hydride addition at Cγ provides a catalytically active alkynyl ruthenium–tin species (Scheme 22).

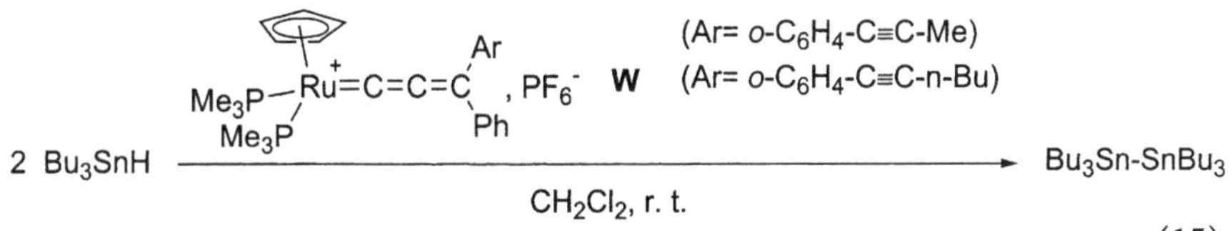

$$2\ Bu_3SnH \xrightarrow[CH_2Cl_2,\ r.t.]{\textbf{W}} Bu_3Sn\text{-}SnBu_3 \quad (15)$$

W: $[(C_5H_5)Ru(PMe_3)_2(=C=C=C(Ar)Ph)]^+PF_6^-$ (Ar= o-C_6H_4-C≡C-Me) (Ar= o-C_6H_4-C≡C-n-Bu)

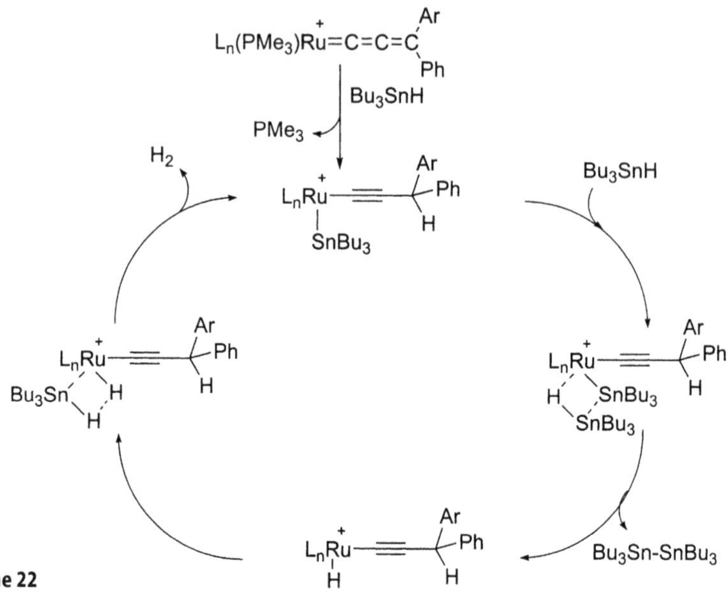

Scheme 22

5.3 Transetherification of Vinyl Ethers

Very recently, the water-soluble binuclear ruthenium allenylidene complex $[\{RuCl(\mu\text{-}Cl)(C{=}C{=}CPh_2)(Ph_2P(2\text{-}OS(O)_2C_6H_4))_2\}_2]$ Na_4 was used to perform selective transetherification of substituted vinyl ethers into acetals and aldehydes according to the solvent (Scheme 23) [110].

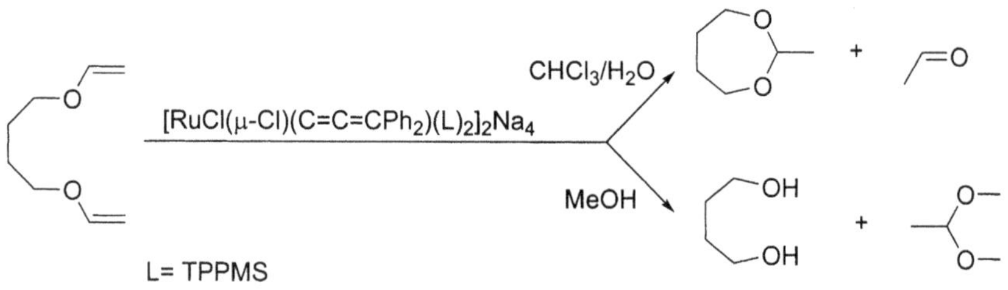

Scheme 23

6 Conclusion

The activation of terminal alkynes and propargylic alcohols by appropriate ruthenium complexes provides general and easy access to ruthenium vinylidene and allenylidene intermediates. These cumulenic systems offer a variety of possibilities in catalysis for selective transformations of acetylenic derivatives, most of them via addition or migration of nucleophilic species onto the α-carbon of vinylidene or γ- and α-carbons of allenylidene intermediates. These new regioselective and often stereoselective catalytic transformations, which proceed under mild conditions, should contribute to increase the creativity of synthetic organic chemists. Stable and isolated ruthenium vinylidene and allenylidene complexes also provide efficient catalyst precursors, especially active in olefin metathesis.

References

1. Bruce MI (1991) Chem Rev 91:197
2. Puerta MC, Valerga P (1999) Coord Chem Rev 193–195:977
3. Bruce MI (1998) Chem Rev 98:2797
4. Touchard D, Dixneuf PH (1998) Coord Chem Rev 178–180:409
5. Silvestre J, Hoffmann R (1985) Helv Chim Acta 68:1461
6. De Angelis F, Sgamellotti A, Re N (2002) Organometallics 21:5944
7. Mahé R, Dixneuf PH, Lécolier S (1986) Tetrahedron Lett 27:6333
8. Mahé R, Sasaki Y, Bruneau C, Dixneuf PH (1989) J Org Chem 54:1518
9. Höfer J, Doucet H, Bruneau C, Dixneuf PH (1991) Tetrahedron Lett 32:7409
10. Fournier J, Bruneau C, Dixneuf PH, Lécolier S (1991) J Org Chem 56:4456
11. Bruneau C, Dixneuf PH (1992) J Mol Catal 74:97
12. Rotem M, Shvo Y (1983) Organometallics 2:1689
13. Mitsudo T, Hori Y, Yamazaki Y, Watanabe Y (1987) J Org Chem 52:2230
14. Bruneau C, Neveux M, Kabouche Z, Ruppin C, Dixneuf PH (1991) Synlett 755
15. Bruneau C, Dixneuf PH (1997) J Chem Soc Chem Commun 507
16. Neveux M, Seiller B, Hagedorn F, Bruneau C, Dixneuf PH (1993) J Organomet Chem 451:133
17. Doucet H, Höfer J, Bruneau C, Dixneuf PH (1993) J Chem Soc Chem Commun 850
18. Doucet H, Martin-Vaca B, Bruneau C, Dixneuf PH (1995) J Org Chem 60:10901
19. Dixneuf PH, Bruneau C, Dérien S (1998) Pure Appl Chem 70:1065
20. Doucet H, Höfer J, Derrien N, Bruneau C, Dixneuf PH (1996) Bull Soc Chim Fr 133:939
21. Doucet H, Derrien N, Kabouche Z, Bruneau C, Dixneuf PH (1997) J Organomet Chem 551:151
22. Gemel C, Trimmel G, Slugovc C, Kremel S, Mereiter K, Schmid R, Kirchner K (1996) Organometallics 15:3998
23. Jimenez Tenorio M, Puerta MC, Valerga P, Moreno-Dorado FJ, Guerra FM, Massanet GM (2001) Chem Commun 2324
24. Melis K, Samulkiewiecz P, Rynkowski J, Verpoort F (2002) Tetrahedron Lett 43:2713
25. Goossen LJ, Paetzold J, Koley D (2003) Chem Commun 706
26. Opstal T, Verpoort F (2002) Synlett 935
27. Picquet M, Bruneau C, Dixneuf PH (1997) J Chem Soc Chem Commun 1201
28. Picquet M, Fernandez A, Bruneau C, Dixneuf PH (2000) Eur J Org Chem 2361
29. Tokunaga M, Wakatsuki Y (1998) Angew Chem Int Ed Engl 37:2867
30. Suzuki T, Tokunaga M, Wakatsuki Y (2001) Org Lett 3:735
31. Tokunaga M, Suzuki T, Koga N, Fukushima T, Horiuchi A, Wakatsuki Y (2001) J Am Chem Soc 123:11917
32. Alvarez P, Bassetti M, Gimeno J, Mancini G (2001) Tetrahedron Lett 42:8467
33. Lo C-Y, Guo H, Lian J-J, Shen F-M, Liu R-S (2002) J Org Chem 67:3930
34. Trost BM, Dyker G, Kulawiec RJ (1990) J Am Chem Soc 112:7809

35. Trost BM, Kulawiec RJ (1992) J Am Chem Soc 114:5579
36. Trost BM, Flygare JA (1994) J Org Chem 59:1078
37. Trost BM, Kulawiec RJ, Hammes A (1993) Tetrahedron Lett 34:587
38. Trost BM, Vidal B, Thommen M (1999) Chem Eur J 5:1055
39. Nishibayashi Y, Takei I, Hidai M (1997) Organometallics 16:3091
40. Trost BM, Rhee YH (1999) J Am Chem Soc 121:11680
41. Trost BM, Rhee YH (2002) J Am Chem Soc 124:2528
42. Fukumoto Y, Dohi T, Masaoka H, Chatani N, Murai S (2002) Organometallics 21:3845
43. Jérôme F, Monnier F, Lawicka H, Dérien S, Dixneuf PH (2003) Chem Commun 696
44. Merlic CA, Pauly ME (1996) J Am Chem Soc 118:11319
45. Wang Y, Finn MG (1995) J Am Chem Soc 117:8045
46. Slugovc C, Mereiter K, Zobetz E, Schmid R, Kirchner K (1996) Organometallics 15:5275
47. Slugovc C, Doberer D, Gemel C, Schmid R, Kirchner K, Winkler B, Stelzer F (1998) Monatsh Chem 129:221
48. Pavlik S, Gemel C, Slugovc C, Mereiter K, Schmid R, Kirchner K (2001) J Organomet Chem 617–618:301
49. Bianchini C, Peruzzini M, Zanobini F, Frediani P, Albinati A (1991) J Am Chem Soc 113:5453
50. Bianchini C, Frediani P, Masi D, Peruzzini M, Zanobini F (1994) Organometallics 13:4616
51. Yi CS, Liu N (1996) Organometallics 15:3968
52. Bassetti M, Marini S, Tortorella F, Cadierno V, Diez J, Pilar Gamasa M, Gimeno J (2000) J Organomet Chem 593–594:292
53. Yi CS, Liu N (1997) Organometallics 16:3910
54. Yi CS, Liu N (1999) Synlett 281
55. Matsuzaka H, Takagi Y, Ishii Y, Nishio M, Hidai M (1995) Organometallics 14:2153
56. Melis K, De Vos D, Jacobs P, Verpoort F (2002) J Organomet Chem 659:159
57. Wakatsuki Y, Yamazaki H, Kumegawa N, Satoh T, Satoh JY (1991) J Am Chem Soc 113:9604
58. Wakatsuki Y, Yamazaki H (1995) J Organomet Chem 500:349
59. Yamazaki H (1976) J Chem Soc Chem Commun 841
60. Bedford RB, Hill AF, Thompsett AR, White AJP, Williams DJ (1996) J Chem Soc Chem Commun 1059
61. Murakami M, Ubukata M, Ito Y (1998) Tetrahedron Lett 39:7361
62. Slugovc C, Mereiter K, Schmid R, Kirchner K (1998) J Am Chem Soc 120:6175
63. Murakami M, Hori S (2003) J Am Chem Soc 125:4720
64. Onitsuka K, Katayama H, Sonogashira K, Ozawa F (1995) J Chem Soc Chem Commun 2267
65. Katayama H, Ozawa F (1998) Chem Lett 67
66. Katayama H, Yoshida T, Ozawa F (1998) J Organomet Chem 562:203
67. Katayama H, Urushima H, Ozawa F (2000) J Organomet Chem 606:16
68. Schwab P, Grubbs RH, Ziller JW (1996) J Am Chem Soc 118:100
69. Katayama H, Yonezawa F, Nagao M, Ozawa F (2002) Macromolecules 35:1133
70. Pineda Contreras A, Masiel Cerda A, Tlenkopatchev MA (2002) Macromol Chem Phys 203:1811
71. Opstal T, Verpoort F (2003) J Mol Catal A 200:49
72. Saoud M, Romerosa A, Peruzzini M (2000) Organometallics 19:4005
73. del Rio I, van Koten G (1999) Tetrahedron Lett 40:1401
74. Louie J, Grubbs RH (2001) Angew Chem Int Ed Engl 40:247
75. Sémeril D, Bruneau C, Dixneuf PH (2002) Adv Synth Catal 344:585

76. Sémeril D, Bruneau C, Dixneuf PH (2001) Helv Chim Acta 84:3335
77. Yi CS, Liu N (1998) Organometallics 17:3158
78. Opstal T, Verpoort F (2002) Tetrahedron Lett 43:9259
79. Trost BM, Flygare JA (1992) J Am Chem Soc 114:5476
80. Trost BM, Flygare JA (1994) Tetrahedron Lett 35:4059
81. Nishibayashi Y, Wakiji I, Hidai M (2000) J Am Chem Soc 122:11019
82. Nishibayashi Y, Inada Y, Hidai M, Uemura S (2002) J Am Chem Soc 124:7900
83. Nishibayashi Y, Wakiji I, Ishii Y, Uemura S, Hidai M (2001) J Am Chem Soc 123:3393
84. Nishibayashi Y, Onodera G, Inada Y, Hidai M, Uemura S (2003) Organometallics 22:873
85. Inada Y, Nishibayashi Y, Hidai M, Uemura S (2002) J Am Chem Soc 124:15172
86. Nishibayashi Y, Yoshikawa M, Inada Y, Hidai M, Uemura S (2002) J Am Chem Soc 124:11846
87. Nishibayashi Y, Inada Y, Hidai M, Uemura S (2003) J Am Chem Soc 125:6060
88. Suzuki T, Tokunaga M, Wakatsuki Y (2002) Tetrahedron Lett 43:7531
89. Yeh K-L, Liu B, Lo C-Y, Huang H-L, Liu R-S (2002) J Am Chem Soc 124:6510
90. Datta S, Chang C-L, Yeh K-L, Liu R-S (2003) J Am Chem Soc 124:9294
91. Fürstner A, Picquet M, Bruneau C, Dixneuf PH (1998) Chem Commun 1315
92. Fürstner A, Müller T (1999) J Am Chem Soc 121:7814
93. Picquet M, Touchard D, Bruneau C, Dixneuf PH (1999) New J Chem 141
94. Fürstner A, Liebl M, Lehmann CW, Picquet M, Kunz R, Bruneau C, Touchard D, Dixneuf PH (2000) Chem Eur J 6:1847
95. Osipov SN, Artyushin OI, Kolomiets AF, Bruneau C, Dixneuf PH (2000) Synlett 1031
96. Osipov SN, Artyushin OI, Kolomiets AF, Bruneau C, Picquet M, Dixneuf PH (2001) Eur J Org Chem 3891
97. Le Gendre P, Picquet M, Richard P, Moise C (2002) J Organomet Chem 643–644:231
98. Fürstner A, Hill AF, Liebl M, Wilton-Ely JDET (1999) Chem Commun 601
99. Harlow KJ, Hill AF, Wilton-Ely JDET (1999) J Chem Soc Dalton Trans 285
100. Jafarpour L, Huang J, Stevens ED, Nolan SP (1999) Organometallics 18:3760
101. Picquet M, Bruneau C, Dixneuf PH (1998) Chem Commun 2249
102. Sémeril D, Le Nôtre J, Bruneau C, Dixneuf PH, Kolomiets AF, Osipov SN (2001) New J Chem 25:16
103. Castarlenas R, Sémeril D, Noels AF, Demonceau A, Dixneuf PH (2002) J Organomet Chem 663:235
104. Alaoui Abdallaoui I, Sémeril D, Dixneuf PH (2002) J Mol Catal A: Chem 182–183:577
105. Akiyama R, Kobayashi S (2002) Angew Chem Int Ed Engl 41:2602
106. Kobayashi S, Akiyama R (2003) Chem Commun 449
107. Sémeril D, Olivier-Bourbigou H, Bruneau C, Dixneuf PH (2002) Chem Commun 146
108. Csihony S, Fischmeister C, Bruneau C, Horvath I, Dixneuf PH (2002) New J Chem 26:1667
109. Maddock SM, Finn MG (2001) Angew Chem Int Ed Engl 40:2138
110. Saoud M, Romerosa A, Manas Carpio S, Gonsalvi L, Peruzzini M (2003) Eur J Inorg Chem 1614

Topics Organomet Chem (2004) 11: 155–171
DOI 10.1007/978-3-540-40027-1

Ruthenium-Promoted Radical Processes Toward Fine Chemistry

Lionel Delaude · Albert Demonceau · Alfred F. Noels (✉)

Center for Education and Research on Macromolecules (CERM), Institut de Chimie (B6a), University of Liège, Sart-Tilman par 4000 Liège, Belgium
af.noels@ulg.ac.be

Abstract Ruthenium holds a prominent position among the many transition metals used in radical chemistry. The dichlorotris(triphenylphosphine) complex [$RuCl_2(PPh_3)_3$] was the first active ruthenium-based catalyst investigated for atom-transfer radical addition processes and has found numerous applications in organic synthesis. Other catalytic systems that have been devised since include (1) neutral or cationic Grubbs-type complexes bearing an alkylidene fragment and either phosphine, *N*-heterocyclic carbene, or Schiff base ligands, (2) half-sandwich ruthenium complexes bearing a cyclopentadienyl, a pentamethylcyclopentadienyl, or an indenyl ligand, and (3) ruthenium complexes bearing anionic carborane–phosphine and dicarbollide ligands. Their activities are discussed and the predictive value of cyclic voltammetry in radical chemistry is questioned. A related example of a ruthenium-catalyzed C–H hydroxylation reaction is also reported.

Keywords Atom-transfer reaction · Catalysis · Kharasch addition · Olefin

1 Introduction

Free-radical reactions are particularly useful to accomplish transformations that are not possible or that are difficult to achieve using ionic pathways. In this respect, the development over the last 15 years of new synthetic methods leading to well-defined and controlled radical reactions has weakened the pessimistic old notion of free-radical processes being difficult to control, owing to the intervention of highly reactive intermediates that usually undergo fast reactions with low selectivity. Recent advances in catalysis directed toward the formation of carbon–carbon bonds via free-radical mediated reactions have added a whole new dimension to the repertoire of synthetic methods available for controlling the precise assembly of small organic molecules and of polymer chains. Numerous radical reactions are now both chemoselective and regioselective, and even stereoselectivity can be achieved with a good understanding of the radical intermediate structures.

Radicals being neutral species tend to react together. Indeed, the most common side reactions in free-radical processes involve the formation of adducts between two radicals, via combination or disproportionation. These unwanted termination steps usually occur much faster than the desired reactions between radicals and substrates. Thus, the key to control in both radical addition and polymerization procedures consists in lowering the concentration of transient radical species. This will minimize the side reactions between radical species, yet the kinetics of the useful reactions will also be affected.

Many different metal-based promoters are available to initiate free-radical reactions. Of particular synthetic importance are organotin compounds, although radicals derived from other group 14 elements, notably silicon, are also attracting significant interest [1]. Yet, transition metals offer a useful alternative to their main-group counterparts for controlled radical transformations, and they have found widespread use in fine chemistry. Among the transition metals employed for generating carbon-centered radicals, manganese, chromium, cobalt, and especially copper have been the most widely studied [2, 3]. Ruthenium-catalyzed radical reactions have also recently emerged as particularly promising and worthy of interest, but their potential in organic synthesis and in polymer chemistry remains largely unexplored so far.

2 Metal-Catalyzed Atom-Transfer Reactions

2.1 Historical Background

Atom-transfer reactions encompass a broad range of radical addition processes in which carbon–heteroatom bonds are added across alkenes, alkynes, or other unsaturated functionalities. The first example of an atom-transfer radical addition (ATRA) involving C–C bond formation and yielding a monomeric product was reported by Kharasch and coworkers in the mid-1940s. Typically, carbon tetrachloride was added to 1-octene in the presence of a radical initiator to afford the anti-Markovnikov addition product [4–7]. Classical initiators such as benzoyl peroxide, azobis(isobutyronitrile), or UV light were employed, and the process took the name of its discoverer in the everyday chemical language. The Kharasch addition reaction was born and has been used in organic synthesis ever since, although its original embodiment suffered from major drawbacks that limited its applicability. These downsides came from competing telomerization and polymerization reactions, which lowered the yields and lengthened the purification procedure.

By the mid-1950s, it was recognized that transition metals were also capable of mediating the free-radical addition of polyhalogenoalkanes onto alkenes while limiting the unwanted side reactions. Minisci [8] was among the first investigators to report that carbon tetrachloride could add to olefins to afford only the corresponding monoadducts in the presence of iron or copper salts. In a number of cases, however, the occurrence of competitive oligomerization and telomerization processes still remains a problem that has not been satisfactorily addressed up to the present day. Although Kharasch additions *stricto sensu* refer to reactions promoted by organic radicals or light – and not by metal complexes – transition-metal-promoted reactions of this type are also commonly designated using the same patronymic nowadays.

2.2
Mechanistic Indications

Although detailed kinetic studies on Kharasch and related additions are sparse, there is agreement that two different (but related to some extent) types of mechanism might be operative in these reactions.

The first type of mechanism involves a redox chain process. As shown in Eqs. (1–3), it begins with the abstraction of a halogen atom from a polyhaloalkane reagent by the metal complex. This generates a radical species that further adds to an olefin. A chain-transfer reaction ensues and yields back the reduced metal species, hence the acronym ATRA, for the sequence.

$$[M^n] + XCCl_3 \rightleftharpoons [M^{n+1}]X + {}^\bullet CCl_3 \qquad X = Br, Cl \tag{1}$$

$$RR'C{=}CH_2 + {}^\bullet CCl_3 \longrightarrow RR'\dot{C}{-}CH_2{-}CCl_3 \tag{2}$$

$$[M^{n+1}]X + RR'\dot{C}{-}CH_2{-}CCl_3 \rightleftharpoons RR'CX{-}CH_2{-}CCl_3 + [M^n] \tag{3}$$

Since metal halides have a much larger chain-transfer constant than $XCCl_3$ (X is Cl or Br), halogen transfer from the oxidized metal is favored over a propagation step. This strongly limits or suppresses the competitive telomerization processes. Only reactions that are rapider than the halogen-atom transfer can occur between the addition and the chain-transfer steps. Moreover, the C–C bond-forming step in ATRA is a discrete step in the productive radical chain. Hence, ATRA reactions tolerate olefins that give slow addition steps. The olefin addition has only to be faster than radical–radical coupling or radical–solvent reactions. Of course, the carbon–heteroatom bond formed in the product must be stronger than the one broken in the initial reactant, and the abstraction of the transferred group needs to be fast (which usually is the case) in order to minimize oligomerization reactions. The selectivity toward a 1:1 addition therefore stems from the controlled chain termination, and the olefin seems to play no role in the rate-determining step of the redox chain mechanism. A similar situation is encountered in some atom-transfer radical polymerization (ATRP) reactions.

The second type of mechanism proceeds via a nonchain pathway and the organic radical results from a single electron transfer (SET) as illustrated in Eqs. (4–6). The species resulting from the SET interaction is a radical that apparently remains caged in the coordination sphere of the metal center (**A** or **B**). Olefin coordination to species **A** or **B** remains a matter of debate and, possibly, depends upon the nature of the metal complex.

$$L_nM\text{-}Cl + XCCl_3 \rightleftharpoons [L_nM(\cdots X \cdots)(\cdots Cl \cdots)\cdot CCl_3] \quad X = Br, Cl \tag{4}$$

$$\mathbf{A}\ [L_nM(\cdots X \cdots)(\cdots Cl \cdots)\cdot CCl_3] \rightleftharpoons [L_n\dot{M}XCl][\cdot CCl_3]\ \mathbf{B} \tag{5}$$

$$[L_n\dot{M}XCl][\cdot CCl_3] + RR'C{=}CH_2 \longrightarrow RR'CX{-}CH_2{-}CCl_3 + L_nM\text{-}Cl \tag{6}$$

2.3
Specificity of Ruthenium-Based Systems

The ability of ruthenium to assume a wide range of oxidation states and coordination geometries provides unique opportunities for catalysis. Indeed, a wide range of mechanistically very different processes are catalyzed by ruthenium complexes [9]. The development of highly efficient ruthenium-based catalysts is also driven by their tolerance toward functional groups, their easy access, and their versatility. Ruthenium holds a prominent position among the many transition metals (Cr, Mn, Fe, Ni, Pd, Cu, etc.) used in radical chemistry, both for synthetic applications [10, 11] and for polymer chemistry [12–14]. In particular, the dichlorotris(triphenylphosphine) complex $[RuCl_2(PPh_3)_3]$ (**1**) was the first active ruthenium-based catalyst precursor for ATRA [15]. The same complex was later used as a promoter for the controlled polymerization of methyl methacrylate (MMA), but in this case further activation by a Lewis acid such as aluminum triisopropoxide was required [16–18].

3
The [$RuCl_2(PPh_3)_3$] Catalytic System

3.1
General Considerations

The Kharasch addition reactions promoted by [$RuCl_2(PPh_3)_3$] are believed to proceed through a redox chain mechanism (Eqs. 1–3) [16]. Their kinetics show a first-order dependence both on the ruthenium complex and on CCl_4. Whereas no clear-cut evidence for alkene coordination to the metal was found with catalyst precursor **1** (which readily loses one phosphine ligand), olefin coordination cannot be excluded because there is a saturation kinetic rate dependence on the alkene. This observation led to the proposal of a reversible step involving olefin coordination to the metal center [16, 19, 20]. Recent work with other ruthenium-based catalysts further supports olefin coordination (see later).

The effect of ring size on the reactivity of C_5–C_{10} *cis*-cycloalkenes in addition reactions with CCl_4 was investigated with a number of metal complexes, including **1**, and relative reactivities were expressed in terms of rate constants compared with those of the corresponding normal alkenes [21]. There were only small differences in the relative reactivities of cycloalkenes when the addition was promoted by a conventional initiator (benzoyl peroxide) or catalyzed by a metal complex (Cu-, Mo-, Co-, Fe-, and Ru-based catalysts were used in the study). The relative rate constants followed the trend observed for alicyclic systems in addition reactions with a number of free radicals, and the relative reactivities of *cis*-cycloalkenes decreased according to the sequence $C_8>C_5>C_7>C_6>C_{10}$. The addition reactions were mainly controlled by *I*-strain in the cycloalkene molecules. A strong catalyst influence on the stereoselectivity of the addition was observed, however, with cyclohexene. The cis-to-trans isomer ratio of the adduct was significantly affected by the ruthenium catalyst when compared with reactions promoted by benzoyl peroxide. Cu-, Mo-, Co-, and Fe-based catalysts also had the same effect, but to a somewhat lesser extent. Here also, a nonchain mechanism involving the coordination of reactants was proposed for the metal-catalyzed reaction, especially in the ruthenium case. Other variations of the mechanism might account for the reaction products as well. For instance, the experimental data did not allow a process involving an oxidative addition of the polyhalogenated molecule to ruthenium(II) and one implying some other interactions between the $^{\bullet}CCl_3$ radical and the metal center to be distinguished. As a matter of fact, many oxidative additions are known to proceed via radical intermediates [22].

Besides promoting the Kharasch addition reaction of polyhalogenated alkanes to MMA, the [$RuCl_2(PPh_3)_3$] complex (**1**) also initiates the controlled polymerization of MMA, provided that the $XCCl_3$ concentration is kept low. Thus, the switch between the polymerization and the 1:1 Kharasch addition reaction depends solely on the relative concentration of the polyhaloalkane ("the initiator" in polymerization reactions) to the metal catalyst. Using near-toequimolar proportions of $XCCl_3$ and complex **1** leads to controlled polymerization, whereas using a large excess of $XCCl_3$ relative to the metal promotes the Kharasch addition. Not all the ruthenium-based catalysts, however, behave in the same way.

3.2 Applications in Organic Synthesis

The synthesis of polyhalogenated alkanes, lactams, lactones, etc., which are versatile intermediates in the synthesis of natural products and of bioactive molecules, has held the attention of chemists for many years. The Ru(II)-catalyzed addition of polychloroacetic acid to terminal olefins affords the corresponding adducts in high yields. Similarly, dichloroacetic and trichloroacetic esters add to a variety of olefins to give the corresponding chloroesters and lactones. Several applications along these lines can be found in Ref. [10]. The same methodology also provides ready access to perfluorinated alkanes as complex **1** catalyzes the reaction of alkenes with perfluoroalkanesulfonyl chlorides at 120 °C. Yields are more than satisfactory (up to 87%) with alkenes and vinylarenes, but poor with cycloolefins [23]. From a mechanistical point of view, the reactions are interesting because the sulfonyl radicals formed by the interaction of the sulfonyl chloride and the ruthenium catalyst release SO_2 to form perfluoroalkyl radicals (Eq. 7).

$$\text{R–CH=CH}_2 + \text{R}_\text{F}\text{–SO}_2\text{Cl} \xrightarrow[\text{C}_6\text{H}_6,\ 120\ ^\circ\text{C}]{[\text{RuCl}_2(\text{PPh}_3)_3]\ (\mathbf{1})} \text{R–CHCl–CH}_2\text{–R}_\text{F} + \text{SO}_2 \tag{7}$$

$$\text{R}_\text{F} = \text{-CF}_3,\ \text{-C}_6\text{F}_{13}$$

Addition of CCl_4 to chloroethene and 1,1-dichloroethene occurred selectively under the influence of a catalytic amount of complex **1** and afforded the 1:1 anti-Markovnikov adducts in 86% and 89% yield, respectively. No reaction was observed with chlorotrifluoroethene [24].

A key intermediate in the synthesis of pretazettine (Eq. 8), an alkaloid that contains a *cis*-3a-arylhydroindole ring system and shows antiviral and anticancer properties, has been synthesized by chlorine-atom transfer cyclization of a chloroacetamide in a highly stereocontrolled manner [25].

[RuCl$_2$(PPh$_3$)$_3$] (**1**)
C$_6$H$_6$, 150 °C

Ar = 3,4-methylenedioxyphenyl

Pretazettine

(8)

Complex **1** also catalyzes the regioselective radical addition of perhalogenoethanes to silyl enol ethers. The primary addition–desilylation products undergo the facile β-elimination of a chloride to afford α,β-unsaturated ketones [26, 27]. For example, CF_2ClCCl_3 adds to the trimethylsilyl enol ether of acetophenone to yield β-chloro-β-(chlorodifluoromethyl)-α,β-acetophenone in 80% yield (Eq. 9).

$$\text{TMSO(Ph)C=CH}_2 + \text{F}_2\text{ClC–CCl}_3 \xrightarrow[\text{DMF, 100 °C}]{[\text{RuCl}_2(\text{PPh}_3)_3]\ (\mathbf{1})} \text{PhCOCH}_2\text{CCl}_2\text{CClF}_2 \xrightarrow{-\text{HCl}} \text{PhCOCH=CClCClF}_2 \quad (9)$$

4 Engineering of New Ruthenium Catalytic Systems

The previous examples have established ruthenium-catalyzed atom-transfer reactions as a valuable addition to the list of synthetic methods available in fine chemistry. The potential of these systems is obvious, but sometimes their applicability is limited by rather poor catalytic activity and/or selectivity, particularly when it comes to the chemoselectivity of the addition and the concurrent formation of telomers. Hence, the need to extend the range of possible substrates and to perform the reactions under milder conditions led to the search for new catalytic systems with improved performances. Yet, the application of ruthenium catalysis to radical reactions remains a relatively unexplored and new field.

Astonishingly, all the reactions described up to 1999 use $[RuCl_2(PPh_3)_3]$ (**1**) as a catalyst precursor, with almost no ligand variation. It is nevertheless evident that the coordination sphere of the ruthenium atom plays a crucial role in tuning the activity of the catalyst for the activation and deactivation steps in the ATRA mechanism. Bulky ligands are expected to reduce the rate of activation by restricting the access to the metal center. Depending on their electronic properties, ligands also modulate the redox potential of the Ru(II)/Ru(III) couple. For instance, basic, strong σ-donating ligands should ease the oxidation step, whereas good π-acceptor ligands are more likely to stabilize the lower oxidation state of the metal center. Developments along these lines have led to new families of ruthenium complexes with quite often improved performances in radical reactions. These new catalyst precursors can be classified among the three following families:

1. Neutral or cationic Grubbs-type complexes bearing an alkylidene fragment and either phosphine, *N*-heterocyclic carbene (NHC), or Schiff base ligands.
2. Half-sandwich ruthenium complexes bearing a cyclopentadienyl (Cp), a pentamethylcyclopentadienyl (Cp*), or an indenyl ligand.
3. Ruthenium complexes bearing anionic carborane–phosphine and dicarbollide ligands.

4.1
Grubbs and Related Complexes as Catalysts for Radical Reactions

Two groups independently reported in 1999 that Grubbs' popular olefin metathesis catalyst $[Cl_2Ru(=CHPh)(PCy_3)_2]$ (**2**), where Cy is cyclohexyl, and related complexes were efficient promoters for Kharasch additions of $CHCl_3$ and CCl_4 across double bonds [28, 29]. Furthermore, the same ruthenium alkylidene complexes also catalyze the controlled ATRP of various monomers [28, 30]. Under the same experimental conditions, complex **2** displays a greater activity in ATRP and affords lesser telomerization than complex **1**. An excess of free phosphine ligand has an inhibitory effect on the Kharasch reactivity, and the presence of radical scavengers severely limits the formation of the addition product without significantly affecting the metathesis activity. This latter result supports the intervention of free radicals in the addition reaction. Furthermore, it has been shown that the Grubbs benzylidene catalyst **2** generates persistent radical anions upon treatment with π-acceptor quinones, and also with dienes and even monoenes. There is evidence that the observed electron paramagnetic resonance signals arise from charge transfer [31, 32]. Thus, charge-transfer complexes with halogenated initiators could enhance the free-radical activity of the ruthenium alkylidene complexes and be responsible for their efficacy.

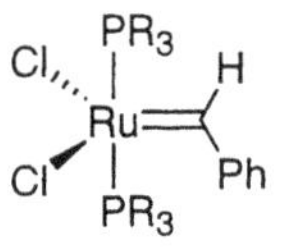

2 (R = Cy)
3 (R = Ph)

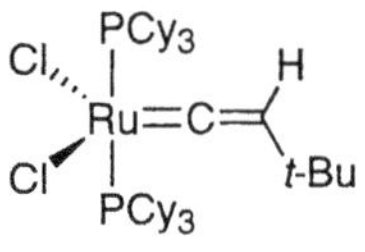

4

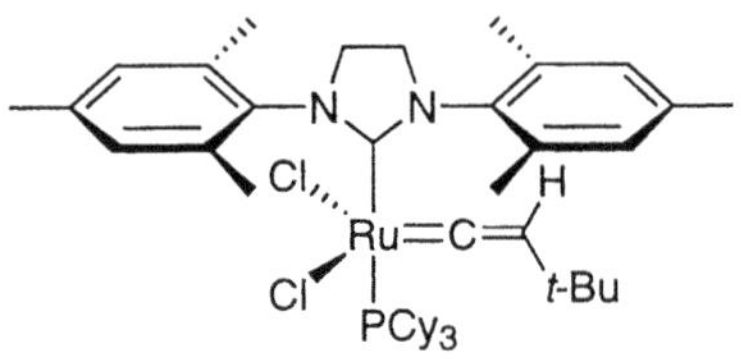

5

Within the series of $[Cl_2Ru(=CHPh)(PR_3)_2]$ complexes tested, the highest ATRA activity for the reaction of CCl_4 with vinyl substrates was obtained with the triphenylphosphine derivative $[Cl_2Ru(=CHPh)(PPh_3)_2]$ (**3**). Quantitative yields of monoadducts were obtained with styrene and MMA, and up to 61% conversion was achieved with 1-octene at a catalyst loading of 2.5 mol % [28]. The corresponding tricyclohexylphosphine complex **2** afforded less active catalytic systems. The reaction worked well with 1,1-disubstituted olefins but could not be extended to their 1,2-disubstituted counterparts. In all cases, yields were significantly lower when chloroform served as reagent instead of carbon tetrachloride [28, 29].

The range of ruthenium alkylidene catalysts active in radical chemistry was further enlarged to the readily accessible vinylidene complexes **4** and **5** [33]. Catalyst precursors **4** and **5** were tested for the ATRA of polyhalogenated alkanes with various olefins. Substitution of one phosphine in **4** by an NHC improves its catalytic efficiency. This is a surprising result given that **3** is more active than **2**, and that NHCs are significantly more basic ligands than alkyl phosphines [34]. It confirms once again that catalyst-tailoring requires a suitable adjustment between the catalyst, monomer, initiator, and atom (or group of atoms) being transferred. The monocationic complexes generated in situ by treating **4** and **5** with silver tetrafluoroborate are less efficient catalysts for ATRA than their neutral parents, although they are more active in ATRP. Since olefin coordination is favored in cationic complexes, this observation may indicate that polymerization reactions with these ruthenium complexes proceed through olefin coordination.

The robust homobimetallic ruthenium complexes **6** also efficiently catalyze the addition of carbon tetrachloride across a variety of C=C double bonds [35]. The influence of various *R* and *R′* substituents on the outcome of the reaction was investigated. Steric bulkiness is needed in the Schiff base moiety to attain reasonable catalytic activity. Electron-withdrawing *R* groups have a detrimental effect. The best combination of steric crowding and electronic balance in the complexes tested is reached when *R*=H and *R′* is the 2,6-dimethyl-4-bromophenyl group. This particular complex stands among the best catalyst precursors for Kharasch additions of carbon tetrachloride to olefins, including acrylonitrile (66% yield at 85 °C). Substitution of the [(*p*-cymene)$RuCl_2$] fragment by PCy_3 or by an NHC ligand yields new ruthenium complexes of the type **7** or **8** which show about the same catalytic efficiency as **6** [36]. The latter family of complexes exhibits poor ATRP activity, although styrene can be polymerized to some extent. These results sharply contrast with those obtained with [(arene)$RuCl_2L$] complexes (*L* is phosphine or NHC), whose activity can be tuned to promote either ATRA or ATRP depending on the exact nature of the ligand. Thus, aliphatic phosphines (typically tricyclohexylphosphine) lead to very good ATRP catalysts, while the presence of aromatic rings (as in triphenylphosphine) allows Kharasch additions to be performed with fairly good selectivity (A.F. Noels, A. Demonceau, unpublished results). Ruthenium-*p*-cymene complexes bearing NHC ligands are liable to even larger variations in the carbene structure, thereby providing a wide range of options for catalyst fine-tuning and engineering in atom-transfer radical reactions [37].

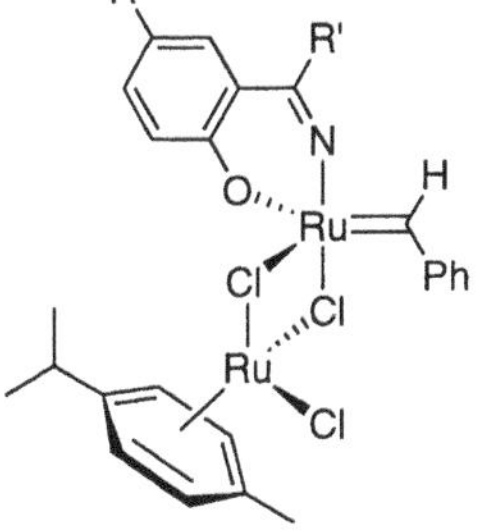

R R′ N O Ru H Ph Cl Cl Ru Cl

6

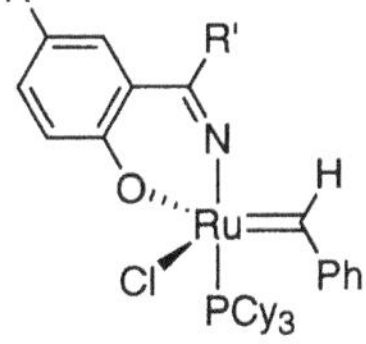

R R′ N O Ru H Ph Cl PCy_3

7

R R′ N O Ru H Ph Cl N N

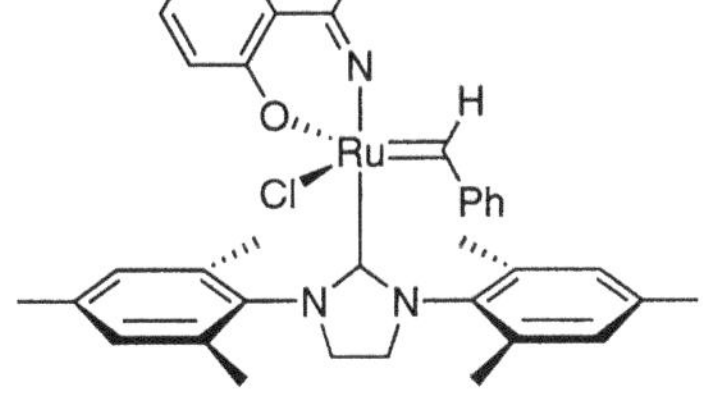

8

4.2
Cp-, Cp*- and Indenyl-Ru(II) Complexes as Catalysts for Radical Reactions

The catalytic activity of half-sandwich ruthenium complexes **9**, **10**, and **11** was investigated in the Kharasch addition of carbon tetrachloride against a set of four representative olefins, viz. *n*-butyl acrylate, MMA, styrene, and 1-decene. Not surprisingly, the outcome of the reaction depended very strongly on the olefin used [38]. Complexes **10** and **11** outperformed **9** in all cases, except with MMA, for which monoadduct production was almost quantitative with all three catalysts within 2 h at 85 °C. With styrene, longer reaction times (5 h) were required to achieve high yields, respectively 95% with **10** and 80% with **11**. *n*-Butyl acrylate, an easily polymerizable substrate, underwent clean addition of CCl_4, yielding 85% of 1,3,3,3-tetrachloropropylbenzene in 4 h with **10**. No telomer formation was observed under these conditions. In all cases 1-decene, a model for the nonfunctionalized α-olefins, was less prone to react. A modest 45% yield of addition product was nevertheless obtained with the indenyl-substituted complex **11** after 24 h at 60 °C.

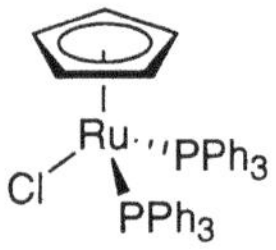

9 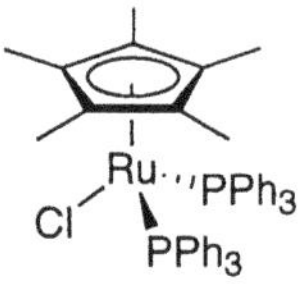 **10** 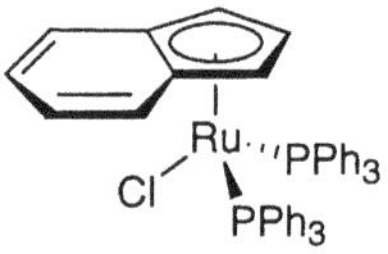 **11**

The high efficiency of Ru-Cp/Cp* catalyst precursors **10** and **11** was preserved when the reaction of styrene and MMA was carried out at 40 °C or even lower temperatures. Indeed, a 90% yield of the styrene–carbon tetrachloride adduct is obtained at room temperature with complex **10** [38]. Hence, it is one of the most efficient catalytic systems reported so far for this reaction. Addition of free triphenylphosphine to the reaction medium strongly depresses the catalytic activity. The same negative trend is observed when the original PPh_3 ligand in **10** is replaced by either tris(4-methoxyphenyl)phosphine or tris(4-trifluoromethylphenyl)phosphine [39]. In fact, the reactivity order correlates well with the ruthenium phosphine bond energy order, i.e., with the relative ease of formation of coordinatively unsaturated 16-electron species through PAr_3 ligand disengagement. A mechanism in which the catalytically active species is generated by release of a phosphine ligand has also been postulated for the Kharasch reaction mediated by **1** [16].

Addition of carbon tetrachloride to complexes **9**–**11** in toluene-d_8 promotes the decomposition of all three molecules into unidentified paramagnetic ruthenium species [38]. The relative rates of decomposition are very different from each other, however, and ^{31}P NMR permitted the correlation of the highest catalytic activity with the highest reactivity toward carbon tetrachloride. All the kinetics data were interpreted in terms of a two-step mechanism, in which phosphine ligand disengagement occurs prior to the activation of the halogenated compound by the ruthenium center.

The family of Ru(II)-Cp* ATRA catalysts was recently extended to novel ruthenium amidinate complexes **12** and **13** [40]. Complex **12** displays two successive one-electron oxidation waves in cyclic voltammetry, assigned to Ru(II)/Ru(III) and Ru(III)/Ru(IV) oxidation steps, respectively. This opens the door to chemical transformations of organic molecules on **12** either by way of one-electron redox processes [i.e., Ru(II) to Ru(III) or Ru(III) to Ru(IV)], or via two-electron processes [i.e., Ru(II) to Ru(IV)].

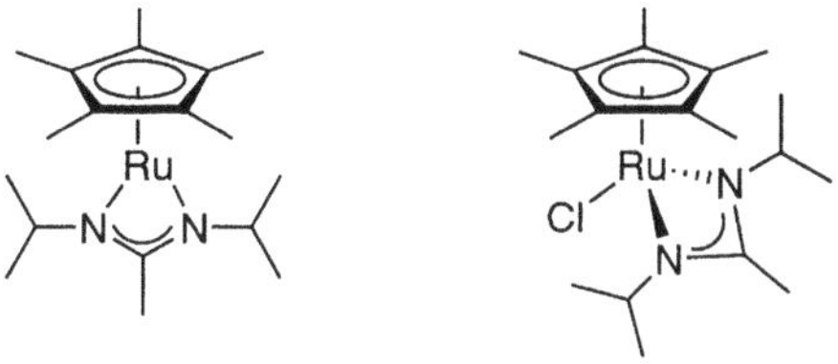

Ru N N Cl Ru N N

12 **13**

Both complexes **12** and **13** are active toward atom-transfer cyclization of *N*-allyltrichloracetamides [40]. Of particular interest is the synthesis of a pyrrozilidine alkaloid skeleton. Much milder conditions are needed when using ruthenium-based catalysts instead of copper derivatives, although a high catalyst loading (30 mol %) is still required. Thus, an *N*-functionalized vinyl pyrrolidine is smoothly converted at room temperature into a bicyclic lactam, in 90% and 85% yield with complexes **12** and **13**, respectively. The product is a precursor of the pyrrozilidine alkaloids trachelantamidine and pseudoheliotridane (Eq. 10).

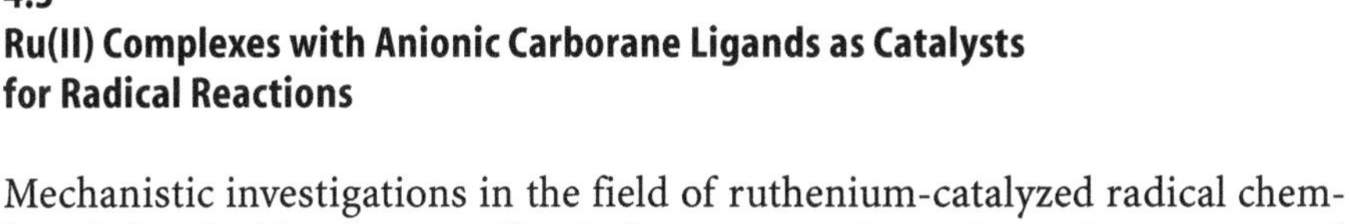

(**12**) or (**13**), C_6H_6, r. t.

Pseudoheliotridane Trachelantamidine

(10)

4.3 Ru(II) Complexes with Anionic Carborane Ligands as Catalysts for Radical Reactions

Mechanistic investigations in the field of ruthenium-catalyzed radical chemistry led to the idea that coordinatively unsaturated metal complexes generated through ligand release from a given precursor might be key intermediates in the catalytic process. It was therefore anticipated that stable, well-defined 14-electron complexes could provide direct access to the catalytic species. Ruthenium carborane complexes in which B–H→Ru agostic bonds are favored meet these criteria. Indeed, boron clusters provide structural and bonding possibilities distinct of conventional organic ligands [41]. Thus, the catalytic activity of three 14-electron ruthenium(II) complexes, **14**, **15**, and **16**, with one or two *nido*-diphosphinocarborane anions was investigated for the addition of carbon tetrachloride onto a set of five representative olefins [42]. The idea behind the catalyst design was borne out to some extent. Yet, the outcome of the reactions dramatically depended both on the complex used and on the substrate. Even if the yields were satisfactory with some olefins and compared favorably with those obtained with $[RuCl_2(PPh_3)_3]$ (**1**), the addition products were always accompanied by oligomers or telomers. Hence, the ruthenium carborane complexes, being less selective, cannot compete with the best catalyst precursors, for example **10**, described so far.

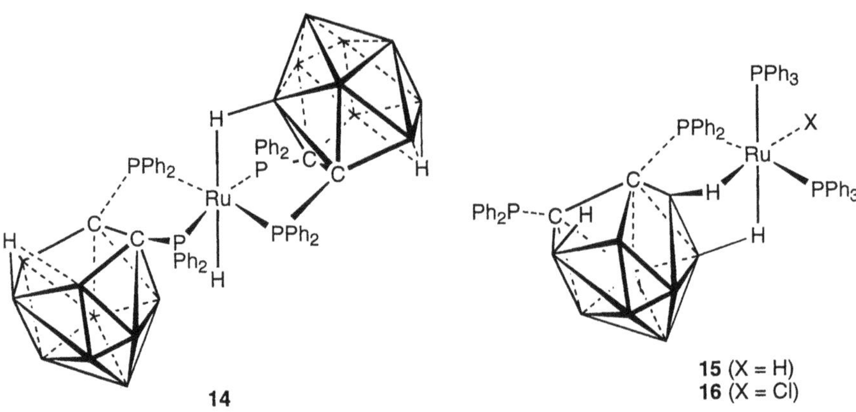

The dicarbollide dianion $[7,8\text{-}C_2B_9H_{11}]^{2-}$ stands among the best studied η^5-boron ligands. The C_2B_3 coordinating motif of this cluster is related to the widely used monoanionic (Cp) ligand, although behind the apparent similarity of the two ring systems, some remarkable differences remain. Among the singularities brought about by the $[7,8\text{-}C_2B_9H_{11}]^{2-}$ ligand prevails its ability to stabilize higher oxidation states than Cp. Moreover, the out-of-plane disposition of the open-face substituents could be beneficial to catalysis, particularly in ATRA reactions. Substitution of one open-face hydrogen by an SR_2 group leads to carbollide monoanions $[X\text{–}R_2S\text{-}7,8\text{-}C_2B_9H_{11}]^-$. The stability of such ligands in coordination complexes and the negative charge dissipation on the bond between substituents of the C_2B_3 open-face ring are well-known. Overall, the monoionic and dianionic boron clusters could behave as an electron pool connected to the metal center and fulfill the catalyst electronic requirements. Ruthenium carbollide complexes **17–22** indeed meet the expectations, and some of them emerge as the most efficient ATRA catalysts described so far for the addition of carbon tetrachloride to styrene and MMA. In particular, structure **19** affords quantitative yields of Kharasch addition products while displaying turnover numbers (TON) of 4,200 and 9,000 with MMA and styrene, respectively, and initial turnover frequencies (TOF) of 1,880 and 1,500 h^{-1} at

40 °C. These values are significantly higher than those recorded with **10** (TON=1,600–1,700, TOF=400 h^{-1} for MMA) [43]. The TON for **19** even surpasses that obtained with the pincer N,C,N-chelating aryldiaminonickel complex reported as the most efficient ATRA catalyst to date (TON=1,731 and TOF=400 h^{-1} for MMA at 25 °C) [44, 45].

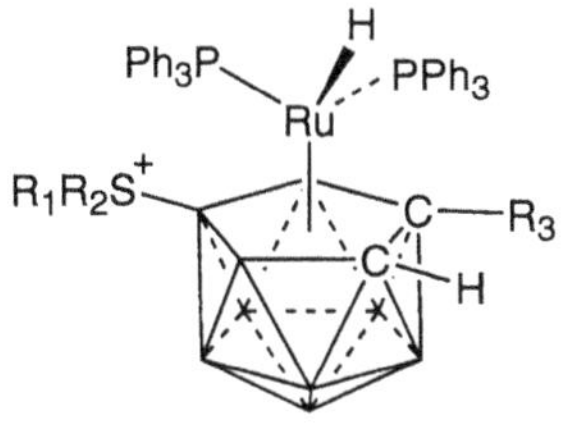

	R_1, R_2	R_3
17	Me, Me	H
18	Et, Et	H
19	Et, Ph	H
20	$(CH_2)_4$	Me
21	Me, Me	Me
22	Et, Et	Me

5 Cyclic Voltammetry as a Probe for Catalyst Efficiency

One expects easy and reversible Ru(II)/Ru(III) redox processes to be crucial for achieving high catalytic efficiency in atom-transfer radical reactions. Electron-transfer properties of some of the most active ruthenium complexes reported so far for controlled radical reactions were determined by cyclic voltammetry in dichloromethane. The investigations commented on hereafter concern the series of closely related complexes described in the preceding sections. They were initiated following the recent disclosure of electrochemical analyses performed with copper-based ATRP catalysts [46, 47]. All potentials are referenced to Fc/Fc^+ (where Fc is ferrocene) and the oxidative response is assigned to the Ru(II)–Ru(III) oxidation. The process is reversible with a peak-to-peak separation (ΔE_p) of about 80–90 mV. Interestingly, the oxidation potentials (E_{ox}) for complexes **17**–**20**, the most active catalyst precursors so far, are close to each other and are centered around –270 mV, whereas E_{ox} values of about –370 mV are observed for the slightly less efficient catalyst precursors **21** and **22**. The oxidation potentials range from –100 to +150 mV for species **9**, **10**, and **11** (+133, –83, and –10 mV, respectively), the former complex (**9**) being by far the less efficient catalyst of the series [43].

A more positive E_{ox} value denotes a more stabilized Ru(II) state, and a more negative E_{ox} value a more stabilized Ru(III) state. The fact that the highest catalytic activities correlate to E_{ox} values lying between the edges suggests that neither Ru(II) nor Ru(III) species should be too stabilized within the same ligand framework to display a good turnover. Accordingly, a complex displaying a relatively high oxidation potential should possess a rather inert divalent metal in its coordination sphere and is not expected to display an outstanding activity in ATR reactions. This observation is fostered when comparing the efficacy of the [(arene)$RuCl_2(PR_3)$] family of complexes with their arene-tethered analogues [η^1:η^6-(phosphinoarene)$RuCl_2$]. Both series of complexes undergo a one-electron reversible oxidation, which occurs at a significantly higher oxidation potential for the strapped complexes [48,49]. Even though some of the former ruthenium arene complexes are outstanding ATRP catalysts but poor ATRA catalysts (depending on the nature of the phosphine ligand), none of the tethered complexes show any significant activity in radical reactions (A.F. Noels, A. Demonceau, unpublished results).

The predictive value of cyclic voltammetry data in radical chemistry seems, however, rather limited. An adequate redox potential of the metal complex is required for obtaining efficient catalysts, and comparison of the values recorded within a family of closely related species can allow meaningful forecasts, but other parameters need to be considered. Atom-transfer processes give rise to an expansion of the metal coordination sphere, and the Ru(II)/Ru(III) equilibrium is affected not only by the redox couple, but also by the energetics of the X–Ru^{n+1} bond. Steric hindrance, ligand oxidation, and a host of other parameters also play a crucial role in the overall process. The fact that electrochemical measurements are carried out on catalyst precursors, and not necessarily on the genuine active species, using an experimental setting quite different from the actual reaction conditions in terms of temperature and solvent (a supporting electrolyte, usually a tetrabutylammonium salt, is required) may also influence the results.

6 Ruthenium Porphyrin Complexes as Catalysts for C–H Hydroxylation

Applications of controlled radical reactions – including oxidation – deal almost exclusively with C=C double bonds. Indeed, a multitude of examples have been reported for the selective transformation of this functional group. Contrasting with this situation, only a very limited number of selective ("stereocontrolled") radical reactions involving sp^3-hybridized C–H bonds are known. Particularly useful functionalizations along these lines include the hydroxylation and the acyloxylation of alkyl chains. The reason for their limited success is of course due to the high stability of the C–H bond compared with that of the olefinic C=C unit: most electrophilic reagents which readily add to unsaturated substrates are not able to oxidize a C–H bond.

Iron-containing cytochrome P-450 constitutes the most famous example of a selective C–H bond oxidizer. Although the exact nature of the mechanism remains controversial, the reaction most likely proceeds through radical intermediates [2]. The hydroxylation of activated C–H bonds has also been carried out in the presence of synthetic porphyrin complexes. In these biomimetic processes, ruthenium plays a relatively minor role when compared with iron. Zhang et al. [50], however, recently reported the enantioselective hydroxylation of benzylic C–H bonds using ruthenium complexes supported by a D_4-symmetric porphyrin bearing a crafted chiral cavity. Thus, complex **23** reacts in a stoichiometric manner with ethylbenzene to give phenethyl alcohol with a

45% ee. The same hydroxylation proceeds catalytically (72% ee) using **24** as the catalyst and 2,6-dichloropyridine *N*-oxide as a terminal oxidant. Other acyclic alkylarenes are converted into alcohols with rather good enantioselectivity, but the reaction of cyclic substrates takes place only with modest selectivity. In all the cases, chemical yields are modest to poor.

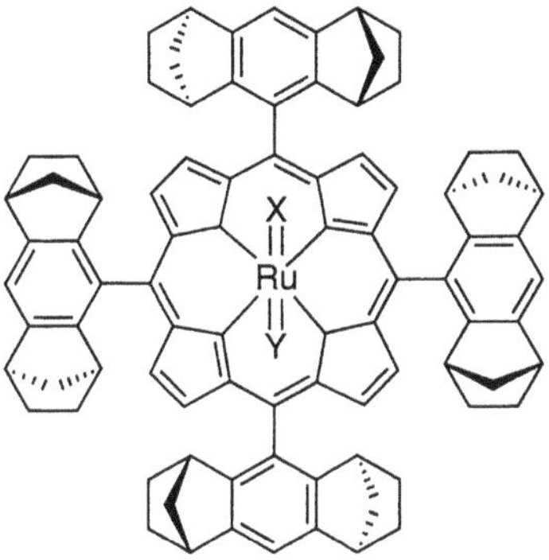

23 (X = Y = O)
24 (X = CO, Y = EtOH)

7
Conclusion and Perspectives

Currently, the number of synthetically useful ruthenium-catalyzed atom-transfer reactions remains rather limited. In view of the versatility and potential utility of these reactions in fine chemistry, it seems likely that further applications and extensions of known reactions will appear in the near future. Suffice it to recall that the recent breakthroughs in the field are due to catalyst engineering and only occurred after 1999. Thus, impressive progress has already been made in a short period of time.

The development of ruthenium complexes for other applications in radical chemistry is still in its infancy, but seems well suited to future expansion, thanks to the versatility of ruthenium as a catalytically active center. Large avenues have not been explored yet and remain open to research. For instance, the development of methodologies for the asymmetric functionalization of C–H bonds remains a challenge. The Kharasch–Sosnovsky reaction [51,52], in which the allylic carbon of an alkene is acyloxylated, its asymmetric counterpart, and the asymmetric version of the Kharasch reaction itself are practically terra incognita to ruthenium chemistry, and await the discovery of improved catalysts.

Thanks to the development of the Grubbs benzylidene catalyst (**2**) and other related ruthenium complexes, olefin metathesis has experienced spectacular advances over the past 10 years. The various incarnations of the reaction (acyclic diene metathesis, ring-closing metathesis, ring-opening metathesis polymerization, etc.) have now acquired first rank importance in synthesis. Clearly, the emergence of a similar, generic, efficient catalytic system for controlled radical reactions would contribute enormously to their popularity among the community of organic chemists. This will presumably follow from a better understanding of the mechanisms of these highly complex reactions.

Acknowledgement Financial support of the EU through contracts HPRN-CT-2000-00010 "Polycat" and G5RD-CT-2001-00554 "Dentalopt" is gratefully acknowledged.

References

1. Gilbert BC, Parsons AF (2002) J Chem Soc Perkin Trans 2 367
2. Renaud P, Sibi MP (2001) (eds) Radicals in organic synthesis, vols 1 and 2. Wiley-VCH, Weinheim
3. Curran DP, Porter NA, Giese B (1996) (eds) Stereochemistry in radical reactions. Wiley-VCH, Weinheim
4. Kharasch MS, Engelman S, Mayo FR (1938) J Org Chem 2:288
5. Kharasch MS, Jenson EV (1945) Science 102:128
6. Kharasch MS, Skell PS, Fisher P (1948) J Am Chem Soc 70:1055
7. Kharasch MS, Jensen EV, Urry WH (1945) J Am Chem Soc 67:1626
8. Minisci F (1975) Acc Chem Res 8:165
9. Trost BM, Toste FD, Pinkerton AB (2001) Chem Rev 101:2067
10. Iqbal J, Bhatia B, Nayyar N (1994) Chem Rev 94:519
11. Bellus D (1985) Pure Appl Chem 57:1827
12. Matyjaszewski K, Xia J (2001) Chem Rev 101:2921
13. Sawamoto M, Kamigaito M (1999) CHEMTECH June:30
14. Matyjaszewski K (1999) Chem Eur J 5:3095
15. Matsumoto H, Nakano T, Nagai Y (1973) Tetrahedron Lett 51:5147
16. Bland WJ, Davis R, Durrant JLA (1985) J Organomet Chem 280:397
17. Kato M, Kamigaito M, Sawamoto M, Higashimura T (1995) Macromolecules 28:1721
18. Ando T, Kato M, Kamigaito M, Sawamoto M (1996) Macromolecules 29:1070
19. Bland WJ, Davis R, Durrant JLA (1984) J Organomet Chem 267:C45
20. Davis R, Furze JD, Cole-Hamilton DJ, Pogorzelec P (1992) J Organomet Chem 440:191
21. Davis R, Stephens K, Hajek M (1994) J Mol Catal 92:269
22. Crabtree RH (2001) The organometallic chemistry of the transition metals, 3rd edn. Wiley, New York
23. Kamigata N, Fukushima T, Terekawa Y, Yoshida M, Sawada H (1991) J Chem Soc Perkin Trans 1 627
24. Kotora M, Hajek M (1992) J Mol Catal A 77:51
25. Ishibashi H, Nakatani H, Iwami S, Sato T, Nakamura N, Ikeda M (1989) J Chem Soc Chem Commun 1767
26. Okano T, Shimizu T, Sumida K, Egushi S (1993) J Org Chem 58:5163
27. Okano T, Uekawa T, Sawaki H, Egushi S (1990) Synlett 403
28. Simal F, Demonceau A, Noels AF (1999) Tetrahedron Lett 40:5689
29. Tallarico JA, Malnick L, Snapper ML (1999) J Org Chem 64:344
30. Simal F, Demonceau A, Noels AF (1999) Angew Chem Int Ed Engl 38:538
31. Amir-Ebrahimi V, Hamilton JG, Nelson J, Rooney JJ, Thompson JM, Beaumont AJ, Rooney AD, Harding CJ (1999) J Chem Soc Chem Commun 1621
32. Amir-Ebrahimi V, Hamilton JG, Nelson J, Rooney JJ, Rooney AD, Harding CJ (2000) J Organomet Chem 606:84
33. Opstal T, Verpoort F (2002) Tetrahedron Lett 43:9259

34. Bourissou D, Guerret O, Gabbaï FP, Bertrand G (2000) Chem Rev 100:39
35. De Clercq B, Verpoort F (2002) Tetrahedron Lett 43:4687
36. De Clercq B, Verpoort F (2003) J Organomet Chem 672:11
37. Richel A, Delfosse S, Cremasco C, Delaude L, Demonceau A, Noels AF (2003) Tetrahedron Lett 44: 6011
38. Simal F, Wlodarczak L, Demonceau A, Noels AF (2001) Eur J Org Chem 2689
39. Simal F, Wlodarczak L, Demonceau A, Noels AF (2000) Tetrahedron Lett 41:6071
40. Nagashima H, Gondo M, Masuda S, Kondo H, Yamagushi Y, Matsubara K (2003) J Chem Soc Chem Commun 442
41. Xie Z (2002) Coord Chem Rev 231:23
42. Simal F, Sebille S, Demonceau A, Noels AF, Nunez R, Abad M, Teixidor F, Vinas C (2000) Tetrahedron Lett 41: 5347
43. Tutusaus O, Vinas C, Nunez R, Teixidor F, Demonceau A, Delfosse S, Noels AF, Mata I, Molins E (2003) J Am Chem Soc 125: 11830
44. van de Kuil LA, Grove DM, Gossage MA, Zwikker JW, Jenneskens LW, Drenth W, van Koten G (1997) Organometallics 16:4985
45. Gossage RA, van de Kuil LA, van Koten G (1998) Acc Chem Res 31:423
46. Qiu J, Matyjaszewski K, Thouin L, Amatore C (2000) Macromol Chem Phys 201:1625
47. Matyjaszewski K, Göbelt B, Paik H-J, Horwitz CP (2001) Macromolecules 34:430
48. Bennett MA, Edwards AJ, Harper JR, Khimyak T, Willis AC (2001) J Organomet Chem 629:7
49. Ghebreyessus KY, Nelson HJ (2000) Organometallics 19:3387
50. Zhang R, Yu W-Y, Lai T-S, Che C-M (1999) J Chem Soc Chem Commun 1791
51. Eames J, Watkinson M (2001) Angew Chem Int Ed Engl 40:3567
52. Andrus MB, Lashley JC (2002) Tetrahedron 58:845

Topics Organomet Chem (2004) 11: 173–195
DOI 10.1007/978-3-540-40027-1

Selective Carbonylations with Ruthenium Catalysts

Naoto Chatani (✉)

Department of Applied Chemistry, Faculty of Engineering, Osaka University, Suita, Osaka 565-0871, Japan
chatani@chem.eng.osaka-u.ac.jp

Abstract Ruthenium-catalyzed carbonylation reactions are described. The purpose of this chapter is to show how ruthenium complexes as catalysts are important in the recent development of carbonylation reactions. This review does not present a complete, historical coverage of ruthenium-catalyzed carbonylation reactions, but presents the most significant developments of the last 10 years. The emphasis is on novel and synthetic transformations of genuine value to organic chemists. Especially, this review will focus on carbonylative cycloadditions and carbonylation of C–H bonds. The review is generally organized according to the nature of the reaction.

Keywords Carbonylation · Ruthenium · Carbon monoxide · Carbonylative cycloaddition · C–H bond activation

1 Introduction

Carbonylation reactions are recognized as useful and reliable transformations for the preparation of a variety of carbonyl compounds [1]. It is well known that various transition metals, such as nickel, cobalt, palladium, platinum, iron, and rhodium, catalyze a variety of carbonylation reactions. In contrast, ruthenium is not commonly used as a catalyst in carbonylation reactions. Ruthenium offered few advantages over other transition-metal complexes in most carbonylation reactions. Carbonylation or related reactions that are catalyzed by ruthenium complexes were restricted to a few reactions such as hydroformylation, the water-gas shift reaction, and reductive carbonylations of nitrobenzene derivatives catalyzed by $Ru_3(CO)_{12}$ or its derivatives. However, compared with the commonly used palladium, nickel, and rhodium catalysts, ruthenium has some new and unique catalytic characteristics. This becomes clear when generalized catalytic cycles are compared. For example, palladium typically operates within catalytic cycles shuttling between the 0 and II oxidation states. Most palladium-catalyzed reactions involve only some elementary reactions, such as oxidative addition, carbopalladation, transmetallation, β-hydride elimination, and reductive elimination. For this reason, the mechanism of catalysis using palladium is relatively easily understood. In contrast, the oxidation states of ruthenium can vary between –2 and +8. As a result, a wide range of mechanistically different processes may be included in ruthenium-catalyzed reactions. These characteristics make the design of reactions using ruthenium difficult and complicated. However, this diversity represents a unique and a characteristic feature of ruthenium. In the past 10 years dramatic growth has occurred in the use of ruthenium catalysts in synthetically important organic transformations. Similarly, a variety of novel types of carbonylation reactions catalyzed by ruthenium have also been discovered. In particular, $Ru_3(CO)_{12}$ or its derivatives show a high catalytic activity for most of the carbonylation reactions reported thus far. The aim of this chapter is to review recent developments in ruthenium carbonylation reactions and does not present a complete, historical coverage of the reactions. This review will focus on novel reactions that have recently been developed and on synthetically important carbonylation reactions, especially carbonylative cycloadditions and direct carbonylation reactions at C–H bonds. Reactions which are known to be catalyzed by other transition-metal complexes are not included in this review.

2 Carbonylative Cycloadditions

Transition-metal-catalyzed cycloaddition reactions provide a new synthetic method for the construction of a numerous types of ring systems, which are achieved with difficulty by thermally promoted cycloaddition reactions [2]. In particular, the utilization of carbon monoxide as a one-carbon unit provides a new and attractive strategy for the construction of cyclic carbonyl compounds. In this section, such catalytic carbonylative cycloadditions will be classified by topology, i.e., the numbers in square brackets refer to the number of atoms that constitute the cycloadducts.

Metal complexes enable one to employ molecules that are thermally unreactive toward cycloadditions by taking advantage of their ability to be activated through complexation. Most of the molecules activated by transition-metal complexes involve C–C unsaturated bonds such as alkynes, alkenes, 1,3-dienes, allenes, and cyclopropanes. In contrast, carbonyl functionalities such as aldehydes, ketones, esters, and imines seldom participate in transition-metal-catalyzed carbonylative cycloaddition reactions. Recently, such a transformation was reported via the use of ruthenium complexes.

2.1 [2+2+1] Cycloadditions of an Alkyne, an Alkene, and CO: The Pauson–Khand-Type Reaction

Among the carbonylative cycloaddition reactions, the Pauson–Khand (P–K) reaction, in which an alkyne, an alkene, and carbon monoxide are condensed in a formal [2+2+1] cycloaddition to form cyclopentenones, has attracted considerable attention [3]. Significant progress in this reaction has been made in this decade. In the past, a stoichiometric amount of $Co_2(CO)_8$ was used as the source of CO. Various additive promoters, such as amines, amine *N*-oxides, phosphanes, ethers, and sulfides, have been developed thus far for a stoichiometric P–K reaction to proceed under milder reaction conditions. Other transition-metal carbonyl complexes, such as $Fe(CO)_4$(acetone), $W(CO)_5$(tetrahydrofuran), $W(CO)_5F^-$, $Cp_2Mo_2(CO)_4$, where Cp is cyclopentadienyl, and $Mo(CO)_6$, are also used as the source of CO in place of $Co_2(CO)_8$. There has been significant interest in developing catalytic variants of the P–K reaction. Rautenstrauch et al. [4] reported the first catalytic P–K reaction in which alkenes are limited to reactive alkenes, such as ethylene and norbornene. Since 1994 when Jeong et al. [5] reported the first catalytic intramolecular P–K reaction, most attention has been focused on the modification of the cobalt catalytic system [3]. Recently, other transition-metal complexes, such as Ti [6], Rh [7], and Ir complexes [8], have been found to be active for intramolecular P–K reactions.

In 1997, Murai et al. [9] reported the first use of $Ru_3(CO)_{12}$ as a catalyst for the intramolecular P–K reaction of 1,6-enynes (Eq. 1). The scope of the reaction with respect to the alkene is limited to enynes with no substituent on the olefinic carbon. The metallacycle **1** is proposed to be involved as an intermediate.

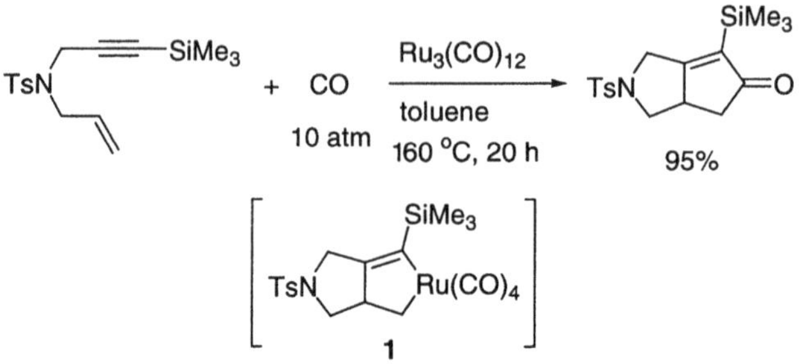

(1)

Mitsudo et al. [10] independently reached the same conclusion (Eq. 2). *N,N*-dimethylacetamide (DMA) was found to be the solvent of choice. This solvent system is applicable to enynes bearing a substituent at the olefinic part.

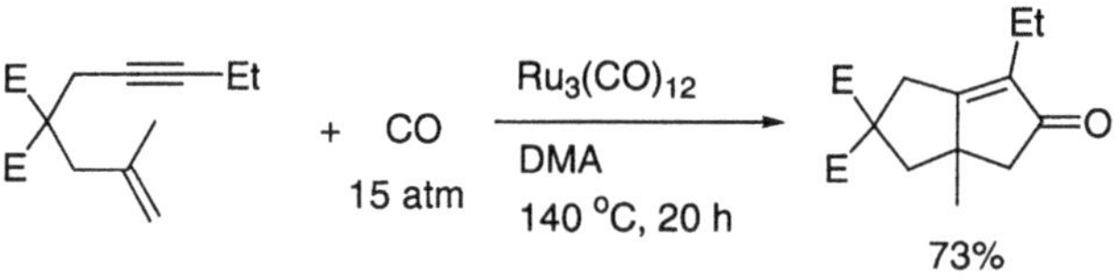

(2)

Although cobalt-catalyzed P–K reactions have been extensively studied, one remaining problem to be solved is the narrow scope of alkenes in intermolecular variants. The reaction is restricted to ethylene or strained alkenes such as norbornene [4]. Ruthenium complexes also failed to catalyze the intermolecular P–K reaction. Recently, such problems were overcome by taking advantage of the coordination of a heteroatom to a metal center. Itami et al. [11] reported that dimethyl(2-pyridyl)silyl group functions as a removable directing group in intermolecular P–K reactions (Eq. 3). The coordination of the pyridine nitrogen in the vinylsilane promotes an oxidative cyclization of alkynes and alkenes to ruthenium to give a metalacycle **2**. The insertion of CO followed by reductive elimination affords a silylcyclopentenone that undergoes protodesilylation to give cyclopentenones. Although the issue of regioselectivity remains in the case of the synthesis of highly substituted cyclopentenones,

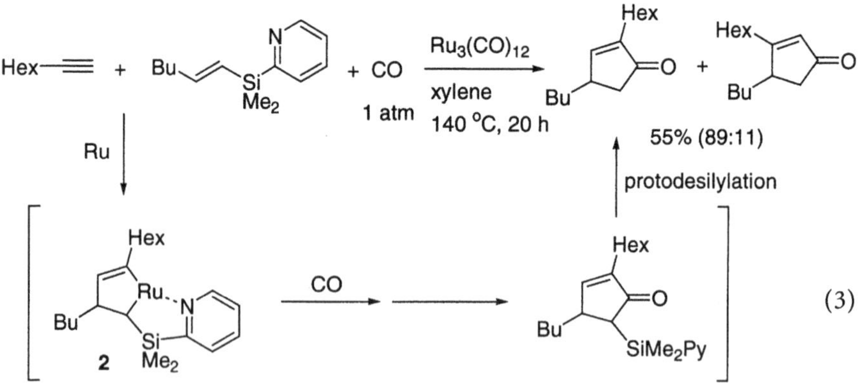

(3)

the reaction is the first success in catalytic intermolecular P–K reactions using alkenes other than ethylene and norbornene.

2.2 [2+2+1] Cycloadditions of an Alkyne, a Carbonyl Functionality, and CO: Hetero-Pauson–Khand-Type Reaction

As already described, the P–K reaction is a useful transformation to give cyclopentenones. If one replaces one olefinic carbon with an oxygen or a nitrogen atom, the formation of α,β-unsaturated lactones or lactams would be expected (Scheme 1). The so-called hetero-P–K-type reaction, which involves the [2+2+1] cycloaddition of an alkyne, an aldehyde (or a ketone), and CO was found to proceed in the presence of a catalytic amount of $Ru_3(CO)_{12}$. The first attempt at the hetero-P–K-type reaction was reported by Crowe and Vu [12], in which a stoichiometric amount of $Cp_2Ti(PMe_3)_2$ reacts with an yne-aldehyde to give a metallacycle, which did not undergo carbonylation, even with heating. The failure can be attributed to the strong Ti–O bond and the Ti–vinyl bond. Buchwald et al. [13] also reported a similar reaction system in which the reaction of a yne-ketone reacts with a stoichiometric amount of $Cp_2Ti(PMe_3)_2$ under 1 atm of CO at 70 °C to give the expected α,β-unsaturated lactone, albeit in low yield.

C≡O

X = O, N

Scheme 1

Murai et al. [14] found that $Ru_3(CO)_{12}$ shows a high catalytic activity for the intramolecular hetero-P–K-type reaction of yne-aldehydes (Eq. 4). A variety of substituents on the acetylenic moiety can be tolerated, and the application to cyclohexane-fused bicyclic systems is also feasible. Although the mechanism of this catalysis remains elusive, two pathways have been proposed as the initial step for the reaction in Eq. (4); via the oxidative cyclization of yne-aldehydes to a ruthenium center, leading to a metallacycle **3**, or via the oxidative addition of an aldehyde C–H bond to ruthenium, leading to **4**.

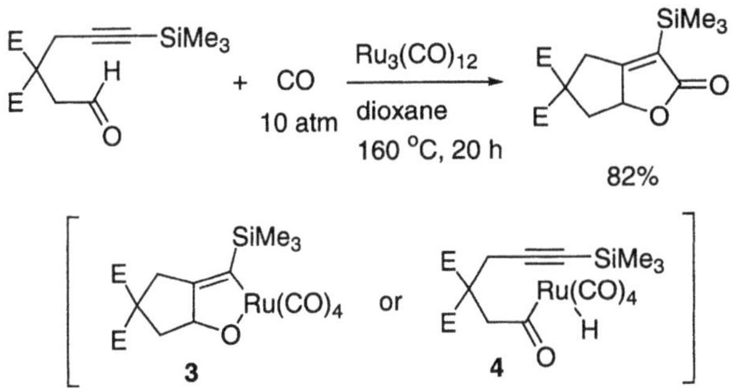

SiMe3 + CO 10 atm; $Ru_3(CO)_{12}$, dioxane, 160 °C, 20 h; 82% (4)

SiMe3, Ru(CO)4, **3** or Ru(CO)4, H, **4**

The reaction was extended to the cyclocarbonylation of yne-imines, leading to bicyclic α,β-unsaturated lactams, indicating that the reaction might proceed via the intermediacy of metallacycles because examples of oxidative additions of imine C–H bonds to transition metals are rare (Eq. 5) [15].

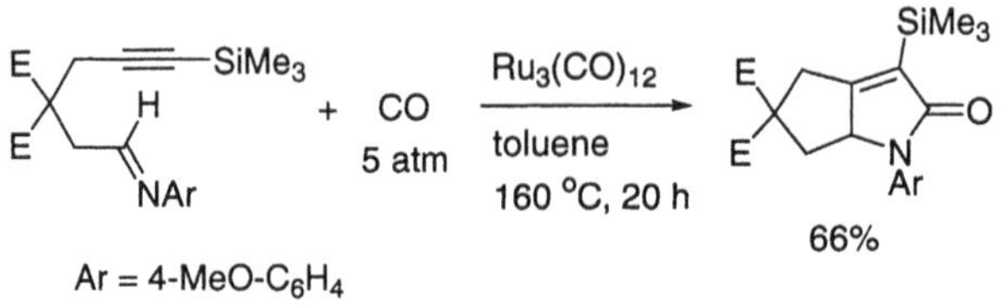

(5)

The reaction can be extended to allene-aldehydes. Kang et al. [16] reported on the $Ru_3(CO)_{12}$-catalyzed cyclocarbonylation of allene-aldehydes or allene-ketones leading to α-methylene-γ-butyrolactones (Eq. 6). The fact that a ketone moiety also functions as a two-atom unit indicates that the reaction includes metallacycle **5** as an intermediate.

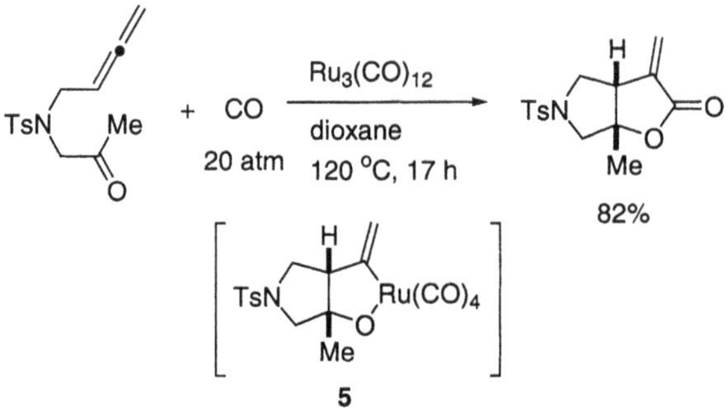

(6)

2.3
[2+2+1] Cyloadditions of an Alkene, a Carbonyl Functionality, and CO

If alkenes are used in place of alkynes in Scheme 1, saturated five-membered lactones or lactams would be expected to form (Scheme 2). The first example of this type of catalytic reaction was achieved under CO by the use of ene-aldehydes or ene-ketones as the substrates and $Cp_2Ti(PMe_3)_2$ as the catalyst precursor [13, 17]. $Cp_2Ti(CO)(PMe_3)$ has been proposed as a key catalytic species and oxatitanocyclopentanes are invoked as key intermediates. Trimethylphosphine may play a role in decreasing the Lewis acidity of the catalytic species and facilitating the ligand-induced reductive elimination from the metallacycle. However, the system is applicable only to intramolecular reactions.

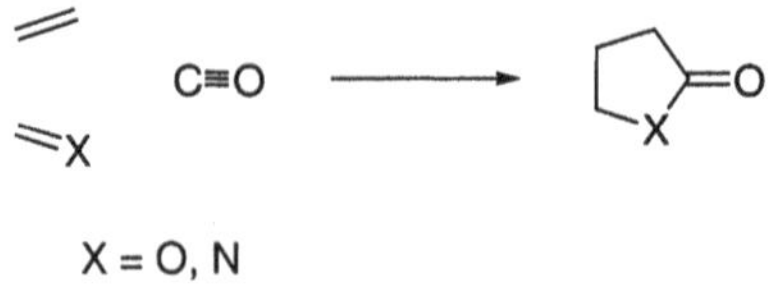

Scheme 2

Chatani and coworkers [18] reported that the intermolecular coupling of ketones (aldehydes), alkenes, and CO proceeds when $Ru_3(CO)_{12}$ is used as a catalyst (Eqs. 7, 8). However, simple aldehydes or ketones, such as benzaldehyde, acetophenone, and cyclohexanone, failed to react. Instead, 1,2-dicarbonyl compounds, including α-keto esters, α-keto amides, α-diketones, and pyridylketones participate in the intermolecular coupling reaction, indicating that the coordination of the substrates to ruthenium by chelation through an oxygen or nitrogen atom, as in **6** or **7**, is essential for the reaction to proceed. The addition of a weakly basic phosphine ligand, $P(4\text{-}CF_3C_6H_4)_3$, in the reaction in Eq. (7) dramatically increases the rate of the reaction of α-dicarbonyl compounds. Unpolarized terminal and cyclic olefins as well as some internal alkynes can be successfully used in the synthesis of a diverse array of γ-lactone derivatives. The reaction mechanism is not clear, but the mechanism was proposed on the basis of a stoichiometric reaction previously reported by Frühauf et al. [19]. On the basis of the results obtained by control experiments, the rate-determining step is different in the reactions in Eqs. (7) and (8). In the reaction in Eq. (7), the formation of metallacycle **8** through the oxidative cyclization of an alkene, a carbonyl group to a ruthenium center is the rate-determining step. On the other hand, the insertion of CO in the metallacycle (related to **8**) or the reductive elimination of the resulting CO-inserted complex (related to **9**) is the

OMe, Ph + CO (5 atm) + $H_2C{=}CH_2$ —[$Ru_3(CO)_{12}$, $P(C_6H_4CF_3)_3$, toluene, 160 °C, 20 h]→ OMe, Ph lactone 94% (7)

[OMe, Ph, Ru—O, $(CO)_3$ **6**]

Ph pyridyl ketone + CO (5 atm) + cyclopentene —[$Ru_3(CO)_{12}$, toluene, 160 °C, 20 h]→ N, Ph lactone 99% (8)

[N, Ph, Ru—O, $(CO)_3$ **7**]

rate-determining step because of the facile formation of the metallacycle via the coordination of the sp^2 nitrogen to a ruthenium center in the reaction in Eq. (8) (Scheme 3).

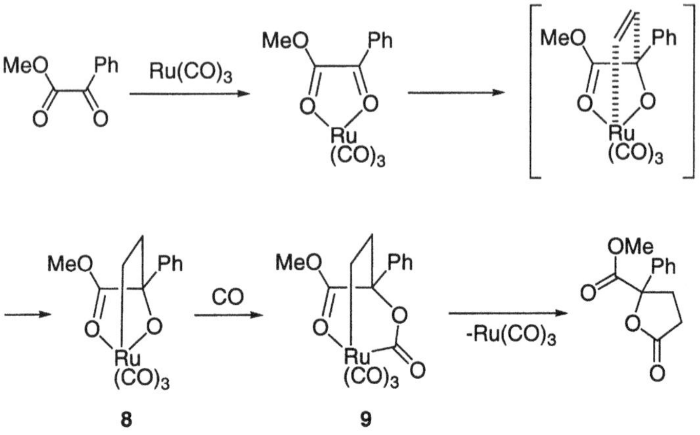

Scheme 3

Iminopyridines or 1,2-iminoesters also underwent a [2+2+1] cycloaddition to give the expected lactams (Eq. 9) [20].

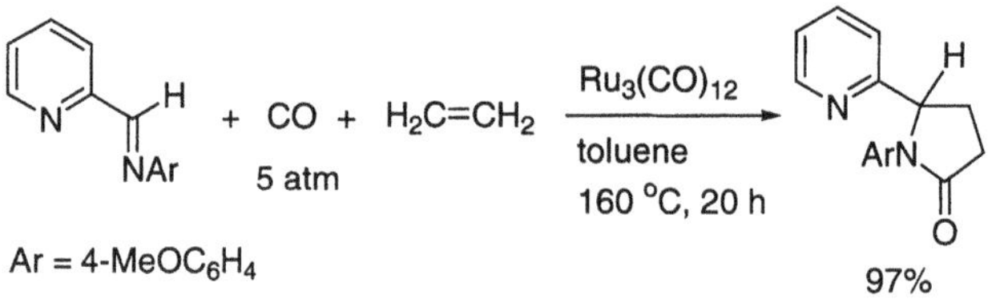

(9)

Göbel and Imhof [21] reported that cyclic diimines also react with CO and alkenes in the presence of $Ru_3(CO)_{12}$ or $Fe_2(CO)_9$ to give spiro lactams (Eq. 10) [21].

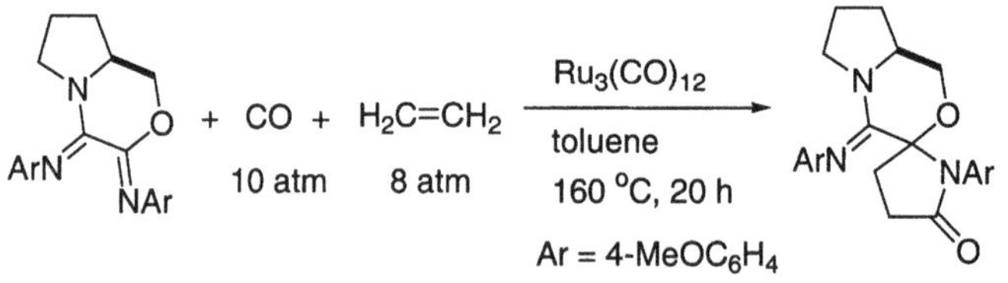

(10)

Imhof et al. [22] studied the reaction mechanism of the [2+2+1] cycloaddition reactions of diimines, CO, and ethylene catalyzed by iron carbonyl complexes on the basis of density functional theory (Scheme 4). The catalytic reaction does not start when CO dissociates from **10** followed by the addition of ethylene, but instead the associative pathway to **11** is proposed. In addition, it can be concluded that the insertion of CO in **11** takes place into a C–Fe bond but not a N–Fe bond, which is different from the fact the *N*-carbonylated complex related to **12** was isolated by themselves et al. [19].

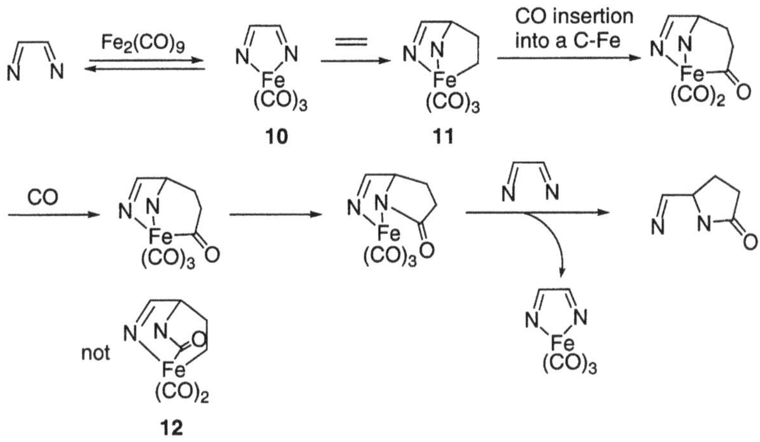

Scheme 4 A proposed reaction mechanism by Imhof et al. [22]

An ester carbonyl group is known to be generally less reactive than an aldehyde or a ketone carbonyl group. As a result, cycloaddition reactions of esters under thermal conditions are very rare. In a unique case, Chatani et al. [23] found that an ester functionality also participates in the carbonylative cycloaddition reaction of α-ketolactones (Eq. 11). The presence of a bulky group next to the keto carbonyl group is required for this selective reaction.

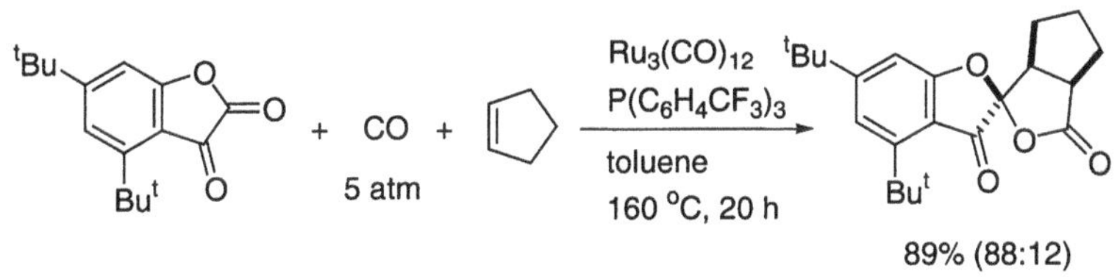

(11)

2.4
[4+1] Cycloadditions of α,β-Unsaturated Imines and CO

In terms of the construction of a five-membered ring system, the [4+1] mode, in which conjugated systems act as four-atom assembling units, represents an attractive and straightforward approach (Scheme 5). In general, the carbonylative [4+1] cycloaddition of a simple 1,3-conjugated system, such as 1,3-butadiene and α,β-unsaturated ketones, is a difficult process, but 1,3-conjugated systems containing cumulated double bonds, such as vinylallenes [24], diallenes [25], and allenylaldehydes [26], have been involved. Recently, a [4+1] cycloaddition of 1,2-diazadiene catalyzed by $Pd(PPh_3)_4$ was reported by Boeckman et al. [27].

Scheme 5

X = O, N

Murai et al. [28] found that the reaction of α,β-unsaturated imines with CO results in a [4+1] cycloaddition to give unsaturated γ-lactams (Eq. 12). For the reaction of imines which contain a β-hydrogen, the initially produced β,γ-unsaturated γ-lactams are isomerized to the stabler α,β-unsaturated isomers. This success can be attributed to the facile coordination of the sp^2 nitrogen of the substrates to a ruthenium center that assembles the substrates to the ruthenium complex.

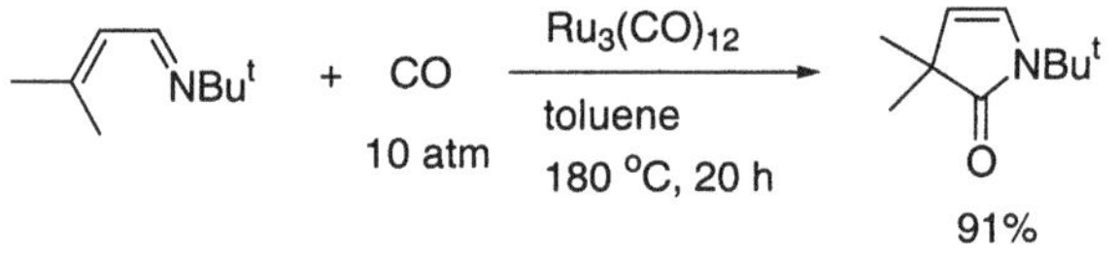

NBu^t + CO (10 atm) $\xrightarrow{Ru_3(CO)_{12}}$ toluene, 180 °C, 20 h; NBu^t, O, 91% (12)

2.5 [5+1] Cycloadditions of Cyclopropylimines and CO

Ring strain present in three-membered ring compounds facilitates ring-opening thus permitting them to serve as a five-atom assembling unit in cycloaddition reactions, when the ring is conjugated with an unsaturated bond. The [4+1] cycloaddition described in previous section was extended to a [5+1] cycloaddiiton when cyclopropyl imines were used as substrates [29] (Eq. 13).

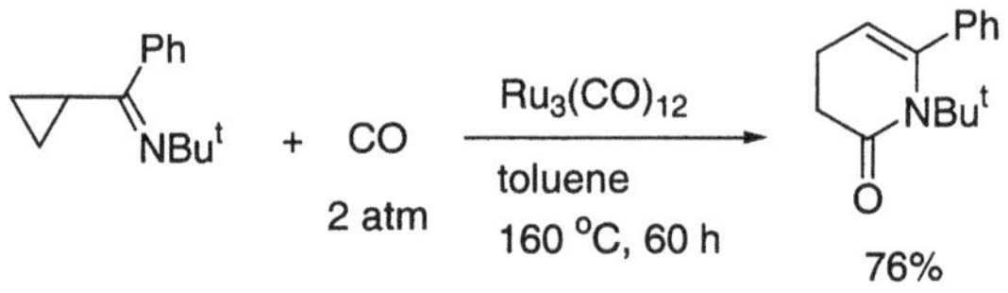

Ph, NBu^t + CO (2 atm) $\xrightarrow{Ru_3(CO)_{12}}$ toluene, 160 °C, 60 h; Ph, NBu^t, O, 76% (13)

2.6 [2+2+1+1] Cycloadditions

Carbonylation of the parent acetylene via stoichiometric or catalytic reactions involving transition-metal carbonyl complexes has been extensively studied. Various types of carbonylation reactions of acetylene were discovered. In 1968, Pino et al. [30] reported on the synthesis of hydroquinone via a $Ru_3(CO)_{12}$-catalyzed carbonylation of acetylene with H_2 or H_2O. The product formally consisted of two molecules of acetylene and CO, and one molecule of H_2 (Eq. 14). To achieve a good yield of hydroquinone, the H_2 pressure must be kept under

10 atm. Otherwise, high-boiling or solid polymeric products of unknown structure are formed.

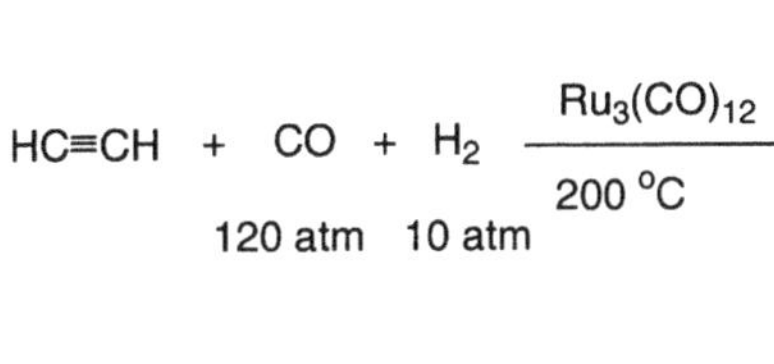

(14)

Some carbonylation reactions of alkynes related to the Pino reaction have been reported. The reaction of 1,6-diynes with a hydrosilane and CO in the presence of $Ru_3(CO)_{12}$ resulted in a [2+2+1+1] cycloaddition to give catecol derivatives, in which two molecules of CO are incorporated (Eq. 15) [31]. The transformation is unique in that it represents a rare example of the successive incorporation of two molecules of CO into diynes. A novel method for incorporating CO via an oxycarbyne complex **13** was proposed, as shown in Scheme 6. The oxidative addition of a hydrosilane to a ruthenium center followed by a 1,3-silyl shift from ruthenium to the oxygen atom of the CO ligand gives the siloxycarbyne complex **13**. The complex reacts with CO to afford dioxyalkyne **14**, which undergoes a [2+2+2] cycloaddition to give monosilyl catechols as the primary products, which were silylated under the reaction conditions used.

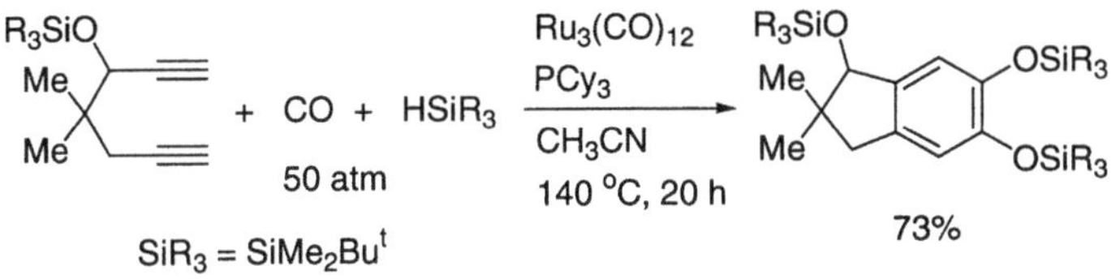

(15)

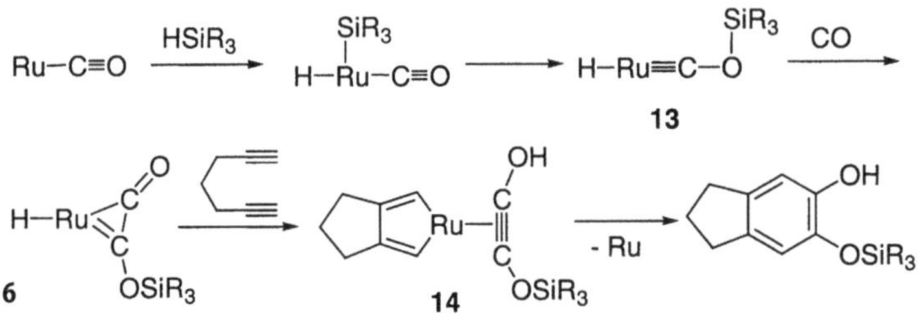

Scheme 6

Mitsudo et al. [32] found that hydroquinones can be obtained by the reaction of internal alkynes with norbornene and CO using *N*-methylpiperidine as a solvent in the presence of $Ru_3(CO)_{12}$ (Eq. 16). The reaction is proposed to proceed via the maleoyl ruthenium complex **15**, which is generated from an alkyne, two molecules of CO, and ruthenium. Norbornene is inserted into this complex to give the quinone, which undergoes reduction to the final product under the reaction conditions used. When other common solvents were used in place of *N*-methylpiperidine, the Reppe-type reaction took place to give a quinone **16** as a byproduct. The selective reaction was attained when *N*-methylpiperidine was used as the solvent.

Pr–C≡C–Pr + CO (60 atm) + norbornene —[$Ru_3(CO)_{12}$, N-methylpiperidine, 140 °C, 20 h]→ hydroquinone product, 85% (16)

[**15**] **16**

3
Carbonylation Reactions with Ring Cleavage

Mitsudo et al. [33] found that the treatment of cyclopropenones with CO in the presence of $Ru_3(CO)_{12}$ and Et_3N results in a carbonylative dimerization to give pyranopyrandiones, in which two molecules of cyclopropenone and two molecules of CO are formally incorporated, in high yield (Eq. 17). Some other ruthenium complexes also show catalytic activity, but $Ru_3(CO)_{12}$ gave the best yield. Labeled experiments using ^{13}CO suggest that three molecules of CO are incorporated into the products, indicating that one of the carbonyl groups of cyclopropenone exchanged with the external CO.

dibutylcyclopropenone + CO (20 atm) —[$Ru_3(CO)_{12}$, Et_3N, toluene, 150 °C, 20 h]→ tetrabutylpyranopyrandione, 86% (17)

The reaction was extended to the cross-carbonylation of cyclopropenones and internal alkynes, leading to unsymmetrically substituted pyranopyrandiones (Eq. 18) [33].

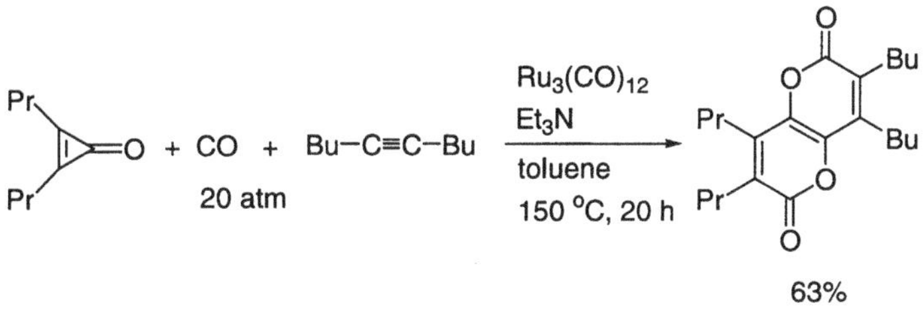

(18)

Mitsudo et al. [34] then examined the reaction of cyclobutenedione with CO and norbornene (Eq. 19). The reaction requires a low pressure of CO. Regioselective C–C bond cleavage followed by decarbonylation takes place to give the four-membered metallacyle **17**, which reacts with norbornene to give the final product.

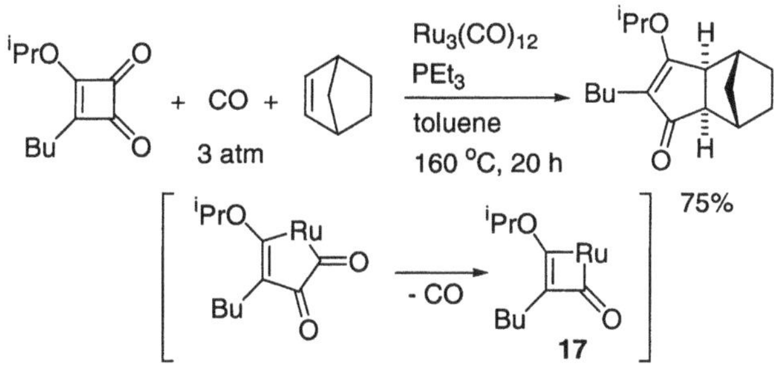

(19)

4 Carbonylative Cylizations

The reactions described in this section are not unique to ruthenium catalysis. These transformations can also be achieved using a palladium or a nickel catalyst. Since carbonylative cyclizations leading to cyclic carbonyl compounds are useful transformations in organic synthesis, these reactions are included in this section.

Because unsaturated lactones and lactams are of importance as biologically active compounds, the carbonylative cyclization of alkynyl alcohols, alkynyl amines, and their allenyl derivatives has been extensively studied using various transition-metal complexes. Takahashi et al. [35] reported that $Ru_3(CO)_{12}$ also catalyzes the cyclocarbonylation of allenyl alcohols to five- to eight-membered unsaturated lactones (Eq. 20).

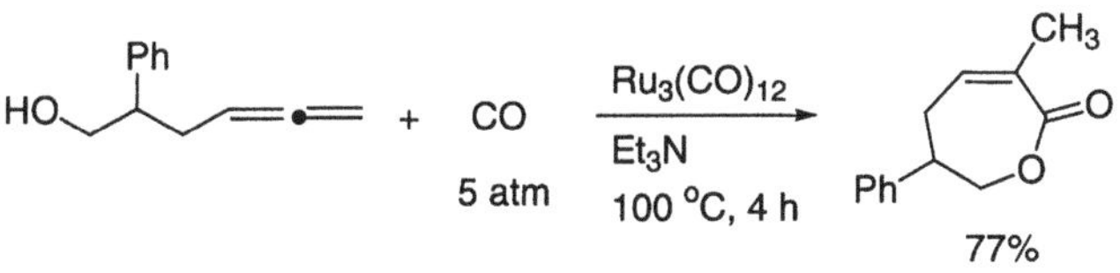

(20)

Kang et al. [36] found that allenyl sulfonamides also undergo cyclocarbonylation catalyzed by $Ru_3(CO)_{12}$ to give α,β-unsaturated γ-lactams (Eq. 21).

Hex, NHTs + CO (20 atm) — $Ru_3(CO)_{12}/Et_3N$, dioxane, 100 °C, 9 h → Me, Hex, N, Ts, O; 91% (21)

The three-component coupling reaction of allyl carbonates, CO, and norbornene leading to cyclopentenones with high exo selectivity was reported by Mitsudo et al. [37] (Eq. 22). Interestingly, $Ru_3(CO)_{12}$ was ineffective, but $[RuCl_2(CO)_2]_2/Et_3N$ was active. The nature of the amines used and the pressure of the CO had significant effects on the efficiency of the reaction. The fact that the same products were obtained from regioisomeric substrates indicates the intermediacy of a common intermediate, the π-allyl ruthenium complex.

Ph, OCO_2Me / OCO_2Me, Ph + CO (3 atm) + norbornene — $[RuCl_2(CO)_3]_2/Et_3N$, THF, 120 °C, 3 h → Ph, O; 72-79% (22)

5
Carbonylation Reactions at C–H Bonds

The activation of C–H bonds (functionalization of C–H bonds) catalyzed by transition-metal complexes has been a topic of interest in organic and organometallic chemistry [38]. This topic is treated in the chapter on Activation of inert C–H bonds. The direct catalytic formation of a C–C bond from C–H bonds has been extensively studied because it provides a promising synthetic approach to the waste-free construction of structurally diverse C–C skeletons. Recently, a series of a three-component coupling reaction of C–H bonds in *N*-heterocyclic componds, CO, and alkenes was found to be catalyzed by $Ru_3(CO)_{12}$. In all cases, the coordination of an sp^2 nitrogen to ruthenium is a key step for the reaction to proceed. The coordination is responsible both for the efficiency and for the site-selectivity of the reaction. Thus, the carbonylation reactions reported can be classified into four types, depending on the position where the carbonylation takes place: (1) α to an sp^2 nitrogen atom, (2) β to an sp^2 nitrogen atom, (3) γ to an sp^2 nitrogen atom, and (4) δ to an sp^2 nitrogen atom.

5.1
α-Carbonylation

In 1992, Moore et al. [39], in a pioneering study, reported that the reaction of pyridine, alkenes, and CO catalyzed by $Ru_3(CO)_{12}$ results in the selective cleavage of the C–H bond α to the pyridine nitrogen to give an acylated pyridine (Eq. 23). Although a variety of transtion-metal carbonyl complexes were examined for their ability to catalyze this new acylation reaction, only ruthenium carbonyl complexes showed catalytic activity. Monophosphine complex, $Ru_3(CO)_9(PPh_3)_3$ was less active. A trinuclear ruthenium cluster **18**, formed by the coordination of the pyridine nitrogen to the ruthenium catalyst followed by specific activation of a C–H bond next to the nitrogen, was proposed as the key catalytic species.

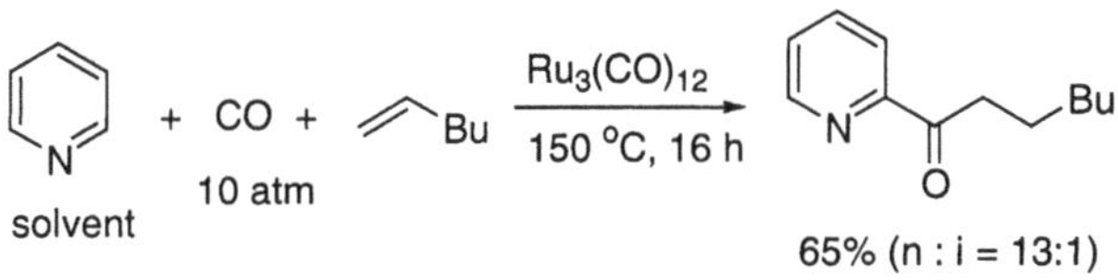

pyridine (solvent) + CO (10 atm) + ═╲Bu → [$Ru_3(CO)_{12}$, 150 °C, 16 h] 2-pyridyl–C(=O)–CH₂CH₂–Bu; 65% (n : i = 13:1) (23)

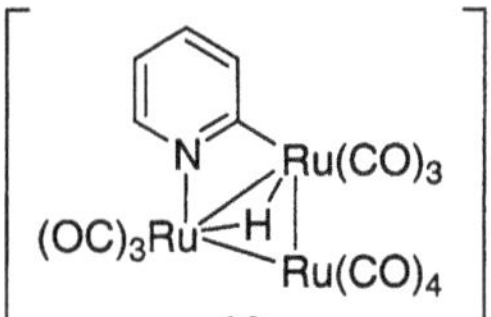

[N, $Ru(CO)_3$, $(OC)_3Ru$, H, $Ru(CO)_4$] **18**

Later, Murai and coworkers [40, 41] reported that the reaction is also applicable to five-membered *N*-heterocycles, such as imidazoles, thiazoles, oxazoles, and pyrazoles (Eq. 24). Functional group compatibility was extensively studied, and it was found that various functional groups, such as ketone, ester, cyano,

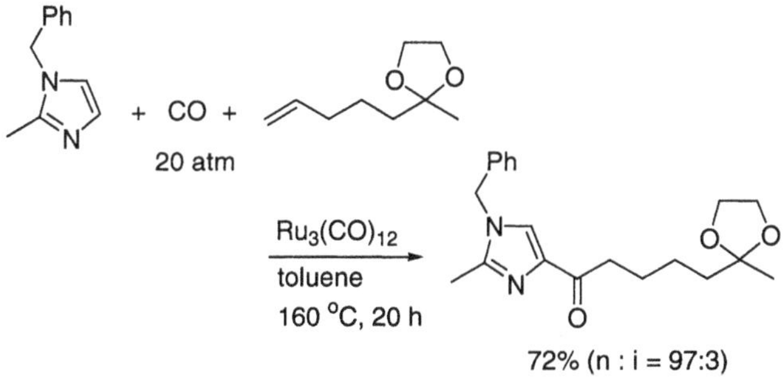

Ph, N, N + CO (20 atm) + O, O → [$Ru_3(CO)_{12}$, toluene, 160 °C, 20 h] Ph, N, N, O, O, O; 72% (n : i = 97:3) (24)

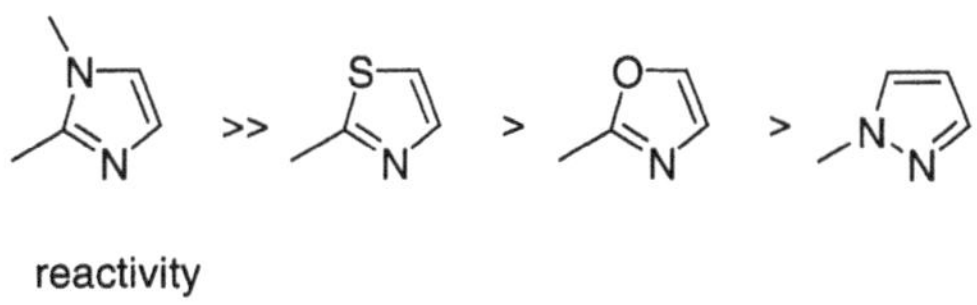

N, N >> S, N > O, N > N, N

reactivity

acetal, *N,O*-acetal, ketal, and silyl groups, were tolerated under the reaction conditions, indicating that C–H bond activation reactions have now reached a satisfactory level in organic synthesis. They observed that the reactivity of the substrates increased with increasing basicity of the *N*-heterocycle according to the series: imidazole>thiazole>oxazole>pyrazole [41]. This indicates that the coordination of the substrates by the sp^2 nitrogen to the ruthenium center is a key step in the carbonylation of C–H bonds in *N*-heterocycles. The substrates must compete with CO in order to coordinate with ruthenium. In fact, the reaction of pyrazole, which has a lower basicity, proceeded effectively only when the reaction was carried out under a lower CO pressure (3 atm 46%, 20 atm trace). This observation highlights the importance of the coordination of a nitrogen to ruthenium for the reaction to proceed.

5.2 *β*-Carbonylation

Carbonylation at a C–H bond *β* to the sp^2 ring nitrogen can also be achieved by a $Ru_3(CO)_{12}$ catalyst. The $Ru_3(CO)_{12}$-catalyzed reaction of 1,2-dimethylbenzimidazole with an alkene and CO provides the corresponding *β*-acylated product in high yield with complete site-selectivity [42] (Eq. 25). A trinuclear ruthenium cluster **19** is proposed as the key catalytic species. A similar basicity-dependent reactivity of substrates as described in the *α*-carbonylation was observed in the case of the carbonylation at C–H bond *β* to the sp^2 nitrogen.

+ CO (5 atm) + CH2=CHBu^t — $Ru_3(CO)_{12}$, toluene, 160 °C, 20 h → 77%

[**19** : Ru(CO)$_3$, H, Ru(CO)$_4$, Ru (CO)$_3$] (25)

reactivity: > >> > >

5.3
γ-Carbonylation

When the reaction of 2-phenylpyridine with CO (20 atm) and ethylene is conducted at 160 °C, the ortho C–H bond (γ to the sp^2 nitrogen) in the benzene ring undergoes carbonylation (Eq. 26) [43]. Carbonylation takes place selectively at a C–H bond γ to the sp^2 nitrogen (ortho C–H bond). C–H bonds in the pyridine ring and meta and para C–H bonds in the benzene ring are completely unreactive. In the reaction of meta-substituted substrates, carbonylation takes place exclusively at the less hindered C–H bond, irrespective of the electronic nature of the substituents, indicating that site-selectivity is determined by steric factors. A wide functional group compatibility was also observed. In sharp contrast to the α- and β-carbonylation described earlier, the reaction is restricted to ethylene as the alkene partner. Thus, the use of 1-hexene resulted in no reaction.

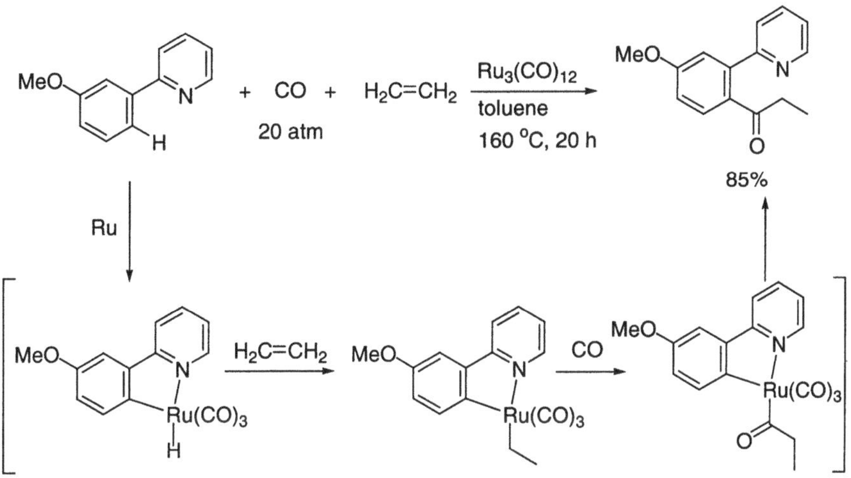

(26)

Some other directing groups which involve the sp^2 nitrogen can also function as a directing group in place of the pyridine ring. The reaction of aromatic imines with CO and ethylene in the presence of $Ru_3(CO)_{12}$ did not stop at the carbonylation step, but rather indenone derivatives were the final products and were formed via an intramolecular aldol-type reaction of the expected carbonylation products in situ (Eq. 27) [44]. The treatment of the reaction mixture with silica gel selectively afforded indenones in good yields.

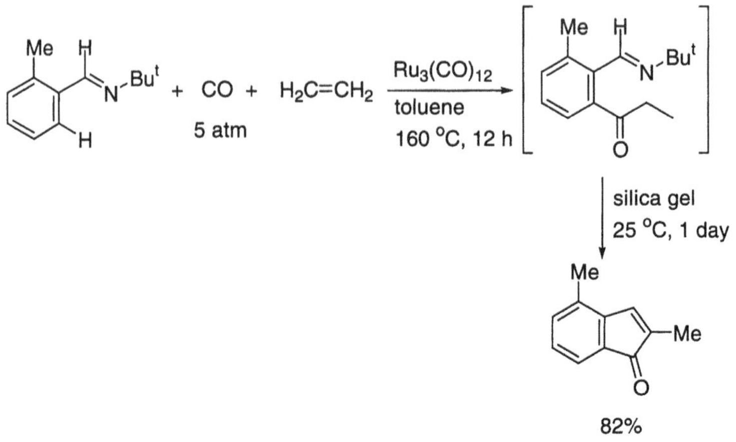

(27)

An oxazoline ring is also an effective directing group for the γ-carbonylation at the C–H bond in the benzene ring (Eq. 28) [45]. In contrast to a pyridine ring, the oxazoline serves as a suitable directing group for further useful transformations because it is readily converted to other functional groups, such as carboxylic acids, esters, and aldehydes.

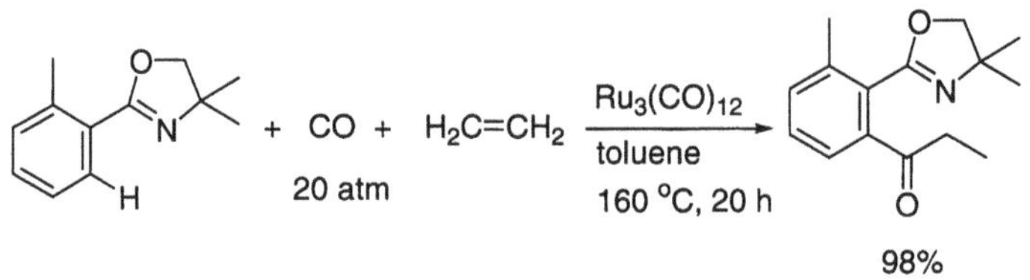

(28)

A pyrazole ring also serves as a directing group (Eq. 29) [46]. The reactivity of *N*-phenylpyrazole is much higher than expected on the basis of the basicity of the pyrazole.

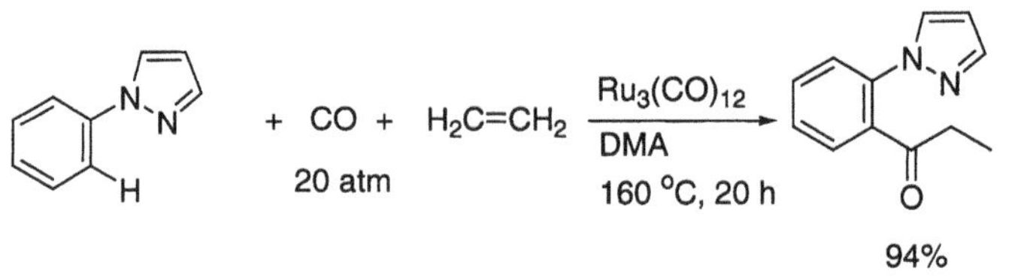

(29)

Imhof et al. [47] reported that the reaction of α,β-unsaturated imines with CO and alkenes in the presence of $Ru_3(CO)_{12}$ gives γ-lactam derivatives (Eq. 30). It was proposed that an aldehyde **19** formed by the direct carbonylation at the C–H bond in the 3-position is the key intermediate.

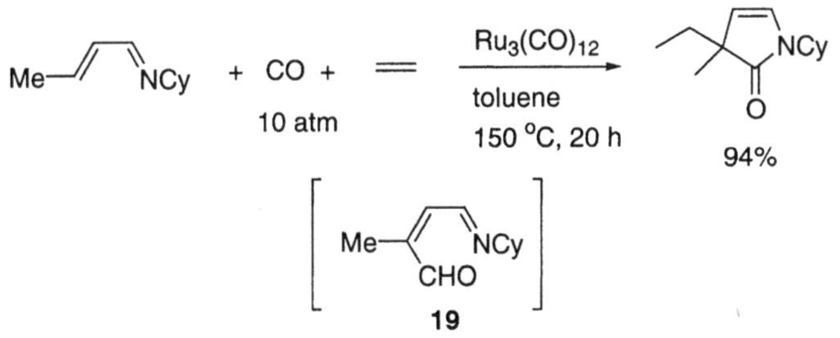

(30)

Chatani et al. [48] also reported that the $Ru_3(CO)_{12}$-catalyzed reaction of α,β-unsaturated imines with CO and ethylene results in a three-component coupling reaction to give unsaturated γ-lactams (Eq. 31). Unlike Imhof and coworkers, they proposed that the reaction proceeds via a two-step sequence involving the initial three-component coupling reaction at the olefinic C–H bonds, leading to **20**. In fact, the corresponding ethyl ketones were isolated in some cases.

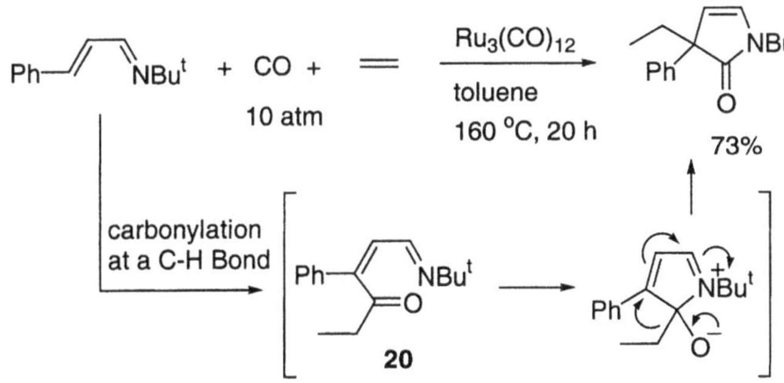

(31)

5.4 δ-Carbonylation

Chatani et al. [49] reported that the carbonylation of the C–H bond at the position δ to the sp^2 nitrogen also proceeds in the presence of a $Ru_3(CO)_{12}$ catalyst (Eq. 32). The reactivity seemed to be sensitive to the polarity of the solvent. The choice of DMA as the solvent is crucial for the reaction to proceed efficiently. The available substrates are extremely limited to an indoline skeleton.

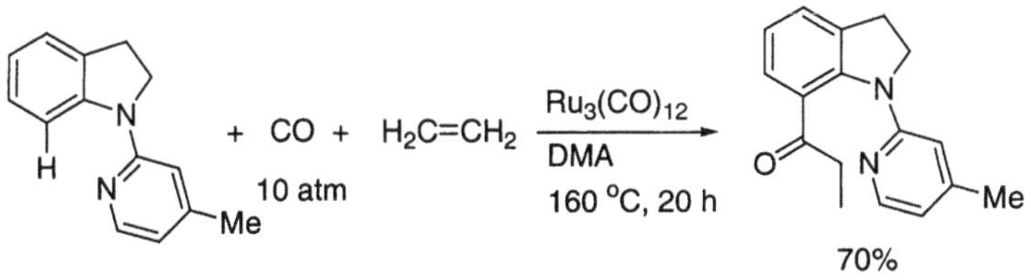

(32)

5.5 Combinatorial Chemistry in Carbonylation at C–H Bonds

Several high-throughput protocols have recently been reported for determining optimal reaction conditions and applicable substrates. Electrospray ionization mass spectrometry has drawn increasing attention for the analysis of combinatorial libraries. Recently, Ellman et al. [50] applied this method to exploit $Ru_3(CO)_{12}$-catalyzed carbonylation at C–H bonds in *N*-heterocycles. The high-throughput strategy for optimization of the carbonylation and the discovery of new products are shown in Scheme 7. A mixture consisting of aromatic *N*-heterocycles (33 different compounds) and *tert*-butylethylene was subjected to the carbonylation at C–H bonds catalyzed by $Ru_3(CO)_{12}$ (40 mol %) under CO (20 atm) at 160 °C. The reaction mixture was treated with a peptide label, $H_2NOGlyArg_4$, to give oxime derivatives, which were then analyzed by electrospray ionization mass spectrometry.

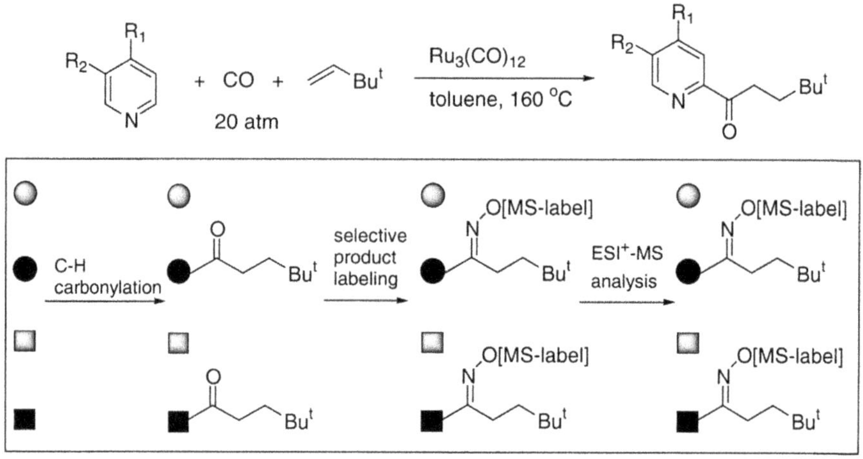

Scheme 7

6 Hydroformylation

The hydroformylation of alkenes to give linear aldehydes constitutes the most important homogeneously catalyzed process in industry today [51]. The hydroformylation of propene is especially important for the production of *n*-butyraldehyde, which is used as a starting material for the manufacture of butanol and 2-ethylhexanol. Catalysts based on cobalt and rhodium have been the most intensively studied for the hydroformylation of alkenes, because they are industrially important catalysts. While ruthenium complexes have also been reported to be active catalysts, ruthenium offers few advantages over cobalt or rhodium. A review on hydroformylation catalyzed by ruthenium complexes has appeared [52].

7 Reductive Carbonylation of Nitro Compounds

The reactions described previously involve carbonylation accompanied by C–C bond formation. In this section, the reactions in which C–C bond formation is not involved and CO functions as a reducing agent will be described. Reductive carbonylation of nitro compounds constitutes an intense field of research owing to the fact that industrially important compounds can be obtained in a single step (Eq. 33). Although a variety of transition-metal complexes show catalytic activity for the reductive carbonylation of nitro compounds, ruthenium complexes have been extensively studied. Among the most important products that can be obtained by this approach are isocyanates, ureas, and carbamates. Although the reaction mechanism is not clear, the probable intermediates in the reductive carbonylation of nitro compounds are the corresponding nitroso and isocyanates. Thus, in the absence of alcohols or H_2O, isocyanates are obtained. The presence of an alcohol leads to urethane formation, and the addition of amines to the system generates ureas. The ruthenium-catalyzed reductive carbonylation of organic nitro compounds has recently been reviewed [53].

$$PhNO_2 + CO \longrightarrow PhNCO \begin{cases} \xrightarrow{ROH} PhNHC(O)OR \\ \xrightarrow{RNH_2} PhNHC(O)NHR \end{cases} \tag{33}$$

8 Conclusion

Ruthenium is not an effective catalyst in many catalytic reactions; however, it is becoming one of the most novel and promising metals with respect to organic synthesis. The recent discovery of C–H bond activation reactions [38] and alkene metathesis reactions [54] catalyzed by ruthenium complexes has had a significant impact on organic chemistry as well as other chemically related fields, such as natural product synthesis, polymer science, and material sciences. Similarly, carbonylation reactions catalyzed by ruthenium complexes have also been extensively developed. Compared with other transition-metal-catalyzed carbonylation reactions, ruthenium complexes are known to catalyze a few carbonylation reactions, such as hydroformylation or the reductive carbonylation of nitro compounds. In the last 10 years, a number of new carbonylation reactions have been discovered, as described in this chapter. We expect the exploitation of a wide variety of reactions catalyzed by ruthenium complexes in the future.

References

1. (a) Falbe J (1989) (ed) New syntheses with carbon monoxide. Springer, Berlin Heidelberg New York; (b) Colquhoun HM, Thompson DJ, Twigg MV (1991) (eds) Carbonylation: direct synthesis of carbonyl compounds. Plenum, New York
2. (a) Schore NE (1988) Chem Rev 88:1081; (b) Lautens M, Klute W, Tam W (1996) Chem Rev 96:49; (c) Ojima I, Tzamarioudaki M, Ki Z, Donovan RJ (1996) Chem Rev 96:635
3. (a) Schore N (1991) Org React 40:1; (b) Chung YK (1999) Coord Chem Rev 188:297; (c) Brummond KM, Kent JL (2000) Tetrahedron 56:3263; (d) Gibson SE, Stevenazzi A (2003) Angew Chem Int Ed Engl 42:1800
4. Rautenstrauch V, Mégard P, Conesa J, Küster W (1990) Angew Chem Int Ed Engl 29:1413
5. Jeong N, Hwang SH, Lee Y, Chung YK (1994) J Am Chem Soc 116:3159
6. Hicks FA, Kalaoui NM, Buchwald SL (1996) J Am Chem Soc 118:9450
7. Koga Y, Kobayashi T, Narasaka K (1998) Chem Lett 249. Jeong N, Lee S, Sung BK (1998) Organometallics 17:3642
8. Shibata T, Takagi K (2000) J Am Chem Soc 122:9852
9. Morimoto T, Chatani N, Fukumoto Y, Murai S (1997) J Org Chem 62:3762
10. Kondo T, Suzuki N, Okada T, Mitsudo T (1997) J Am Chem Soc 119:6187
11. Itami K, Mitsudo K, Yoshida J (2002) Angew Chem Ind Ed Engl 41:3481
12. Crowe WE, Vu AT (1996) J Am Chem Soc 118:1557
13. (a) Kablaoui NM, Hicks FA, Buchwald SL (1996) J Am Chem Soc 118:5818; (b) Kablaoui NM, Hicks FA, Buchwald SL (1997) J Am Chem Soc 119:4424
14. Chatani N, Morimoto T, Fukumoto Y, Murai S (1998) J Am Chem Soc 120:5335
15. Chatani B, Morimoto T, Kamitani A, Fukumoto Y, Murai S (1999) J Organomet Chem 579:177
16. Kang SK, Kim KJ, Hong YT (2002) Angew Chem Int Ed Engl 41:1584
17. Mandal SK, Amin SR, Crowe WE (2001) J Am Chem Soc 123:6457
18. (a) Chatani N, Tobisu M, Asaumi T, Fukumoto Y, Murai S (1999) J Am Chem Soc 121:7160; (b) Tobisu M, Chatani N, Asaumi T, Amako K, Ie Y, Fukumoto Y, Murai S (2000) J Am Chem Soc 122:12663
19. van Wijnkoop M, de Lange PPM, Frühauf HW, Vrieze K, Smeets WJJ, Spek AL (1995) Organometallics 14:4781
20. Chatani N, Tobisu M, Asaumi T, Murai S (2000) Synthesis 925
21. (a) Göbel A, Imhof W (2001) Chem Commun 593; (b) Imhof W, Gobel A (2003) J Mol Catal A 197:15
22. Imhof W, Anders E, Göbel A, Görls H (2003) Chem Eur J 9:1166
23. Chatani N, Amako K, Tobisu M, Asaumi T, Murai S (2003) J Org Chem 68:1591
24. Murakami M, Itami K, Ito Y (1999) J Am Chem Soc 121:4130 and references cited therein
25. Sigman MS, Eaton BE (1996) J Am Chem Soc 118:11783
26. Sigman MS, Eaton BE, Heise JD, Kubiak CP (1996) Organometallics 15:2829
27. Boeckman RK Jr, Reed JE, Ge P (2001) Org Lett 3:3651
28. Morimoto T, Chatani N, Murai S (1999) J Am Chem Soc 121:1785
29. Kamitani A, Chatani N, Morimoto T, Murai S (2000) J Org Chem 65:9230
30. Pino P, Braca G, Sbrana G, Cuccuru A (1968) Chem Ind 1732
31. Chatani N, Fukumoto Y, Ida T, Murai S (1993) J Am Chem Soc 115:11614
32. Suzuki N, Kondo T, Mitsudo T (1998) Organometallics 17:766

33. Kondo T, Kaneko Y, Toguchi Y, Nakamura A, Okada T, Shiotsuki M, Ura Y, Wada K, Mitsudo T (2002) J Am Chem Soc 124:6824
34. Kondo T, Nakamura A, Okada T, Suzuki N, Wada K, Mitsudo T (2000) J Am Chem Soc 122:6319
35. (a) Yoneda E, Kaneko T, Zhang AW, Onitsuka K, Takahashi S (2000) Org Lett 2:441; (b) Yoneda E, Zhang SW, Onitsuka K, Takahashi S (2001) Tetrahedron Lett 42:5459
36. Kang SK, Kim KJ, Yu CM, Hwang JW, Do YK (2001) Org Lett 3:2851
37. Morisaki Y, Kondo T, Mitsudo T (2000) Org Lett 2:949
38. Kakiuchi F, Murai S (1999) In: Murai S (ed) Activation of unreactive bond. Topics in organometallic chemistry, vol 3. Springer, Berlin Heidelberg New York, pp 47–79
39. Moore EJ, Pretzer WR, O'Connell TJ, Harris J, LaBounty L, Chou R, Grimmer SS (1992) J Am Chem Soc 114:5888
40. Chatani N, Fukuyama T, Kakiuchi F, Murai S (1996) J Am Chem Soc 118:493
41. Chatani N, Fukuyama T, Tatamidani H, Kakiuchi F, Murai S (2000) J Org Chem 65:4039
42. Fukuyama T, Chatani N, Taysumi J, Kakiuchi F, Murai S (1998) J Am Chem Soc 120:11522
43. Chatani N, Ie Y, Kakiuchi F, Murai S (1997) J Org Chem 62:2604
44. Fukuyama T, Chatani N, Kakiuchi F, Murai S (1997) J Org Chem 62:5647
45. Ie Y, Chatani N, Ogo T, Marshall DR, Fukuyama T, Kakiuchi F, Murai S (2000) J Org Chem 65:1475
46. Asaumi T, Chatani N, Matsuo T, Kakiuchi F, Murai S (2003) J Org Chem 68:7538
47. (a) Berger D, Imhof W (1999) Chem Commun 1457; (b) Berger D, Imhof W (2000) Tetrahedron 56:2015; (c) Göbel A, Imhof W, (2001) Chem Commun 593; (d) Imhof W, Berger D, Kötteritzsch M, Rost M, Schönecker B (2001) Adv Synth Catal 343:795
48. Chatani N, Kamitani A, Murai S (2002) J Org Chem 67:7014
49. Chatani N, Yorimitsu S, Asaumi T, Kakiuchi F, Murai S (2002) J Org Chem 67:7557
50. Szewczk JW, Zuckerman RL, Bergman RG, Ellman JA (2001) Angew Chem Int Ed Engl 40:216
51. van Leeuwen PWNM, Claver C (2000) (eds) Rhodium catalyzed hydroformylation. Kluwer, Dordrecht
52. Kalck P, Peres Y, Jeck J (1991) Adv Organomet Chem 32:121
53. (a) Tafesh AM, Weiguny J (1996) Chem Rev 96:2035; (b) Cenini S, Ragaini F (1997) (eds) Catalytic reductive carbonylation of organic nitro compounds. Kluwer, Dordrecht
54. Fürstner A (1998) (ed) Alkene methasesis in organic synthesis. Topics in organometallic chemistry, vol 1. Springer, Berlin Heidelberg New York

Topics Organomet Chem (2004) 11: 197–248
DOI 10.1007/978-3-540-40027-1

Synthesis of Silicon Derivatives with Ruthenium Catalysts

Bogdan Marciniec (✉) · Cezary Pietraszuk

Department of Organometallic Chemistry, Faculty of Chemistry, Adam Mickiewicz University, Grunwaldzka 6, 60-780 Poznań, Poland
marcinb@main.amu.edu.pl

Abstract Methods for synthesis of saturated and unsaturated organosilicon compounds as well as other silicon derivatives based on ruthenium-catalyzed reactions have been reviewed. All the catalytic processes discussed lead to formation of either novel carbon–silicon bonds or novel carbon–carbon bonds affected by a substituted silyl group at carbon. Saturated organosilicon products can be prepared via hydrosilylation of alkenes and activation of the C–H bond of arenes, alkenes and triethylsilane. Linear unsaturated organosilicon compounds are usually prepared via hydrosilylation of alkynes, alkene self-metathesis and cross-metathesis and related metathetical transformations as well as silylative coupling of alkenes with vinylsilanes and coupling of arenes and alkenes with silylalkynes. On the other hand, cyclic unsaturated organosilicon compounds have been described as having been synthesized via intramolecular hydrosilylation of silylalkynes as well as ring-closing metathesis reactions and condensation of silicon-containing dienes (and enynes). Other silicon derivatives containing mostly Si–*X*–C bonds (where *X* is O or N) can be successfully prepared by ruthenium-catalyzed reactions of hydrosilylation, silylformylation, silylcarbonylation and dehydrocondensation of the respective initial silicon compounds. The final subchapter contains a brief overview of catalytic methods for synthesizing organosilicon polymers.

Keywords Homogeneous catalysis · Ruthenium complexes · Organosilicon compounds · Organic synthesis

1 Introduction

In contrast to organic chemistry, where numerous transition-metal-catalyzed reactions (e.g., olefin metathesis, olefin oxidation as in the Wacker process, hydroformylation, hydrogenation) have been developed in the last 50 years, in organosilicon chemistry only hydrosilylation has gained considerable attention and has become widely applied both in laboratory preparations and in industry [1–3]. However, in the last 20 years a number of transformations of silicon compounds have been revealed and some of them spectacularly developed. Among them are dehydrogenative silylation and double silylation of alkenes and alkynes by hydrosilanes, silylative coupling of alkenes with vinylsilanes, metathesis of silicon-containing alkenes, the coupling of C–H of alkenes and arenes with olefins including unsaturated organosilicon compounds, dehydrocoupling of hydrosilanes as well as silylformylation and silylcarbonylation of a variety of organic compounds [1–7]. These catalytic methods have been applied in various strategies for the synthesis of organosilicon compounds via either direct formation of carbon–silicon bonds or formal creation of novel carbon–carbon bonds affected by silyl groups substituted at carbon. Although the hydrosilylation of alkenes and alkynes predominantly uses platinum catalysts and silylcarbonylation uses cobalt group catalysts, most of these processes may also proceed effectively in the presence of ruthenium complexes as catalysts. The aim of this chapter is to describe the applications of ruthenium-catalyzed reactions in the synthesis of molecular organosilicon compounds and related silicon derivatives. No (or very limited) mechanistic implications are introduced. Processes resulting in the formation of macromolecular organosilicon products are briefly summarized in the final subchapter and are not discussed extensively.

2 Saturated Organosilicon Products

2.1 Hydrosilylation of Alkenes

Ruthenium complexes including carbonyl derivatives used in the hydrosilylation of alkenes exceptionally give regular saturated products, but predominantly lead to unsaturated silyl olefins, which are the products of dehydrogenative silylation [1–4].

Unlike trichlorosilane, dichlorosilane is very effective in ruthenium-complex-catalyzed addition to 1-alkenes (Eq. 1) [8].

$$Cl_2SiH_2 + CH_2{=}CH{-}R \xrightarrow{[Ru]} R{-}CH_2CH_2{-}SiHCl_2 \tag{1}$$

On the other hand, trialkoxysilanes undergo efficient addition to 1-alkenes in the presence of $RuCl_2(PPh_3)_3$ in air (Eq. 2) [9].

$$(R'O)_3SiH + CH_2{=}CH{-}R \xrightarrow{RuCl_2(PPh_3)_3} (R'O)_3Si{-}CH_2CH_2{-}R \tag{2}$$

Recently $Ru_3(CO)_{12}$ has been reported to be an effective catalyst for hydrosilylation of 1-octene [10] and of allyl chloride (Eq. 3) [11] by triethoxysilane. The latter process is of great importance for production of the main intermediate in manufacturing silane coupling agents.

$$(EtO)_3SiH + CH_2{=}CHCH_2{-}Cl \xrightarrow{Ru_3(CO)_{12}} (EtO)_3Si{-}CH_2CH_2CH_2{-}Cl \tag{3}$$

$Ru(CO)_3(PPh_3)_2$ appeared to be an effective catalysts for direct production of the most common silane coupling agents, i.e., 3-aminopropyltrialkoxysilane (Eq. 4) [12]

$$(RO)_3SiH + CH_2{=}CHCH_2{-}NH_2 \xrightarrow{Ru(CO)_3(PPh_3)_2} (RO)_3Si{-}CH_2CH_2CH_2{-}NH_2 \tag{4}$$

2.2 Activation of the C–H Bond

2.2.1 Dehydrocoupling of Triethylsilane

A quite unusual functionalization of the C–H bonds catalyzed by (*p*-cymene)$Ru(H)_2(SiEt_3)_2$ or [(*p*-cymene)$RuCl_2]_2$ was found by Berry group (Eq. 5) [13]. The complex catalyzes the dehydrocoupling of Et_3SiH to a carbosilane dimer in the presence of hydrogen acceptors such as alkenes

$$2\,Et_3SiH + R'CH{=}CH_2 \xrightarrow{[Ru]} HSiEt_2CHMeSiEt_3 + R'CH_2CH_3 \tag{5}$$

The crucial step of this new example of Si–C formation is a β-hydrogen elimination from a silyl ligand followed by an insertion of η^2-silene coordinated to the metal in the Ru–Si bond (Eq. 6) [14]. Only traces of hydrosilylation and dehydrogenative silylation reactions are observed.

$$[Ru](SiR_2CH_3)(SiR_3) \xrightarrow{\beta\text{-H elimination}} [Ru](H)(SiR_3)(\eta^2\text{-}R_2Si{=}CH_2) \longrightarrow [Ru](H)SiR_2CH_2SiR_3 \tag{6}$$

2.2.2 Dehydrogenative Silylation of Arenes with Trialkylsilanes

A very attractive method for the dehydrogenative silylation of arenes via the C–H bond cleavage has been recently reported by Kakiuchi et al. [15]. The reaction of aryloxazolines with trialkylsilanes (mostly triethylsilane) catalyzed by Ru complexes results in the formation of ortho-silylated aryloxazolines in good-to-excellent yields (Eq. 7).

$$\text{R-aryl-oxazoline (4,4-dimethyl)} + Et_3SiH + CH_2{=}CH{}^tBu \xrightarrow{[Ru]} \text{R-(2-}SiEt_3\text{)aryl-oxazoline} + CH_3CH_2{}^tBu \tag{7}$$

Yield [%] (time)

R	$Ru_3(CO)_{12}$	$RuH_2(CO)(PPh_3)_3$
2-F	85 (48 h)	0 (72 h)
3-NMe_2	25 (48 h)	97 (60 h)
3-OMe	33 (20 h)	97 (65 h)
3-Me	39 (20 h)	97 (65 h)
3-F	12 (48 h)	8 (72 h)

This is the first example of the direct silylation at the C–H bond using a hydrosilane as the source of the silyl group. Of key importance in this efficient process is the use of an olefin as the scavenger of the two hydrogen atoms, which effectively concludes the catalytic cycle.

2.2.3
Coupling of Arenes and Cycloaalkenes with Vinylsilanes

In 1993 Murai et al. [16] reported a rare example of the catalytic C–C bond formation between aromatic ketones or acyl-substituted heteroaromatics and cycloalkenes via activation of the C–H bond in the ortho position to the ketone functionality. These reactions are described in the chapter "Activation of Inert C–H Bonds".

2.2.4
Cyclopropanation

Terminal alkenes and cycloalkenes have been found to react in the presence of the polymeric ruthenium complex $[Ru_2(CO)_4(\mu\text{-OAc})_2]_n$ with methyl diazo(trimethylsilyl)acetate (Eq. 8) [17].

$$R^1CH{=}CHR^2 + N_2{=}C(SiMe_3)(COOMe) \xrightarrow[-N_2,\ CH_2Cl_2]{[Ru_2(CO)_4(\mu\text{-OAc})_2]_n} \text{cyclopropane (}SiMe_3\text{, }COOMe\text{)} + \text{cyclopropane (}COOMe\text{, }SiMe_3\text{)} \quad (8)$$

R^1	R^2	Yield [%]	E/Z (*anti/syn*)
H	H	70	1.8/1
H	H	89	3.4/1
$-(CH_2)_4-$		54	7/1

The same complex is a suitable catalyst for the cyclopropanation of 1,1- and 1,2-disubstituted alkenes with trimethylsilyldiazomethane (Eq. 9) [18]. High exo selectivity was obtained when cyclohexene was used as an olefin.

$$R^1R^3C{=}CHR^2 + N_2{=}CHSiMe_3 \xrightarrow[-N_2,\ CH_2Cl_2]{[Ru_2(CO)_4(\mu\text{-OAc})_2]_n} \text{cyclopropanes (}SiMe_3\text{ anti/syn)} + Me_3SiCH{=}CHSiMe_3 \quad (9)$$

R^1	R^2	R^3	Yield [%]	E/Z (*anti/syn*)
Ph	H	H	64	7/3
EtO	H	H	68	6/4
Ph	H	Me	61	6/4
$-(CH_2)_4-$		H	48	9/1

3 Linear Unsaturated Organosilicon Compounds

Alkenylsilanes, mainly vinyl silanes and allyl silanes or related compounds, being widely used intermediates for organic synthesis can be efficiently prepared by several reactions catalyzed by transition-metal complexes, such as dehydrogenative silylation of alkenes, hydrosilylation of alkynes, alkene metathesis, silylative coupling of alkenes with vinylsilanes, and coupling of alkynes with vinylsilanes [1–7]. Ruthenium complexes have been used for chemoselective, regioselective and stereoselective syntheses of unsaturated products.

3.1 Dehydrogenative Silylation of Alkenes

As we have already mentioned, ruthenium complexes predominantly catalyze the dehydrogenative silylation of alkenes but competitively with the hydrosilylation so the reaction usually gives a mixture of the dehydrogenative silylation and hydrosilylation products. $Ru_3(CO)_{12}$ appears to be a very active catalyst for the dehydrogenative silylation of styrene, para-substituted styrenes [19, 20], trifluoropropene and pentafluorostyrene [21] by trialkyl-, phenyldialkylsilanes (but also triethoxysilane) (Eq. 10).

$$HSiR_3 + 2\ \text{=}\!\!\diagup^{R'} \xrightarrow{Ru_3(CO)_{12}} R'\text{-CH=CH-}SiR_3 + \text{—}\!\!\diagup^{R'} \qquad (10)$$

Alkenes having a hydrogen atom at the allylic position (1-hexene, allylbenzene, 3-phenoxyprop-1-ene) form mixtures of vinylsilanes and allylsilanes [20].

A highly selective dehydrogenative silylation of ethylene proceeds in the presence of $RuH_2(H_2)_2(PCy_3)_2$, where Cy is cyclohexyl, as a catalyst precursor to yield vinylsilane under very mild conditions (Eq. 11). The formation of vinylsilane is promoted by high olefin-to-silane ratios [22].

$$\text{=} + HSiR_3 \xrightarrow[\substack{20\ \text{bar}\ C_2H_4 \\ 17^{o}C,\ 15\ \text{min.}}]{RuH_2(H_2)_2(PCy_3)_2} \text{=}\!\!\diagup^{SiR_3} + \text{—}\!\!\diagup^{SiR_3} \qquad (11)$$

Vinylchlorodimethylsilane has been recently reported as a product of dehydrogenative silylation of ethylene by chlorodimethylsilane with the same catalyst precursor. The product is obtained with 46–92% selectivity of unsaturated-to-saturated (ethyldimethylchlorosilane) products [23].

3.2 Hydrosilylation of Alkynes

Hydrosilylation of 1-alkynes catalyzed by transition-metal complexes is still the most versatile method for preparing 1-alkenylsilanes but the reaction often gives a mixture of three possible isomers (Eq. 12)

$$R'—\equiv \; + \; HSiR_3 \xrightarrow{\text{catalyst}} \underset{E}{R'CH{=}CHSiR_3} \; + \; \underset{Z}{R'CH{=}CHSiR_3} \; + \; \underset{\alpha}{CH_2{=}C(R')SiR_3} \quad (12)$$

Ruthenium complexes are known to be generally less reactive in hydrosilylation reactions when compared with platinum and rhodium ones. However, very selective ruthenium-based catalytic systems have been recently developed. The hydrosilylation of terminal alkynes generally tends to proceed through cis addition, resulting in trans adducts as the main products.

$RuHCl(CO)(P\textit{i}\text{-}Pr_3)_2$ was found to be the first highly efficient catalyst for the hydrosilylation of phenylacetylene with $HSiEt_3$ in $ClCH_2CH_2Cl$ (60 °C), yielding (*Z*)-$PhCH{=}CHSiEt_3$ with 100% selectivity [24]. Other ruthenium complexes, for example, $RuH(XY)(CO)(P\textit{i}\text{-}Pr_3)_2$, where *XY*=Cl, acetate, acetylacetate [25], $[RuCl_2(p\text{-cymene})]_2$ [26] and $Ru(SiMe_2Ph)Cl(CO)(P\textit{i}\text{-}Pr_3)_2$ [27] have been reported subsequently for the stereoselective production of *Z*-styrylsilanes and other Z products (Eq. 13, Table 1).

$$R'—\equiv \; + \; HSiR_3 \xrightarrow[25^\circ C]{[Ru]} R'CH{=}CHSiR_3 \quad (13)$$

Table 1

R'	Si*R*$_3$	Yield (%)	Z/E	[Ru]	Ref.
Ph	$SiEt_3$	100	Z	$RuHCl(CO)(P\textit{i}\text{-}Pr_3)_2$	[25]
Ph	$SiEt_3$	100	97/2	$RuH(acetate)(CO)(P\textit{i}\text{-}Pr_3)_2$	[25]
Ph	$SiEt_3$	100	95/3	$RuH(acac)(P\textit{i}\text{-}Pr_3)_2$	[25]
Ph	$SiEt_3$	94	96/4	$[RuCl_2(p\text{-cymene})]_2$	[26]
n-Bu	$SiPh_3$	98	98/2	$[RuCl_2(p\text{-cymene})]_2$	[26]
4-Me-C_6H_4	$SiPh_3$	83	95/5	$[RuCl_2(p\text{-cymene})]_2$	[26]
$Cl(CH_2)_3$	$SiPh_3$	87	96/4	$[RuCl_2(p\text{-cymene})]_2$	[26]
$PhCH_2OCH_2$	$SiPh_3$	78	98/2	$[RuCl_2(p\text{-cymene})]_2$	[26]
$PhCH_2O(CO)(CH_2)_2$	$SiPh_3$	85	96/4	$[RuCl_2(p\text{-cymene})]_2$	[26]
Ph	$Si(OEt)_3$	88	96/4	$RuCl(CO)(SiMe_2Ph)(P\textit{i}\text{-}Pr_3)_2$	[27]
p-Tolyl	$SiMe_2Ph$	93	99/1	$RuCl(CO)(SiMe_2Ph)(P\textit{i}\text{-}Pr_3)_2$	[27]
Cyclohexyl	$SiMe_2Ph$	80	95/5	$RuCl(CO)(SiMe_2Ph)(P\textit{i}\text{-}Pr_3)_2$	[27]
n-Hexyl	$SiMe_2Ph$	98	91/9	$RuCl(CO)(SiMe_2Ph)(P\textit{i}\text{-}Pr_3)_2$	[27]

Table 2

R'	SiR_3	Yield (%)	E/Z
Ph	$SiMe_2Ph$	94	99/1
Ph	$SiMe_2Ph$	100	>99/1[a]
Ph	$Si(OEt)_3$	96	95/5
Ph	$SiMe_2[C_6H_3\text{-}3,5\text{-}(CF_3)_2]$	98	>99/1
p-Tolyl	$SiMe_2[C_6H_3\text{-}3,5\text{-}(CF_3)_2]$	100	>99/1
Cyclohexyl	$SiMe_2[C_6H_3\text{-}3,5\text{-}(CF_3)_2]$	94	99/1

[a] Five-fold excess of $HSiR_3$ relative to alkyne.

In contrast to the silyl–ruthenium complexes including triisopropylphosphine, the ruthenium hydride complex with triphenylphosphine directs the stereoselectivity toward the E product according to Eq. (14) (Table 2) [27].

$$R'\text{—}\equiv \;+\; HSiR_3 \xrightarrow[25^{\circ}C]{RuHCl(CO)(PPh_3)_3} R'\text{CH=CH}SiR_3 \tag{14}$$

It has been also found that the platinum–ruthenium cluster complex $[Pt_3Ru_6(CO)_{20}(\mu_3\text{-}PhC_2Ph)(\mu_3\text{-}H)(\mu\text{-}H)]$ is an effective catalyst precursor for the highly selective catalytic hydrosilylation of diphenylacetylene with triethoxysilane (Eq. 15). The true catalyst is actually the decarbonylated species.

$$Ph\text{—}\equiv\text{—}Ph \;+\; HSiEt_3 \xrightarrow[30^{\circ}C,\ 2\ h]{Pt\text{-}Ru\ cluster} (Ph)(H)C\text{=}C(Ph)(SiEt_3) \;+\; PhCH\text{=}CHPh \tag{15}$$

yield = 60%, TOF = 29 h^{-1}, 100% E; yield = 6%

Most transformations occur at the ruthenium atoms but the high activity is due to some form of synergetic enhancement by platinum because this activity is not observed in the absence of this metal [28].

The rigorous stereocontrol to yield either (*E*)-1-silyl-1-alkene or (*Z*)-1-silyl-1-alkene has been achieved using ruthenium complexes suppressing the formation of the internal adduct in both cases. On the other hand, recent reports also describe the selective formation of the internal adduct in the ruthenium-catalyzed hydrosilylation of alkynes either by the necessary functional group directed addition of trialkyl and trialkoxysilanes [26, 27] or by a more general Markovnikov-type one [29]. If alkynes having a hydroxyl group at the β-position to the triple bond are employed as a substrate, then α-vinylsilanes are generated with excellent selectivity (Eq. 16, Table 3) [26].

$$R'\text{—}\equiv \;+\; HSiR_3 \xrightarrow[CH_2Cl_2,\ 45^{\circ}C,\ 3\ h]{[RuCl_2(p\text{-}cymene)]_2} \underset{\alpha}{CH_2\text{=}C(R')SiR_3} \;+\; \underset{\beta(Z)}{R'CH\text{=}CHSiR_3} \tag{16}$$

Table 3

R'	Yield (%)	α/β(Z)
$HOCH_2CH_2$	47	98/2
$PhCH_2OCH_2CH_2$	89	>99/1
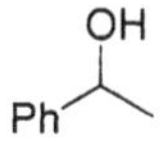 OH, Ph	60	87/13
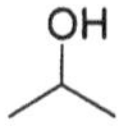 OH	59	98/2
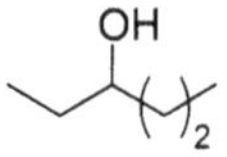 OH, 2	53	92/8
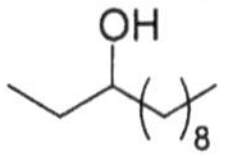 OH, 8	86	98/2

The cationic ruthenium complex $[CpRu(MeCN)_3]^+(PF_6)^-$ catalyses the hydrosilylation of alkyne with triethylsilane in good yield and with very limited formation of linear 1,2-disubstituted products (Eq. 17, Table 4).

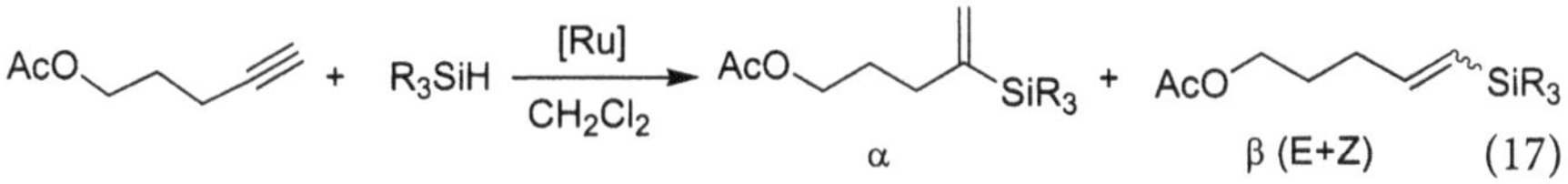

AcO + R_3SiH $\xrightarrow[CH_2Cl_2]{[Ru]}$ AcO SiR_3 (α) + AcO SiR_3 (β (E+Z)) (17)

Table 4

[Ru]	Si*R*$_3$	Yield (%)	α/β
$[CpRu(MeCN)_3]^+(PF_6)^-$	$SiEt_3$	89	20/1
$[CpRu(MeCN)_3]^+(PF_6)^-$	$SiMe(OEt)_2$	85	6/1
$[Cp^*Ru(MeCN)_3]^+(PF_6)^-$	$SiMe(OEt)_2$	88	13/1

The hydrosilylation with diethoxymethylsilane also gives the α-product but with a lower selectivity. The use of a more sterically demanding catalyst involving a pentamethylcyclopentadienyl (Cp*) ligand has improved the selectivity to 13:1 [29].

Recent reports on the ruthenium(IV) complex also including the electron-donating Cp* ligand $[Cp^*RuH_3(PPh_3)]$ as a catalyst have shown that the hydrosilylation of 1-alkynes by methyldichlorosilane proceeds with novel regioselectivity to afford preferentially internal (α) adducts (Eq. 18) [30].

$$\text{R}{-}{\equiv}\text{X} + \text{Cl}_2\text{MeSiH} \xrightarrow{\text{Cp*RuH}_3(\text{PPh}_3)} \underset{\alpha}{\text{CH}_2{=}\text{C(R)SiCl}_2\text{Me}} + \underset{\beta}{\text{RCH}{=}\text{CHSiCl}_2\text{Me}}$$

yield ($\alpha + \beta$) = 68-89% $\alpha/\beta > 99/1 \sim 89/11$

R = n-C_5H_{11}, n-C_6H_{13}, $AcO(CH_2)_3$, $AcO(CH_2)_2$, $PhOCH_2$ (18)

The latter complex catalyzes the hydrosilylation of a wide variety of alkynes under mild conditions with good functional group tolerance and complete selectivity for the unusual trans addition. Trost et al. [31] have coupled the process with protodesilylation to achieve a two-step trans reduction of alkynes (Eq. 19, Table 5).

$$\text{R}^1{-}{\equiv}{-}\text{R}^2 + (\text{EtO})_3\text{SiH} \xrightarrow[\text{CH}_2\text{Cl}_2,\ 0^\circ\text{C - r.t.}]{[\text{Cp*Ru(MeCN)}_3]^+(\text{PF}_6)^-} \left[\text{R}^1\text{CH}{=}\text{C}(\text{Si(OEt)}_3)\text{R}^2 + \text{R}^1\text{C}(\text{Si(OEt)}_3){=}\text{CHR}^2\right] \xrightarrow[\text{THF}]{\text{CuI, TBAF}} \text{R}^1\text{CH}{=}\text{CHR}^2 \quad (19)$$

The hydrosilylation of internal alkynes with triethoxysilane proceeds clearly to afford vinylsilane as a mixture of (Z) regioisomers and is followed by protodesilylation. The latter cuprous iodide initiated process allows a neat *trans*-alkyne reduction of a wide variety of alkynes. In the experiment, the interme-

Table 5

Alkyne	Product	Yield (%)
$CH_3(CH_2)_8C{\equiv}CCH_2CH_2C(O)CH_3$	$CH_3(CH_2)_8CH{=}CHCH_2CH_2C(O)CH_3$	96
$Cl(CH_2)_3C{\equiv}C(CH_2)_3CH_3$	$Cl(CH_2)_3CH{=}CH(CH_2)_3CH_3$	59
$PhCH(OH)C{\equiv}CCH_3$	$PhCH(OH)CH{=}CHCH_3$	76
$(C_2H_5O)_2CHC{\equiv}CCH_3$	$(C_2H_5O)_2CHCH{=}CHCH_3$	67

diate vinylsilanes were not isolated but were filtered to remove the ruthenium catalyst and then concentrated under reduced pressure to remove excess silane [31].

Catalytic hydrosilylation of enyne (buten-3-yne) with three kinds of hydrosilanes ($HSiMeH_2$, $HSiMe_2Ph$ and $HSiEt_3$) in the presence of $RuHCl(CO)(PPh_3)_3$ has given five types of reaction products (Eq. 20), the isolation has not been achieved.

$$\mathbf{1} + HSiR_3\ (\mathbf{2}) \xrightarrow[CDCl_3,\ 30^\circ C]{RuHCl(CO)(PPh_3)_3} \mathbf{3} + \mathbf{4} + \mathbf{5} + \mathbf{6} + \mathbf{7} \tag{20}$$

Dienylsilanes **3** and **6** are the cis and trans addition products of hydrosilanes across the triple bond of **1**, respectively. Allylsilane is the product formed by hydrogenation followed by hydrosilylation of **1**. Alkenylsilane **5** is the 1,4-adduct of hydrosilane across the enyne skeleton of **1**, and bis(silyl)butenyne **7** is the dehydrogenative silylation product of **1** [32].

3.3 Silylative Coupling (trans-Silylation) of Alkenes with Vinylsilanes

In 1984, the first very effective example of metathesis (disproportionation) of vinylsubstituted silicon compounds catalyzed by ruthenium complexes was reported [33]. It opened a new route of great synthetic importance and has allowed synthesis of a series of unsaturated silicon compounds according to the following equations, with the yield predominantly higher than 70%. Numerous reports on vinylsilane disproportionation (Eq. 21) [5, 33–38] and its co-disproportionation with olefins (Eq. 22) [37–44] have been published.

$$2\,R_3SiCH{=}CH_2 \xrightarrow[-\ CH_2{=}CH_2]{[Ru]} R_3SiCH{=}CHSiR_3 + (R_3Si)_2C{=}CH_2 \tag{21}$$

R = Me, Ph, OR'

$$R_3SiCH{=}CH_2 + CH_2{=}CH(CH_2)_nR' \xrightarrow[-\ CH_2{=}CH_2]{[Ru]} R_3SiCH{=}CH(CH_2)_nR' + CH_2{=}C(SiR_3)(CH_2)_nR' \tag{22}$$

R = Me, Ph, OR"
R' = Me, n = 3-15; R' = Ph, n = 0

The following ruthenium complexes appeared to be active in this reaction: $RuCl_2(PPh_3)_3$, $RuHCl(PPh_3)_3$, $[RuCl_2(CO)_3]_2$, $Ru_3(CO)_{12}$, $[RuCl_2(p\text{-cymene})]_2$, $RuCl_3$ with $HSiEt_3$, $HSi(OEt)_3$, $HSiPh_3$ as well as $LiAlH_4$ used as co-catalysts.

All experimental results as well as the report of Seki et al. [45] on catalysis of disproportionation of vinylsilanes by $Ru_3(CO)_{12}/HSiPh_3$ and $RuHCl(CO)(PPh_3)_3$ have provided convincing evidence supporting the thesis that formation of the Ru–H bond is a crucial stage in the initiation of catalytically active species.

In all reactions of vinyl-substituted silicon compounds performed in the presence of precursors containing no Ru–H (or Ru–Si) bond, the generation of catalytically active species can occur as follows [36].

$$\underset{[Ru]}{\overset{|}{H_2C{=}CHSiR_3}} \rightleftharpoons \underset{[Ru]-H}{\overset{|}{HC{=}CHSiR_3}} \longrightarrow \underset{[Ru]}{\overset{||}{CH{-}CH_2SiR_3}} \tag{23}$$

However, at that time the results obtained did not permit distinction between the reaction mechanism involving ruthenium carbene intermediates, being classical catalysts of the metathesis, and the non-metallacarbene mechanism.

Evidence for the migratory insertion of ethylene [46] and vinylsilane [47] into the Ru–Si bond yielding vinylsilane and two bis(silyl)ethene regioisomers [*E*-1,2-bis(silyl)ethene and 1,1-bis(silyl)ethene], respectively, has proved that in the reaction referred to as the "metathesis" of vinylsilanes and their "co-metathesis" with olefins, instead of the C=C bond cleavage formally characterizing alkene metathesis (Eq. 24a), a new type of olefin conversion that is a silylative coupling of olefins with vinylsilanes occurs (Eq. 24b).

$$\text{a)}\quad CH_2{=}CHSiR_3 + CH_2{=}CHR' \xrightleftharpoons{\text{M(carbene)}} CH_2{=}CH_2 + R'CH{=}CHSiR_3\ (E{+}Z)$$

$$\text{b)}\quad CH_2{=}CHSiR_3 + H_2C{=}CHR' \xrightleftharpoons{\text{[Ru]-H or [Ru]-SiR}_3} CH_2{=}CH_2 + R'CH{=}CHSiR_3 + CH_2{=}C(R')SiR_3 \tag{24}$$

A mechanistic scheme of this new type of silyl olefin conversion involves the migratory insertion of the olefin into the Ru–Si bond and vinylsilane into the *M*–H bond followed by β-hydrogen and β-silicon elimination to give 1,2-bis(silyl)ethenes, 1,1-bis(silyl)ethenes and ethylene (Scheme 1) [46, 47].

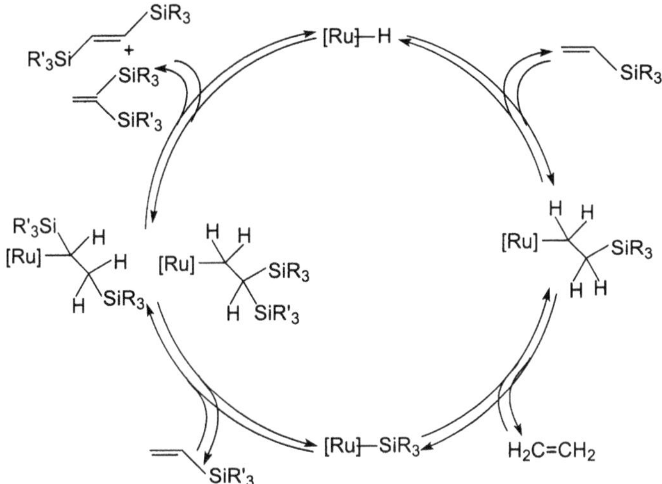

Scheme 1

The most synthetically effective results of homocoupling of vinyltrisubstituted silanes were compiled (Eq. 25, Table 6).

$$CH_2{=}CHSiR_3 \xrightarrow[-\ C_2H_4]{[Ru]} \underset{\mathbf{1}}{R_3SiCH{=}CHSiR_3} + \underset{\mathbf{2}}{CH_2{=}C(SiR_3)_2} \tag{25}$$

As we have already mentioned, substituted vinylsilanes $R'CH{=}CHSiR_3$ and $R'(SiR_3)C{=}CH_2$ constitute a very important class of organosilicon reagents. Synthetic and catalytic studies of the heterocoupling of 1-alkenes and para-substituted styrenes with vinylsilanes have led to new methods for regioselec-

Table 6

SiR_3	Catalyst or catalytic system	Conversion (Yield **1**+**2**) (%)	Reference
$SiMe_3$	$Ru_3(CO)_{12}/HSiPh_3$	Up to 75	[45]
	Various Ru complexes	50–70	[36, 38]
	$Ru(SiMe_3)Cl(CO)(PPh_3)_2$	80	[47]
$Si(OMe)_3$	$RuCl_2(PPh_3)_3$	(58)	[33]
$Si(OEt)_3$	$RuCl_2(PPh_3)_3$	(82)	[44]
	$Ru(SiMe_3)Cl(CO)(PPh_3)_2$	90 (85)	[47]
$SiMe_2(OEt)$	$RuHCl(CO)(PPh_3)_3$	(49)	[46]
$Si(Oi\text{-}Pr)_3$	$RuCl_3$(hydrate)/$HSi(OEt)_3$	(45)	[44]
$SiMe(OSiMe_3)_2$	Various Ru complexes	20–92 (16–88)	[49]
$SiMe_2Ph$	Various Ru complexes	(50–90)	[35]
	$Ru(SiMe_3)Cl(CO)(PPh_3)_2$	92 (87)	[47]

tive preparation of 1-silyl-1-alkenes and stereoselective and regioselective synthesis of 1-phenyl-2-ethenes. The most synthetically efficient data have been compiled (Eqs. 26, 27, Tables 7, 8).

$$CH_2{=}CHR' + CH_2{=}CHSiR_3 \xrightarrow[-\,CH_2{=}CH_2]{[Ru]} R'CH{=}CHSiR_3 + R_3SiCH{=}CHSiR_3 \tag{26}$$

$$X{-}C_6H_4{-}CH{=}CH_2 + CH_2{=}CHSiR_3 \xrightarrow[-\,CH_2{=}CH_2]{[Ru]} X{-}C_6H_4{-}CH{=}CH{-}SiR_3 \tag{27}$$

Many ruthenium complexes have been tested in the silylative coupling reaction. In the synthetic procedure the absence of by-products of the homocoupling of vinylsilanes is required so an excess of the olefin has usually been used. However, the screening tests performed at the 1:1 ratio of styrene and phenyldimethylvinylsilane with a variety of ruthenium catalysts have shown that pentacoordinated monocarbonyl bisphosphine complexes appear to be the most active and selective catalysts of which $RuHCl(CO)(PCy_3)_2$ has shown high catalytic activity under conditions of catalyst loadings as low as 0.05 mol % [55]. Cuprous salts (chloride, bromide) have recently been reported to be very successful co-catalysts of ruthenium phosphine complexes, markedly increasing the rate and selectivities of all ruthenium phosphine complexes [54].

A series of 1-silyl-2-N(O or B)-substituted ethenes (with high preference of isolated E isomers) were synthesized in the presence of ruthenium complexes (Eq. 28) [51–53] (B. Marciniec, M. Jankowska, M. Zaidlewicz, J. Cytarska, unpublished results). The reaction opens a general synthetic route for 1-silyl-2-heteroatom-substituted ethenes. The compounds of this kind, for example, β-alkoxysubstituted vinylsilanes are difficult to synthesize by other transition-metal-catalyzed reactions.

Table 7

SiR_3	R′ (C_nH_{2n+1})	[Ru]	Yield (%)	E/Z (E/self)	Reference
$SiMe_3$	n=1	$RuHCl(CO)(PPh_3)_3$	74	76/24	[46]
	n=4–8, 16	$RuCl_2(PPh_3)_3$	46–60	(45/1)	[43]
$Si(OEt)_3$	n=8	$RuCl_2(PPh_3)_3$	81	(70/11)	[42]
$SiMePh_2$	n=8	$RuCl_2(PPh_3)_3$	70	50/1	[40]
$SiMe_2Ph$	n=5–8, 10, 12, 16	$RuCl_2(PPh_3)_3$	55–75	20–50/1	[40]
	n=8	$[RuCl_2(CO)_3]_2$	85		[40]
	n=8	$Ru(acac)_3$	50		[40]

Table 8

SiR_3	X	[Ru]	Yield (%)	Reference
$SiMe_3$	H	Various Ru complexes[a]	58–90	[48]
$SiMe_2OEt$	H	$RuHCl(CO)(PPh_3)_3$	72	[46]
$SiMe_2Ph$	H	Various Ru complexes[a]	56–75	[48]
$Si(OEt)_3$	H	Various Ru complexes[a]	60–75	[48]
$Si(OEt)_3$	H, Cl, Br, Me, OMe	$RuHCl(CO)(PPh_3)_3$/CuCl	95–100	[54]
$Si(OMe)_3$	H, Cl, Br, Me	$RuHCl(CO)(PPh_3)_3$/CuCl	90–99	[54]
$Si(OSiMe_3)_3$	H, Cl, Br, Me	$RuHCl(CO)(PPh_3)_3$/CuCl	83–99	[54]
$SiMe_2Ph$	H, Cl, Br	$RuHCl(CO)(PPh_3)_3$/CuCl	99	[54]
$Si(OEt)_3$	H, Cl, Br	$RuHCl(CO)(PCy_3)_2$/CuCl	81–96	[54]
$Si(OMe)_3$	H, Cl, Br	$RuHCl(CO)(PCy_3)_2$/CuCl	99	[54]
$Si(OSiMe_3)_3$	H	$RuHCl(CO)(PCy_3)_2$/CuCl	92	[54]
$SiMe_2Ph$	H, Me	$RuHCl(CO)(PCy_3)_2$/CuCl	60–95	[54]
$SiMe_3$	H, Cl, Br, Me	$RuHCl(CO)(PCy_3)_2$/CuCl	100	[54]

[a] $RuHCl(CO)(PPh_3)_3$, $RuCl(SiMe_3)(CO)(PPh_3)_2$, $RuCl[Si(OEt)_3](CO)(PPh_3)_2$, $RuCl(SiMe_2Ph)(CO)(PPh_3)_2$.

$$R_3Si\text{–CH=CH}_2 + \text{CH}_2\text{=CH–R'} \xrightarrow[-\,\text{CH}_2\text{=CH}_2]{[Ru]} R_3Si\text{–CH=CH–R'} \tag{28}$$

[Ru]	SiR_3	R'	Yield [%]	E/Z	Ref.
$RuHCl(CO)(PPh_3)_3$ $RuCl(SiMe_3)(CO)(PPh_3)_2$ $RuHCl(CO)(PCy_3)_2$ $RuCl_2(PPh_3)_3$	$SiMe_3$, $SiMe_2Ph$, $Si(OEt)_3$	OEt, OPr, OBu, O^tBu, OC_6H_{11}, $SiMe_3$	56-99	2/1-5/1	[51]
$RuHCl(CO)(PPh_3)_3$	$SiMe_3$, $SiMe_2(OEt)$	OBu, COOMe	59-84	4/1-13/1	[46]
$RuHCl(CO)(PPh_3)_3$	$Si(OEt)_3$	COOMe			[52]
$RuHCl(CO)(PCy_3)_2$	$SiMe_3$, $SiMe_2Ph$, $Si(OEt)_3$	N-pyrrolidinone, N-phthalimide	12-97	E	[53]
$RuHCl(CO)(PCy_3)_2$	$SiMe_3$, $SiMe_2Ph$,	B(O–)(O–) cyclic boronate	72-78	3/1	[a]

[a] [B. Marcimec et al., unpubl.]

In view of recent reports, it seems prospective to use the reaction of trans silylation for the synthesis of other types of unsaturated compounds, particularly for the synthesis of novel organosilicon starburst compounds having a silicon-bridged π-conjugated structure which is expected to have potential optoelectronic properties. The effective functionalization of 1,3,5-tris(dimethylvinylsilyl)benzene by the respective reactions with 1,4-divinylbenzene leads to the formation of a dendrimer (Fig. 1a) [55].

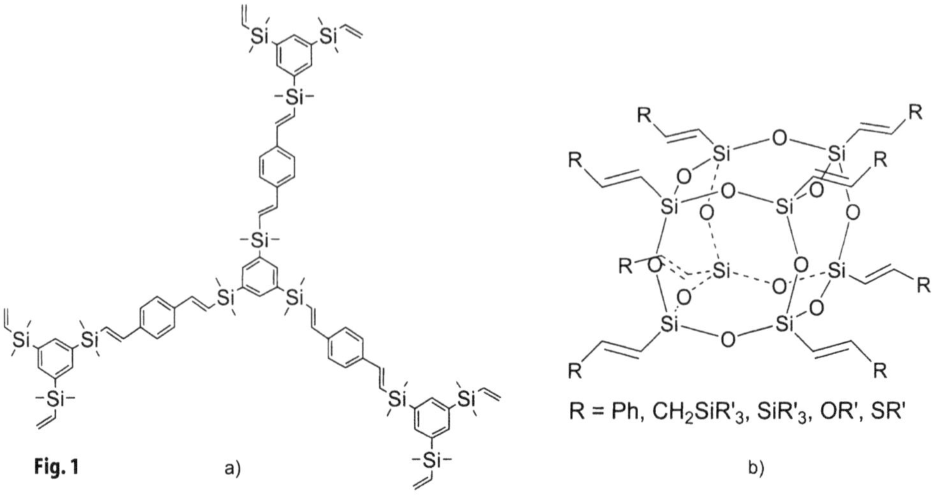

Fig. 1 a) b)

The reactions of vinylcyclosiloxanes and vinylcyclosilazanes with styrene in the presence of $RuHCl(CO)(PCy_3)_2$ opened a new route to functionalized monomers for the ring-opening polymerization of cyclosiloxanes and cyclosilazanes (Eq. 29) [56].

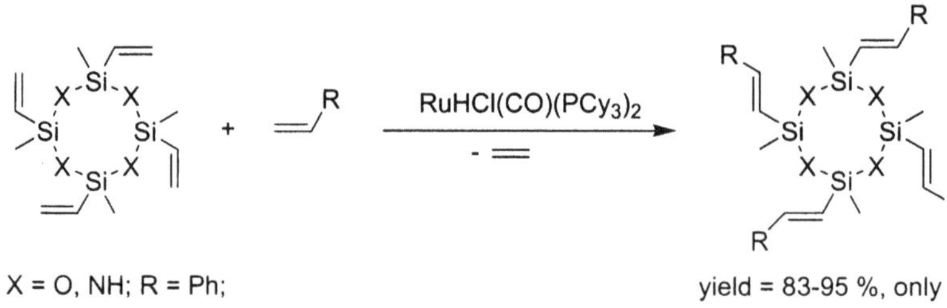

(29)

Organosubstituted octasilsesquioxanes (Fig. 1b) have also been prepared by cross-metathesis (CM) and silylative coupling of vinylsilsesquioxane with olefins in the presence of the ruthenium carbene complex $Cl_2(PCy_3)_2Ru(=CHPh)$ (Grubbs catalyst) and Ru–H (Ru–Si) complexes, for example, $RuHCl(CO)(PCy_3)_2$, respectively [57].

Untypical conversion of allyltrisubstituted silanes has been proved to occur via preliminary isomerization of 1-propenyl-trisubstituted silanes, followed by heterocoupling with parent allylsilanes to finally yield the (E+Z) isomers of bis(silyl)propene and propene [50, 58].

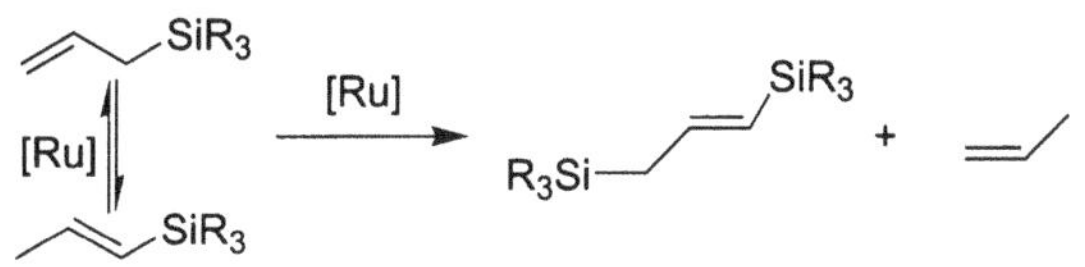

(30)

[Ru] = $RuHCl(CO)(PPh_3)_3$, $RuCl_2(PPh_3)_3$
SiR_3 = $SiMe_3$, $Si(OEt)_3$

The cross-coupling of allylsilanes with alkenes [50] and styrene [58] also occurs via their preliminary isomerization followed by the reaction of 1-propenylsilane with exemplary olefin-1-decene resulting in 1-(triethoxysilyl)-1-decene as a product (Eq. 31). If the cross-coupling takes place, an expected product of this reaction is 1-silyl-2-undecene, which is not detected.

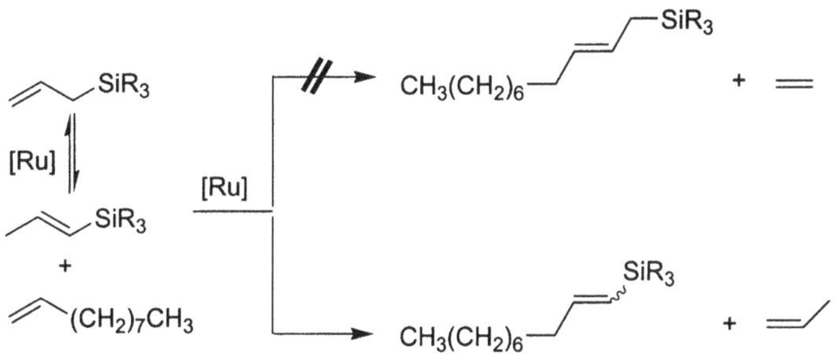

(31)

Our recent synthetic examinations have confirmed the non-metallacarbene mechanism for Ru–H and Ru–Si complex catalyzed reactions of vinyl alkyl ethers with vinylsilanes, yielding a mixture of (*E*+*Z*)-1-silyl-2-(alkoxy)ethenes (Eq. 32). Interestingly, 1-silyl-1-(alkoxy)ethene has not been found among the products [51].

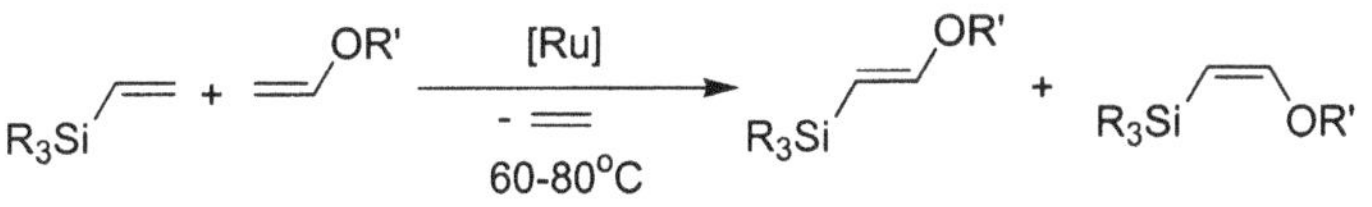

(32)

3.4 Self-Metathesis and Cross-Metathesis of Unsaturated Organosilicon Derivatives

Prior to the late 1980s very little information on the effective metathesis conversion of unsaturated organosilicon compounds had been reported [59]. Remarkable developments over the last 10–15 years in the synthesis of well-defined, functional-group-tolerant metal carbene complexes, i.e., molybdenum Schrock-type catalyst $[(CF_3)_2MeCO]_2Mo(NAr)(=CHCMe_2Ph)$, ruthenium first-generation $Cl_2(PCy_3)_2Ru(=CHPh)$ (Fig. 2a) and second generation $Cl_2(PCy_3)(IMesH_2)Ru(=CHPh)$, where IMes is imidazol-2-ylidene, (Fig. 2b) Grubbs catalyst as well as the Hoveyda–Grubbs catalyst (Fig. 2c), have opened new opportunities in organic [60] and organosilicon chemistry [6, 7].

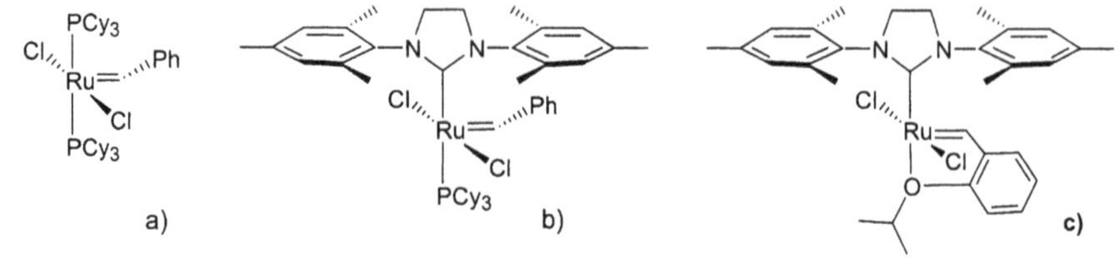

a) First generation Grubbs catalyst

b) Second generation Grubbs catalyst

c) Hoveyda - Grubbs catalyst

Fig. 2

3.4.1 Self-Metathesis and Cross-Metathesis of Vinylsilanes

Vinylsilicon compounds exhibit specific behavior in metathesis owing to the strong steric and electronic influence of silicon on the double bond. Vinyl derivatives of organosilicon compounds do not undergo effective self-metathesis. The formation of a small amount of *E*-1,2-bis(dichloromethylsilyl)ethene observed as an accompanying product during the CM of dichloromethylvinylsilane with some olefins [61] is the only reported example of vinylsilane self-metathesis occurring in the presence of a well-defined ruthenium catalyst. On the other hand, metathesis transformation of vinylsilanes such as CM with olefins, ring-closing metathesis (RCM) and ring-opening metathesis (ROM)/CM) have been reported. The first example of an effective CM of vinylsilane in the presence of a first-generation Grubbs catalyst was the stereoselective and regioselective synthesis of silylstyrenes (Eq. 33) [62, 63]. Recently, chlorosubstituted vinylsilanes have also been shown to react selectively with styrene in the presence of a second-generation Grubbs catalyst (Eq. 33) [61].

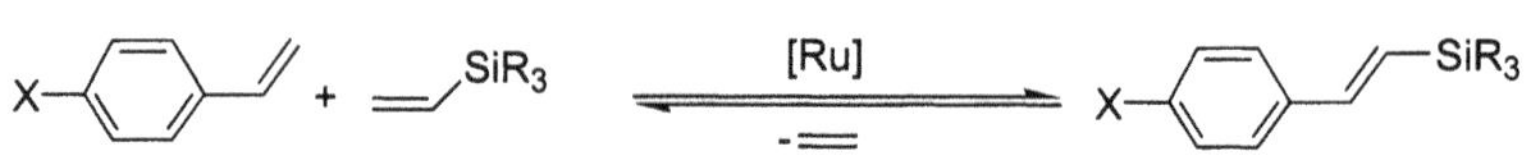

[Ru]	X	R	Yield [%]	Ref.
$Cl_2(PCy_3)_2Ru(=CHPh)$	H	OMe, OEt, $OSiMe_3$	72-90	[62]
$Cl_2(PCy_3)_2Ru(=CHPh)$	H, Cl, OMe, Me	OEt	56-99	[63]
$Cl_2(PCy_3)(IMesH_2)Ru(=CHPh)$	H	Cl	83	[61]

(33)

The CM of vinylsilanes with alkenes and allyl-substituted heteroorganic compounds in the presence of Grubbs catalysts gives alkenylsilanes and 1-silyl-3-*N*(*O* or *S*)-substituted propenes, respectively, in moderate-to-very high yields (D. Chadyniak, W. Prukala, B. Marciniec, unpublished results) (Eq. 34).

$$R_3Si\text{-}CH{=}CH_2 + CH_2{=}CH\text{-}CH_2\text{-}Y \xrightleftharpoons[-\,CH_2=CH_2]{Cl_2(PCy_3)_2Ru(=CHPh)} R_3Si\text{-}CH{=}CH\text{-}CH_2\text{-}Y \tag{34}$$

Y	R	Yield [%]	E/Z	Ref.
C_3H_7	OMe, OEt, $OSiMe_3$	60-75	9/1-10/1	[63]
C_3H_7	Cl	100*	20/1	[61]
C_3H_7	OCOMe	100*	20/1	[61]
$SiMe_3$, $Si(OEt)_3$	OEt	71, 95	*E*, 15/1	[63]
$SiMe_3$	Cl	100*	25/1	[61]
$SiMe_3$	OCOMe	100*	25/1	[61]
OEt, OBu, OC_6H_{11}, OPh, OCH_2Ph, $OSiMe_3$ glycydyloxyl,	OMe, OEt, $OSiMe_3$	50-95	5/1-12/1	[64]
OEt,	Cl	92*	8/1	[61]
OCOMe, OCOEt, OCOPr	OEt	87-89	4/1-14/1	[52]
NMePh, N(COMe)Ph	OEt	57, 63	6/1, traces *Z*	[53]

*) $Cl_2(PCy_3)(IMesH_2)Ru(=CHPh)$

Oligo(phenylene,vinylene)s carrying terminal vinyl groups can react with vinyltriethoxysilane in the presence of a Grubbs catalyst [65].

(35)

Reaction scheme: oligo(phenylene vinylene) bearing OC_8H_{17} groups, $[\ldots]_n$, with terminal vinyl + $CH_2{=}CH\text{-}R$ → ($Cl_2(PCy_3)_2Ru(=CHPh)$, $-\,CH_2{=}CH_2$, benzene, r.t.) → product with terminal $CH{=}CH\text{-}R$

R = $CH_2Si(OEt)_3$, $Si(OEt)_3$
n = 1, 2, 3

yield = 5-48 %

The products being highly fluorescent molecules, oligophenylene vinylene chromophores, rigidly connected to hydrolyzable alkoxysilane moieties, are interesting for electrical and optical applications.

Low conversions of octavinylsilsesquioxanes have been observed when reacting them with pent-4-en-1-one and 5-bromopent-1-ene in the presence of $Cl_2(PCy_3)_2Ru(=CHPh)$ (Eq. 36) [66]. Recent reexamination of the reaction with styrene, 1-hexene and allyltrimethylsilane has succeeded. The X-ray structures of the trans-styryl- and trans-3-trimethylsilyl-1-propenyl-substituted silsesquioxanes have also been obtained [57].

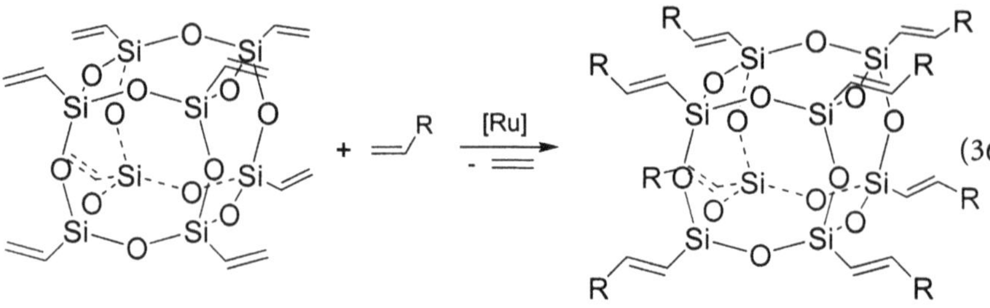

(36)

[Ru] = $Cl_2(PCy_3)_2Ru(=CHPh)$ R = Ph, Bu, CH_2SiMe_3; yield = 69-96%

[Ru] = $Cl_2(PCy_3)_2(IMesH_2)Ru(=CHPh)$ R = S^tBu; yield = 95%

CM of 1,9-decadiene with an excess of trialkoxy- and trisiloxy-substituted vinylsilanes results in the formation of *E,E*-1,10-bis(silyl)deca-1,9-dienes (Eq. 37) [67].

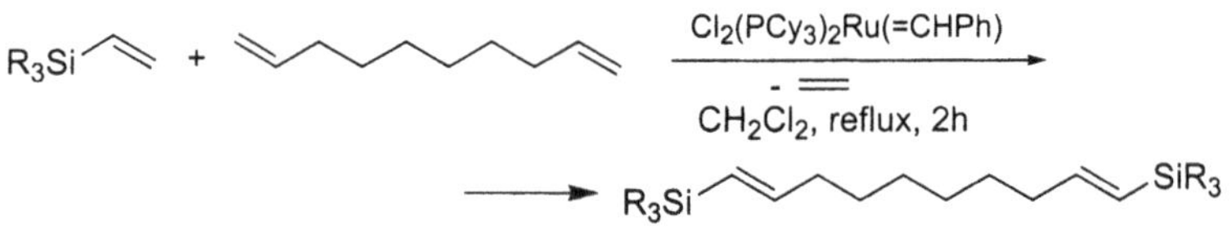

$Cl_2(PCy_3)_2Ru(=CHPh)$; CH_2Cl_2, reflux, 2h (37)

SiR_3 = $Si(OMe)_3$, $Si(OEt)_3$, $Si(OSiMe_3)_3$; yield = 45-73%

3.4.2
Self-Metathesis and Cross-Metathesis of Allylsilanes

In some systems self-metathesis of allylsilanes has been observed to accompany the CM. The CM of allylsilane with unsaturated compounds is a convenient method of introduction of the silyl group into the olefin. Moreover, a chemoselective run of CM has been demonstrated (Eq. 38) [68].

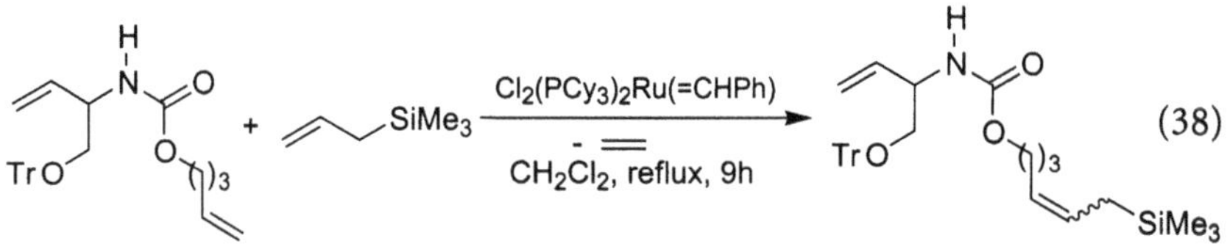

$Cl_2(PCy_3)_2Ru(=CHPh)$; CH_2Cl_2, reflux, 9h (38)

yield = 50%, *E/Z* = 1.5/1

CM has been successfully applied to carbohydrate derivatives [70, 71] to generate a wide range of modified neoglyco conjugates of great potential in glycobiology. $Cl_2(PCy_3)_2Ru(=CHPh)$ offers the advantage of being compatible with most protecting groups normally utilized in carbohydrate chemistry. *O*-Allyl-α-D-galactopyranoside has been effectively reacted with $H_2C{=}CHCH_2SiMe_3$ (Eq. 39).

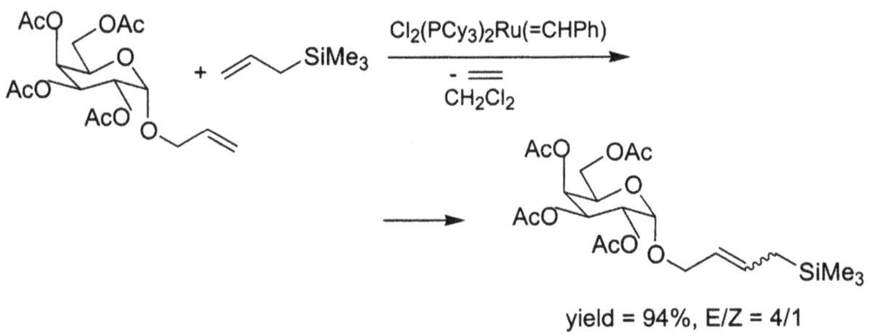

$Cl_2(PCy_3)_2Ru(=CHPh)$, CH_2Cl_2

yield = 94%, E/Z = 4/1 (39)

CM of allyltrimethylsilane with homoallylic alcohols containing both *anti*-allylic and *syn*-allylic substituents has been reported to proceed effectively and display enhanced E selectivity [72, 73]. The reaction has been proposed to be a step in the synthetic route leading to vinylcyclopropanes (Eqs. 40, 41).

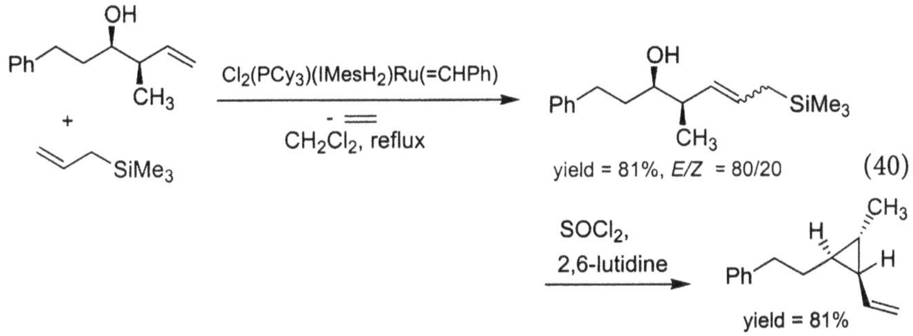

$Cl_2(PCy_3)(IMesH_2)Ru(=CHPh)$, CH_2Cl_2, reflux

yield = 81%, *E/Z* = 80/20 (40)

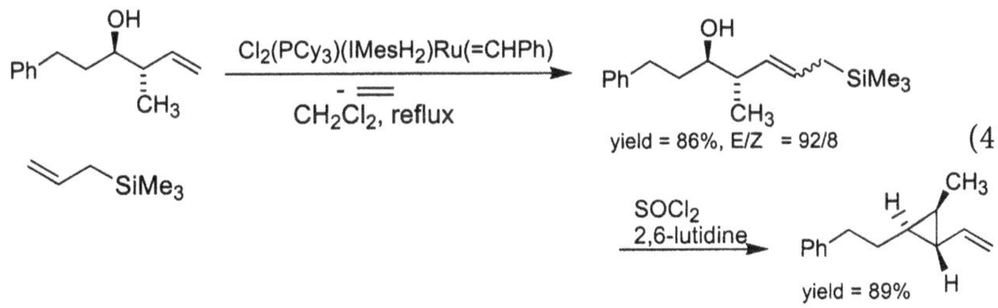

$SOCl_2$, 2,6-lutidine

yield = 81%

$Cl_2(PCy_3)(IMesH_2)Ru(=CHPh)$, CH_2Cl_2, reflux

yield = 86%, E/Z = 92/8 (41)

$SOCl_2$, 2,6-lutidine

yield = 89%

Effective CM of allyltrimethylsilane with allyl alcohols has been reported by Cossy et al. [74] and was achieved with the Hoveyda–Grubbs catalyst (Eq. 42).

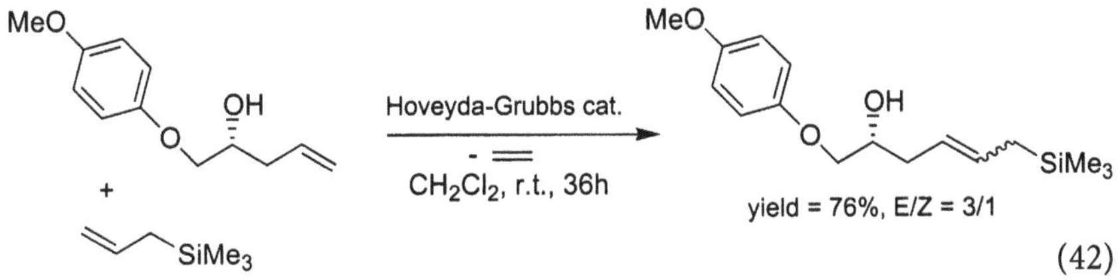

Hoveyda-Grubbs cat., CH_2Cl_2, r.t., 36h

yield = 76%, E/Z = 3/1 (42)

The effective reaction of allyltrimethylsilane with alkenylepoxides leads to the products being potentially useful synthetic building blocks for Lewis-acid-mediated cyclization and condensation reactions (Eq. 43) [75].

$$\text{epoxide-}(CH_2)_n\text{-CH=CH}_2 + CH_2\text{=CHCH}_2SiMe_3 \xrightarrow[-\,CH_2=CH_2;\; CH_2Cl_2,\ 40^{\circ}C]{Cl_2(PCy_3)_2Ru(=CHPh)} \text{epoxide-}(CH_2)_n\text{-CH=CHCH}_2SiMe_3 \quad (43)$$

n = 2, 3, 4, 6; yield = 30-45%, *E/Z* = 1/1

A series of CMs of α,β-unsaturated carbonyl compounds with allylsilanes have been tested in the presence of the Hoveyda–Grubbs catalyst (Eq. 44) [76]. Functionalized allylsilanes have been obtained in moderate-to-good yield and very good stereoselectivity in favour of the E isomer. Examples have proved a high functional group tolerance of the Hoveyda–Grubbs catalyst.

$$R^1C(O)CH\text{=CH}_2 + CH_2\text{=CHCH}_2SiR^2_3 \xrightarrow[-\,CH_2=CH_2;\; CH_2Cl_2\ \text{r.t.}]{\text{Hoveyda-Grubbs cat.}} RC(O)CH\text{=CHCH}_2SiR^2_3 \quad (44)$$

R^1 = OEt, Et, H, OH; R^2 = Me, Ph; yield = 37 - 86 %

Vinyl-substituted cyclic acetals are another group of reagents that undergo effective CM with allylsilane [77].

CM has been reported to provide a synthetic tool for immobilization of reagents. Polymer-supported synthesis with an allylsilyl unit as a linker was developed. Divinylbenzene cross-linked allyldimethylsilylpolystyrene has been reported to undergo highly efficient ruthenium-catalyzed CM with functionalized terminal alkenes (Eq. 45) [78]. Products have been liberated by protodesilylation with trifluoroacetic acid.

$$\text{(P)-C}_6H_4\text{-SiMe}_2\text{-CH}_2CH\text{=CH}_2 + CH_2\text{=CHR} \xrightarrow[-\,CH_2=CH_2]{Cl_2(PCy_3)_2Ru(=CHPh)} \text{(P)-C}_6H_4\text{-SiMe}_2\text{-CH}_2CH\text{=CHR} \quad (45)$$

3.5 Tandem Ring-Opening Metathesis/Cross-Metathesis

The term ROM/CM refers to the processes described by the following equation:

$$CH_2\text{=CHR} + \text{cycloalkene} \xrightarrow{\text{cat.}} RCH\text{=CH-}\frown\text{-CH=CH}_2 \quad (46)$$

Opening of a strained ring system and the subsequent coupling with an acyclic alkene results in the formation of diene products. Because of many metathesis pathways available in the systems containing a cyclic and a linear olefin, the reaction proceeds selectively only when the reacting partners are properly chosen.

ROM/CM of norbornene derivatives with allyltrimethylsilane in the presence of various ruthenium catalysts has been reported by Blechart et al. (see Eq. 47 and Refs. [79–82]).

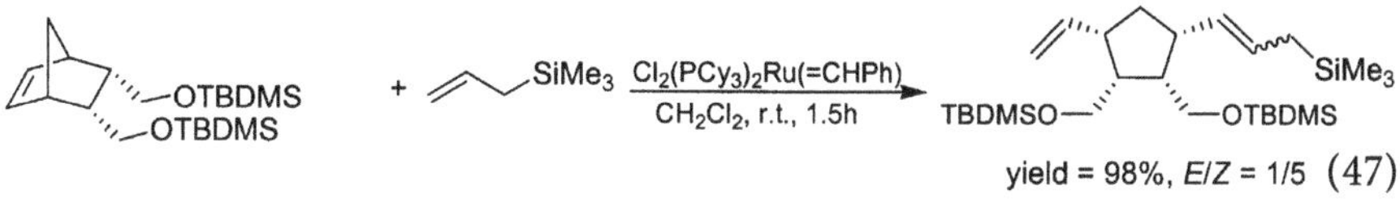

OTBDMS OTBDMS + $SiMe_3$ $Cl_2(PCy_3)_2Ru(=CHPh)$ CH_2Cl_2, r.t., 1.5h TBDMSO OTBDMS $SiMe_3$

yield = 98%, E/Z = 1/5 (47)

ROM/CM of unsymetrical norbornenes with olefins gives a mixture of regioisomers and stereoisomers. The effect of *R* in exo-2- and endo-2-substituted norbornenes on the regioisomer distribution has been examined [83] (Eq. 48).

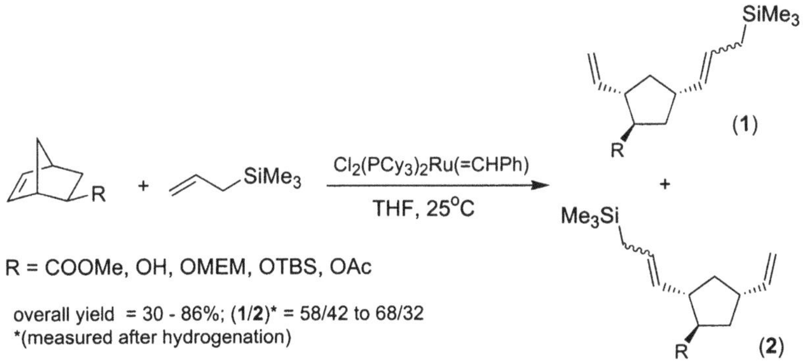

R + $SiMe_3$ $Cl_2(PCy_3)_2Ru(=CHPh)$ THF, 25°C → $SiMe_3$ R (**1**) + Me_3Si R (**2**) (48)

R = COOMe, OH, OMEM, OTBS, OAc

overall yield = 30 - 86%; (**1**/**2**)* = 58/42 to 68/32
*(measured after hydrogenation)

Silylsubstituted norbornenes undergo ROM with ethylene in the presence of $Cl_2(PCy_3)_2Ru(=CHPh)$ or the Noels catalyst [84] (generated in situ) to form α,ω-dialkenes, giving quantitative yield when the Grubbs complex was used as a catalyst [85]. The reactions with 2,3-disubstituted norbornadiene have been found to be highly chemoselective with ROM occurring on the less substituted or sterically less hindered double bond regardless of the nature of substituents (Eq. 49) [83].

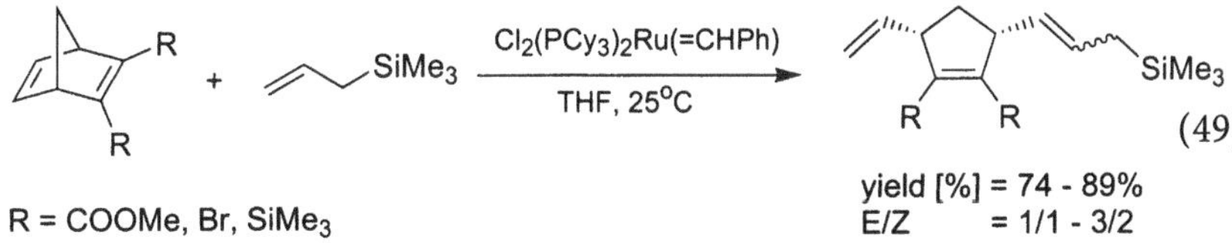

R R + $SiMe_3$ $Cl_2(PCy_3)_2Ru(=CHPh)$ THF, 25°C → R R $SiMe_3$ (49)

R = COOMe, Br, $SiMe_3$

yield [%] = 74 - 89%
E/Z = 1/1 - 3/2

ROM/CM of linear olefins with cyclooctene and cyclobutene derivatives provides an exciting route to asymmetric acyclic dialkenes. Cyclobutene derivatives have been reported to undergo ROM/CM with allylsilane (Eq. 50) [86]. For other examples see Refs. [79, 87].

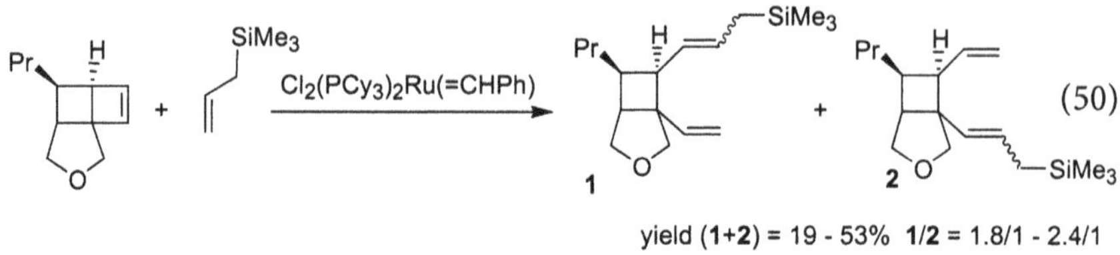

$Cl_2(PCy_3)_2Ru(=CHPh)$ (50)

yield (**1**+**2**) = 19 - 53% **1**/**2** = 1.8/1 - 2.4/1

Reactions of cyclopropenone ketal with terminal alkenes afford 1,4-divinyl ketone ketals in good yields (Eq. 51) [88].

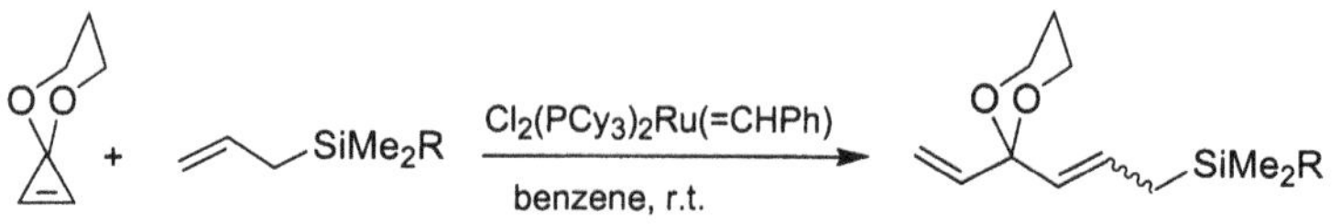

$SiMe_2R$ $Cl_2(PCy_3)_2Ru(=CHPh)$ benzene, r.t. $SiMe_2R$ (51)

R	Yield [%]	E/Z
Me	86	95/5
$CH{=}CH_2$	52	95/5

The only example of effective ruthenium-catalyzed ROM/CM involving the vinylsilane moiety is the reaction of cyclooctene with trialkoxy- and trisiloxy-substituted vinylsilanes leading to the formation of *E*,*E*-1,10-bis(silyl)deca-1,9-dienes (Eq. 52) [67].

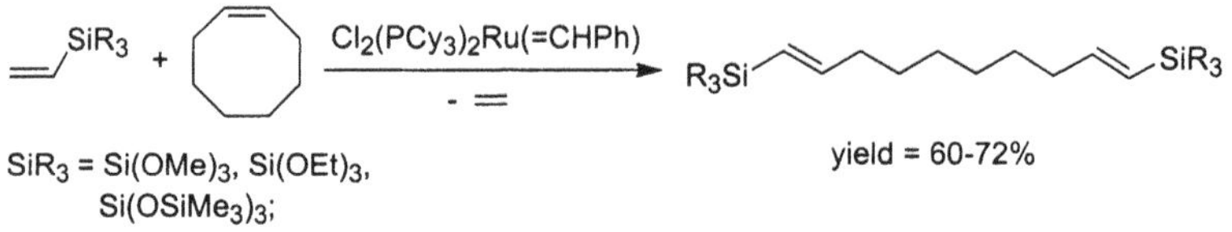

SiR_3 + $Cl_2(PCy_3)_2Ru(=CHPh)$ - = R_3Si SiR_3 (52)

$SiR_3 = Si(OMe)_3, Si(OEt)_3, Si(OSiMe_3)_3$; yield = 60-72%

Stragies and Blechert [89] have described the synthesis of [*n*.3.0]bicyclic derivatives with different ring sizes and functional groups in a single domino process. With an excess of allyltrimethylsilane, a bicyclic product was obtained in high yield (Eq. 53).

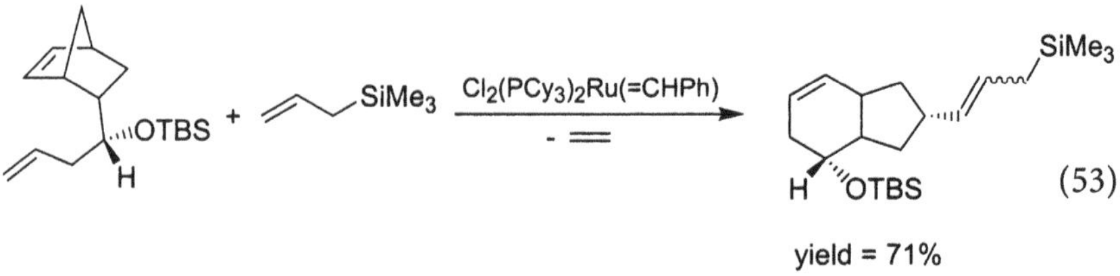

OTBS H + $SiMe_3$ $Cl_2(PCy_3)_2Ru(=CHPh)$ - = $SiMe_3$ H OTBS (53)

yield = 71%

Formally the reaction combines the ring opening of norbornene, RCM with the terminal double bond and CM with a second alkene. The domino metathesis involving an allylsilyl derivative has been used as a step in the synthesis of (–)-halosaline [69] and (–)-indolizidine [90].

3.6 Intermolecular Enyne Metathesis

Intermolecular enyne (ene-yne) metathesis combines an alkene and an alkyne into a 1,3-diene (Eq. 54).

$$R^1\text{—}\equiv \; + \; \text{=\!\!\diagup}R^2 \xrightarrow{\text{cat.}} R^1\text{C(=CH}_2\text{)CH=CH}R^2 \tag{54}$$

Important progress has been made in the study of the process since the discovery and development of well-defined ruthenium carbene complexes. The interest in the reaction as a synthetic tool for organic synthesis has been growing, among others, because of its atom economy. For recent reviews on enyne metathesis see Refs. [91–93]. For a review on the application of enyne metathesis in organosilicon chemistry see Ref. [6].

Terminal alkynes and terminal alkenes can be selectively transformed into disubstituted dienes in the presence of ruthenium carbene complexes [94–96]. Reactions of a series of alkynes with $H_2C{=}CHCH_2SiMe_3$ lead to silyl-substituted conjugated dienes.

$$R^1\text{—}\equiv\text{—}R^2 \; + \; H_2C{=}CHCH_2SiMe_3 \xrightarrow[CH_2Cl_2,\ r.t.]{[Ru]} Me_3SiCH_2CH{=}C(R^1)C(R_2){=}CH_2$$

R^1	R^2	[Ru]	Yield [%]	E/Z	Ref.
CH_2OTHP	H	$Cl_2(PCy_3)_2Ru(=CHPh)$	81	1/1	[94]
CH_2OAc	H	$Cl_2(PCy_3)_2Ru(=CHPh)$	90	1/1	[94]
CH(Me)OAc	H	$Cl_2(PCy_3)_2Ru(=CHPh)$	63	1/1	[94]
$SiMe_3$	H	$Cl_2(PCy_3)_2Ru(=CHPh)$	22	9/1	[94]
$SiMe_3$	H	$Cl_2(PCy_3)(IMesH_2)Ru(=CHPh)$	77	6/1	[95]
Cy	H	$Cl_2(PCy_3)(IMesH_2)Ru(=CHPh)$	69	1.9/1	[95]
$TBPSOCH_2$	H	$Cl_2(PCy_3)(IMesH_2)Ru(=CHPh)$	79	1.3/1	[96]
CH_2OAc	CH_2OAc	$Cl_2(PCy_3)(IMesH_2)Ru(=CHPh)$	89	1/2.5	[95]

(55)

The reaction offers a convenient way of synthesis of conjugated allylsilanes. The second-generation Grubbs catalyst shows a significantly increased activity in the process, when compared with the first-generation Grubbs catalyst, especially in the ene-yne CM involving sterically hindered alkenes [95]. The reaction of alkynes with ethylene is interesting and is potentially of great use [97], and an example with an organosilicon derivative is presented in Eq. (56) [98].

$$4\text{-MeOC}_6H_4CH_2C\equiv CSiMe_3 \; + \; H_2C{=}CH_2 \xrightarrow[\text{toluene, }80^{o}C]{Cl_2(PCy_3)(IMesH_2)Ru(=CHPh)} 4\text{-MeOC}_6H_4CH_2C(=CH_2)C(SiMe_3){=}CH_2 \tag{56}$$

yield = 87%

Polymer-supported synthesis of 1,3-dienes by efficient ruthenium-catalyzed intermolecular enyne metathesis has been reported by Schürer and Blechert [99]. The polystyrene resin, containing a propargyl ester moiety, was reacted with functionalized alkene in the presence of $Cl_2(PCy_3)_2Ru(=CHPh)$. The dienes obtained were cleaved from the polymer support using a paladium-catalyzed reaction with different nucleophiles (Eq. 57).

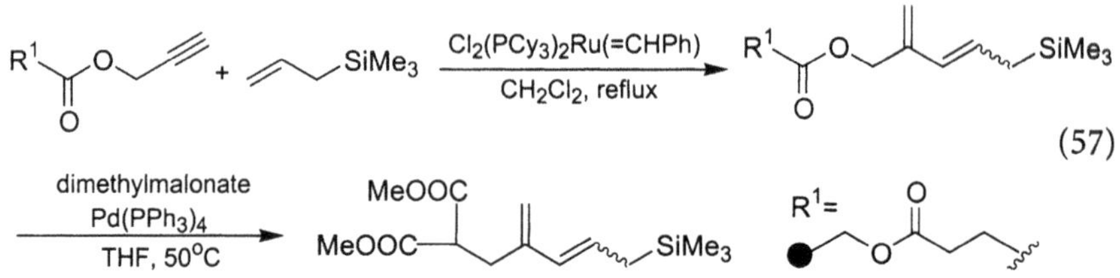

$Cl_2(PCy_3)_2Ru(=CHPh)$, CH_2Cl_2, reflux; dimethylmalonate, $Pd(PPh_3)_4$, THF, 50°C

(57)

An approach involving the use of polystyrene with attached allyldimethylsilyl fragments, which react with a free acetylene derivative, has also been described [100].

3.7 Coupling of Substituted Aromatic and Olefinic Reagents with Silylalkynes

Ortho vinylations of various aromatic ketones with alkynylsilanes catalyzed by $RuH_2(CO)(PPh_3)_3$ proceed regioselectively with high preference for the E geometry (Eq. 58) [101, 102].

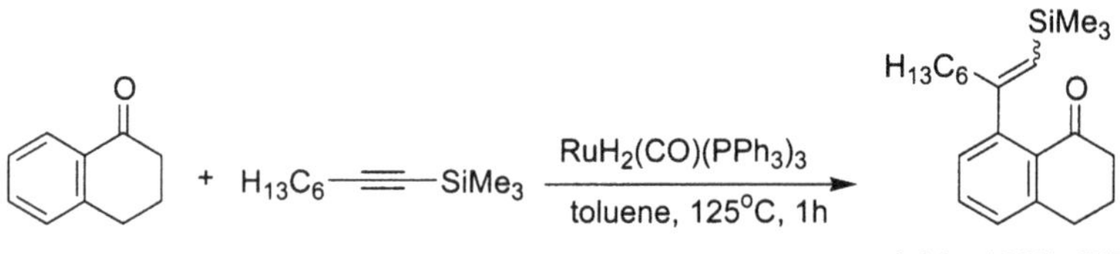

$RuH_2(CO)(PPh_3)_3$, toluene, 125°C, 1h

yield = 100%, E/Z = 11/1 (58)

Many aromatic ketones have been tested in the reaction, for example, (Eq. 59) [102].

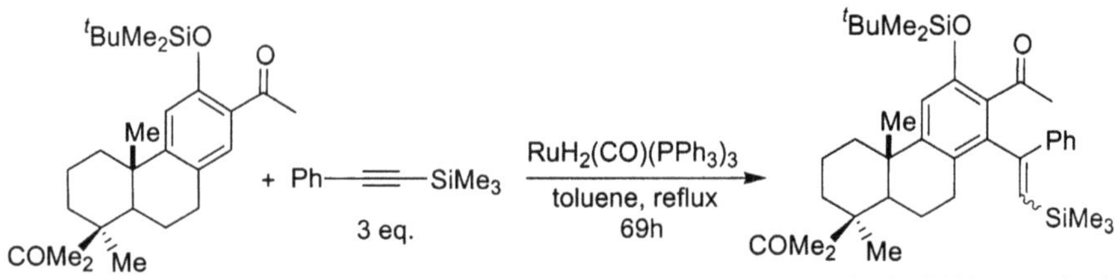

$RuH_2(CO)(PPh_3)_3$, toluene, reflux, 69h; 3 eq.

yield = 79% (59)

Heteroaromatic ketones (Eq. 60) [101] and enones (Eq. 61) [103] with an activated olefinic C–H bond have also been found to undergo effective coupling with 1-phenyl-2-silylacetylene.

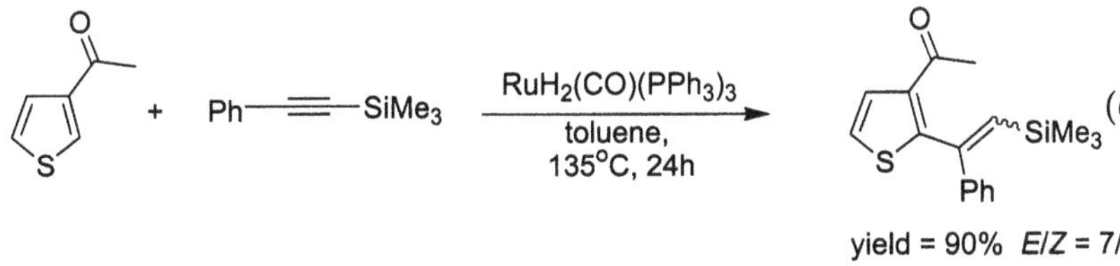

$RuH_2(CO)(PPh_3)_3$, toluene, 135°C, 24h (60)

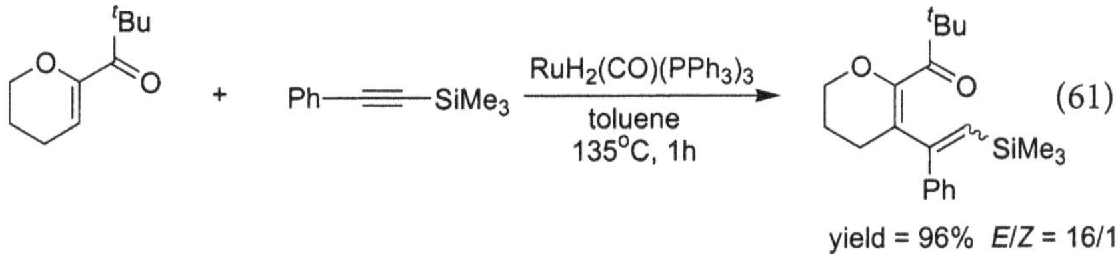

yield = 90% E/Z = 7/1

$RuH_2(CO)(PPh_3)_3$, toluene, 135°C, 1h (61)

yield = 96% E/Z = 16/1

3.8 Dimerization of Trimethylsilylethyne

Dimerization of terminal acetylenes is a convenient route to an unsaturated C4 skeleton. The process was recently overviewed briefly [104]. Dimerization of trimethylsilylacetylene proceeds in the presence of $[(PP_3)RuH(H_2)]BPh_4$, where $PP_3{=}P(CH_2CH_2PPh_2)_3$ or $[(PP_3)RuH(N_2)]BPh_4$ [105].

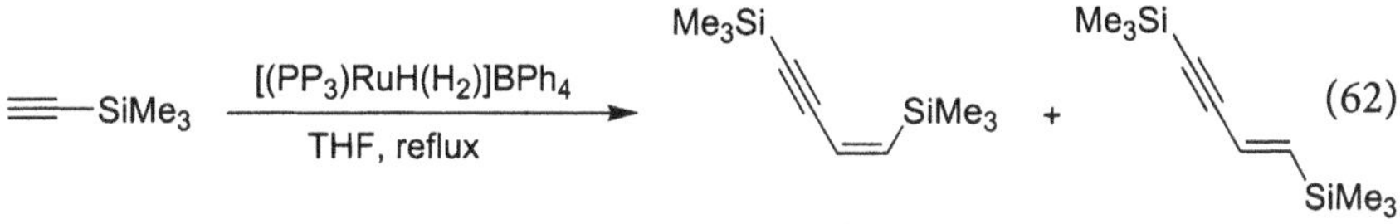

$[(PP_3)RuH(H_2)]BPh_4$, THF, reflux (62)

Different stereoselectivities and regioselectivities were observed when Cp^*RuH_3L, where L is PPh_3, PCy_3 or PMe_3, or (tris(pyrazol-1-yl)borate)-$RuCl(PPh_3)_2$ were used as catalysts (Eq. 63) [106, 107]

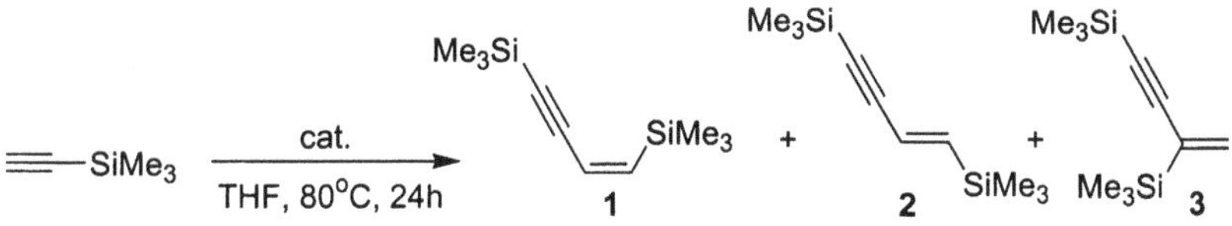

cat., THF, 80°C, 24h → **1** + **2** + **3**

Cat.	Yield **1+2+3** [%] (conversion)	**1/2/3**	Ref.
$Cp^*RuH_3(PPh_3)$	100	2/98 (no 3)	[106]
$Cp^*RuH_3(PCy_3)$	58	1/19 (no 3)	[106]
$Cp^*RuH_3(PMe_3)$	83	1/3/6	[106]
$TpRuCl(PPh_3)_2$	(94)	82/-/15 (no 2)	[107]

(63)

In the presence of Ru(cyclooctadiene)(cyclooctatetraene)/6 PPh_3, dimerization of HC≡$CSiMe_3$ leads to effective formation of 1,4-bis(trimethylsilyl)butatriene (Eq. 64) [108].

$$\equiv\!-SiMe_3 \xrightarrow[-10^{\circ}C,\ 30h]{Ru(cod)(cot)\ /6\ PPh_3} \mathbf{1} + \mathbf{2} + \mathbf{3} \tag{64}$$

yield (**1**+**2**+**3**) = 92% **1/2/3** = 1 / 4 / 15

Silylacetylenes were reported to undergo cross-coupling with internal alkynes (Eq. 65) [109].

$$\equiv\!-SiMe_3 + R^1\!-\!\equiv\!-R^2 \xrightarrow[Et_3N,\ C_6H_6,\ r.t.]{Cp^*Cl(PPh_3)Ru(=C=CHPh)} Me_3Si\!-\!\equiv\!-C(R^1)=CH(R^2) \tag{65}$$

R^1 = Me, Et; R^2 = COOEt, COMe

yield = 85-90%

3.9 Catalytic Double Addition of Trimethylsilyldiazomethane to Alkynes

Synthesis of substituted 1,4-bis(trimethylsilyl)buta-1,3-dienes was achieved via the reaction of terminal or internal alkynes with N_2=$CHSiMe_3$ (Eq. 66) [110].

$$R^1\!-\!\equiv\!-R^2 + 2\ N_2{=}CHSiMe_3 \xrightarrow[dioxane,\ 60^{\circ}C]{Cp^*RuCl(cod)} Me_3SiCH{=}C(R^1)\!-\!C(R^2){=}CHSiMe_3 \tag{66}$$

R^1	R^2	Yield [%]	E/Z
H	Ph	72	7/3
Ph	Me	56	E/E
Ph	Ph	30	E/E
Et	Et	72	E/E
CH_2OH	Me	95	E/E

4 Cyclic Unsaturated Organosilicon Compounds

4.1 Intramolecular Hydrosilylation of Silylalkynes

The reaction of organosilicon substrates containing a C≡C bond has been used for synthesizing predominantly cyclic products having an exocyclic rather than an endocyclic double bond [111–114]. Platinum complexes have been mostly used as catalysts; however, recently a highly regioselective and stereoselective intramolecular hydrosilylation of alkenyl silyl ethers catalyzed by a ruthenium arene complex has been developed (Eq. 67) [115].

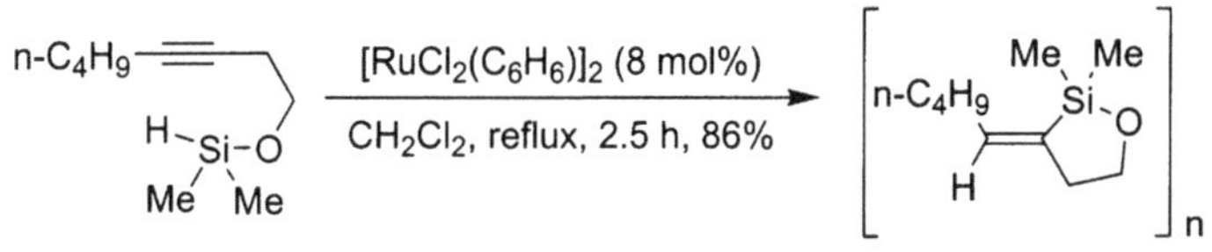

n-C_4H_9 H–Si–O Me Me
$[RuCl_2(C_6H_6)]_2$ (8 mol%)
CH_2Cl_2, reflux, 2.5 h, 86%
[n-C_4H_9 H Me Me Si O]$_n$ (67)

The product (*Z*)-alkylidenosilacyclopentane is efficiently coupled with aryl or alkenyl halides in the presence of butylammonium fluoride and a palladium(0) complex (Eq. 68).

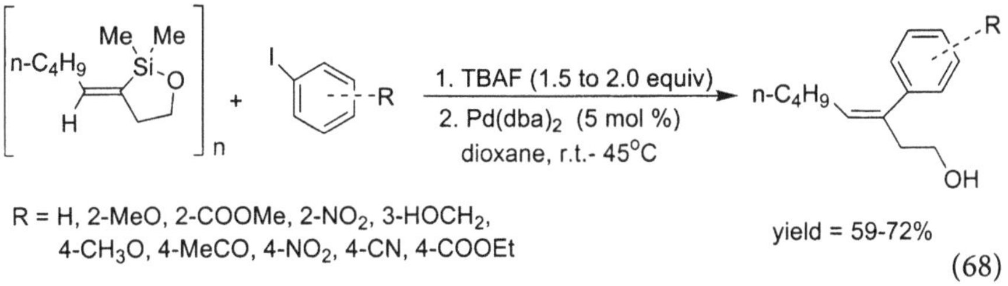

[n-C_4H_9 H Me Me Si O]$_n$ + I–C_6H_4–R
1. TBAF (1.5 to 2.0 equiv)
2. $Pd(dba)_2$ (5 mol %)
dioxane, r.t.- 45°C
n-C_4H_9 R OH

R = H, 2-MeO, 2-COOMe, 2-NO_2, 3-$HOCH_2$, 4-CH_3O, 4-MeCO, 4-NO_2, 4-CN, 4-COOEt

yield = 59-72%

(68)

The cationic $[Cp^*Ru(MeCN)_3]^+(PF_6)^-$ complex, reported as a stereoselective catalyst for trans hydrosilylation of internal alkynes, has been successfully used in intermolecular endo-dig hydrosilylation of propargyloxyhydrosilanes synthesized in situ via silylation of propargylic alcohols by tetramethyldisilazane [116].

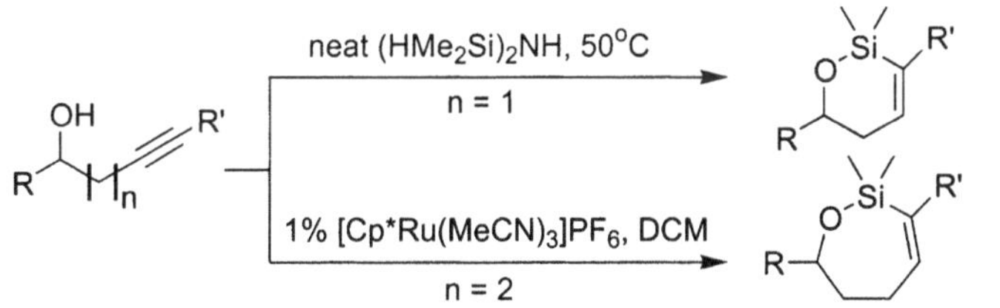

OH R R' n
neat $(HMe_2Si)_2NH$, 50°C
n = 1
1% $[Cp^*Ru(MeCN)_3]PF_6$, DCM
n = 2
O Si R' R
O Si R' R
(69)

In contrast to the process catalyzed by platinum complexes the ruthenium-catalyzed reaction gives endo-dig cyclization in this case to provide seven-membered ring products even though the exo-dig process would form the more favorable six-membered ring. In all reactions only a single product isomer is observed [116].

4.2 Silylative Coupling Condensation of Silicon-Containing Dienes

In the presence of ruthenium, rhodium and cobalt complexes that initially contain or generate *M*–H and *M*–Si bonds, the divinylderivatives of silicon compounds undergo silylative (poly)condensation to yield a mixture of oligomers and cyclic unsaturated siloxanes (silanes, silazanes, Eq. 70) [6, 117, 118].

2 [Si] —catalyst, - ═→ [Si] [Si] ; [Si] [Si] → polymers; [Si] [Si] ; [Si] [Si] [Si] (70)

where [Si] = -SiMe$_2$ or -Me$_2$Si–O–SiMe$_2$- or -Me$_2$Si–N(H)–SiMe$_2$- or -Me$_2$Si–(CH$_2$)$_n$–SiMe$_2$-, n = 1 - 4

In the presence of $[RuCl_2(CO)_3]_2$ as a catalyst, *trans*-bis(vinylsilyl)ethenes are exclusively formed but $[(\text{cyclooctadiene})RhX]_2$ (where X is Cl or $OSiMe_3$) catalyzes mostly the formation of gem-dimeric products. Ruthenium phosphine complexes give both products [119]; the gem products subsequently undergo intramolecular ring closure to yield cyclotetrasiloxane [120], cyclotetrasilazane and cyclohexacarbosilanes [117], respectively.

Cyclization has also been reported to furnish cyclocarbosilanes with one exocyclic methylene group (Eqs. 71, 72) [121].

RuHCl(CO)(PPh$_3$)$_3$, - ═, toluene, 80°C, 24 h; Me$_2$Si, SiMe$_2$; yield = 87% (71)

RuHCl(CO)(PPh$_3$)$_3$, - ═, toluene, 80°C, 25 h; SiMe$_2$, SiMe$_2$; yield = 97% (72)

Such a ring-closing reaction has also been observed for divinyldisilyl ethers

O–SiMe$_2$, O–SiMe$_2$; RuHCl(CO)(PPh$_3$)$_3$, - ═, 80°C, 24h; O–SiMe$_2$, O–SiMe$_2$; yield = 85% (73)

The compound can be a very useful intermediate for the production of other organosilanes, for example, 1,1-bis(silyl)ethenes as well as the previously mentioned exocyclic siloxane (P. Pawluc, Y. Itami, B. Marciniec, unpublished results).

4.3 Ring-Closing Metathesis of Silicon-Containing Dienes

In this subchapter the most important synthetic aspect of the RCM of unsaturated organosilicon compounds will be discussed. For a general review on RCM see Refs. [59, 122–124]. For a review on RCM of silyl dienes see Refs. [6, 7]. A number of valuable synthetic applications have been proposed over the last few years. As follows from the data available, from the point of view of the application of RCM of vinyl and allylsilanes, the most important are silicon-tethered processes. For a general review on the silicon tethered processes see Refs. [125, 126].

The molybdenum complex $[(CF_3)_2MeCO]_2Mo(NAr)(=CHCMe_2Ph)$ has been observed to be a more efficient catalyst for cyclization of vinyl silyl ether dienes than the ruthenium complex $Cl_2(PCy_3)_2Ru(=CHPh)$, probably because this type of alkene is sterically more demanding (than allyl derivatives) and therefore requires a catalyst less sensitive to steric bulkiness near the reaction center. However, some examples of the RCM of substituted vinylsilanes catalyzed by ruthenium complexes have been reported [127, 131] (Eq. 74). For more examples see Ref. [127].

Ph SiMe3 O → $Cl_2(PCy_3)(IMesH_2)Ru(=CHPh)$ / $- =$ / CH_2Cl_2, reflux, 1h → Ph SiMe3 O, yield = 90% (74)

Grubbs has reported a synthetic strategy leading to highly functionalized organic molecules, involving RCM of silicon-tethered unsaturated organic fragments and oxidative cleavage of the cycles formed. Acyclic silyl ether dienes, easily available via silylation of secondary alcohols in the presence of molybdenum $[(CF_3)_2MeCO]_2Mo(NAr)(=CHCMe_2Ph)$, ruthenium vinylcarbene $Cl_2(PCy_3)_2Ru(=CHCH=CPh_2)$ and ruthenium benzylidene $Cl_2(PCy_3)_2Ru(=CHPh)$ complexes give the RCM products (Eq. 75) [128]. The six-, seven-, eight- and ten-membered rings obtained undergo oxidative cleavage to give hydroxy alkenes.

Me2 Si O R (n, m) → cat. / $- =$ → Me2 Si O R (n, m) → $KF/KHCO_3$, H_2O_2 → HO (n) = (m) OH R

n = 0, 1, 2; m = 0, 1, 2, 4; R = Cl, OCH_2Ph (75)

A series of substituted tetrahydrofurans and tetrahydropyrans have been synthesized via RCM of silyl ether dienes followed by a modified Sakurai reaction (Eq. 76) [129, 130].

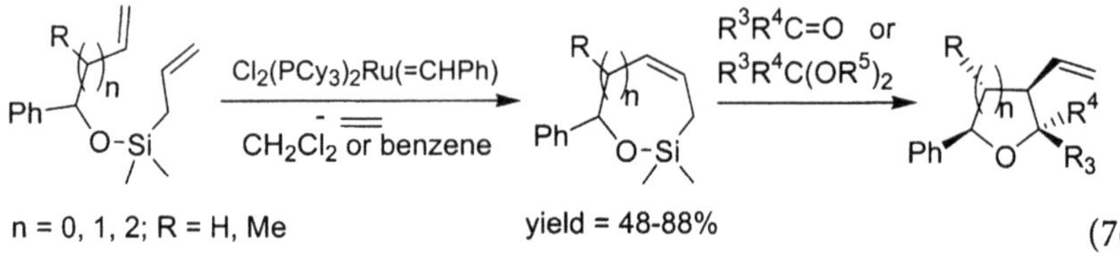

$Cl_2(PCy_3)_2Ru(=CHPh)$, CH_2Cl_2 or benzene; $R^3R^4C=O$ or $R^3R^4C(OR^5)_2$

n = 0, 1, 2; R = H, Me   yield = 48-88%   (76)

A practical and efficient route for the stereoselective conversion of homoallylic alcohols to diastereomerically pure substituted cyclopropanes has been developed by Taylor et al. [73] (Eqs. 77, 78).

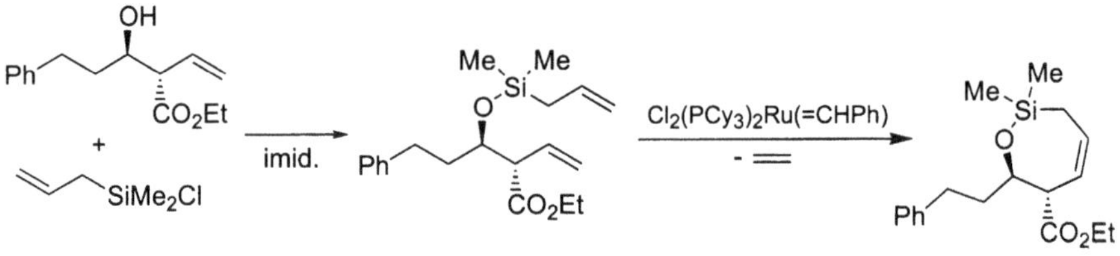

$SiMe_2Cl$, imid.; $Cl_2(PCy_3)_2Ru(=CHPh)$; CO_2Et (77)

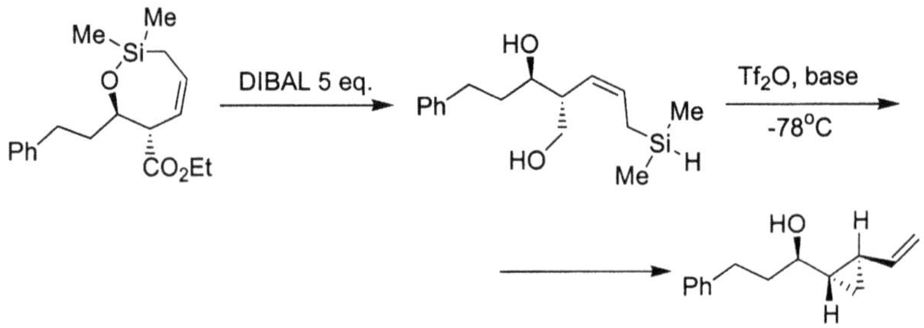

DIBAL 5 eq.; Tf_2O, base, -78°C   (78)

Stereoselective synthesis of functionalized carbocyclic and heterocyclic compounds via tandem ester enolate Claisen rearrangement/RCM has been reported (Eq. 79) [131, 132].

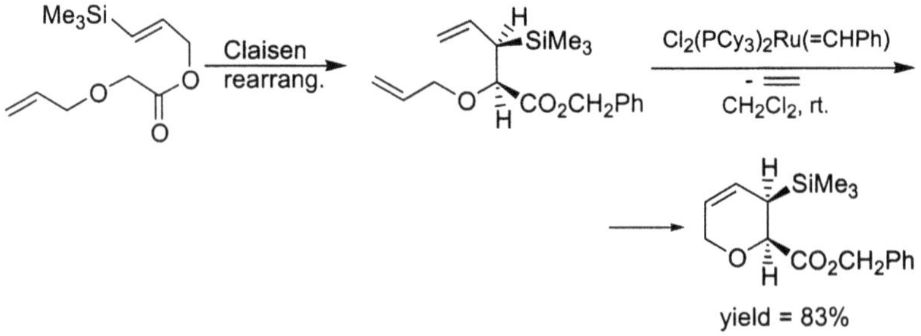

Me_3Si; Claisen rearrang.; $SiMe_3$, CO_2CH_2Ph; $Cl_2(PCy_3)_2Ru(=CHPh)$, CH_2Cl_2, rt.   (79)

yield = 83%

The effective RCM of dienes, catalyzed by ruthenium allenylidene salts $[Cl(PCy_3)(p\text{-cymene})\,Ru(=C=C=CPh_2)]^+(OCF_3SO_2)^-$, has been observed in ionic liquid (1-butyl-3-methylimidazolium salts) (Eq. 80) [133].

2.5 mol% [Ru]; 80°C, 4h (80)

Silaketals constitute another group of convenient starting materials for silicon-tethered syntheses involving RCM [134–136]. The example in Eq. (81) [134] illustrates the synthetic potential of the reaction. For more examples see Ref. [136].

$Cl_2(PCy_3)_2Ru(=CHPh)$, CH_2Cl_2; 1. OsO_4, NMO, r.t. 2. TBAF, then Ac_2O, pyr; NaOMe, MeOH, r.t.; D-altritol (81)

Silacycloalkenes of different size can be easily obtained via RCM of respective siladienes. Successful synthesis of 5–7-membered rings has been reported [137]. An exemplary reaction is presented in Eq. (82).

$Cl_2(PCy_3)_2Ru(=CHPh)$, benzene, 20°C; yield = 70% (82)

The relative ease of ring formation from dienes decreases in the order six->seven->five-membered rings. Hoshi et al. [138] have recently reported the synthesis of a number of disilacycloalkenes via RCM of bis(allyldimethylsilyl)substituted compounds, with the ruthenium vinylcarbene complex $Cl_2(PCy_3)_2Ru(=CHCH=CPh_2)$ used as a catalyst. Successful formation of seven- and eight-membered rings has been achieved under mild reaction conditions.

A series of novel alkenylidene-bridged silsesquioxanes have been synthesized via RCM (Eq. 83) [139]. The molecules obtained are interesting building blocks for new materials or are a platform for macrocyclic hosts.

$Cl_2(PCy_3)_2Ru(=CHPh)$, CH_2Cl_2, 65°C, 36h (83)

4.4 Intramolecular Enyne Metathesis

The term intramolecular enyne metathesis describes two types of processes. One involves a [2+2] cycloaddition of a multiple bond and a transition-metal carbene complex and the other is an oxidative cyclization catalyzed by low-valent transition-metal complexes, for example, Pt, Pd and Ru. The latter reaction is also called a skeletal reorganization. Both processes lead to similar products (Eq. 84).

cat. (84)

The enyne metathesis has been recently reviewed [91–93]. For an overview of the enyne metathesis involving organosilicon compounds see Ref. [6].

Ruthenium-catalyzed skeletal reorganization involving organosilicon compounds have been reported by Chatani et al. [140] (Eq. 85). The reaction provides an efficient method for converting enynes to vinylcycloalkenes.

$[RuCl_2(CO)_3]_2$, toluene, 80°C, CO, 6h; yield = 86% (85)

A facile approach to polycyclic heterocycles via a cascade series of enyne metathesis reactions has been proposed. Norbornene derivatives bearing propargyloxy substituents undergo a series of enyne metathesis reactions leading to heterocyclic dienes [141]. The products have been shown to undergo stereoselective Diels–Alder coupling giving a range of heteropolycycles (Eq. 86). A similar approach has been used for the synthesis of cyclic silyl ethers [142, 143].

$Cl_2(PCy_3)_2Ru(=CHPh)$; maleic anhydride (86)

An easy-to-use, in situ generated catalyst resulting from a combination of commercially available, stable reagents has been reported to display a high activity in the process. The products can be easily transformed further to heterobicyclic derivatives, diols or tetrasubstituted alkenes (Eq. 87) [142].

$[RuCl_2(p\text{-cymene})]_2$ / Mes–N(+)N–Mes Cl^- / Cs_2CO_3 (87)

The ring-closing enyne metathesis has been used for the synthesis of alkenyl substituted six- and seven-membered cyclic enol ethers (Eq. 88) [144]. The reaction has been proposed as an element of the strategy for preparation of subunits of brevetoxins and ciguatoxins.

$Cl_2(PCy_3)_2Ru(=CHPh)$, CH_2Cl_2, reflux; yield = 20% (88)

An efficient method for the preparation of highly functionalized conjugated dienes with the use of silicon-tethered ring-closing enyne metathesis in the presence of first- and second-generation Grubbs catalysts has been described by Yao [145] (Eq. 89).

$Cl_2(PCy_3)_2Ru(=CHPh)$, CH_2Cl_2, reflux, 3h; not isolated; KF, H_2O_2, $KHCO_3$, MeOH/THF; total yield = 88% (89)

5 Silicon Products Containing Si–X–C Bonds, where X is O or N

Organic derivatives of silicon compounds particularly those involving Si–O–C and Si–N(*R*)–C bonds are of great synthetic importance as intermediates which, when suitably reacted (usually via solvolysis), lead to the desired organic fine chemicals [146a]. These classes of compounds may be effectively synthesized via some catalytic reactions such as the hydrosilylation of compounds containing C=O and C=N, C≡N bonds catalyzed predominantly by numerous rhodium complexes, silylformylation of many unsaturated organic compounds usually catalyzed by cobalt group (Co, Rh, Ir) carbonyls and other complexes, as well as a dehydrogenative condensation (e.g., alcohols) catalyzed by many transition-metal and main group metal compounds [1–7].

5.1 Hydrosilylation of Ketones, Amides, Imides and Related Compounds

Although ruthenium-based catalysts, mainly ruthenium clusters, are generally less active than rhodium or cobalt clusters and other complexes in the hydrosilylation of ketones, imides and amides, they can be interesting in view of their specific catalytic activity and selectivity leading finally via hydrosilylation to reduction of the organic substrates [1]. The triruthenium carbonyl cluster bearing the acenaphthylene ligand (μ_3-η^2:η^3:η^5-acenaphthylene)$Ru_3(CO)_7$ appeared to be an active catalyst in the efficient reduction of aldehydes and ketones [147], as well as carboxylic acids, esters and amides [148] via hydrosilylation with trialkylsilane. Prior activation of the catalyst by hydrosilanes dramatically accelerated the reactions. Sterically small trialkylsilanes such as $HSiMe_2Et$ and $HSiMeEt_2$ are very effective in the catalytic reduction of many organic compounds. Hydrosilylation of esters with $EtMe_2SiH$ followed by hydrolytic cleavage of the Si–C bonds gives two types of organic products (Eq. 90) [147, 148].

$$\mathrm{RC(O)OR'} + \mathrm{EtMe_2SiH} \xrightarrow[\text{1,4-dioxane, } 20^\circ\mathrm{C}]{(\mu_3\text{-acenaphthylene})\mathrm{Ru_3(CO)_7}} \xrightarrow[\mathrm{H_2O}]{\mathrm{H^+}} \underset{\mathrm{A}}{\mathrm{RCH_2OH}} + \underset{\mathrm{B}}{\mathrm{RCH_2OR'}} \tag{90}$$

R	R'	Yield A [%]	Yield B [%]
$PhCH_2CH_2$	Me	97	0
	Et	55	11
	^{t}Bu	80	7
$C_{11}H_{23}$	Et	17	75
C_6H_5	Me	52	0
$Br(CH_2)_5$	Et	47	40

Also carboxylic acids (Eq. 91) and tertiary amides (Eq. 92) undergo reduction via hydrosilylation to give the corresponding alcohols and amines, respectively [148].

$$\mathrm{RC(O)OH} + \mathrm{EtMe_2SiH} \xrightarrow[\text{1,4-dioxane, r.t.}]{(\mu_3\text{-acenaphthylene})\mathrm{Ru_3(CO)_7}} \xrightarrow[\mathrm{H_2O}]{\mathrm{H^+}} \mathrm{RCH_2OH} \tag{91}$$

$$\mathrm{RC(O)NR'_2} + \mathrm{EtMe_2SiH} \xrightarrow[\text{1,4-dioxane, r.t.}]{(\mu_3\text{-}\eta^2:\eta^3:\eta^5\text{-acenaphthylene})\mathrm{Ru_3(CO)_7}} \xrightarrow[\mathrm{H_2O}]{\mathrm{H^+}} \mathrm{RCH_2NR'_2} \quad (92)$$

The hydrosilylation reaction of aldehydes and ketones (Eq. 93) catalyzed by (μ_3-η^2:η^3:η^5-acenaphthylene)$Ru_3(CO)_7$ proceeds at room temperature to form the corresponding silyl ethers in good yield (Eq. 93) [147].

$$\mathrm{O{=}CR^1R^2} + \mathrm{HMe_2SiCH_2CH_2SiMe_2H} \xrightarrow[\text{benzene, r.t., 1h}]{(\mu_3\text{-}\eta^2:\eta^3:\eta^5\text{-acenaphthylene})\mathrm{Ru_3(CO)_7}} \mathrm{R^1R^2CHOSiMe_2CH_2CH_2SiMe_2OCHR^1R^2} \xrightarrow{\mathrm{H^+}} \mathrm{HO{-}CHR^1R^2} \quad (93)$$

The ruthenium cluster $Ru_3(CO)_{12}$ and the $RuCl_2(CO)_2(PPh_3)_2$ system with ethyl or methyl iodide and additionally with diethyl amine as cocatalyst(s) have shown high catalytic activity in facile transformation of cyclic and acyclic amides to amines via hydrosilylation with many trisubstituted silanes (Eq. 94) [149].

$$\text{1-acetylpiperidine} + \mathrm{R_3SiH} \xrightarrow[\text{toluene, 100°C, 16 h}]{\mathrm{RuCl_2(CO)_2(PPh_3)_2}} \text{1-ethylpiperidine} \quad (94)$$

SiR_3	Co-catalyst	Yield [%]
$SiMe_2Ph$	EtI	90
$SiMe_2{}^tBu$	EtI	70
$SiMe_2(OEt)$	EtI	93
$SiMe(OEt)_2$	EtI	91
$SiEt_3$	EtI	96
$SiEt_3$	MeI	98
$SiEt_3$	I_2	94

The diruthenium complex having Ru–H–Si interactions $\{Ru(CO)_2SiTol_2H)\}_2$(μ-dppm)(μ-η^2:η^2-H_2SiTol_2), where dppm is bis(diphenylphosphine)methane and Tol is *p*-tolyl, has been recently reported to be very effective in the hydrosilylation of various ketones (Eq. 95) and imines (Eq. 96) with dihydrosilanes [150].

$$R^1C(=O)R^2 + (p\text{-Tol})_2SiH_2 \xrightarrow[C_6D_6,\ 50^\circ C]{[(CO)_2(HR_2Si)Ru(\mu\text{-}H_2SiR_2)(\mu\text{-dppm})Ru(CO)_2(SiR_2H)],\ R = p\text{-tol}} H(p\text{-Tol})_2Si\text{-}O\text{-}CHR^1R^2 \quad (95)$$

R^1	R^2	Yield [%]
Me	Me	96
Me	$CH{=}CMe_2$	88
$(CH_2)_5$	$(CH_2)_5$	92

$$R^1C(=NR^3)R^2 + (p\text{-Tol})_2SiH_2 \xrightarrow[C_6D_6,\ 50^\circ C]{[(CO)_2(HR_2Si)Ru(\mu\text{-}H_2SiR_2)(\mu\text{-dppm})Ru(CO)_2(SiR_2H)],\ R = p\text{-tol}} H(p\text{-Tol})_2Si\text{-}N(R^3)\text{-}CHR^1R^2 \quad (96)$$

R^1	R^2	R^3	Yield [%]
Ph	Ph	H	74
^{t}Bu	Ph	H	82
Ph	Ph	Me	82

The ruthenium carbene complex (Grubbs catalyst) which has shown high efficiency in alkene methathesis and related processes, since it displays tolerance toward a wide variety of common functional groups, has also appeared of synthetic utility in the hydrosilylation of ketones to yield silyl ethers–one of the most widely used classes of protecting groups in synthetic chemistry (Eq. 97) [151]. The reaction requires temperatures above 50 °C, which generate a slightly increased amount of silylated by-products.

$$R^1C(=O)R^2 + R'_3SiH \xrightarrow[\text{neat, 50-80}^\circ C]{Cl_2(PCy_3)_2Ru(=CHPh)} R^1CH(OSiR'_3)R^2 \quad (97)$$

R^1	R^2	SiR'_3	Yield (%)
Ph	H	$SiMe_2Ph$	95
Ph	H	$SiEt_3$	95
Ph	Me	$SiPh_2H$	95
(structure: alkenyl)	H	$Si(CH_2CH{=}CH_2)Me_2$	95
(structure: alkenyl)	Me	$SiEt_3$	85

Recently, the use of carbon dioxide as a carbon building block [152] has attracted increasing attention. The hydrosilylation of carbon dioxide catalyzed preferably by ruthenium complexes leads to the synthesis of silyl formate esters (Eq. 98) [153]. Results of the reaction of hydrosilylation in supercritical carbon dioxide as a solvent and substrate have recently been reported [154].

$$R_3SiH + O{=}C{=}O \xrightarrow[\text{200-220atm, 90°C}]{RuH_2(PMe_3)_4} R_3SiO{-}CH{=}O \tag{98}$$

A new complex $[Ru^{II}Cl(MeCN)_5][Ru^{III}Cl_4(MeCN)_2]$ prepared by the reaction of $RuCl_3$ (hydrate) in MeCN has appeared as an effective catalyst for the hydrosilylation of CO_2 with n-$(H_{13}C_6)_3SiH$, Me_2PhSiH and some diorganosilanes Et_2SiH_2, Ph_2SiH_2, p-$C_6H_4(SiMe_2H)_2$ [155].

Asymmetric hydrosilylation of prochiral carbonyl compounds and imides has been studied but traditionally a rhodium catalyst system is used. The groups of Uemura and Hidai have developed the ruthenium(II)-catalyzed asymmetric hydrosilylation of imines by using $RuCl_2(PPh_3)$(oxazolinylferrocenyl)diphenylphosphine as chiral ligands and obtained the corresponding secondary amines with high enantioselectivities after acid hydrolysis (up to 89% ee) [156].

The same complexes have been found to be effective catalysts for asymmetric hydrosilylation of ketoximes to give the corresponding amines with high enantioselectivities (up to 89% ee) after acid hydrolysis (Eq. 99) [157].

$$\text{(benzocycloalkanone oxime, } n = 1, 2, 3\text{)} + Ph_2SiH_2 \xrightarrow[\text{AgOTf, THF, r.t.}]{RuCl_2(PPh_3)L} \xrightarrow{H^+} \text{(benzocycloalkyl amine, } NH_2\text{)} \tag{99}$$

n = 1, 2, 3

5.2 Reactions of Organosilicon Compounds with Carbon Monoxide

5.2.1 Silylcarbonylation

Formal silylcarbonylation and silylformylation reactions are mainly catalyzed by cobalt and rhodium complexes (clusters); yet, Chatani et al. [158] have found a new type of carbonylation of diynes with trialkylsilanes leading to catechols (Eq. 100).

$$\text{diyne} + HSi^tBuMe_2 + CO \xrightarrow[50\ atm,\ 140^\circ C,\ 20\ h]{Ru_3(CO)_{12}/PCy_3} \text{product (OSiBu}^t\text{Me}_2)_2 \quad (100)$$

Diyne	Product	Yield [%]
$(MeC(O))_2C(CH_2C\equiv CH)_2$	OSi^{t}BuMe$_2$ / OSi^{t}BuMe$_2$	71
$CH_2(CH_2C\equiv CH)_2$	OSi^{t}BuMe$_2$ / OSi^{t}BuMe$_2$	45
$O(CH_2C\equiv CH)_2$	OSi^{t}BuMe$_2$ / OSi^{t}BuMe$_2$	40

5.2.2 Carbonylation

Chatani and coworkers reported the effective carbonylation of the C–H bond in the aromatic ring via $Ru_3(CO)_{12}$-catalyzed reaction of olefins and CO with heteroaromatics (Eq. 101) [159] and substituted benzene (Eq. 102) [160]. For more examples of the acylation of five-membered heteroaromatic compound see Ref. [161]. The reaction is closely related to the process of the ortho alkylation of substituted aromatic compounds and involves an additional step of CO insertion.

$$\text{1,2-dimethylimidazole} + H_2C{=}CHSiMe_3 \xrightarrow[\text{20 atm CO, toluene, }160^\circ C,\ 20h]{Ru_3(CO)_{12}} \text{imidazolyl–C(O)CH}_2\text{CH}_2\text{SiMe}_3 \quad \text{yield = 66\%} \quad (101)$$

$$\text{2-phenylpyridine} + H_2C{=}CHSiMe_3 \xrightarrow[\text{20 atm CO, toluene, }160^\circ C,\ 20h]{Ru_3(CO)_{12}} \text{ortho-C(O)CH}_2\text{CH}_2\text{SiMe}_3 \quad (102)$$

The coupling of α,β-unsaturated imines, CO and $H_2C{=}CHSiMe_3$ results in the formation of α,α-disubstituted-β,γ-unsaturated butyrolactams (Eq. 103) [162].

$$\text{Ph–CH=CH–CH=N}^t\text{Bu} + H_2C{=}CHSiMe_3 \xrightarrow[\text{10atm CO, toluene, }160^\circ C,\ 20h]{Ru_3(CO)_{12}} \text{Me}_3\text{SiCH}_2\text{CH}_2\text{, Ph-lactam (N}^t\text{Bu)} \quad \text{yield = 72\%} \quad (103)$$

Analogous reactions of pyridylolefins (Eq. 104) [163] and 2-phenyloxazolines [164] lead to a mixture of products.

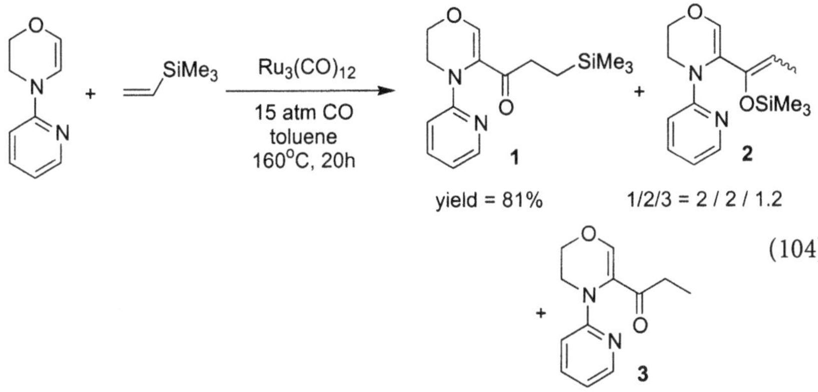

SiMe3 + Ru$_3$(CO)$_{12}$, 15 atm CO, toluene, 160°C, 20h → 1 + 2 + 3

yield = 81% 1/2/3 = 2 / 2 / 1.2

(104)

5.2.3
Cyclocarbonylation

Ruthenium carbonyl complexes have been shown to catalyze a number of carbonylation processes. The ruthenium-catalyzed intramolecular Pauson–Khand reaction was found to proceed in the presence of $Ru_3(CO)_{12}$ (Eq. 105) [165, 166]. The reaction is a valuable tool for selective organic synthesis.

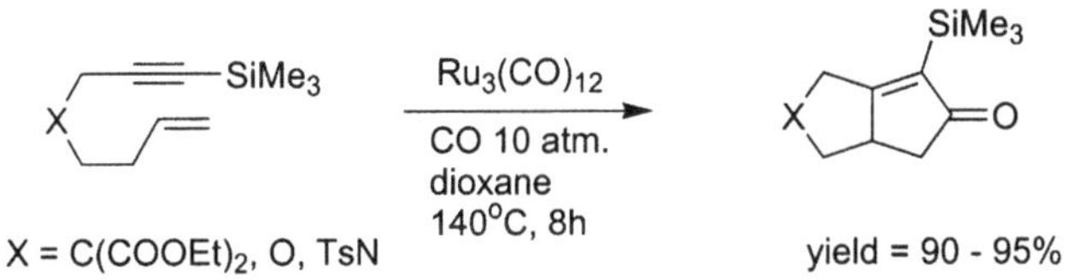

SiMe3; Ru$_3$(CO)$_{12}$, CO 10 atm., dioxane, 140°C, 8h

X = C(COOEt)$_2$, O, TsN yield = 90 - 95%

(105)

Chatani et al. [167] have developed new cyclocarbonylation of internal acetylenes containing terminal aldehyde group (Eq. 106).

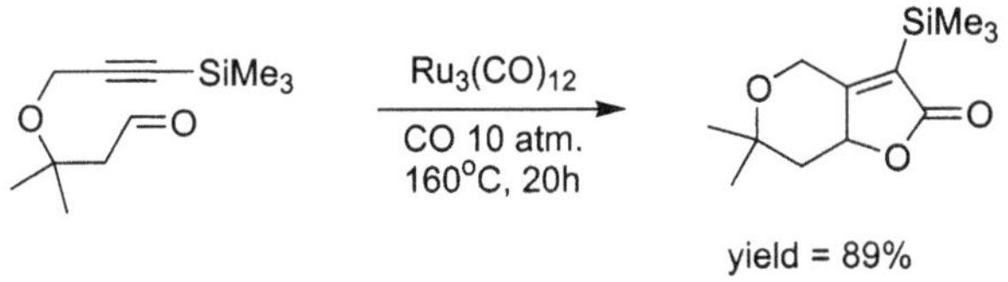

SiMe3; Ru$_3$(CO)$_{12}$, CO 10 atm., 160°C, 20h

yield = 89%

(106)

The reaction can also be carried out intermolecularly, i.e., between internal acetylene, ketone and CO. This [2+2+1] cycloaddition in the presence of $Ru_3(CO)_{12}$ leads to the formation of unsaturated five-membered lactones (Eq. 107) [168, 169].

$$\text{di(2-pyridyl) ketone} + Ph-C\equiv C-SiMe_3 \xrightarrow[\text{CO 5 atm (initially), 160 °C, 20 h}]{Ru_3(CO)_{12}} \text{product} \quad (107)$$

The treatment of internal diyne in the presence of a catalytic amount of $Ru_3(CO)_{12}$ under CO pressure causes novel carbonylation involving 1,2-silyl migration and leads to the formation of a five-membered ring with relatively low yield (Eq. 108) [170].

$$\xrightarrow[\text{CO 10 atm., 150°C, 24h}]{Ru_3(CO)_{12}} \quad \text{yield} = 36\% \quad (108)$$

5.3
Dehydrogenative Condensation Reactions

As we have already mentioned, the silane solvolysis (alcoholysis) is an exceptionally efficient method performed under mild conditions for protecting hydroxyl functions by trialkyl or aryl silyl groups. Many transition-metal complexes as well as other catalysts have been used for this purpose. Among them, ruthenium complexes e.g. monomeric $RuCl_2(CO)_2(PMe_3)_2$ [171] used in organic solvent and dimeric $Ru_2(\mu\text{-}Cl_2)Cl_2(CO)_4(PMe_3)_4$ [172] used in polar solvent were found to be catalytically active in the reaction. However, the Grubbs complex is a much more active and selective catalyst in the alcoholysis of trisubstituted silanes by various alcohols to yield silyl ethers with high yield (above 95%) (Eq. 109, Table 9) [151].

$$R\text{-}OH + R_3'SiH \xrightarrow{Cl_2(PCy_3)_2Ru(=CHPh)} ROSiR'_3 \quad (109)$$

Diphenylsilane reacts with 1,1′-ferrocenedimethanol in the presence of $RuCl_2(PPh_3)_3$ to give a cyclic product [173].

$$\text{Fe}(C_5H_4CH_2OH)_2 + H_2SiPh_2 \xrightarrow{RuCl_2(PPh_3)_3} \text{Fe}(C_5H_4CH_2O)_2SiPh_2 \quad (110)$$

Table 9

Alcohol	Silane	Temp [°C]/time [h]	Silyl ether	Yield [%]
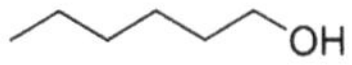 OH	Me_2PhSiH	25 / 0.25	 $OSiMe_2Ph$	> 95
	$^tBuMe_2SiH$	45 / 6	 $OSiMe_2{}^tBu$	> 95
	$(EtO)_3SiH$	35 / 5.5	 $OSiEt_3$	> 95
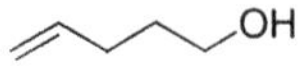 OH	Et_3SiH	25 / 0.5	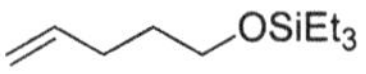 $OSiEt_3$	75
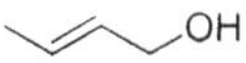 OH	Me_2PhSiH	25 / 2	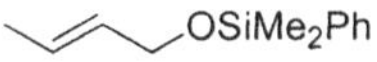 $OSiMe_2Ph$	> 95
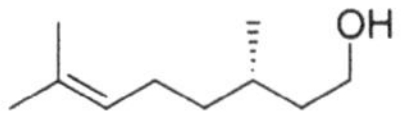 OH	Me_2PhSiH	25 / 2	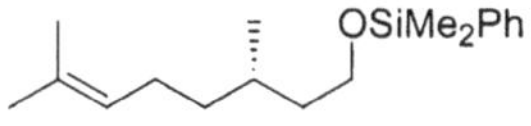 $OSiMe_2Ph$	> 95
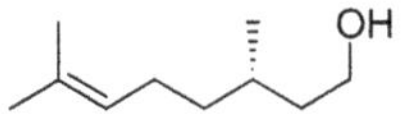 OH	$^tBuMe_2SiH$	45 / 8	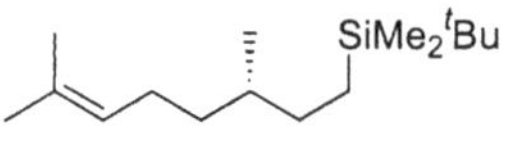 $SiMe_2{}^tBu$	
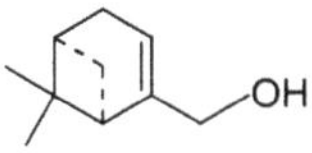 OH	Me_2PhSiH	25 / 1	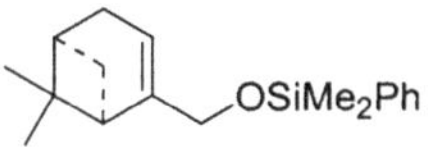 $OSiMe_2Ph$	> 95

6 Synthesis of Organosilicon Polymers

The challenging chemistry of organosilicon derivatives is not limited to molecular compounds. There are numerous examples of catalyzed reactions leading to macromolecular products of different composition, structures and properties. Organosilicon oligomers and polymers prepared by such methods have found wide application as adhesives, membranes, materials of special electrophysical, optical and thermal properties as well as precursors for ceramics [2, 174]. In this subchapter we summarize the most important ruthenium-catalyzed processes leading to silicon-containing macromolecules. They are based on catalytic reactions leading to molecular products presented in the preceding subchapter.

It is well known that hydrosilylation processes usually catalyzed by Pt and Rh complexes can be efficiently applied in polymer chemistry. $Ru_3(CO)_{12}$ was effectively used for the functionalization of polysiloxanes via hydrosilylation of allyl derivatives with polymethylhydrosiloxanes [175]. On the other hand, polymerization via coupling of activated aromatics with dienes occurs mostly in the presence of ruthenium complexes as catalysts (Eq. 111). For representative references see Ref. [176] and papers cited therein.

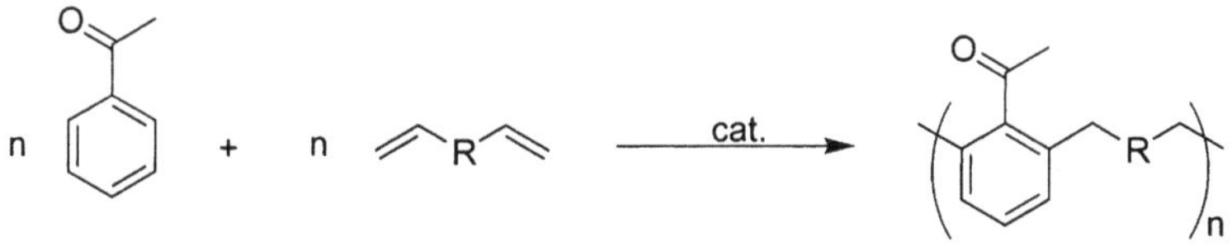

n + n R —cat.→ R n (111)

In the silylative coupling reactions of olefins and dienes with vinylsubstituted silanes, ruthenium catalysts, containing initially or generating in situ Ru–H/Ru–Si bonds, catalyze polycondensation of divinylsubstituted silicon compounds to yield unsaturated silylene (siloxylene, silazanylene)–vinylene–alkenylene (arylene) products (Eq. 112). For recent results see Refs. [177, 178] and for reviews see Refs. [6, 7, 117, 118].

$$(n+1)\ \text{CH}_2{=}\text{CH}\text{–[Si]–CH}{=}\text{CH}_2 \xrightleftharpoons[]{[\text{Ru}],\ -n\ \text{CH}_2{=}\text{CH}_2} \text{CH}_2{=}\text{CH}\text{–}\left[\text{[Si]–CH}{=}\text{CH}\right]_n\text{–[Si]–CH}{=}\text{CH}_2 \tag{112}$$

where [Si] = $-Si(CH_3)_2-$, $-(CH_3)_2Si-O-Si(CH_3)_2-$, $-(C_2H_5O)_2Si-O-Si(OC_2H_5)_2-$, $-(CH_3)_2Si-NH-Si(CH_3)_2-$, $-(CH_3)_2Si-C_6H_4-Si(CH_3)_2-$, $-(CH_3)_2Si-C_6H_4-$

[Ru] = $[RuCl_2(CO)_3]_2$ or $[Ru(H)(OAc)(CO)(PPh_3)_2]$

In the presence of a ruthenium complex, divinylsilicon compounds can also undergo co-polycondensation with dienes, for example, 1,4-divinylbenzene [178, 179]. For recent reviews on the silylative coupling (co)polycondensation see Refs. [6, 7, 117, 118].

Remarkable development over the last 10–15 years in the synthesis of well-defined functional-group-tolerant ruthenium carbenes (Grubbs-related catalysts) also caused real development of the metathesis-based reactions in organosilicon polymers. For recent reviews on metathesis of organosilicon compounds see Refs. [6, 7]. Unsaturated organosilicon polymers can be synthesized via ruthenium carbene catalyzed ring-opening metathesis polymerization (ROMP) of silylsubstituted cycloalkenes (Eq. 113).

$$\text{norbornene}(R^1,R^2) \xrightarrow{\text{cat.}} \left[\text{CH}{=}\text{CH–}c\text{-C}_5\text{H}_7(R^1R^2)\right]_n \tag{113}$$

For general reviews concerning ROMP see Refs. [59, 180, 181], and for synthesis from organosilicon monomers.

$$\text{CH}_2{=}\text{CH}(\text{CH}_2)_m\text{–SiR}_2\text{–}(\text{CH}_2)_n\text{CH}{=}\text{CH}_2 \xrightarrow[-\ \text{CH}_2{=}\text{CH}_2]{\text{cat.}} \left[\text{CH}{=}\text{CH}(\text{CH}_2)_m\text{–SiR}_2\text{–}(\text{CH}_2)_n\right]_x \tag{114}$$

R - Me, Ph
m,n > 0

$$\text{CH}_2{=}\text{CH}(\text{CH}_2)_m\text{–SiMe}_2\text{–Z–SiMe}_2\text{–}(\text{CH}_2)_n\text{CH}{=}\text{CH}_2 \xrightarrow[-\ \text{CH}_2{=}\text{CH}_2]{\text{cat.}} \left[\text{CH}{=}\text{CH}(\text{CH}_2)_m\text{–SiMe}_2\text{–Z–SiMe}_2\text{–}(\text{CH}_2)_n\right]_x \tag{115}$$

Z - atom or group of atoms
m,n > 0

For general reviews on ADMET polymerization see Refs. [59, 180, 183] and for reviews on ADMET polymerization of organosilicon monomers (Eqs. 114, 115) see Refs. [6, 7, 59, 182].

Divinylsilicon compounds similar to monovinyl derivatives do not undergo homometathesis but they react with dienes to give co-polymers according to Eq. (116) [179, 184].

$$\mathrm{CH_2{=}CH\text{-}Si(OEt)_2\text{-}O\text{-}Si(OEt)_2\text{-}CH{=}CH_2} \xrightarrow[\mathrm{CH_2Cl_2,\ reflux,\ 18h}]{\mathrm{Cl_2(PCy_3)_2Ru(=CHPh)}}$$

with 1,4-divinylbenzene: $\mathrm{\left[Si(OEt)_2\text{-}O\text{-}Si(OEt)_2\text{-}CH{=}CH\text{-}C_6H_4\text{-}CH{=}CH\right]_n}$ — linear and *trans*-regular copolymer

with 1,9-decadiene: $\mathrm{\left[Si(OEt)_2\text{-}O\text{-}Si(OEt)_2\text{-}CH{=}CH\text{-}(CH_2)_6\text{-}CH{=}CH\right]_n}$

(116)

Ruthenium carbonyl complexes, for example, $Ru_3(CO)_{12}$ can be used as catalysts for the synthesis of oligosilazanes via dehydrocoupling of Si–H bonds with H–N bonds (Eq. 117) [185].

$$\mathrm{n\ R_2SiH_2 + n\ NH_3 \xrightarrow[THF,\ 60^\circ C]{Ru_3(CO)_{12}} H\text{-}[R_2SiNH]_n\text{-}SiR_2 + 2n\ H_2} \tag{117}$$

$(Me_3P)_3RuH_3(SiMe_3)$ appeared to be the most active catalyst for the dehydrocoupling of alkylsilanes (Eq. 118) [186].

$$\mathrm{HSiMe_3 \xrightarrow[-\,H_2;\ C_6H_{12},\ 150^\circ C]{(Me_3P)_3RuH_3(SiMe_3)} H\text{-}[SiMe_2CH_2]_n\text{-}SiMe_3} \tag{118}$$

Finally, $RuCl_2[P(C_6H_4Me\text{-}p)_3]_3$ was reported to effectively catalyze desilanative polycondensation of bis(trihydrosilyl)benzene (Eq. 119) [187].

$$\mathrm{H_3Si\text{-}C_6H_4\text{-}SiH_3 \xrightarrow[-SiH_4]{RuCl_2[P(C_6H_4Me\text{-}\mathit{p})_3]_3} \left(C_6H_4\text{-}SiH_2\right)_n} \tag{119}$$

7 Conclusions

1. Organometallic catalysis constitutes at present the most valuable tool in organic and heteroorganic syntheses. Although hydrosilylation remains the most often commercially used process in production of organosilicon compounds, other attractive catalytic conversions of silicon derivatives have been revealed and developed over the last 2 decades.

2. While platinum and rhodium are predominantly used as efficient catalysts in the hydrosilylation and cobalt group complexes are used in the reactions of silicon compounds with carbon monooxide, in the last couple of years the chemistry of ruthenium complexes has progressed significantly and plays a crucial role in catalysis of these types of processes (e.g., dehydrogenative silylation, hydrosilylation and silylformylation of alkynes, carbonylation and carbocyclisation of silicon substrates).
   The development of well-defined metal carbene complexes in the last decade has made olefin metathesis one of the most useful catalytic methods for organic synthesis. Although the reactivity of molybdenum and tungsten carbenes is higher than that of ruthenium ones in the field of organosilicon synthesis, the greater tolerance of the functional groups, water and oxygen of ruthenium ones, has made the Grubbs-type ruthenium carbene complexes the most significant catalysts for all metathetical conversion, for example, CM, RCM, tandem ROC/CM, also involving vinyl-substituted silicon compounds. In the presence of ruthenium complexes containing or generating in situ Ru–H/Ru–Si bonds, the latter class of silicon derivatives undergo a novel transformation with alkenes, called the silylative coupling or trans silylation. Ruthenium hydride complexes are fundamental for activation of the C–H bond in alkenes and arenes in the reaction with vinylsilicon and alkynylsilicon compounds.
3. All the reactions overviewed lead to formation of either a direct carbon–silicon bond or a new carbon–carbon bond affected by silyl groups. From the mechanistic point of view, the hydrosilylation and silylformylation reactions as well as the silylative coupling and dehydrogenative condensation occur via intermediates containing a Ru–Si bond (i.e., silicometallics) and Ru–H bonds. On the other hand, all metathesis-based transformations as well as the coupling of arenes and alkenes with vinylsilanes and silylacetylenes proceed via intermediates including ruthenium–carbon bond, for example ruthenium carbene and ruthenium carbyl (aryl) – (i.e., organometallics) and Ru–H bonds.
4. The two different mechanistic pathways of catalysis as well as particularly different activation of C–H by Ru–H complexes in the presence of vinylsilanes with or without β-migration of the silyl group, i.e., the silylative coupling, characteristic only of silicon compounds versus Murai coupling, common in organic syntheses, implies a combination of these reactions leading to quite novel catalytic processes for silicon-based initial compounds. Besides, the collection of experimental material on catalysis by ruthenium complexes in silicon chemistry can be regarded as an initial step in searching for catalytic methods for the synthesis of other p-block elements (e.g., B, Ge, Sn, P) – carbon-bond-containing compounds occurring via intermediates involving a ruthenium–p-block element bond (i.e., inorganometallics).

References

1. (a) Ojima I, Li Z, Zhu J (1998) In: Rappoport Z, Apeloig Y (eds) The chemistry of organic silicon compounds. Wiley, Chichester, chap 29; (b) Marciniec B (ed) (1992) Comprehensive handbook on hydrosilylation. Pergamon, Oxford
2. Brook MA (2000) Silicon in organic organometallic and polymer chemistry. Wiley, New York, chaps 4–6, 9, 12
3. (a) Marciniec B (2002) In: Cornils B, Herrmann WA (eds) Applied homogeneous catalysis with organometallic compounds. Wiley-VCH, Weinheim, chap 2.6; (b) Marciniec B (2002) Silicon Chem 1:155
4. (a) Reichl JA, Berry DH (1999) Adv Organomet Chem 43:197; (b) Marciniec B (2000) Appl Organomet Chem 14:527
5. Marciniec B (1997) New J Chem 21:815
6. Marciniec B, Pietraszuk C (2003) Curr Org Chem 7:691
7. Marciniec B, Pietraszuk C (2003) In: Grubbs R (ed) Handbook of metathesis. Wiley-VCH, Weinheim, chap 2.13
8. Watanabe H, Aoki M, Sakurai N, Watanabe K, Nagai Y (1978) J Organomet Chem 160:C1
9. (a) Marciniec B, Gulinski J (1981) J Mol Catal 10:123; (b) (1978) Polish Patent PL 111 276
10. Hilal HS, Khalaf S, Jondi W (1993) J Organomet Chem 452:167
11. Tanaka M, Hayashi T, Mi Z-Y (1993) J Mol Catal 81:207
12. Takatsuna K, Shiozawa K, Okumura Y (1990) US Patent 4,927,953
13. Djurovich PI, Dolich AR, Berry DH (1994) J Chem Soc Chem Commun 1897
14. Procopis LJ, Mayer B, Plossi D, Berry H (1992) Polym Prepr Am Chem Soc Div Polym Chem 33:1241
15. Kakiuchi F, Igi K, Matsumoto M, Chatani N, Murai S (2001) Chem Lett 422
16. Murai S, Kakiuchi F, Sekine S, Tanaka Y, Kamatani A, Sonoda M, Chatani N (1993) Nature 366:529
17. Maas G, Werle T, Alt M, Mayer D (1993) Tetrahedron 49:881
18. Maas G, Seitz J (2001) Tetrahedron Lett 42:6137
19. Nesmeyanov AN, Friedlina RK, Chukovskaya EC, Petrov RG, Belyavsky AB (1962) Tetrahedron 17:61
20. Seki Y, Takeshita K, Kawamoto K, Murai S, Sonoda N (1986) J Org Chem 51:3890
21. Ojima I, Fuchikami T, Yatabe M (1984) J Organomet Chem 260:335
22. Christ ML, Sabo-Etienne S, Chaudret B (1995) Organometallics 14:1082
23. Lachaize S, Sabo-Etienne S, Donnadieu B, Chaudret B (2003) Chem Commun 214
24. Esteruelas MA, Herrero J, Oro LA (1993) Organometallics 12:2377
25. Martin M, Solo E, Lahoz FJ, Oro LA (2002) Organometallics 21:4027
26. Na Y, Chang S (2000) Org Lett 2:1887
27. Katayama H, Taniguchi K, Kobayashi M, Sagawa T, Minami T, Ozawa F (2002) J Organomet Chem 645:192
28. Adams RD, Barnard TS (1998) Organometallics 17:2567
29. Trost BM, Ball ZT (2001) J Am Chem Soc 123:12726
30. Kawanami Y, Sonoda Y, Mori T, Yamamoto K (2002) Org Lett 4:2825
31. Trost BM, Ball ZT, Joege T (2002) J Am Chem Soc 124:7922
32. Maruyama Y, Yamamura K, Nakayama I, Yoshiuchi K, Ozawa F (1998) J Am Chem Soc 120:1421
33. Marciniec B, Gulinski J (1984) J Organomet Chem 266:C19
34. Marciniec B, Maciejewski H, Gulinski J, Rzejak L (1989) J Organomet Chem 362:273
35. Marciniec B, Pietraszuk C (1991) J Organomet Chem 412:Cl
36. Marciniec B, Pietraszuk C, Foltynowicz Z (1994) J Organomet Chem 474:83

37. Marciniec B, Rzejak L, Gulinski J, Foltynowicz Z, Urbaniak W (1988) J Mol Catal 46:329
38. Marciniec B, Foltynowicz Z, Pietraszuk C, Gulinski J, Maciejewski H (1994) J Mol Catal 90:213
39. Foltynowicz Z, Marciniec B (1989) J Organomet Chem 376:15
40. Marciniec B, Pietraszuk C (1993) J Organomet Chem 447:163
41. Foltynowicz Z, Marciniec B, Pietraszuk C (1991) J Mol Catal 65:113
42. Marciniec B, Pietraszuk C, Foltynowicz Z (1992) J Mol Catal 76:307
43. Foltynowicz Z, Marciniec B, Pietraszuk C (1993) Appl Organomet Chem 7:539
44. Foltynowicz Z, Marciniec B (1997) Appl Organomet Chem 11:667
45. Seki Y, Takeshita K, Kawamoto K (1989) J Organomet Chem 369:117
46. Wakatsuki Y, Yamazaki H, Nakano M, Yamamoto Y (1991) J Chem Soc Chem Commun 703
47. Marciniec B, Pietraszuk C (1995) J Chem Soc Chem Commun 2003
48. Marciniec B, Pietraszuk C (1997) Organometallics 16:4320
49. Marciniec B, Pietraszuk C, Kujawa M (1998) J Mol Catal A 133:41
50. Marciniec B, Foltynowicz Z, Lewandowski M (1994) J Mol Catal 90:125
51. Marciniec B, Kujawa M, Pietraszuk C (2000) Organometallics 19:1677
52. Kujawa-Welten M, Marciniec B (2002) J Mol Catal A 190:79
53. Marciniec B, Chadyniak D, Krompiec S (2004) Tetrahedron Lett 45:0000 (in press)
54. Pietraszuk C, Marciniec B, Jankowska M (2002) Polish Patent PL 355 875
55. Itami Y, Marciniec B, Majchrzak M, Kubicki M (2003) Organometallics 22:1835
56. Itami Y, Marciniec B, Kubicki M (2003) Organometallics 22:3717
57. Itami Y, Marciniec B, Kubicki M (2004) Chem Eur J 10:1239
58. Kakiuchi F, Yamada A, Chatani N, Murai S, Furukawa N, Seki Y (1999) Organometallics 18:2033
59. Ivin KJ, Mol JC (1997) Olefin metathesis and metathesis polymerization. Academic, San Diego
60. (a) Gibson S (2001) In: Fürstner A (ed) Alkene metathesis in organic synthesis. Springer, Berlin Heidelberg New York, p 155; (b) Blechert S, Connon SJ (2003) Angew Chem 115:1944; (c) Blechert S, Connon SJ (2003) Angew Chem Int Ed Engl 42:1900
61. Pietraszuk C, Marciniec B, Fischer H (2003) Tetrahedron Lett 44:7121
62. Pietraszuk C, Marciniec B, Fischer H (2000) Organometallics 19:913
63. Pietraszuk C, Fischer H, Kujawa M, Marciniec B (2001) Tetrahedron Lett 42:1175
64. Kujawa-Welten M, Pietraszuk C, Marciniec B (2002) Organometallics 21:840
65. Sugiono E, Metzroth T, Detert H (2001) Adv Synth Catal 343:351
66. Feher FJ, Soulivong D, Eklund AG, Wydham KD (1997) Chem Commun 1185
67. Pietraszuk C, Marciniec B, Jankowska M (2002) Adv Synth Catal 344:789
68. Brümmer O, Rückert A, Blechert S (1997) Chem Eur J 3:441
69. Stragies R, Blechert S (1999) Tetrahedron 55:8179
70. Roy R, Das SK, Dominique R, Trono MC, Hernandez-Mateo F, Santoyo-Gonzalez F (1999) Pure Appl Chem 71:565
71. Roy R, Dominique R, Das SK (1999) J Org Chem 64:5408
72. Engelhardt FC, Schmitt MJ, Taylor RE (2001) Org Lett 3:2209
73. Taylor RE, Engelhardt FC, Schmitt MJ, Yuan H (2001) J Am Chem Soc 123:2964
74. (a) Cossy J, BouzBouz S, Hoveyda AH (2001) J Organomet Chem 624:327; (b) (2001) J Organomet Chem 634:216
75. Langer P, Holtz E (2002) Synlett 110
76. BouzBouz S, DeLemos E, Cossy J (2002) Adv Synth Catal 344:627
77. Maishal TK, Sinha-Mahapatra DK, Paranjape K, Sarkar A (2002) Tetrahedron Lett 43:2263

78. Schuster M, Lucas N, Blechert S (1997) Chem Commun 823
79. (a) Schneider MF, Lucas N, Velder J, Blechert S (1997) Angew Chem 109:257; Schneider MF, Lucas N, Velder J, Blechert S (1997) (b) Angew Chem Int Ed Engl 36:257
80. (a) Wakamatsu H, Blechert S (2002) Angew Chem 114:2509; (b) Wakamatsu H, Blechert S (2002) Angew Chem Int Ed Engl 41:2403
81. Connon SJ, Rivard M, Zaja M, Blechert S (2003) Adv Synth Catal 345:572
82. (a) Connon SJ, Dunne AM, Blechert S (2002) Angew Chem 114:3989; (b) Connon SJ, Dunne AM, Blechert S (2002) Angew Chem Int Ed Engl 41:3835
83. Mayo P, Tam W, (2002) Tetrahedron 58:9513
84. Delaude L, Demonceau A, Noels AF (1999) Macromolecules 32:2091
85. Karlou-Eyrisch K, Müller BKM, Herzig C, Nuyken O (2000) J Organomet Chem 606:3
86. Tallarico JA, Randall ML, Snapper ML (1997) Tetrahedron 53:16511
87. Feng J, Szeimies G (2002) Eur J Org Chem 2942
88. Michaut M, Parrain J-L, Santelli M (1998) Chem Commun 2567
89. Stragies R, Blechert S (1998) Synlett 169
90. Zaminer J, Stapper C, Blechert S (2002) Tetrahedron Lett 43:6739
91. Poulsen CS, Madsen R (2003) Synthesis 1
92. Mori M (2001) In: Fürstner A (ed) Alkene metathesis in organic synthesis. Springer, Berlin Heidelberg New York, p 133
93. Diver ST, Giessert AJ (2004) Chem Rev 104:1317
94. (a) Stragies R, Schuster M, Blechert S (1997) Angew Chem 109:2628, (b) Stragies R, Schuster M, Blechert S (1997) Angew Chem Int Ed Engl 36:2518
95. Stragies R, Voigtmann V, Blechert S (2000) Tetrahedron Lett 41:5465
96. Lee H-Y, Kim BG, Snapper ML (2003) Org Lett 5:1855
97. Kinoshita A, Sakakibara N, Mori M (1997) J Am Chem Soc 119:12388
98. Tonogaki K, Mori M (2002) Tetrahedron Lett 43:2235
99. Schürer SC, Blechert S (1998) Synlett 166
100. Schuster M, Blechert S (1998) Tetrahedron Lett 39:2295
101. Kakiuchi F, Yamamoto Y, Chatani N, Murai S (1995) Chem Lett 681
102. Harris PWR, Rickard CEF, Woodgate PD (1999) J Organomet Chem 589:168
103. Kakiuchi F, Uetsuhara T, Tanaka Y, Chatani N, Murai S (2002) J Mol Catal A 182–183: 511
104. Bruneau C, Dixneuf PH (1999) Acc Chem Res 32:311
105. Bianchini C, Peruzzini M, Zanobini F, Frediani P, Albinati A (1991) J Am Chem Soc 113:5453
106. Yi ChS, Liu N (1996) Organometallics 15:3968
107. Slugovc C, Mereiter K, Zobetz E, Schmid R, Kirchner K (1996) Organometallics 15: 5275
108. Wakatsuki Y, Yamazaki H, Kumegawa N, Johar PS (1993) Bull Chem Soc Jpn 66:987
109. Yi C, Liu N (1998) Organometallics 17:3158
110. Le Paih J, Derien S, Oezdemir I, Dixneuf PH (2000) J Am Chem Soc 122:7400
111. Steinmetz MG, Udayakumar BS (1989) J Organomet Chem 378:1
112. Steinmetz MG, Mayes RT, Udayakumar BS (1987) J Chem Soc Chem Commun 759
113. Tamao K, Maeda K, Tanaka T, Ito Y (1988) Tetrahedron Lett 29:6955
114. Marshall JA, Yanik MM (2000) Org Lett 2:2173
115. Denmark SE, Pan W (2002) Org Lett 4:4163
116. Trost BM, Ball ZT (2003) J Am Chem Soc 125:30
117. Marciniec B (2000) Mol Cryst Liq Cryst 354:173
118. Marciniec B (2002) In: Khosravi E, Szymanska-Buzar T (eds) Ring opening metathesis polymerization and related chemistry. Kluwer, Dordrecht, p 331

119. Marciniec B, Lewandowski M (1997) Tetrahedron Lett 38:3777
120. Marciniec B, Lewandowski M, Bijpost E, Malecka E, Kubicki M, Walczuk-Gusciora E (1999) Organometallics 18:3968
121. Mise T, Takaguchi Y, Umemiya T, Shimizu S, Wakatsuki Y (1998) Chem Commun 699
122. Fürstner A (1998) In: Fürstner A (ed) Alkene metathesis in organic synthesis. Springer, Berlin Heidelberg New York, p 37
123. Grubbs RH, Chang S (1998) Tetrahedron 54:4413
124. Armstrong SK (1998) J Chem Soc Perkin Trans 1:371
125. Bols M, Skrydstrup T (1995) Chem Rev 95:1253
126. Fensterbank L, Malacria M, Sieburth SMCN (1997) Synthesis 813
127. Shuman M, Gouverneur V (2002) Tetrahedron Lett 43:3513
128. Chang, S, Grubbs RH (1997) Tetrahedron Lett 38:4757
129. Meyer C, Cossy J (1997) Tetrahedron Lett 38:7861
130. Cassidy JH, Marsden SP, Stemp G (1997) Synlett 1411
131. Miller JF, Termin A, Koch K, Piscopio AD (1998) J Org Chem 63:3158
132. Burke SD, Ng RA, Morrison JA, Alberti MJ (1998) J Org Chem 63:3160
133. Semeril D, Olivier-Bourbigou H, Bruneau C, Dixneuf PH (2002) Chem Commun 146
134. Evans PA, Murthy VS (1998) J Org Chem 63:6768
135. Harrison BA, Verdine GL (2001) Org Lett 3:2157
136. (a) Postema MHD, Piper JL (2002) Tetrahedron Lett 43:7095; (b) Boiteau J-G, Van de Weghe P, Eustache J (2001) Tetrahedron Lett 42:239; (c) Briot A, Bujard M, Gouverneur V, Nolan SP, Mioskowski CH (2000) Org Lett 2:1517; (d) Hoye TR, Promo MA (1999) Tetrahedron Lett 40:1429
137. Ahmad I, Falck-Pedersen ML, Undheim K (2001) J Organomet Chem 625:160
138. Hoshi T, Yasuda H, Sanji T, Sakurai H (1999) Bull Chem Soc Jpn 72:821
139. Wada K, Izuhara D, Yamada K, Shiotsuki M, Kondo T, Mitsudo T (2001) Chem Commun 1802
140. Chatani N, Morimoto T, Muto T, Murai S (1994) J Am Chem Soc 116:6049
141. Banti D, North M (2002) Adv Synth Catal 344:694
142. Semeril D, Cleran M, Perez A J, Bruneau C, Dixneuf PH (2002) J Mol Catal A 190:9
143. Semeril D, Cleran M, Bruneau C, Dixneuf PH (2001) Adv Synth Catal 343:184
144. Clark JS, Trevitt GP, Boyall D, Stammen B (1998) Chem Commun 2629
145. Yao Q (2001) Org Lett 3:2069
146. (a) Hevesi L (1995) In: Katritzky AR, Meth-Cohn O, Rees ChW (eds) Comprehensive organic functional group transformations. Pergamon, Oxford, chap 2.18; (b) Colvin EW (1988) Silicon reagents in organic synthesis. Academic, London
147. Nagashima H, Suzuki A, Iura T, Ryu K, Matsubara K (2000) Organometallics 19:3579
148. Matsubara K, Iura T, Maki T, Nagashima H (2002) J Org Chem 67:4985
149. Igarashi M, Fuchikami T (2001) Tetrahedron Lett 42:1945
150. Hashimoto H, Aratani I, Kabuto C, Kira A (2003) Organometallics 22:2199
151. Maifeld SV, Miller RL, Lee D (2002) Tetrahedron Lett 43:6363
152. Leitner W (1996) Coord Chem Rev 153:257
153. (a) Koinuma H, Kawakami F, Kato H, Hirai H (1981) J Chem Soc Chem Commun 213; (b) Süss-Fink, Reiner J (1981) J Organomet Chem 221:C36
154. Jessop PG, Ikariya T, Noyori R (1999) Chem Rev 99:475
155. Jansen A, Goerls H, Pitter S (2000) Organometallics 19:135
156. (a) Nishibayashi Y, Takei I, Uemura S, Hidai M (1998) Organometallics 17:3420; (b) Takei I, Nishibayashi Y, Arikawa Y, Uemura S, Hidai M (1999) Organometallics 18:2271

157. Takei I, Nishibayashi Y, Ishii Y, Mizobe Y, Uemura S, Hidai M (2001) Chem Commun 2360
158. Chatani N, Fukumoto Y, Ida T, Murai S (1993) J Am Chem Soc 115:11614
159. Chatani N, Fukuyama T, Kakiuchi F, Murai S (1996) J Am Chem Soc 118:493
160. Chatani N, Ie Y, Kakiuchi F, Murai S (1997) J Org Chem 62:2604
161. Chatani N, Fukuyama T, Tatamidani H, Kakiuchi F, Murai S (2000) J Org Chem 65: 4039
162. Chatani N, Kamitani A, Murai S (2002) J Org Chem 67:7014
163. Chatani N, Ishii Y, Ie Y, Kakiuchi F, Murai S (1998) J Org Chem 63:5129
164. Ie Y, Chatani N, Ogo T, Marshall DR, Fukuyama T, Kakiuchi F, Murai S (2000) J Org Chem 65:1475
165. Morimoto T, Chatani N, Fukumoto Y, Murai S (1997) J Org Chem 62:3762
166. Kondo T, Suzuki N, Okada T, Mitsudo T (1997) J Am Chem Soc 119:6187
167. Chatani N, Morimoto T, Fukumoto Y, Murai S (1998) J Am Chem Soc 120:5335
168. Tobisu M, Chatani N, Asaumi T, Amako K, Ie Y, Fukumoto Y, Murai S (2000) J Am Chem Soc 122:12663
169. Chatani N, Amako K, Tobisu M, Asaumi T, Fukumoto Y, Murai S (2003) J Org Chem 68:1591
170. Onitsuka K, Katayama H, Sonogashira K, Ozawa F (1995) J Chem Soc Chem Commun 2267
171. Oehmichen U, Singer H (1983) J Organomet Chem 243:199
172. Chung MK, Ferguson G, Robertson V, Schlaf M (2001) Can J Chem 79:949
173. Yamaguchi I, Ishii H, Sakano T, Osakada K, Yamamoto T (2001) Appl Organomet Chem 15:197
174. (a) Zeldin M, Wynne KJ, Allcock HR (1988) Inorganic and organometallic polymers: Macromolecules containing silicon, phosphorus and other inorganic elements. American Chemical Society, Washington, DC; (b) Zeigler JM, Fearon FWG (eds) (1990) Silicon-based polymer science. American Chemical Society, Washington, DC
175. Marciniec B, Gulinski J, Kopylova L, Maciejewski H, Grundwald-Wyspianska M, Lewandowski M (1997) Appl Organomet Chem 11:843
176. (a) Londergan TM, Weber WP (1997) Macromol Rapid Commun 18:207; (b) Wang G, Guo H, Weber WP (1996) J Organomet Chem 521:351; (c) Guo H, Wang G, Tapsak MA, Weber WP (1995) Macromolecules 28:5686
177. (a) Majchrzak M, Itami Y, Marciniec B, Pawluc P (2000) Tetrahedron Lett 41:10303; (b) Marciniec B, Małecka E, Ścibiorek M (2003) Macromolecules 36:5545; (c) Marciniec B, Małecka E (2003) J Phys Org Chem 16:818; (d) Marciniec B, Itami Y, Majchrzak M (2004) Mol Cryst Liq Cryst (in press)
178. Majchrzak M, Itami Y, Marciniec B, Pawluc P (2001) Macromol Rapid Commun 22: 202
179. (a) Marciniec B, Małecka E, Majchrzak M, Itami Y (2001) Macromol Symp 174:137, (b) Marciniec B, Majchrzak M (2003) J Organomet Chem 686:228
180. (a) Buchmeiser MR (2000) Chem Rev 100:1565
181. (a) Frenzel U, Nuyken O (2002) J Polym Sci Part A Polym Chem 40:2895; (b) Khosravi E, Grubbs RH (1999) In: Schlütter AD (ed) Synthesis of polymers. Wiley-VCH, Weinheim, chap 3
182. (a) Finkel'shtein ES, Marciniec B (1995) In: Marciniec B, Chojnowski J (eds.) Progress in organosilicon chemistry. Gordon and Breach, Amsterdam, p 445, (b) Finkel'shtein ES (1995) Polym Sci Ser B 37:185; (c) Finkel'shtein ES, Portnykh EB, Ushakov NV, Marciniec B (1995) In: Jones RG (ed) Silicon-containing polymers. Royal Society of Chemistry, Cambridge, p 27

183. (a) Schwendeman JE, Church AC, Wagener KB (2002) Adv Synth Catal 344:597; (b) Tindall D, Pawlow JH, Wagener KB (2001) In: Fürstner A (ed) Alkene metathesis in organic synthesis. Springer, Berlin Heidelberg New York, p 183; (c) Davidson TA, Wagener KB (1999) In: Schlütter AD (ed) Synthesis of polymers. Wiley-VCH, Weinheim, chap 4
184. Malecka E, Marciniec B, Pietraszuk C, Church AC, Wagener KB (2002) J Mol Catal 190: 27
185. Biran C, Blum YD, Glaser R, Tse DS, Youngdahl KA, Laine RM (1988) J Mol Catal 48: 183
186. Procopio LJ, Mayer B, Plossi K, Berry DH (1992) Polym Prepr Am Chem Soc Div Polym Chem 33:1241
187. Sakakura T, Kumberger O, Tan RP, Arthur M-P, Tanaka M (1995) J Chem Soc Chem Commun 193

Topics Organomet Chem (2004) 11: 249–276
DOI 10.1007/978-3-540-40027-1

Ruthenium-Catalyzed Synthesis of Heterocyclic Compounds

Yoshihiko Yamamoto[1] · Kenji Itoh[2] (✉)

[1] Department of Applied Chemistry, Graduate School of Engineering, Nagoya University, Chikusa, Nagoya 464-8603, Japan
yamamoto@apchem.nagoya-u.ac.jp

[2] Department of Molecular Design and Engineering, Graduate School of Engineering, Nagoya University, Chikusa, Nagoya 464-8603, Japan
itohk@apchem.nagoya-u.ac.jp

Abstract Heterocyclic compounds have been synthesized by means of stoichiometric reagents under acidic or basic conditions. Recent progress of homogeneous transition-metal catalysis realizes a highly selective and atom-economical methodology for synthesis of heterocycles under neutral and mild conditions. This review highlights the recent advances in the area of ruthenium-catalyzed syntheses of heterocyclic compounds.

Keywords Ruthenium catalysis · Heterocyclic compounds · Cyclization · Carbon–heteroatom bond formation · Carbon–carbon bond formation

1 Introduction

The occurrence of heterocyclic compounds in nature is widespread, and their use for pharmaceuticals and electronic devices such as conductors, sensors, and light-emitting diodes is becoming more and more important [1]. In this respect, the development of new efficient strategies for the synthesis of heterocyclic compounds with a variety of structural diversity is a continuing challenge in synthetic organic chemistry. Heterocyclic compounds have generally been synthesized by means of condensation reactions in acidic or basic media, which produce considerable amounts of salt waste. The formation of undesirable byproducts is also a problem to be avoided in the conventional methods using stoichiometric reagents. Transition-metal catalysis offers a powerful solution to synthesize heterocyles in atom-economical manner under neutral and mild reaction conditions. The judicious control of chemoselectivity, regioselectivity, and stereoselectivity is also an important merit of transition-metal catalysis.

This review outlines the recent advances in the synthesis of heterocyclic compounds utilizing ruthenium catalysts. The first part is devoted to the synthesis of heterocycles via carbon–heteroatom bond formations. Heterocyclic frameworks are also constructed by ring closure of heteroatom-tethered acyclic molecules. The second part covers the ruthenium-catalyzed carbon–carbon bond forming cyclizations yielding heterocycles. Other examples, in which ruthenium catalysis indirectly participates in heterocycle formation, are collected in the final section. Although a heterocyclic ring was formed without catalysis, ruthenium-catalyzed processes play pivotal roles in such examples.

2 Synthesis of Heterocycles via Carbon–Heteroatom Bond Forming Cyclizations

2.1 Cyclization of Unsaturated Alcohols and Amines

Saturated heterocyclic compounds with various ring sizes can be synthesized by cyclization of an ω-substituted amine, alcohol, or thiol via an intramolecular nucleophilic substitution. This method is quite general, but a leaving group must be discarded as organic and inorganic waste. From the viewpoint of atom-economy [2], cyclization via simple addition is highly desirable. Related intramolecular additions of a heteroatom functionality to an alkene or alkyne through π complexes with cations such as Br^+, I^+, and Hg^+ are useful methods giving rise to heterocycles with a functionalized side chain, although they call for stoichiometric amounts of an electrophilic promoter. In this context, transition-metal-catalyzed cyclizations of an unsaturated amine or alcohol have been developed as an environmentally benign process. A ruthenium(II) arene complex, $RuCl_2(PPh_3)$(p-cymene), has proved to convert catalytically hydroxy enynes **1** to highly substituted furans **2** upon heating (Eq. 1) [3]. Such a cycloisomerization was considered to proceed via intramolecular addition of the hydroxy group to the terminal alkyne moiety activated by a cationic ruthenium species followed by a 1,5-proton shift.

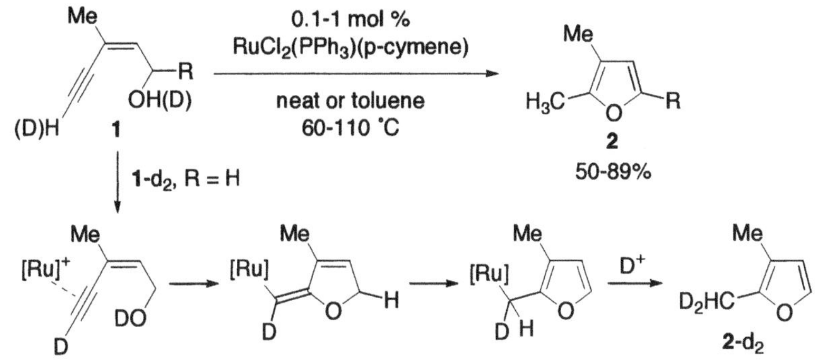

(1)

The same research group has also developed the ruthenium-catalyzed intermolecular addition of carboxylic acids to carbon–carbon triple bonds [4]. When α-hydroxy acids **3** were employed with terminal alkynes, 1,3-dioxolan-4-ones **5** were synthesized via cyclization of enol ester intermediates **4** (Eq. 2) [5].

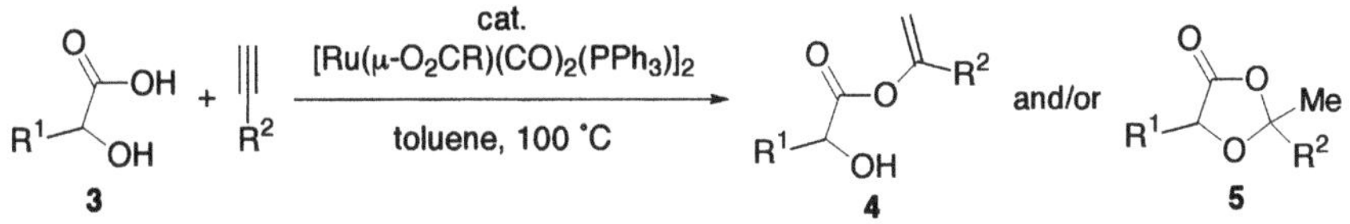

(2)

The 5-endo and 6-endo cyclizations of α,ω-alkynols leading to dihydrofurans and dihydropyrans have been achieved with molybdenum and tungsten catalysis [6]. Transition-metal vinylidene intermediates have been claimed to be involved in these cycloisomerizations [7]. Related cyclizations of bis-homopropargyl alcohols were recently developed using ruthenium catalysis as shown in Eq. (3) [8]. In the presence of the sodium salt of *N*-hydroxysuccinimide **9**,

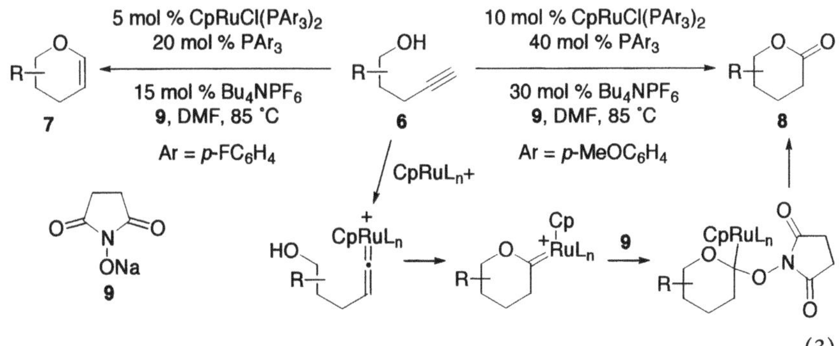

(3)

cationic ruthenium species derived from $CpRuCl(PAr_3)_2$, where Cp is cyclopentadienyl, and Bu_4NPF_6 converted alkynols **6** into dihydropyrans **7** and lactones **8**. When a more electron-accepting triarylphosphine was used, **7** was obtained as a major product. On the other hand, the less electron-accepting phosphine ligand predominantly gave rise to the oxidation product **8**. In a similar manner, the cycloisomerization–oxidation of homopropargyl alcohol afforded γ-butyrolactones [9]. Remarkably, a related cationic ruthenium catalyst system transformed alkynyldiols **10** and allyl alcohols into oxygen heterocycles **12** (Eq. 4) [10]. Such a reconstructive coupling process proceeds via ruthenium allenylidene intermediates **11**.

10 mol % $CpRuCl(PPh_3)_2$

20 mol % NH_4PF_6
100 °C

(4)

As shown in Eqs. (1–3), the cyclization of hydroxyalkynes gave unsaturated oxygen heterocycles in atom-economical manners. If readily available hydroxyalkenes undergo similar cyclization with concomitant elimination of H_2, such an oxidative cyclization would also afford an unsaturated oxygen heterocycle. In fact, 1,1-disubstituted 4-penten-1-ols **13** gave 2,3-dihydrofurans **14** in good yields upon being heated at 160 °C in the presence of $Ru_2(CO)_3$–PPh_3 catalyst, allyl acetate, and K_2CO_3 under 5 atm CO (Eq. 5) [11]. Allyl acetate is imperative as a hydrogen acceptor.

cat $Ru_2(CO)_3$-PPh_3

5 atm CO, K_2CO_3, allyl acetate
toluene, 160 °C

14 62->99%

(5)

Intramolecular addition of amine N–H bonds to carbon–carbon multiple bonds would afford nitrogen heterocycles. To realize catalytic cyclization of α,ω-aminoalkenes or aminoalkynes, various catalytic systems have been developed especially with early transition metals such as titanium, zirconium, lanthanide metals, and actinide metals [12]. Late-transition-metal catalysis based on Ni, Pd, and Rh has also proved to be efficient [12]. Recently, the ruthenium-catalyzed intramolecular hydroamination of aminoalkynes **15** was reported to afford 5–7-membered ring products **16** in various yields (Eq. 6) [13]. Among various precatalysts tested, ruthenium complexes with a π-acidic ligand, especially $Ru_3(CO)_{12}$ and $(\eta^3\text{-}C_3H_5)RuBr(CO)_3$, showed high catalytic activity. Both an internal and a terminal alkyne substrate gave cyclic imines **16** as sole products, although 7-phenylhept-6-ynl-1-amine (**15** *R*=Ph, *n*=3) gave a saturated compound **17** as well as the corresponding normal cyclic imine. Using this protocol, indole was obtained in 54% isolated yield via 5-endo cyclization of 2-ethynylbenzene **18** (Eq. 7) [13].

15 → **16**, **17**: cat $Ru_3(CO)_{12}$, diglyme, 110-140 °C; R = H, Ph, Me; n = 1-3 (6)

18 → indole 54%: cat $Ru_3(CO)_{12}$, diglyme, 110 °C (7)

Similar cyclic imines can be synthesized by intramolecular oxidative amination of aminoalkenes, which are less expensive than aminoalkynes. In the presence of catalytic amounts of $[RuCl_2(CO)_3]_2$/1,3-bis(diphenylphosphino)propane and excess K_2CO_3/allyl acetate, various aminoalkenes **19** possessing substituent(s) β to the amino group afforded five- and six-membered cyclic imines **20** in moderate-to-excellent yields (Eq. 8) [14].

19 → **20** 40 - 87%: cat $[RuCl_2(CO)_3]_2$ / dppp, K_2CO_3, allyl acetate, *N*-methylpiperidine, 140 °C, 8 h (8)

R^1 = H, Ph, Me, Et; R^2, R^3 = H, Ph, Me; n = 1-2

2.2 Co-Cyclotrimerizations of Alkynes with Carbon–Heteroatom Multiple Bonds and Related Cycloaddition

The catalyzed cycloadditions have received growing attention because they form multiple carbon–carbon and carbon–heteroatom bonds simultaneously. In other words, a catalytic cycloaddition strategy is a convergent and highly atom-economical approach [2]. Heterocycle synthesis using such a multicomponent coupling process is highly beneficial in terms of sustainable chemistry. Catalyzed [2+2+2] co-cyclotrimerizations of alkynes with carbon–heteroatom multiple bonds are one of the most straightforward and atom-efficient method to assemble a complex heterocyclic framework from readily available acyclic components [15].

Although transition-metal-catalyzed co-cyclotrimerization of two alkyne molecules with a nitrile is a viable route to a substituted pyridine, such useful pyridine annulations were almost always confined to cyclopentadienylcobalt catalysts until the efficient ruthenium-catalyzed cycloaddition of 1,6-diynes with electron-deficient nitriles was developed by these authors under extremely moderate conditions [16]. The cobalt-catalyzed cycloaddition of diynes with nitriles has been reported to furnish bicyclic pyridines. Electron-deficient nitriles such as ethyl cyanoformate and pentafluorobenzonitrile, however, gave the desired pyridine only in poor yields, less than 10% [17]. In contrast, 1,6-diynes derived from dimethyl malonate **21** [*X* is $C(CO_2Me)_2$] reacted with a variety of nitriles directly connected with an electron-withdrawing group at 60–80 °C in the presence of 2–10 mol % (pentamethylcyclopentadienyl)RuCl(cyclooctadiene) [Cp*RuCl(cod)] to give the bicyclic pyridines **22** in 31–90% yields (Eq. 9) [16]. When *N,N*-dipropargyl tosylamide and dipropargyl ether (*X* is *N*-tosyl and O, respectively) were employed, interesting 3-pyrroline-fused and 2,5-dihydrofuran-fused pyridines were obtained.

$$\mathbf{21} + \text{Ewg}-\text{C}\equiv\text{N} \xrightarrow[\text{ClCH}_2\text{CH}_2\text{Cl, 60-80 °C}]{\text{2-10 mol \% Cp*RuCl(cod)}} \mathbf{22}\ \text{31-90\%} \tag{9}$$

X = $C(CO_2Me)_2$, NTs, O; Ewg = CO_2Et, COPh, COMe, Ts, C_6F_5, CCl_3

This ruthenium catalysis is limited to the electron-deficient nitriles, and acetonitrile or benzonitrile hardly produced the corresponding bicyclic pyridines. On the other hand, several dicyanides having two proximate cyano groups, malononitrile, succinonitrile, *o*-phthalonitrile, and fumaronitrile, were found to participate into a novel ruthenium-catalyzed pyridine annulation (Eq. 10) [18]. Especially, the reaction of malonate-derived 1,6-diyne with 1.5 equiv. malononitrile proceeded even at room temperature in the presence of 5 mol % of Cp*RuCl(cod) to afford a cyanomethyl-substituted pyridine in 95% yield. It is interesting to note that one of two cyano groups remains intact after the completion of the reaction. When unsymmetrical 1,6-diynes ($R^1 \neq H$) were used, one of the two possible regioisomers, the 2,3,4,6-substituted one, was obtained almost exclusively. Utilizing such an excellent regioselectivity, a symmetrical 2,2′-bipyridine derivative **24** was obtained in only a single operation from a linear tetrayne substrate **23** and malononitrile in 95% isolated yield (Eq. 11).

$$\text{diyne} + \text{Z(CN)}_2 \xrightarrow[\text{ClCH}_2\text{CH}_2\text{Cl, rt-80 °C; 46-97\%}]{\text{2-10 mol \% Cp*RuCl(cod)}} \text{major} + \text{minor} \tag{10}$$

X = $C(CO_2Me)_2$, NTs, O; R^1 = H, Me, Ph, $SiMe_3$

23 → 10 mol % Cp*RuCl(cod), $ClCH_2CH_2Cl$, 80 °C, 20 h; E = CO_2Me → **24** 95% (11)

Related co-cyclotrimerizations of two alkyne molecules with limited isocyanates have also been achieved using cobalt and nickel catalysts. With respect to intramolecular versions, two examples of the cobalt(I)-catalyzed cycloaddition of α,ω-diynes with isocyanates have been reported to afford bicyclic pyridones only in low yields, although 2,3-dihydro-5(1*H*)-indolizinones were successfully obtained from isocyanatoalkynes and several silylalkynes with the same cobalt catalysis [19]. On the other hand, the ruthenium catalysis using Cp*RuCl(cod) as a precatalyst effectively catalyzed the cycloaddition of 1,6-diynes **21** with 4 equiv. of isocyanates in refluxing 1,2-dichloroethane to afford bicyclic pyridones **25** in 58–93% yield (Eq. 12) [20]. In this case, both aryl and aliphatic isocyanates can be widely employed.

21 + O=C=NR → 5 mol % Cp*RuCl(cod), $ClCH_2CH_2Cl$, 90 °C → **25** 58-93% (12)

X = $C(CO_2Me)_2$, CH_2, NTs, O; R = Ph, 1-naphthyl, 2-furyl, benzyl, cyclohexyl

In contrast to isocyanates, isothiocyanates have hardly been examined as cycloaddition components, because the strong coordination of organosulfur compounds frequently deactivates a catalytic species [21]. Some organoruthenium complexes, however, recently proved to be efficient catalysts for the formation of carbon–sulfur bonds [21]. The catalytic cycloaddition of diynes with isothiocyanates was also successfully achieved using Cp*RuCl(cod) as a precatalyst [22]. Importantly, the cycloaddition took place at the C=S double bonds of the isothiocyanates to afford thiopyranimines **26** (Eq. 13). This reaction requires 10 mol % of the precatalyst as well as the diynes possessing a quarternary carbon center at the 4-position. When excess amounts of carbon disulfide were also employed in place of the isothiocyanates, a bicyclic dithiopyrone **26** [*X* is $C(CO_2Me)_2$, *Z* is S] was obtained in 50% yield.

21 + S=C=Z → 10 mol % Cp*RuCl(cod), $ClCH_2CH_2Cl$, 90 °C → **26** 35-88% (13)

X = $C(CO_2Me)_2$, $C(COMe)_2$, $C(CN)_2$, etc; Z = NPh, NCO_2Et, NCOPh, NCy, S

A highly electron-deficient carbon–oxygen double bond can also participate in the co-cyclotrimerization with alkynes under the ruthenium catalysis. The cycloaddition of commercially available diethyl ketomalonate with the diynes **21** proceeded at 90 °C in the presence of 5–10 mol % Cp*RuCl(cod). The expected fused 2*H*-pyrans **27**, however, underwent thermal electrocyclic ring-opening to produce cyclopentene derivatives **28** (Eq. 14) [23].

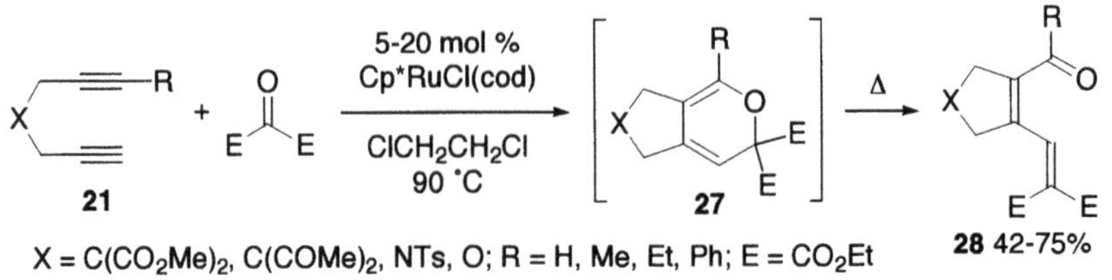

X = $C(CO_2Me)_2$, $C(COMe)_2$, NTs, O; R = H, Me, Et, Ph; E = CO_2Et **28** 42-75%

(14)

In addition to these intramolecular [2+2+2] cycloadditions, intramolecular [4+2] cycloaddition of yne-enones **29** leading to fused pyrans **30** has been achieved by means of the ruthenium catalysis with a cationic complex, $CpRu(MeCN)_3PF_6$ (Eq. 15) [24]. Such hetero Diels–Alder cycloaddition was considered to proceed via an oxaruthenacycle **31**.

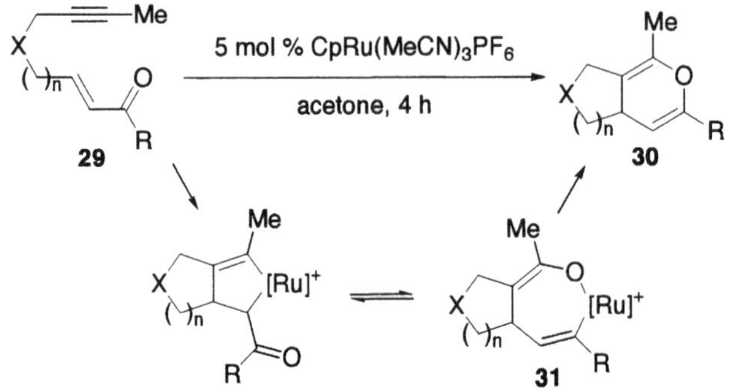

(15)

2.3 Cyclocarbonylations of Unsaturated Molecules

Transition-metal-catalyzed carbonylation reactions are useful one-carbon homologation techniques in organic synthesis, involving industrially important processes, for example, Fisher–Tropsch reaction, Monsanto acetic acid process, and hydroformylation (oxo reaction) [25].

Intramolecular condensation of ω-hydroxycarboxylic acids is a standard method to prepare lactones. Acid catalysts or more elaborate mediators are usually required as well as continuous removal of water. Transition-metal-catalyzed cyclocarbonylation of unsaturated alcohols is a fascinating alternative, which proceeds under neutral conditions [26]. Intramolecular hydroesterification of propargyl alcohols gives 2(5*H*)-furanones, which are ubiquitous lactones in natural products. In this respect, an alternative oxidative cyclocarbonylation is advantageous because readily available allyl alcohols can be employed as substrates. In the presence of 5 mol % of $RuCl_2(PPh_3)_3$ and excess allyl acetate and K_2CO_3, 1,1-disubstituted allyl alcohols **32** were converted into substituted 2(5*H*)-furanones **33** in 15–77% yields under 10 kg cm^{-2} CO (Eq. 16) [27]. Similarly, a phthalide **35** was synthesized in 49% yield from a 1,1-disubstituted benzyl alcohol **34** (Eq. 17) [27].

5 mol % $RuCl_2(PPh_3)_3$; 10 kg cm^{-2} CO; K_2CO_3, allyl acetate; THF, 200 °C — **32** → **33** 15-77% (16)

5 mol % $RuCl_2(PPh_3)_3$; 10 kg cm^{-2} CO; K_2CO_3, allyl acetate; THF, 200 °C — **34** → **35** 49% (17)

Substituted 2(5*H*)-furanones were also synthesized from allenyl alcohols **36** (Eq. 18) [28]. In the presence of 1 mol % of $Ru_3(CO)_{12}$, Et_3N and 10 atm CO, **36** underwent cyclocarbonylation to afford the butenolides **37** in excellent yields. Remarkably, the ruthenium(0)-catalyzed cyclocarbonylation protocol is applicable to the syntheses of 6–8-membered α,β-unsaturated lactones **38–40** in good yields (Eq. 19) [28, 29].

1 mol % $Ru_3(CO)_{12}$; Et_3N, 10 atm CO; dioxane, 100 °C — **36** → **37** 91-99% (18)

1 mol % $Ru_3(CO)_{12}$; 5-10 atm CO; Et_3N/dioxane or *N*-methylpiperidine or Et_3N, 100 °C — **38** n = 1, 95%; **39** n = 2, 80%; **40** n = 3, 81% (19)

Similarly, ruthenium(0)-catalyzed cyclocarbonylation of allenic sulfonamides **41** yielded γ- and δ-unsaturated lactams **42** (Eq. 20) [30]. The lactam formation is claimed to start with the oxidative addition of a $Ru(CO)_4$ fragment into the N–H bond of **41**. Subsequent syn addition of the resultant Ru–H species to the terminal C–C double bond followed by the carbonyl insertion into the Ru–N bond and the reductive elimination would give **42**. This mechanism is supported by the cyclization of a deuterated substrate (R=n-hexyl, n=0), in which the deuterium label was selectively transferred onto the α methyl substituent.

1 mol % $Ru_3(CO)_{12}$, Et_3N, 20 atm CO, dioxane, 100 °C

41 n = 0, 1 → **42** 70-95%

(20)

The transition-metal-mediated cyclocoupling reaction of an alkyne, an alkene, and CO, the Pauson–Khand reaction, is a powerful method to construct a cyclopentenone skeleton, which is a useful building block in organic synthesis [31]. Recently, such a powerful cyclocarbonylation reaction was extended to intramolecular variants of enynes catalyzed by several late-transition-metal complexes, including ruthenium a one [32]. Furthermore, the ruthenium-catalyzed cyclocarbonylation was successfully applied to ynals **43** to obtain interesting bicyclic α,β-unsaturated γ-butyrolactones **44** in good yields (Eq. 21) [33]. The butenolide formation is claimed to start from the oxidative addition of the formyl C–H bond of **43** to the ruthenium center. The next step is the intramolecular insertion of the pendant alkyne into the generated acyl–ruthenium bond to give *exo*-2-(ruthenamethylene)cyclopentanone **45**, which induces CO insertion to result in the second acylruthenium intermediate. Its cyclization onto the C=O double bond is followed by reductive elimination to give the final product **44**. In the case of R=Me, α-ethylidenecyclopentane **46** was formed by the reductive elimination from **45**.

cat $Ru_3(CO)_{12}$, 10 atm CO, toluene, 160-180 °C

43 → **44** 62-93%; **45**; **46**

(21)

Related cyclocarbonylations of enones or enals leading to γ-butyrolactones were achieved using stoichiometric and catalytic amounts of titanocene and its derivative [34]. In addition to those partially intramolecular reactions, their completely intermolecular version was recently developed by means of ruthenium catalysis [35,36]. The ruthenium-catalyzed [2+2+1] γ-butyrolactone annulation employed α-dicarbonyl compounds **47** (α-diketones, α-ketoesters, and α-ketoamides), ethylene, and CO (Eq. 22). As a ketone substituent R^2, electron-deficient groups such as $4\text{-}CF_3C_6H_4$ or CF_3 are superior to the corresponding less electron-deficient ones. When an unsymmetrical α-diketone was used as the dicarbonyl component, two possible regioisomers of **48** were obtained. On the other hand, α-ketoesters usually reacted selectively at their ketone carbonyl groups to form ester-substituted lactones. Benzofuran-2,3-diones were, however, found to react at both ketone and ester carbonyl groups with various ratios depending on the substituents on the phenyl ring [37]. The presence of two neighboring carbonyl groups in **47** was essential in order to form chelated intermediates **49** (Eq. 22). In accord with this assumption, heteroaryl groups including pyridine, pyrazine, thiazole, and oxazole can also be used as directing functionalities instead of the carbonyl groups in **47**. In fact, di-2-pyridyl ketone is successfully coupled with CO and various alkenes and alkynes (Eq. 23) [36].

47 → **48**; cat $Ru_3(CO)_{12}$ / $P(4\text{-}CF_3C_6H_4)_3$, 3 atm ethylene, 5 atm CO, toluene, 160-180 °C; **49** (22)

cat $Ru_3(CO)_{12}$ / with or without $P(4\text{-}CF_3C_6H_4)_3$, 5 atm CO, toluene, 160 °C (23)

In a similar manner, imines **50** with various ancillary groups *X*, such as ethoxycarbonyl, 2-pyridyl, or 2-thiazolyl, are also converted into lactams **51** in moderate-to-good yields (Eq. 24) [38]. The [2+2+1] lactam formation using a chiral substrate **52**, ethylene, and CO quantitatively furnished the spiro lactam **53** (Eq. 25) [39]. The cycloaddition exclusively took place at the carbon–nitrogen double bond next to the oxazine oxygen atom, although **53** was obtained as a diastereomeric mixture.

50 (Ar = *p*-$MeOC_6H_4$) → **51**; cat $Ru_3(CO)_{12}$ / with or without $P(4\text{-}CF_3C_6H_4)_3$, 5 atm CO, toluene, 160 °C (24)

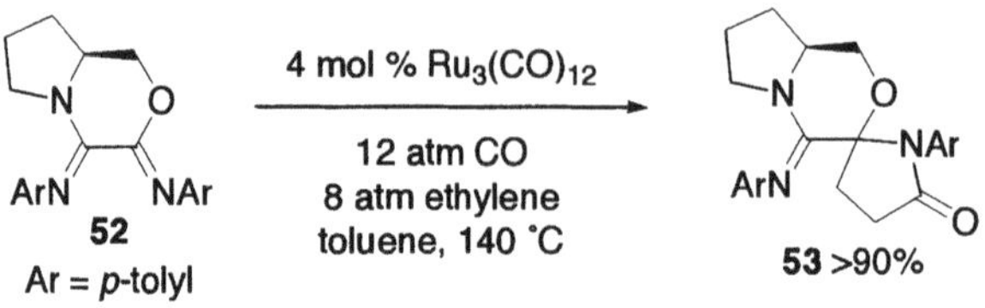

(25)

The cyclocarbonylation of α,β-unsaturated imines **54** also gave five-membered lactams **55** or **56** with or without **57** depending on the substitution patterns of the starting materials (Eq. 26) [40]. Such a useful [4+1] cycloaddition was applied to a cyclic imine **58** conjugated with a cyclohexenyl group to afford an interesting aza-tricycle **59** in 65% yield (Eq. 27). In the presence of an alkene such as ethylene, norbornene, and vinyltrimethylsilane, a similar cyclocarbonylation of α,β-unsaturated imines **60** gave rise to α-alkylated β,γ-unsaturated lactam **61** in various yields (Eq. 28) [41, 42]. Several plausible mechanisms are proposed for the formation of **61**, but the elucidation of the entire mechanism calls for further work.

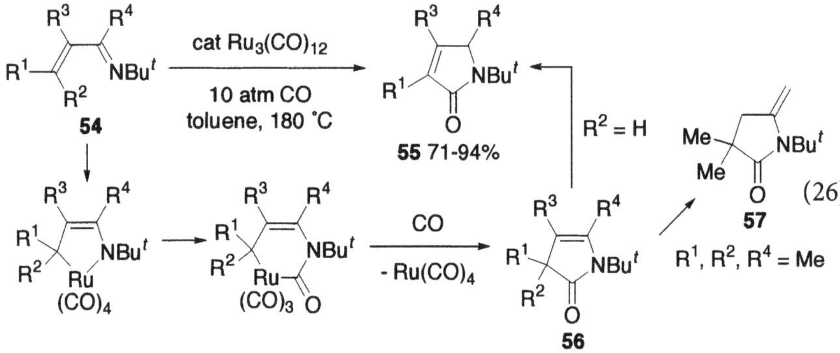

(26)

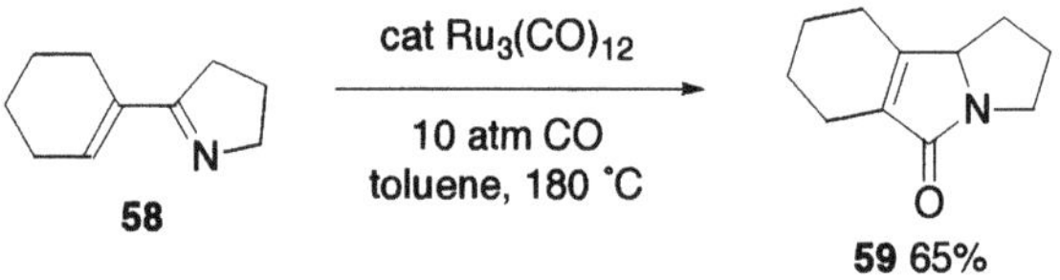

(27)

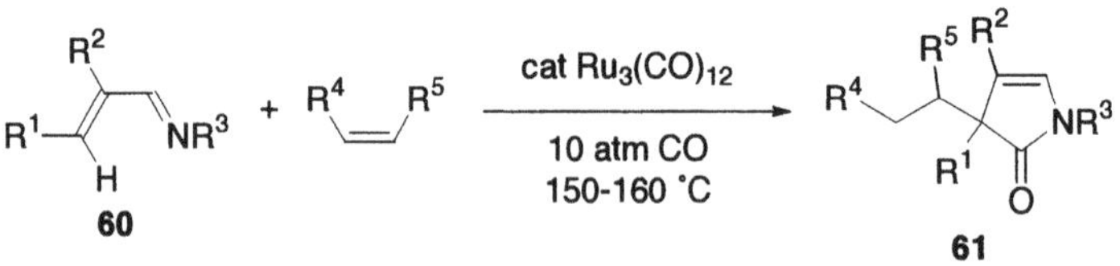

(28)

Cleavage of carbon–carbon bonds by transition-metal catalysts is one of the major challenges in organic and organometallic chemistry [43]. For that purpose, strained small-ring ketones are useful substrates. Recently, some synthetic transformations of cyclobutanes via C–C bond fission were developed using rhodium catalysts [43]. With respect to ruthenium catalysis, an interesting catalytic reaction of cyclobutenediones with alkenes was reported, in which highly substituted cyclopentenones were synthesized in a reconstructive manner with CO extrusion ([4+2–1]) [44]. This remarkable cycloaddition was further extended to a novel reconstructive carbonylation of cyclopropenones **62** (Eq. 29) [45]. As a result, pyranopyrandiones **63** were formed in good yields from both molecules of cyclopropenone and CO. Unsymmetrical pyranopyrandiones **64** were also obtained by a three-component coupling of **62**, internal alkynes, and CO. Quite a complex mechanism was proposed for the formation of the pyranopyrandiones on the basis of isotope-labeling experiments with ^{13}CO.

cat $Ru_3(CO)_{12}/Et_3N$, 15 atm CO, THF, 140 °C; **63** 70-87% ← **62** → cat $Ru_3(CO)_{12}/Et_3N$, R'C≡CR', 20 atm CO, toluene, 150 °C; **64** 54-82%

(29)

2.4 Heterocycle Formations via Oxygenation and Related Reactions

Epoxides, one of the smallest heterocyclic molecules, are useful building blocks in organic synthesis. Especially, chiral epoxides have played important roles in asymmetric synthesis. In this context, the development of an efficient protocol to prepare enantiopure epoxides has been a continuing target (Eq. 30) [46]. One of the most promising methods is the Katsuki–Sharpless asymmetric epoxidation of allyl alcohols using chiral titanium reagents [47]. Toward the asymmetric epoxidation of unfunctionalized alkenes, various chiral catalysts have been developed based on metal(salen) or metal(porphyrin) motifs [48]. Ruthenium catalysts have also been examined with respect to the stoichiometric and catalytic epoxidations of unfunctionalized alkenes [49]. Catalytic asymmetric epoxidations were achieved to some extent using a chiral porphyrin complex **65** [50] and bis(oxazolinyl)pyridine complexes **66** [51] (Fig. 1). Most catalytic methods require stoichiometric oxidants such as pyridine *N*-oxides, iodosobenzene, or iodosobenzene diacetate, but the use of molecular oxygen is ideal from both economical and environmental points of view. Catalytic aerobic epoxidations were carried out with ruthenium porphyrin complexes [52], 1,10-phenanthroline complexes [53], 2-pyridinecarboxyamide complex [54], and a ruthenium-substituted polyoxometalate [55].

$R^1R^2C{=}CR^3R^4$ —cat→ epoxide/aziridine (X)

X = O: epoxide
X = NR: aziridine

(30)

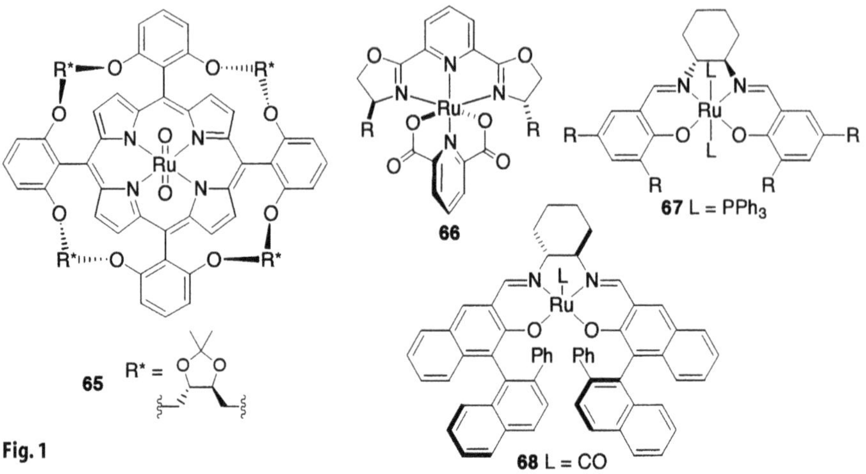

Fig. 1

In addition to epoxides, three-membered nitrogen heterocycles, aziridines, can be obtained by means of catalytic asymmetric aziridinations (Eq. 30). To this aim, chiral ruthenium(salen) complexes **67** [56] and **68** [57] were useful (Fig. 1). The former phosphine complexes **67** gave the aziridine from two cycloalkenes with 19–83% ee [56]. On the other hand, terminal alkenes selectively underwent aziridination in the presence of the latter carbonyl complex **68** with 87–95% ee [57]. In these examples, *N*-tosyliminophenyliodinane or *N*-tosyl azide were used as nitrene sources. Quite recently, catalytic intramolecular amidation of saturated C–H bonds was achieved by the use of a ruthenium(porphyrin) complex (Eq. 31) [58]. In the presence of the ruthenium catalyst and 2 equiv iodosobenzene diacetate, sulfamate esters **69** were converted into cyclic sulfamidates **70** in moderate-to-good yields.

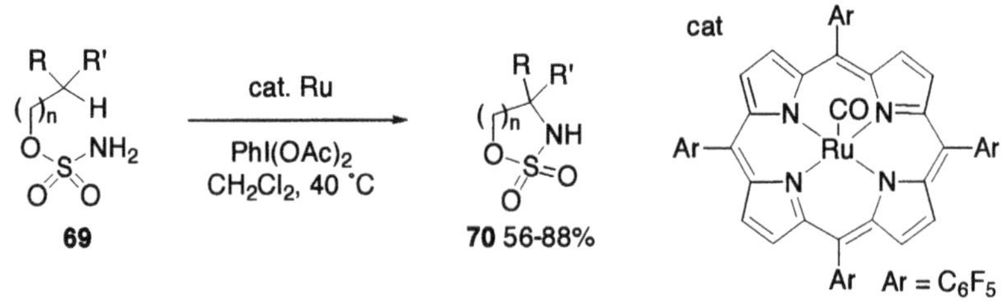

(31)

The oxidation of alkenes with ruthenium tetraoxide generally gives ketones, aldehydes, or carboxylic acids via C=C bond scission. An improved procedure for the oxidative cleavage of alkenes using a catalytic amount of RuO_4 and a stoichiometric amount of sodium metaperiodate in the biphasic solvent system $CCl_4/CH_3CN/H_2O$, however, has been applied to geranyl acetate **71** to produce tetrahydrofuran derivatives **72** and **73** (Eq. 32) [59]. Later, the generality of this oxidative cyclization reaction was confirmed using a simple 1,5-diene **74** and

(*S*)-(–)-limonene **75** (Eq. 33) [60]. These reactions gave tetrahydrofuran derivatives as single diastereomers.

71 AcO — cat RuO_4 / $NaIO_4$, $CCl_4/CH_3CN/H_2O$, 0 °C, 10 min → **72** *cis* = 39%, *trans* = 12% + **73** 12% (32)

74 → HO ... OH 50%; **75** → OH ... OH 45%; cat $RuO_2{\cdot}H_2O$ / $NaIO_4$, AcOEt/acetone/H_2O (33)

3 Synthesis of Heterocycles via Carbon–Carbon Bond Forming Cyclizations

With the aid of transition-metal catalysis, heterocycle formations can be achieved not only by carbon–heteroatom bond forming cyclizations of an acyclic molecule with a terminal group such as alcohols and amines, but also by intramolecular carbon–carbon bond forming reactions of an acyclic precursor containing one or more heteroatoms in its tether moiety. This section will briefly survey heterocycle synthesis via carbon–carbon bond formations. For details of ruthenium-catalyzed C–C bond formations, see other chapters of this book.

During the past decade, ring-closing metathesis (RCM) has emerged as a powerful method to synthesize cyclic molecules [61]. Especially with the evolution of well-defined ruthenium catalysts combining high activity with an excellent functional group tolurance [62], RCM has been extensively applied to the construction of complex heterocycles. The introduction of sterically demanding saturated or unsaturated *N*-heterocyclic carbene ligands into the standard Grubbs catalyst **76** dramatically improves the reactivity without loss of the functional group compatibility [62] (Fig. 2). As summarized in Table 1, the second-generation catalysts **77** and **78** exhibited excellent reactivities toward the synthesis of a tetrasubstituted *N*-heterocycloalkene (entry 1) [63], a cyclic disulfide (entry 2) [64], a seven-membered phosphorus heterocycle (entry 3) [65], and a seven-membered sulfone (entry 4) [66]. In all these cases, using the first-generation catalyst **76** instead of **77** or **78** resulted in lower yields. Cascade metathesis reactions are powerful tools to assemble a complex polycyclic framework from a simple acyclic precursor in only a single operation. The second-generation catalyst **78** also expanded the scope of the cascade metathesis technology. An acrylate derivative was converted into a bicyclic lac-

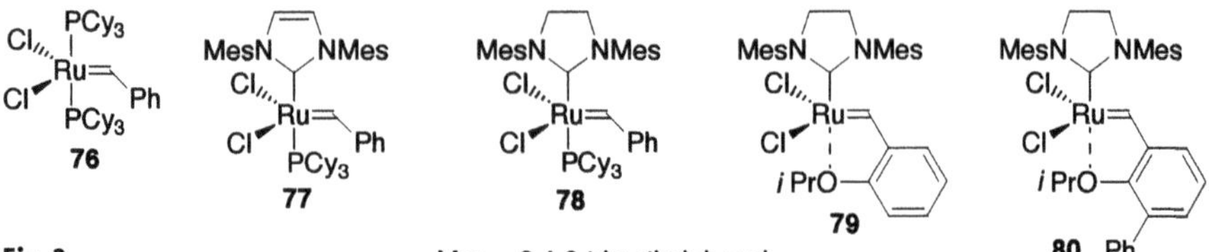

Fig. 2 Mes = 2,4,6-trimethylphenyl

Table 1 Syntheses of heterocycles using ruthenium-catalyzed RCM

Entry	Substrate	Product	Conditions yield	Ref.
1	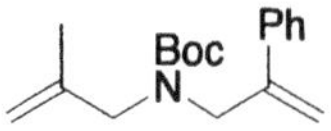	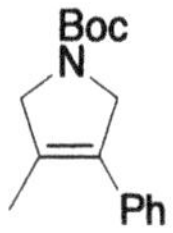	14 mol % **77** CH_2Cl_2, reflux 83%	[63]
2	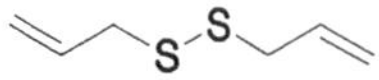		5 mol % **77** CD_2Cl_2, reflux quant	[64]
3	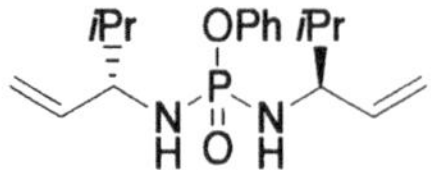	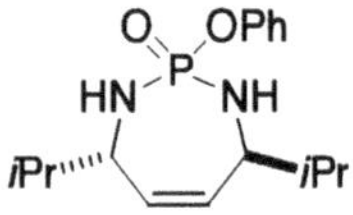	5 mol % **78** CH_2Cl_2, reflux 88%	[65]
4	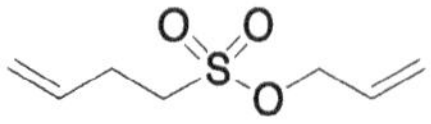	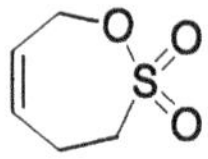	5 mol % **78** C_6H_6, 70 °C 100%	[66]
5	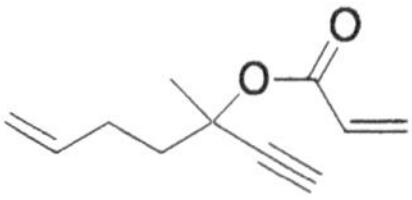	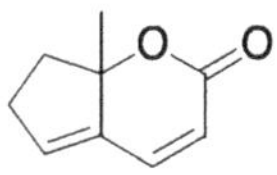	5 mol % **78** CH_2Cl_2, 40 °C 95%	[67]
6	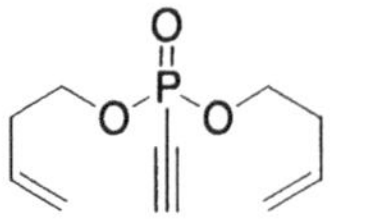	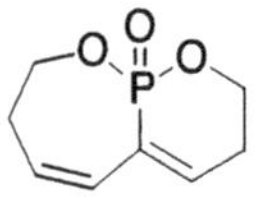	1 mol % **78** CH_2Cl_2, reflux 76%	[68]
7	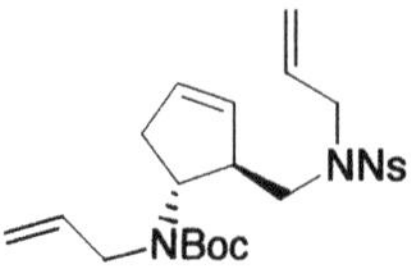	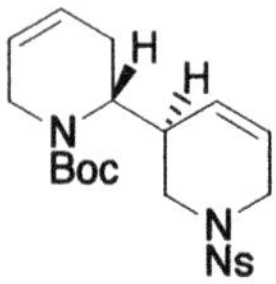	1 mol % **78** CH_2Cl_2, 40 °C 82%	[69]

tone (entry 5) [67]. The conventional catalyst **76** could not incorporate such an α,β-unsaturated carbonyl moiety into the cascade process. Similarly, a phosphonate dienyne and a cyclopentene with two allylamine side chains were cyclized to give a phosphorus bicyclic compound [68] and the precursor of a naturally occurring bipiperidine, astrophylline [69], respectively (entries 6 and 7).

Further modification of the benzilidene moiety of **78** by isopropoxystyrene gives a highly robust complex **79**, which can be easily recovered by silica gel chromatography after RCM [70]. A similar phosphine-free recyclable complex with a biaryl moiety **80** proved to be more active than **76** and **79**; the relative reactivity order is **79**<**78**<**80** [71].

Cycloisomerizations are environmentally benign processes to synthesize cyclic molecules from acyclic precursors via literally isomerization reactions. Transition-metal-catalyzed cycloisomerizations are highly advantageous because they requires no additional reagents except for an appropriate catalyst [72]. Various types of catalyzed cycloisomerizations have been developed to date, and among them, some recent examples of the ruthenium-catalyzed heterocycle formations are compiled in Table 2. There exist several types of cycloisomerization of enyne substrates. The most extensively studied is the intramolecular metathesis of enynes using Grubbs' catalyst **76**, and the resultant products, a conjugated inner–outer ring diene, can be used for Diels–Alder reactions [73]. The second-generation catalyst **78** proved to catalyze the enyne metathesis more effectively than **76**. For instance, an ene-ynamide was effectively converted into a cyclic diene using the thermally robust **78** in toluene at 80 °C for 15 min, although the desired metathesis hardly proceeded using **76** in refluxing dichloromethane for 24 h (entry 1) [74]. An enyne substrate with an internal alkyne gave rise to an unusual six-membered ring product along with a normal metathesis product (entry 2) [75]. In contrast, a ruthenium hydride species generated from Cp*RuCl(cod) and EtOH isomerized an allyl propargyl ether into a 3,4-dialkylidene tetrahydrofuran (entry 3) [76]. Interestingly, a carbapenam analogue was synthesized by means of a similar cycloisomerization using an isolated ruthenium(hydrido) complex (entry 4) [77]. A cationic cyclopentadienylruthenium complex **81** catalyzed an Alder-ene type cyclization of a 1,7-enyne leading to a piperidine ring (entry 5) [78]. The same catalyst system also proved to be effective for the intramolecular [5+2] cycloaddition constructing bicyclo[5.3.0]azacycle (entry 6) [79] and the reconstructive cycloisomerization of a diyne-ol (entry 7) [80]. Catalytic cycloisomerizations of an ene-allene and a diene were performed with an isolated ruthenium(hydride) complex, $RuClH(CO)(PPh_3)_3$, and an in situ formed hydride species from an oligomeric complex $[RuCl_2(cod)]_n$ and *i*-PrOH, respectively, to obtain five-membered heterocycles (entries 8 and 9) [81, 82]. A ruthenium(II) amidinate complex **82** catalyzed the chlorine-atom-transfer radical cyclization of a cyclic trichloroacetoamide leading to a pyrrolizidine framework, even at room temperature (entry 10) [83], while previous ruthenium catalysts required a reaction temperature above 80 °C [84].

Table 2 Syntheses of heterocycles using ruthenium-catalyzed cycloisomerizations

Entry	Precursor	Product yield	Conditions	Ref.
1	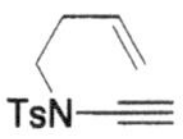 TsN	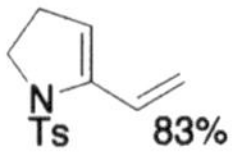 N Ts 83%	5 mol % **78** 1 atm ethylene toluene, 80 °C	[74]
2	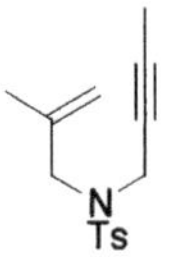 N Ts	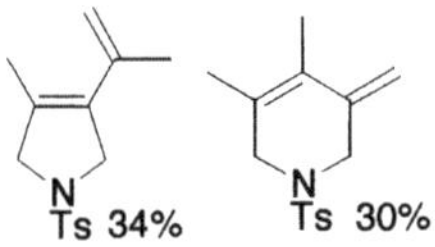 N Ts 34%; N Ts 30%	5 mol % **77** toluene, 80 °C	[75]
3	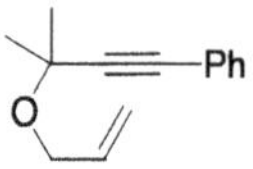 Ph, O	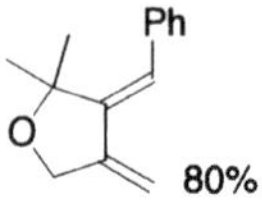 Ph, O 80%	5 mol % Cp*RuCl(cod) EtOH, rt	[76]
4	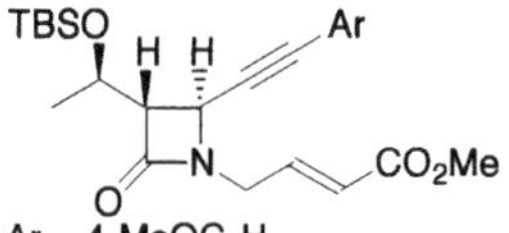 TBSO, H H, Ar, O, N, CO_2Me Ar = 4-$MeOC_6H_4$	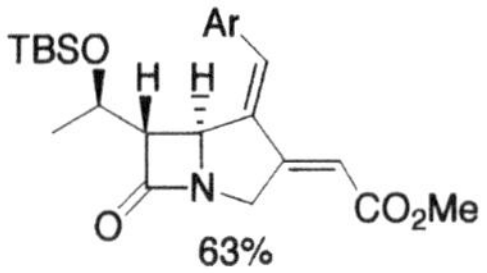 TBSO, Ar, H H, O, N, CO_2Me 63%	10 mol % $RuH_2CO(PPh_3)_3$ toluene, reflux	[77]
5	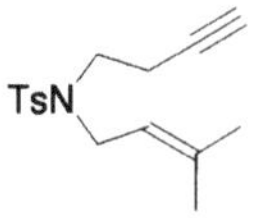 TsN	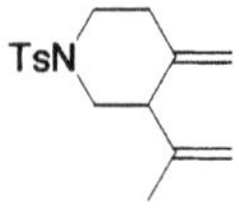 TsN 75%	10 mol % **81** acetone, rt	[78]
6	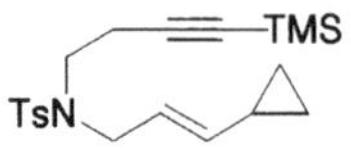 TMS, TsN	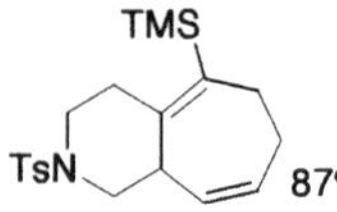 TMS, TsN 87%	15-20 mol % **81** acetone, 50 °C	[79]
7	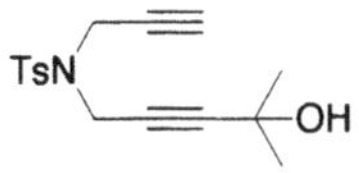 TsN, OH	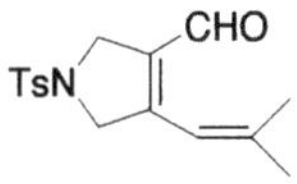 CHO, TsN 99%	3 mol % **81** acetone, rt	[80]
8	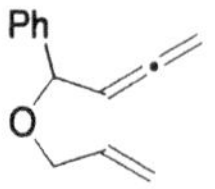 Ph, O	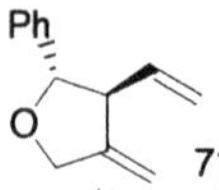 Ph, O 71% (*trans*:*cis* = 11:1)	5 mol % RuClH(CO)$(PPh_3)_3$ toluene, reflux	[81]
9	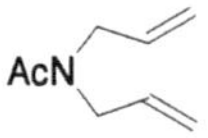 AcN	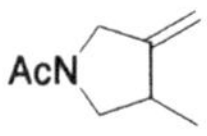 AcN 62% (83% isomeric purity)	5 mol % $[RuCl_2(cod)]_n$ *i*PrOH, reflux	[82]
10	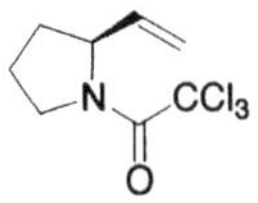 N, CCl_3, O	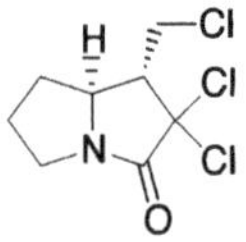 H, Cl, Cl, Cl, N, O 90%	30 mol % **82** benzene, rt	[83]

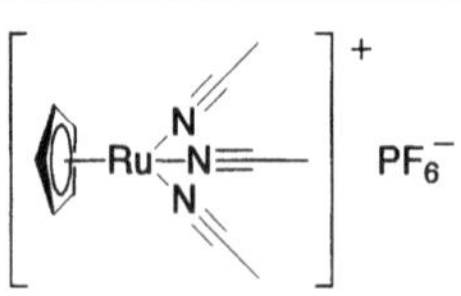

Ru, N, N, N, +, PF_6^- — **81**

Ru, N, N — **82**

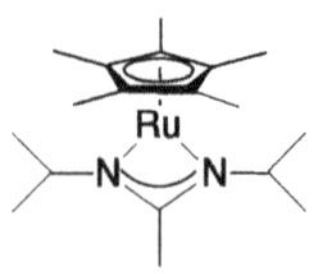

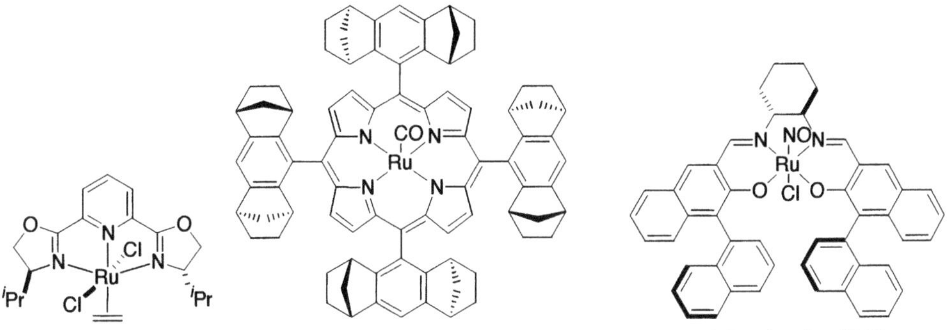

85 5 mol %, rt
93%, 86% ee (1*R*, 5*S*)

86 0.2 mol %, 40 °C
77%, 85% ee (1*S*, 5*R*)

87 5 mol %, hν, 37 °C
54%, 82% ee (1*S*, 5*R*)

Fig. 3

Intramolecular cyclopropanation of allyl diazoacetates gives rise to interesting cyclopropane-fused γ-butyrolactones. A chiral ruthenium bis(oxazolinyl)pyridine complex **85** was employed for the catalytic cyclization of *trans*-cinnamyl diazoacetate **83** at room temperature to obtain an optically active lactone **84** in 93% yield with 86% ee (Eq. 34, Fig. 2) [85]. Chiral porphyrin and salen complexes of ruthenium **86** [86] and **87** [87] also catalyzed the asymmetric intramolecular cyclopropanation of **83** to afford **84** in similar yields and enantiomeric excess.

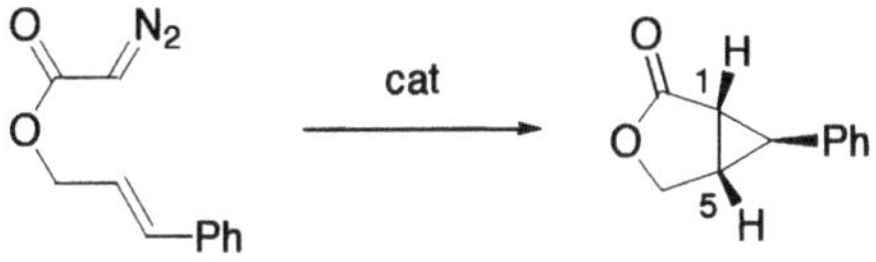

83 —cat→ **84** (34)

Intramolecular [2+2+2] cyclotrimerizations of diynes and triynes possessing heteroatom tethers furnish benzoheterocycles. The cyclization of triynes **88** using the Grubbs catalyst **76** proceeds via cascade metathesis as shown in Eq. (35) to yield a tricyclic product **89** [88]. This novel type of catalytic alkyne cyclotrimerization can be applied to the cycloaddition of 1,6-diynes with monoalkynes [89].

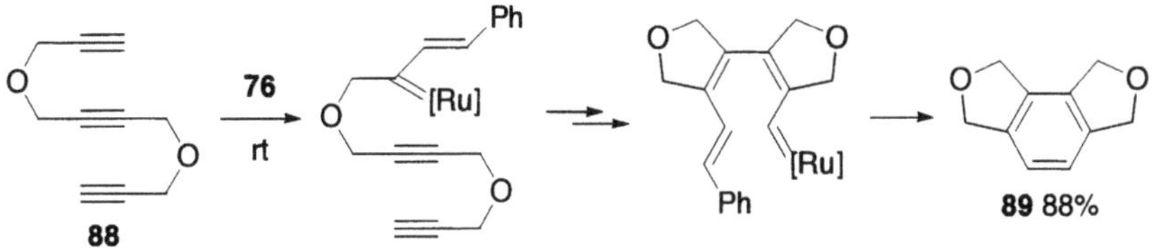

88 —**76**, rt→ [Ru] → → **89** 88% (35)

In addition to the Grubbs carbene complex, organoruthenium complexes possessing a Cp-type planar ligand turned out to be an excellent precatalyst for the cycloaddition of 1,6-diynes with unsaturated molecules (Eq. 36). Using the Ru(II) complex, Cp*RuCl(cod), as a precatalyst, a nitrogen-tethered diyne **90** reacted with 1-hexyne to afford an isoindolin derivative **91** in 80% yield [90]. In contrast, a dinuclear Ru(III) complex, $[Cp^*RuCl_2]_2$, exhibited a superior catalytic activity for the cycloaddition of 1,6-diynes with heterocycloalkenes [91]. An interesting heterotricyclic compound **92** was obtained from **90** and 2,3-dihydrofuran in 70% yield via [2+2+2] alkyne–alkene co-cyclotrimerization. On the other hand, the cycloaddition of 1,6-diynes with a strained alkene such as norbornene gave rise to unusual tandem cyclopropanation products (e.g., **93**) together with the expected co-cyclotrimerization products [91]. Especially, an indenyl complex, $(\eta^5\text{-}C_7H_9)RuCl(PPh_3)_2$, afforded predominantly the tandem cyclopropanation products. These cycloaddition reactions are considered to commonly proceed via a ruthenacyclopentatriene intermediate **94**.

1 mol % Cp*RuCl(cod), rt, ≡—Bu: TsN-diyne **90** → **91** 80%

0.5 mol % $[Cp^*RuCl_2]_2$, rt, 2,3-dihydrofuran: **90** → **92** 70%

5 mol % $(\eta^5\text{-}C_7H_9)RuCl(PPh_3)_2$, 40 °C, norbornene: **90** → **93** 47%

X, [Ru]: **94**

(36)

Quite recently, the alkenylative cyclization of enynes with ethylene was achieved using Cp*RuCl(cod) as a precatalyst at room temperature [92]. This mild and selective transformation was applied to a sulfonamide **95** to produce the corresponding pyrrolidine derivative **96** (Eq. 37). A ruthenacyclopentene intermediate was proposed for this novel cyclization.

5 mol % Cp*RuCl(cod), 1 atm ethylene, rt: TsN **95** → TsN **96** 83%

(37)

An interesting catalytic indole synthesis was realized by the benzilic C–H bond activation on 2,6-xylyl isocyanide (Eq. 38) [93]. A coordinatively unsaturated

Ru(0) active species was generated from the reductive elimination of naphthalene from a naphthylruthenium(hydrido) complex **97**.

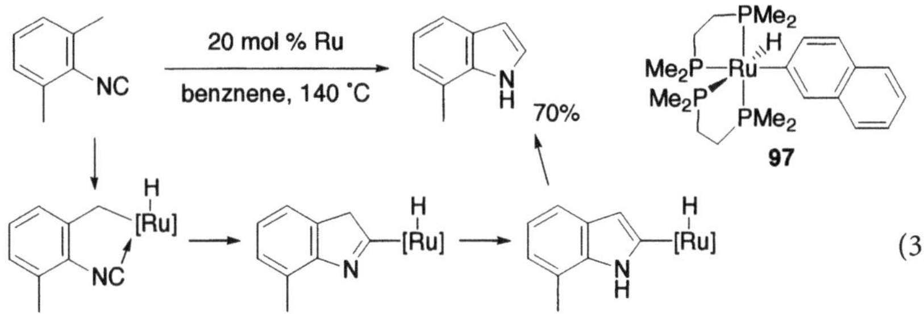

(38)

4
Miscellaneous Heterocyclizations

The ruthenium-catalyzed hydrogen-transfer reaction of primary alcohols gives rise to esters. Such dehydrogenation has been successfully extended to intramolecular variants, lactone formations from diols [94], and lactam formations from aminoalcohols [95] (Eq. 39). Although the heterocyclic ring formations take place without recourse to a ruthenium catalyst in these examples, such heterocyclizations involving hydrogen-transfer processes are becoming more and more important as an environmentally friendly approach to heterocycles.

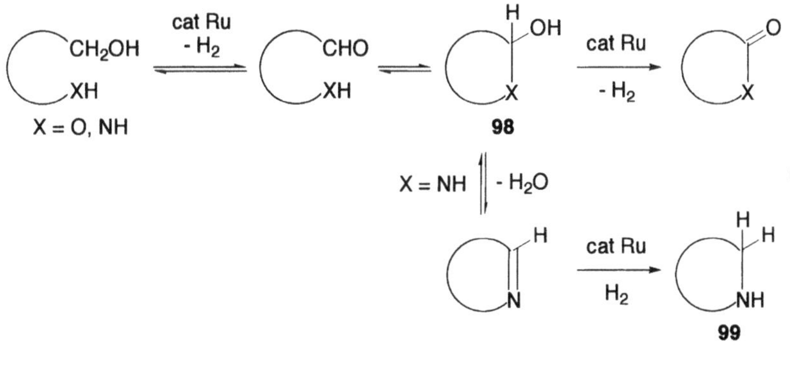

(39)

A related intramolecular N-alkylation leading to saturated nitrogen heterocycles **99** can proceed via dehydration of intermediates **98** (Eq. 39) [96]. Unsaturated nitrogen heterocycles such as pyrroles [97], indoles [98], benzo-azoles [99], 2,3-dihydroimidazol-2-ones [100], and imidazo[1,2-*a*]pyridines [101] were obtained through similar cyclocondensation reactions. Interesting ruthenium-catalyzed syntheses of quinolines have been achieved by means of cyclocondensations of aniline derivatives with propanediols, aminoalcohols, or diamines (Eq. 40) [102, 103]. Similar indole syntheses employed glycols, triethanolamines, or epoxides (Eq. 40) [102, 104]. Substituted quinoline rings were also assembled from anilines and 2 Eq of allylic alcohols (Eq. 41) [105]. Allylic amines can be used as alkyl sources [106] and, surprisingly, even with saturated trialkylamines, the same transformation became possible under ruthenium catalysis [107].

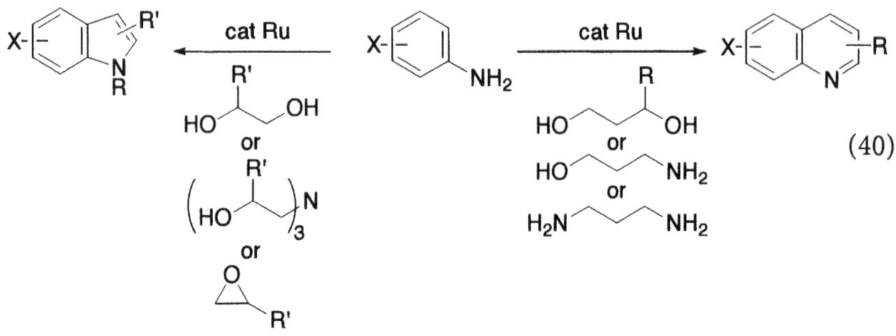

(40)

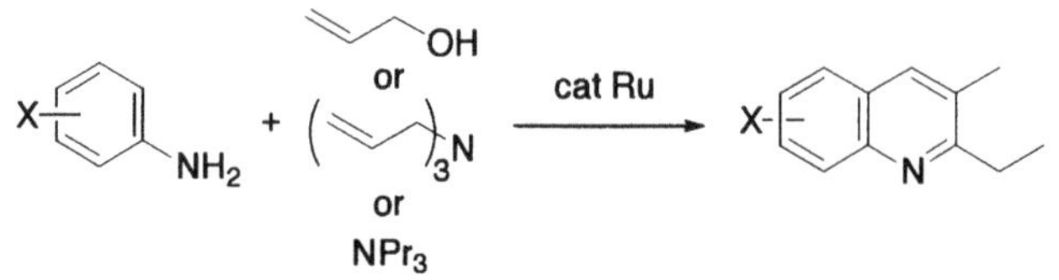

(41)

Quinoline and indole derivatives were also synthesized by cyclocondensation reactions of aniline derivatives with alkynes (Eqs. 42, 43) [108]. These protocols involve the ruthenium-catalyzed intermolecular hydroaminations of terminal alkynes as the initial steps.

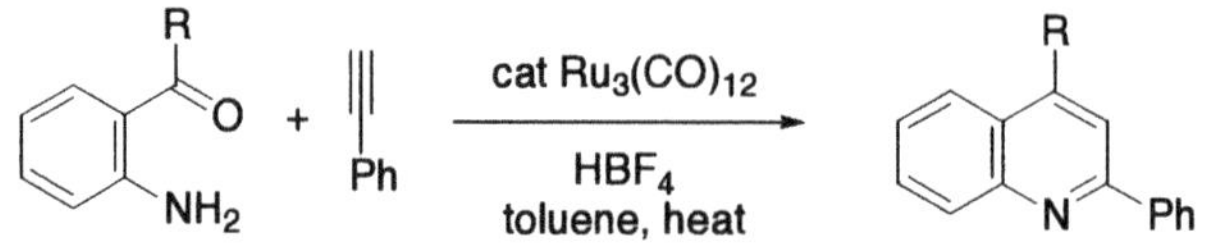

(42)

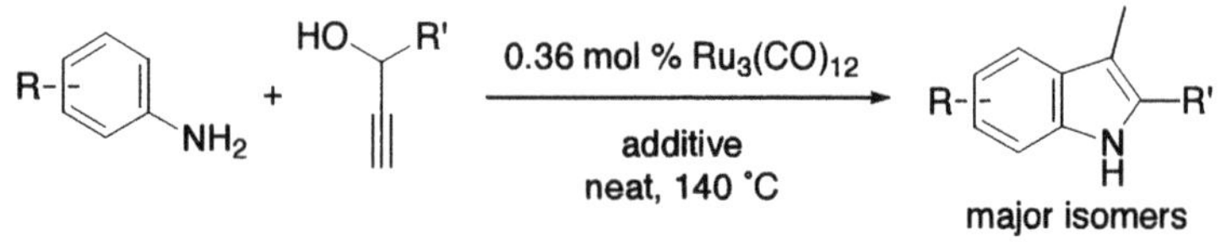

(43)

The reductive annulation of nitrobenzene **100a** with ethynylbenzene also gave a substituted indole **101** in 39% yield in the presence of catalytic amounts of $[Cp^*Ru(CO)_2]_2$ and CO at 170 °C (Eq. 44) [109]. The intermediary nitrosobenzene **100b** was assumed on the basis of following observations [110]. The reaction time was reduced from 48 to 24 h and the yield of **101** was increased to

53% when **100b** was employed instead of **100a**. In addition, **100b** undergoes thermal cycloaddition with ethynylbenzene even in the absence of the ruthenium catalyst to give rise to an *N*-hydroxy indole derivative **102**, which was converted into **101** by the ruthenium catalyst and CO.

cat $[Cp^*Ru(CO)_2]_2$, CO (750 psi), 170 °C; **100a** + PhC≡CH → **101**; **100a** –[Ru]/CO→ **100b**; **100b** –PhC≡CH, Δ→ **102**; **102** –[Ru]/CO→ **101** (44)

Nitriles are important synthetic intermediates in organic synthesis. The transformations of a cyano group into other functionalities, however, generally require acidic or basic media. In this respect, the activation of this functionality under neutral and mild conditions is highly valuable. A ruthenium(hydrido) complex, $RuH_2(PPh_3)_4$, is suitable for this purpose. In the presence of the ruthenium precatalyst, intramolecular condensations of nitriles with alcohols proceeded in good yields [111]. For instance, the cyclization of a hydroxynitrile **103** afforded a naturally occurring butenolide **104** in 91% yield (Eq. 45). The same ruthenium complex catalyzed the hydration of nitriles, leading to amides. Especially, the catalyzed hydration of δ-ketonitriles gave rise to ene-lactams in moderate-to-high yields by way of the cyclization of the resultant ketoamides [112]. The synthetic potential of this method was demonstrated by the application to the total synthesis of (–)-pumiliotoxin C (Eq. 46). The ruthenium-catalyzed hydration/cyclization of **105** gave the desired intermediate **106** with concomitant extrusion of the isopropylidene moiety.

103 –cat $RuH_2(PPh_3)_4$, H_2O, 140 °C→ **104** 91% (45)

105 –3 mol % $RuH_2(PPh_3)_4$, H_2O, DME, 160 °C→ **106** 56% → (-)-pumiliotoxin C

(46)

Taking advantage of the Lewis acidic character of the cationic ruthenium complex **107**, catalytic asymmetric 1,3-dipolar cycloadditions of nitrones have been developed [113]. In the presence of 5 mol % of **107**, the cycloaddition of **108** with crotonaldehyde afforded an isoxazolidine **109** in 80% yield with 66% ee (Eq. 47).

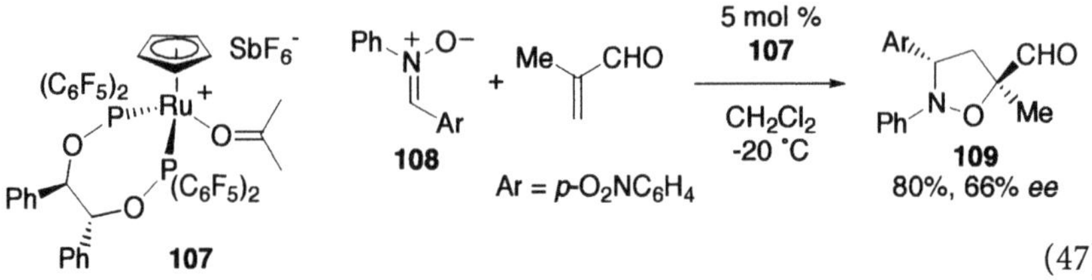

(47)

5 Concluding Remarks

Ruthenium catalysis has been extensively explored during the past decade [114]. Newly developed carbon–carbon bond forming cyclizations such as [2+2+2] cycloaddition, RCMs, and cycloisomerizations have dramatically expanded the scope of heterocycle synthesis. Relatively unexplored catalytic carbon–heteroatom bond formations have also made significant contributions to this area. Further progress in ruthenium catalysis will not only improve the conventional synthetic methodologies, but will also open the way to an unprecedented class of heterocyclic compounds, which might have a significant potential as pharmaceuticals or functional materials.

References

1. Joule JA, Mills K (2000) Heterocyclic chemistry, 4th edn. Blackwell, Oxford
2. (a) Trost BM (1991) Science 254:1471; (b) Trost BM (1995) Angew Chem Int Ed Engl 34:259
3. (a) Seiller B, Bruneau C, Dixneuf PH (1994) J Chem Soc Chem Commun 493; (b) Seiller B, Bruneau C, Dixneuf PH (1995) Tetrahedron 51:13089
4. Bruneau C, Dixneuf PH (1997) Chem Commun 507
5. Neveux M, Seiller B, Hagedorn F, Bruneau C, Dixneuf PH (1993) J Organomet Chem 451:133
6. McDonald FE (1999) Chem Eur J 5:3103
7. Sheng Y, Musaev DG, Reddy KS, McDonald FE, Morokuma K (2002) J Am Chem Soc 124:4149 and references therein
8. Trost BM, Rhee YH (2002) J Am Chem Soc 124:2528
9. Trost BM, Rhee YH (1999) J Am Chem Soc 121:11680
10. Trost BM, Flygare JA (1992) J Am Chem Soc 114:5476
11. Kondo T, Tsunawaki F, Sato R, Ura Y, Wada K, Mitsudo T (2003) Chem Lett 24
12. Müller TE, Beller M (1998) In: Beller M, Bolm C (eds) Transition metals for organic synthesis, vol 2. Wiley-VCH, Weinheim, p 316
13. Kondo T, Okada T, Suzuki T, Mitusdo T (2001) J Organomet Chem 622:149

14. Kondo T, Okada T, Mitsudo T (2002) J Am Chem Soc 124:186
15. (a) Schore NE (1991) In: Paquette LA, Trost BM, Fleming I (eds) Comprehensive organic synthesis, vol 5. Pergamon, Oxford, p 1129; (b) Grotjahn DB (1995) In: Hegedus LS, Abel EW, Stone FGA, Wilkinson G (eds) Comprehensive organometallic chemistry II, vol 12. Pergamon, Oxford, p 741; (c) Bönnemann H, Brijoux W (1998) In: Beller M, Bolm C (eds) Transition metals for organic synthesis, vol 1. Wiley-VCH, Weinheim, p 114
16. Yamamoto Y, Okuda S, Itoh K (2001) Chem Commun 1102
17. Naiman A, Vollhardt KPC (1977) Angew Chem Int Ed Engl 24:248
18. Yamamoto Y, Ogawa R, Itoh K (2001) J Am Chem Soc 123:6189
19. Earl RA, Vollhardt KPC (1984) J Org Chem 49:4786
20. Yamamoto Y, Takagishi H, Itoh K (2001) Org Lett 3:2117
21. Kondo T, Mitsudo T (2000) Chem Rev 100:3205
22. Yamamoto Y, Takagishi H, Itoh K (2002) J Am Chem Soc 124:28
23. Yamamoto Y, Takagishi H, Itoh K (2002) J Am Chem Soc 124:6844
24. Trost BM, Brown RE, Toste FD (2000) J Am Chem Soc 122:5877
25. (a) Thompson DJ (1991) In: Pattenden G, Trost BM, Fleming I (eds) Comprehensive organic synthesis, vol 3. Pergamon, Oxford, p 1015; (b) Bates RW (1995) In: Hegedus LS, Abel EW, Stone FGA, Wilkinson G (eds) Comprehensive organometallic chemistry II, vol 12. Pergamon, Oxford, p 349
26. Ali BE, Alper H (1998) In: Beller M, Bolm C (eds) Transition metals for organic synthesis, vol 1. Wiley-VCH, Weinheim, p 49
27. Kondo T, Kodoi K, Mitsudo T, Watanabe Y (1994) J Chem Soc Chem Commun 755
28. Yoneda E, Kaneko T, Zhang S-W, Onitsuka K, Takahashi S (2000) Org Lett 2:441
29. Yoneda E, Kaneko T, Zhang S-W, Onitsuka K, Takahashi S (2001) Tetrahedron Lett 42:5459
30. Kang S-K, Kim K-J, Yu C-M, Hwang J-W, Do Y-K (2001) Org Lett 3:2851
31. (a) Schore NE (1991) In: Pattenden G, Trost BM, Fleming I (eds) Comprehensive organic synthesis, vol 3. Pergamon, Oxford, p 1037; (b) Schore NE (1995) In: Hegedus LS, Abel EW, Stone FGA, Wilkinson G (eds) Comprehensive organometallic chemistry II, vol 12. Pergamon, Oxford, p 703; (c) Jeong N (1998) In: Beller M, Bolm C (eds) Transition metals for organic synthesis, vol 1. Wiley-VCH, Weinheim, p 560; (d) Chung YK (1999) Coord Chem Rev 188:297
32. (a) Kondo T, Suzuki N, Okada T, Mitsudo T (1997) J Am Chem Soc 119:6187; (b) Morimoto T, Chatani N, Fukumoto Y, Murai S (1997) J Org Chem 62:3762
33. Chatani N, Morimoto T, Fukumoto Y, Murai S (1998) J Am Chem Soc 120:5335
34. (a) Crowe WE, Vu AT (1996) J Am Chem Soc 118:1557; (b) Mandal SK, Amin SR, Crowe WE (2001) J Am Chem Soc 123:6457; (c) Kablaouni NM, Hicks FA, Buchwald SL (1996) J Am Chem Soc 118:5818; (d) Kablaoui NM, Hicks FA, Buchwald SL (1997) J Am Chem Soc 119:4424
35. Chatani N, Tobis M, Asaumi T, Fukumoto Y, Murai S (1999) J Am Chem Soc 121:7160
36. Tobis M, Chatani N, Asaumi T, Amako K, Ie Y, Fukimoto Y, Murai S (2000) J Am Chem Soc 122:12663
37. Chatani N, Amako K, Tobisu M, Asaumi T, Fukumoto Y, Murai S (2003) J Org Chem 68:1591
38. Chatani N, Tobisu M, Asaumi T, Murai S (2000) Synthesis 925
39. Imhof W, Göbel A (2003) J Mol Catal A 197:15
40. Chatani N, Kamitani A, Murai S (1999) J Am Chem Soc 121:1758
41. Chatani N, Kamitani A, Murai S (2002) J Org Chem 67:7014
42. (a) Berger D, Imhof W (1999) Chem Commun 1457; (b) Berger D, Imhof W (2000) Tetrahedron 56:2015; (c) Imhof W, Berger D, Kötteritzsch M, Rost M, Schönecker B (2001) Adv Synth Catal 343:795

43. Murakami M, Ito Y (1999) In: Murai S (ed) Activation of Unreactive bonds and organic synthesis. Springer, Berlin Heidelberg New York, p 97
44. Kondo M, Nakamura A, Okada T, Suzuki N, Wada K, Mitsudo T (2000) J Am Chem Soc 122:6319
45. Kondo T, Kaneko Y, Taguchi Y, Nakamura A, Okada T, Shiotuski M, Ura Y, Wada K, Mitsudo T (2002) J Am Chem Soc 124:6824
46. Johnson RA, Sharpless KB (1991) In: Ley SV, Trost BM, Fleming I (eds) Comprehensive organic synthesis, vol 7. Pergamon, Oxford, p 389
47. Johnson RA, Sharpless KB (2000) In: Ojima I (ed) Catalytic asymmetric synthesis, 2nd edn. Wiley-VCH, New York, p 231
48. Katsuki T (2000) In: Ojima I (ed) Catalytic asymmetric synthesis, 2nd edn. Wiley-VCH, New York, p 287
49. Barf GA, Sheldon RA (1995) J Mol Catal A 102:23
50. Gross Z, Ini S, Kapon M, Cohen S (1996) Tetrahedron Lett 37:7325
51. Nishiyama H, Shimada T, Itoh H, Sugiyama H, Motoyama Y (1997) Chem Commun 1863
52. Groves JT, Quinn R (1985) J Am Chem Soc 107:5790
53. (a) Baily CL, Drago RS (1987) J Chem Soc Chem Commun 179; (b) Goldstein AS, Beer RH, Drago RS (1994) J Am Chem Soc 116:2424
54. Qi JY, Qiu LQ, Lam KH, Yip CW, Zhou ZY, Chan ASC (2003) Chem Commun 1058
55. Neumann R, Dahan M (1997) Nature 388:353
56. Liang J-L, Yu X-Q, Che C-M (2002) Chem Commun 124
57. Omura K, Murakami M, Uchida T, Irie R, Katsuki T (2003) Chem Lett 32:354
58. Liang J-L, Yuan S-X, Huang J-S, Yu W-Y, Che C-M (2002) Angew Chem Int Ed Engl 41:3465
59. Carlsen PHJ, Katsuki T, Martin VS, Sharpless KB (1981) J Org Chem 46:3936
60. Albarella L, Musumeci D, Sica D (2001) Eur J Org Chem 997
61. (a) Fürstner A (2000) Angew Chem Int Ed Engl 39:3012; (b) Phillips AJ, Abell AD (1999) Aldrichimica Acta 32:75; (c) Grubbs RH, Chang S (1998) Tetrahedron 54:4413; (d) Armstrong SK (1998) J Chem Soc Perkin Trans 1 371; (e) Ivin KJ (1998) J Mol Cat A 133:1; (f) Fürstner A (1998) (ed) Alkene metathesis in organic synthesis. Springer, Berlin Heidelberg New York; (g) Fürstner A (1997) Top Catal 4:285
62. (a) Trnka TM, Grubbs RH (2001) Acc Chem Res 34:18; (b) Jafarpour L, Nolan SP (2001) J Organomet Chem 617-618:17
63. Briot A, Bujard M, Gouverneur V, Nolan SP, Mioskowski C (2000) Org Lett 2:1517
64. Spagnol G, Heck M-P, Nolan SP, Mioskowski C (2002) Org Lett 4:1767
65. Sprott KT, McReynolds MD, Hanson PR (2001) Synthesis 612
66. Flohic AL, Meyer C, Cossy J, Desmurs J-R, Galland J-C (2003) Synlett 667
67. Choi T-L, Grubbs RH (2001) Chem Commun 2648
68. Timmer MSM, Ovaa H, Filippov DV, van der Marel GA, van Boom JH (2001) Tetrahedron Lett 42:8231
69. Schaudt M, Blechert S (2003) J Org Chem 68:2913
70. Garber SB, Kingsbury JS, Gray BL, Hoveyda AH (2000) J Am Chem Soc 122:8168
71. Wakamatsu H, Blechert S (2002) Angew Chem Int Ed Engl 41:2403
72. Trost BM, Krische MJ (1998) Synlett 1
73. (a) Poulsen CS, Madsen R (2003) Synthesis 1; (b) Mori M (1998) In: Fürstner A (ed) Alkene metathesis in organic synthesis. Springer, Berlin Heidelberg New York, p 133
74. Saito N, Sato Y, Mori M (2002) Org Lett 4:803
75. (a) Kitamura T, Sato Y, Mori M (2001) Chem Commun 1258; (b) Kitamura T, Sato Y, Mori M (2002) Adv Synth Catal 344:378
76. Paih JL, Rodríguez DC, Dérien S, Dixneuf PH (2000) Synlett 95

77. Mori M, Kozawa Y, Nishida M, Kanamaru M, Onozuka K, Takimoto M (2000) Org Lett 2:3245
78. Trost BM, Toste FD (2000) J Am Chem Soc 122:714
79. Trost BM, Toste FD, Shen H (2000) J Am Chem Soc 122:2379
80. Trost BM, Rudd MT (2002) J Am Chem Soc 124:4178
81. Kang S-K, Ko B-S, Lee D-M (2002) Tetrahedron Lett 43:6693
82. (a) Yamamoto Y, Nakagai Y, Ohkoshi N, Itoh K (2001) J Am Chem Soc 123:6372; (b) Michaut M, Santelli M, Parrain JL (2003) Tetrahedron Lett 44:2157
83. Nagashima H, Gondo M, Masuda S, Kondo H, Yamaguchi Y, Matsubara K (2003) Chem Commun 442
84. Iqbal J, Bhatia B, Nayyar NK (1994) Chem Rev 94:519
85. Nishiyama H, Itoh Y, Sugawara Y, Matsumoto H, Aoki K, Itoh K (1995) Bull Chem Soc Jpn 68:1247
86. Teng P-F, Lai T-S, Kwong H-L, Che C-M (2003) Tetrahedron Asymmetry 14:837
87. Saha B, Uchida T, Katsuki T (2003) Tetrahedron Asymmetry 14:823
88. Peters J-U, Blechert S (1997) Chem Commun 1983
89. Witulski B, Stengel T, Fernández-Hernández JM (2000) Chem Commun 1965
90. (a) Yamamoto Y, Ogawa R, Itoh K (2000) Chem Commun 549; (b) Yamamoto Y, Arakawa T, Ogawa R, Itoh K (2003) J Am Chem Soc 125:12143
91. Yamamoto Y, Kitahara H, Ogawa R, Kawaguchi H, Tatsumi K, Itoh K (2000) J Am Chem Soc 122:4310
92. Mori M, Saito N, Tanaka D, Takimoto M, Sato Y (2003) J Am Chem Soc 125:5606
93. (a) Jones WD, Kosar WP (1986) J Am Chem Soc 108:5640; (b) Hsu GC, Kosar WP, Jones WD (1994) Organometallics 13:385
94. (a) Ishii Y, Osakada K, Ikariya T, Saburi M, Yoshikawa S (1986) J Org Chem 51:2034; (b) Murahashi SI, Naota T, Ito K, Maeda Y, Taki H (1987) J Org Chem 52:4319
95. Naota T, Murahashi SI (1991) Synlett 693
96. (a) Murahashi SI, Kondo K, Hakata T (1982) Tetrahedron Lett 23:229; (b) Tsuji Y, Huh K-T, Ohsugi Y, Watanabe Y (1985) J Org Chem 50:1365; (c) Marsella JA (1991) J Organomet Chem 407:97; (d) Youn YZ, Lee DY, Woo BW, Shim JG, Chae SA, Shim SC (1993) J Mol Catal 79:39; (e) Abbenhuis RATM, Boersma J, van Koten G (1998) J Org Chem 63:4282; (f) Sakano T, Ishii H, Yanaguchi I, Osakada K, Yamamoto T (1999) Inorg Chim Acta 296:176
97. Tsuji Y, Yokoyama Y, Huh K-T, Watanabe Y (1987) Bull Chem Soc Jpn 60:3456
98. Tsuji Y, Kotachi S, Huh K-T, Watanabe Y (1990) J Org Chem 55:580
99. Kondo T, Yang S, Huh K-T, Kobayashi M, Kotachi S, Watanabe Y (1991) Chem Lett 1275
100. Kondo T, Kotachi S, Watanabe Y (1992) J Chem Soc Chem Commun 1318
101. Kondo T, Kotachi S, Ogino S, Watanabe Y (1993) Chem Lett 1317
102. Tsuji Y, Huh K-T, Watanabe Y (1987) J Org Chem 52:1673;
103. Cho CS, Oh BH, Shim SC (1999) J Heterocycl Chem 36:1175
104. (a) Shim SC, Youn YZ, Lee DY, Kim TJ, Cho CS, Uemura S, Watanabe Y (1996) Synth Commun 26:1349; (b) Cho CS, Lim HK, Shim SC, Kim TJ, Choi H-J (1998) Chem Commun 995; (c) Cho CS, Kim JH, Shim SC (2000) Tetrahedron Lett 41:1811; (d) Cho CS, Kim JH, Shim SC (2001) Tetrahedron 57:3321; (e) Cho CS, Kim JH, Choi H-J, Kim T-J, Shim SC (2003) Tetrahedron Lett 44:2975
105. Watanabe Y, Tsuji Y, Ohsugi Y, Shida J (1983) Bull Chem Soc Jpn 56:2452
106. (a) Cho CS, Oh BH, Shim SC (1999) Tetrahedron Lett 40:1499; (b) Cho CS, Oh BH, Shim SC, Oh DH (2000) J Heterocycl Chem 37:1315; (c) Cho CS, Kim JS, Oh BH, Kim T-J, Shim SC, Yoon NS (2000) Tetrahedron 56:7747
107. Cho CS, Oh BH, Kim JS, Kim T-J, Shim SC (2000) Chem Commun 1885

108. (a) Tokunaga M, Eckert M, Wakatsuki Y (1999) Angew Chem Int Ed Engl 38:3222; (b) Tokunaga M, Ota M, Haga M, Wakatsuki Y (2001) Tetrahedron Lett 42:3865
109. Penoni A, Nicholas KM (2002) Chem Commun 484
110. Penoni A, Volkmann J, Nicholas KM (2002) Org Lett 4:699
111. Naota T, Shichijjo Y, Murahashi SI (1994) J Chem Soc, Chem Commun 1359
112. Murahashi SI, Sasao S, Saito E, Naota T (1993) Tetrahedron 49:8805
113. Viton F, Bernardinelli G, Kündig EP (2002) J Am Chem Soc 124:4968
114. (a) Naota T, Takaya H, Murahashi SI (1998) Chem Rev 98:2599; (b) Trost BM, Toste FD, Pinkerton AB (2001) Chem Rev 101:2067

Topics Organomet Chem (2004) 11: 277–320
DOI 10.1007/978-3-540-40027-1

Oxidation Using Ruthenium Catalysts

I. W. C. E. Arends[1] (✉) · T. Kodama[2] · R. A. Sheldon[1]

[1] Laboratory for Biocatalysis and Organic Chemistry, Department of Biotechnology, Delft University of Technology, Julianalaan 136, 2628 Bl Delft, The Netherlands
i.arends@tnw.tudelft.nl

[2] Current adress: Asahi Kasei Corporation, 2767-II Niihama Shionasu Kojima, Kurashihi Okayama, 711-8510 Japan

Abstract Among the catalysts which have potential for use in selective oxidations, ruthenium takes a special position owing to its versatility. Ruthenium can catalyse numerous oxidative transformations: the oxidation of alkanes, the cleavage of double bonds, the asymmetric epoxidation of alkenes, the oxidation of alcohols and ethers and the oxidation of amines and amides. In the field of alcohol, ether and amide oxidation, ruthenium-based catalysts certainly belong to the state of the art in their field and bear great potential for application in fine chemical synthesis. A range of oxidation states can be encountered in the various ruthenium intermediates which are responsible for the respective transformations: Ru(VIII) as in ruthenium tetroxide, Ru(VII) as in perruthenates, Ru(IV)–Ru(VI) in oxo-intermediates and Ru(II) can be applied in (de)hydrogenations. Details regarding the various oxidation mechanisms are discussed. In this chapter the focus is on green technologies and therefore examples of ruthenium-catalysed oxidations using environmentally benign oxidants are emphasized.

Keywords Ruthenium · Ruthenium tetroxide · Catalytic oxidation · Alcohols · Ethers

Abbrevations

bipy	2,2′-Bipyridyl
DMSO	dimethylsulfoxide
dppp	1,3-Bis(diphenylphosphino)propane
H_2amp	*N*-(Hydroxyphenyl) salicyldiimine
HPA	Heteropolyanion
NMO	*N*-Methylmorpholine oxide
PcS	Tetrasodium 2,3-tetrasulfophthalocyanato
POM	polyoxometalate
PSP	Polymer-supported perruthenate
Py	Pyridine
TBAB	Tetra-*n*-butylammonium bromide
TBAP	Tetra-*n*-butylammonium perruthenate
TEMPO	2,2′,6,6′-Tetramethylpiperidinyl-*N*-oxyl
TMP	5,10,15,20-Tetramesitylporphyrinato
TOF	Turnover frequency
TON	Turnover number
TPAP	Tetra-*n*-propylammonium perruthenate
TPFPP	5,10,15,20-tetra(pentafluorephenyl)porphyrinato

1 Introduction

Oxidation catalysis is an important field of research. Basic transformations in oxidations are nowadays in the fine chemical industry still largely performed using stoichiometric amounts of high-valent metal salts, for example, $PyCr_2O_7$ (where Py is pyridine), RuO_4 and $KMnO_4$ [1, 2]. For the oxidation of alcohols the use of high-valent iodine compounds (notably the Dess Martin reagent) or the Swern method (involving the stoichiometric use of dimethylsulfoxide, DMSO) are still very popular among synthetic chemists, but both require large amounts of potentially polluting reagents [3]. Therefore much is to gain using catalysis in this field. Among the catalysts designed, ruthenium takes a special place owing to its versatility. Ruthenium can catalyse numerous oxidative transformations: the oxidation of alkanes, the asymmetric epoxidation of alkenes, the cleavage of double bonds, the oxidation of alcohols and ethers, and the oxidation of amines and amides. In the field of alcohol, ether and amide oxidation, ruthenium-based catalysts belong to the state of the art, and have high potential for applications in the fine chemical industry.

Ruthenium's versatility is due to the large range of accessible oxidation states, –1 to +8 [4], and different types of oxidation mechanisms that are operative, depending on the ruthenium source used (see later). The oldest application is probably the use of RuO_4 for the oxidation of alcohols and ethers.

Modern variations include the in situ, and thus catalytic, use of this high-valent selective reagent, not only for alcohols but also for ethers (see later). Ru(VII) (perruthenate) in the compounds tetra-*n*-butylammonium perruthenate (TBAP) and tetra-*n*-propylammonium perruthenate (TPAP) has found wide application in alcohol oxidation. Ru-oxo complexes with valence states of IV to VI are key intermediates in, for example, the selective oxygen transfer to alkenes, leading to epoxides. On the other hand 16-electron Ru(II) complexes can be used to catalyse hydrogen transfer; thus these are excellent catalysts for oxidative dehydrogenation of alcohols. A separate section is included to describe the different mechanisms in more detail.

In this chapter the focus is on green oxidation technologies using promising or potentially promising ruthenium-based catalysts for fine chemical synthesis. This implies that the choice of oxidant is taken as a criterion when including examples of new technology (see later). For a more exhaustive coverage of the literature up to 1999 we refer to the excellent reviews on ruthenium-catalysed oxidation reactions in organic synthesis by Naota et al. [5] and Murahashi and Komiya [6]. In this chapter a division is made between alkane, alkene and oxygen- and nitrogen-containing compounds. Per category, the newest developments are compared with the state of the art in this particular field. Homogeneous as well as heterogeneous catalysts are described.

From the standpoint of atom efficiency, and clean technology, oxygen and hydrogen peroxide are the oxidants of choice. They both produce water as the sole by-product. However, in the case of ruthenium catalysts a combination of aldehyde and oxygen, to produce peracids in situ, has also been shown to be extremely useful for the oxidation of amides. Other useful oxidants are hypochlorite or bleach (NaOCl) which can be used to generate RuO_4 in situ, *tert*-BuOOH in combination with Ru-oxo complexes (including asymmetric epoxidation), and *N*-methylmorpholine *N*-oxide (NMO), which can be easily recycled oxidation with H_2O_2. Other oxidants which have a higher environmental impact, such as iodine compounds and $KHSO_5$ (oxone), are generally not considered. However in some cases, for example, in the field of asymmetric epoxidation studies, a few examples using other oxidants are discussed.

1.1 Mechanistic Aspects of Ruthenium-Catalysed Oxidations

The rich redox chemistry of ruthenium is dominated by its propensity for the formation of high-valent complexes containing the strongly σ- and π-donating oxo (O^{2-}) ligand. The most well-known ruthenium oxidant is the tetroxide (E_0 for RuO_4/RuO_4^- is +0.99 V), which effects a range of oxidative tranformations, such as ether oxidation and oxidative cleavage of the double bond (see later). Owing to its high oxidation potential, RuO_4 is capable of oxidizing unactivated C–H bonds as exemplified in Fig. 1.

Analogous to chromium(VI)-catalysed C–H oxidations [7], RuO_4 can abstract a hydrogen, leading to a free radical and Ru(VII) in a solvent cage. In this

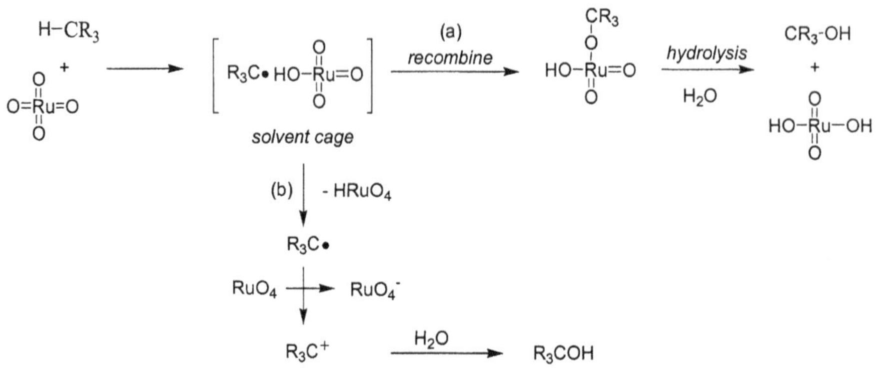

Fig. 1 Mechanism for RuO_4-mediated oxidation of unactivated C–H bonds

solvent cage, recombination might take place giving rise to a ruthenium(VI) ester. Subsequent hydrolysis with ruthenium–oxygen bond cleavage gives an alcohol with the same configuration as the parent hydrocarbon. Alternatively, the free radical can diffuse out of the cage where it will be further oxidised leading to a carbocation (Fig. 1, route b). The alcohols formed under these conditions are further oxidized, leading to the corresponding carbonyl compounds. It is commonly assumed that for alcohol oxidation pathway b, hydrogen abstraction at the α C–H bond followed by one-electron transfer takes place, leading to $R_2C{=}OH^+$ cations and subsequent aldehyde/ketone formation [8]. The active species in RuO_4-catalysed oxidations is pH-dependent. It has been proposed that at pH>8 the active species in reactions with (catalytic amounts of) RuO_4 and NaOCl is probably RuO_4^-, and at pH>12 ruthenate RuO_4^{2-} is the dominant species [8].

For selective oxygen-transfer processes, as in, for example, epoxidation, Ru-oxo species in lower oxidation states have been commonly applied. In general, catalytic systems for oxygen-transfer processes can be divided into two major categories, involving peroxometal and oxometal species as the active oxidant, respectively [1]. The peroxometal mechanism is generally observed with early transition elements whereby high-valent peroxometal complexes of, for example, Mo^{VI}, W^{VI}, V^{V} and Ti^{IV}, are the active oxidants (Fig. 2, pathway a). Catalysis by later and/or many first-row transition elements (Cr, Mn, Fe) on the other hand, involves the intermediacy of high-valent oxometal species, formed via reaction of the metal catalyst with a single oxygen donor (pathway b).

A characteristic feature of this second category is that the olefin epoxidation is often observed in the presence of organic ligands that modulate the activity of the oxometal intermediate, which is certainly the case with ruthenium. In cytochrome P450-dependent monooxygenases, for example, a porphyrin ligand stabilizes a formally oxoiron(V) intermediate [9]. In vivo the active oxoiron(V)porphyrin is formed by reaction of iron(III) with dioxygen in the presence of a sacrificial reductant according to the stoichiometry in Fig. 3. In vitro, the need for a sacrificial reductant can be circumvented by using a single oxy-

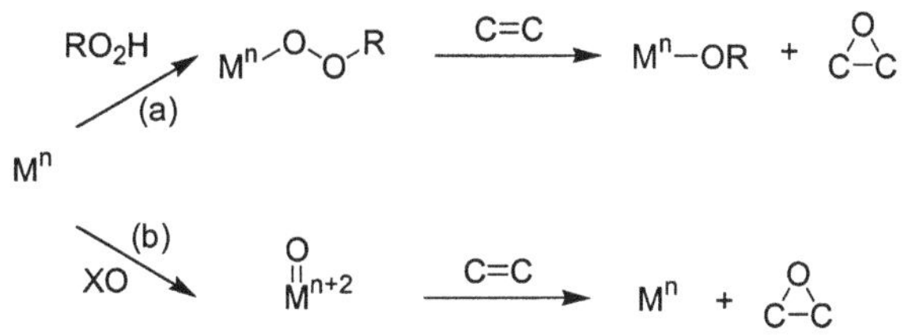

Fig. 2 a Peroxometal and b oxometal mechanisms for olefin epoxidation

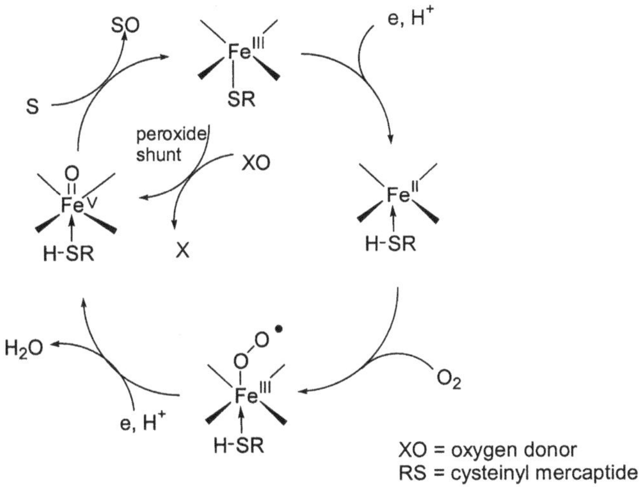

Fig. 3 Mechanism of cytochrome P450 mediated oxidation

gen donor, such as $R\mathrm{O_2H}$, $\mathrm{ClO^-}$, $\mathrm{IO_4^-}$, $R_3\mathrm{NO}$ or PhIO, in the so-called shunt pathway (Fig. 3).

In the presence of electron-donating N, O and P ligands, ruthenium catalyses, for example, olefin epoxidation, with a variety of oxygen donors. However in the case of ruthenium the exact nature of the active oxidant is less clear. The active oxidant is undoubtedly an oxoruthenium complex but various candidates can be envisaged. Moreover oxidative cleavage is generally observed as a competing side reaction and different species may be responsible for epoxidation and oxidative cleavage. Drago [10] has presented evidence in favour of a monooxoruthenium(IV) complex being responsible for epoxidation and a *cis*-dioxoruthenium(VI) species for competing epoxidation and oxidative cleavage. However a ruthenium(V)-oxo intermediate was proposed by Groves et al. [11] in the case of the ruthenium pentafluorophenylporphyrin using $\mathrm{Cl_2PyNO}$ as the oxygen donor. Ruthenium is one of the few examples where these oxoruthenium species can be formed using molecular oxygen as the oxidant. Thus, formally a type of Mars–van Krevelen mechanism is encountered. In 1985 Groves

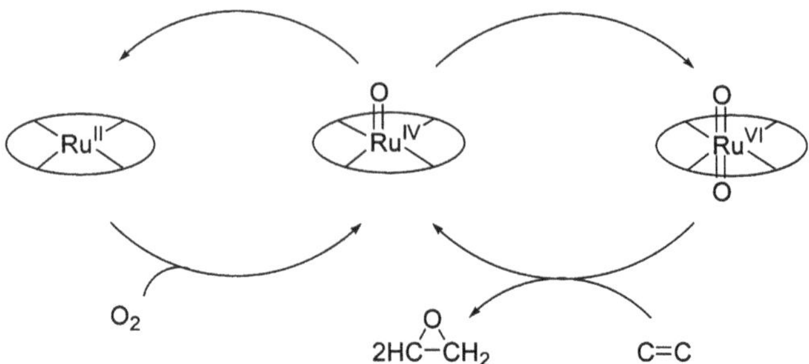

Fig. 4 Proposed mechansim for ruthenium porphyrin catalysed epoxidation in 1985

and Quinn [12] proposed the mechanism shown in Fig. 4 to account for the observed catalytic epoxidation with dioxygen. The ruthenium(IV)oxo species which are formed upon interaction of divalent ruthenium with oxygen can disproportionate to form the Ru(VI)oxo intermediate together with the ruthenium(II) starting compound. A Ru(V)/Ru(III) mechanism was later favoured [11]. The Ru(V) is initially formed by a stepwise oxidation of a ruthenium(II) carbonyl complex.

1.1.1 Ruthenium-Catalysed Alcohol Oxidations

The aerobic oxidation of alcohols is catalysed by both low- and high-valent forms of the metal. In the former case the reaction involves (Fig. 5) the formation of a hydridometal species (or its equivalent), while the latter involves an oxometal intermediate (Fig. 6) which is regenerated by reaction of the reduced form of the catalyst with dioxygen instead of a peroxide. It is difficult to distinguish between the two and one should bear in mind, therefore, that aerobic oxidations with high-valent oxometal catalysts could involve the formation of low-valent species, even the (colloidal) metal, as the actual catalyst.

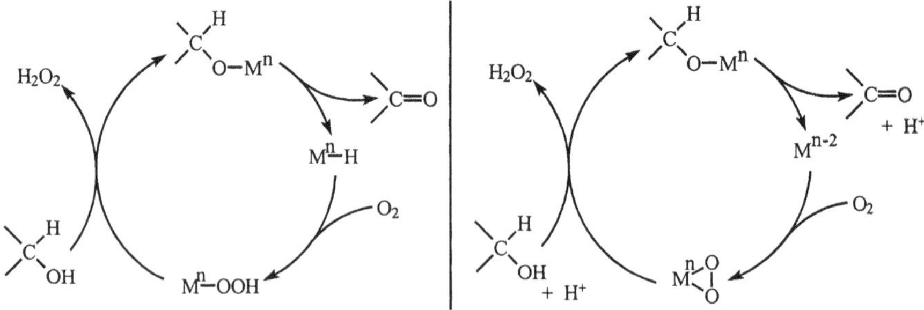

Fig. 5 Hydridometal pathways for alcohol oxidation

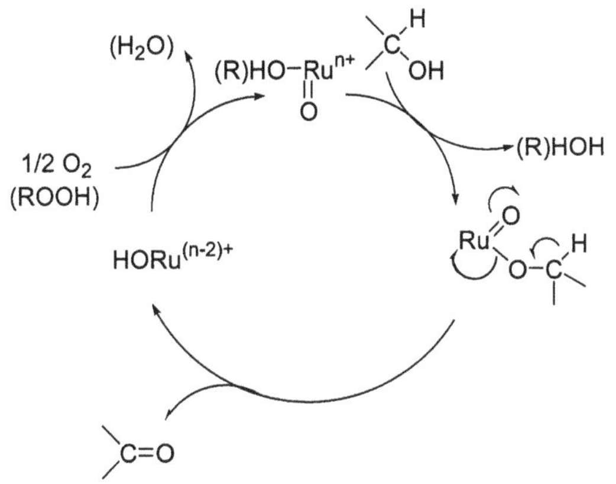

Fig. 6 Oxometal pathway for the oxidation of alcohols

The aerobic oxidation of alcohols catalysed by low-valent late-transition-metal ions, particularly those of group VIII elements, involves an oxidative dehydrogenation mechanism. In the catalytic cycle (Fig. 5) ruthenium can form a hydridometal species by β-hydride elimination from an alkoxymetal intermediate, which is reoxidized by dioxygen, presumably via insertion of O_2 into the M–H bond with formation of H_2O_2. Alternatively, an alkoxymetal species can decompose to a proton and the reduced form of the catalyst (Fig. 5), either directly or via the intermediacy of a hydridometal intermediate. These reactions are promoted by bases as cocatalysts, which presumably facilitate the formation of an alkoxymetal intermediate and/or β-hydride elimination.

2 Ruthenium-Catalysed Oxidation of Alkanes and Arenes

2.1 Ruthenium-Catalysed Oxidation of Alkanes

The selective oxidation of hydrocarbons with dioxygen is of immense industrial importance [1]. A general problem in this area is to obtain high selectivities, particularly at high substrate conversions. The reasons for this are twofold: oxidation can occur at different C–H bonds in a molecule, leading to a low primary selectivity, and the initially formed product is often more reactive than the substrate and is oxidized further, ultimately to carbon dioxide and water, leading to low secondary selectivities. Hence examples of industrial processes tend to involve the oxidation of hydrocarbons in which one particular C–H bond is significantly more reactive, for example, cumene hydroperoxide from cumene, and/or the product is relatively stable towards further oxidation, for example, maleic anhydride from n-butane, phthalic anhydride from o-xylene and benzoic acid from toluene. Even in these cases where one C–H bond is significantly more reactive, for example, cumene, the conversions are kept relatively low (around 20–25%) to avoid competing consecutive processes. And generally speaking, the milder the conditions, the higher the selectivity.

An important goal is, therefore, to develop effective methods for catalytic oxidations with dioxygen, under mild conditions in the liquid phase. Two substrates which are often chosen as models for alkane oxidations are cyclohexane and adamantane. Cyclohexane is of immense industrial importance as its oxidation products – cyclohexanone and adipic acid – are the raw materials for the manufacture of nylon-6 and nylon-6,6. Adamantane is an interesting substrate as the ratio of oxidation at the secondary versus the tertiary C–H bonds is used as a measure of radical versus nonradical oxidation pathways. Industrial processes for the oxidation of cyclohexane, to a mixture of cyclohexanol and cyclohexanone, generally involve low conversions (under 10%). Even at such low conversions, selectivities are modest (70–80%) and substantial amounts of overoxidation products, mostly dicarboxylic acids, are formed.

Most examples of ruthenium-catalysed oxidations of alkanes using porphyrins [11, 13–15] or phthalocyanine [16] complexes as mimics of cytochrome P450 dependent mono-oxygenases, or a variety of other complexes [17–25] as catalysts, have been reported. However, these systems give low turnover numbers (TONs) and/or involve the use of oxygen donors, such as PhIO, *tert*-BuOOH, *m*-chloroperbenzoic acid, oxone ($KHSO_5$) or H_2O_2, as the primary oxidant and are not viable for the industrial oxidation of simple alkanes. They can contribute, however, to an understanding of the role of the various oxoruthenium complexes in oxidation mechanisms.

One example which deserves special mention is the use of a percarboxylic acid such as peracetic acid, generated in situ by autoxidation of the corresponding aldehyde, developed by Murahashi and coworkers, see Eq. (1) [25–27]. These reactions are generally considered to involve high-valent oxoruthenium complexes, generated by reaction of the percarboxylic acid with the ruthenium catalyst, as the active oxidant.

cyclohexane (excess) + CH_3CHO (1 eq.) $\xrightarrow[\text{1 atm } O_2\text{, EtOAc, 70°C}]{Ru^{II}(TPFPP)(CO)}$ cyclohexanone (54% yld) + cyclohexanol (8.1% yld) TON =1410 (1)

TPFPP = tetra(pentafluorophenyl)porphyrinato

Few examples involve the use of dioxygen alone as the primary oxidant. The use of a Ru(III) ethylenediaminetetraacetate complex has been described [28] but this almost certainly involves a free-radical autoxidation pathway and offers no advantages. Following the initial report by Neumann et al. [29] on the use of $[WZnRu_2(OH)(H_2O)(ZnW_9O_{34})_2]^{11-}$ attention has been focused on the use of ruthenium-containing polyoxometalates (POMs) as catalysts for the aerobic oxidation of alkanes. POMs are potentially attractive because their acidic and redox properties can be readily controlled and they constitute "soluble" metal oxides for liquid-phase oxidations. Furthermore they are stable towards oxidizing conditions, in contrast with most organic ligands. Recently, Yamaguchi and Mizuno [30] reported the use of $[(n\text{-}C_4H_9)_4N]_4H[SiW_{11}Ru^{III}(H_2O)O_{39}]\cdot 2H_2O$ as a heterogeneous catalyst for the aerobic oxidation of, for example, adamantane and cyclohexane. High TONs (above 1,000) were claimed (Eq. 2, Table 1).

$[(n\text{-}C_4H_9)_4N]_4H[SiW_{11}Ru^{III}(H_2O)O_{39}].2H_2O$ (Heterogeneous Cat.)

O_2, isobutylacetate, 110°C

59% sel. 2% sel.

25% sel. 14% sel.

Yield of sum of products : 64%
TON : 1282 (2)

Table 1 Oxidation of alkanes with molecular oxygen catalysed by ruthenium-substituted-silico tungstate **1**

Substrate	Time (h)	Temp. (°C)	Yield (%)[a]	TON[b]	Products	Sel. (%)
Adamantane	72	100	64	1282	1-Adamantanol	59
					2-Adamantanol	2
					2-Adamantanone	14
					1,3-Adamantanediol	25
Cyclohexane[c]	48	100	3	1110	Cyclohexanol	33
					Cyclohexanone	67
Cyclooctane	96	110	12	242	Cyclooctanol	13
					Cyclooctanone	87
n-Octane	86	110	3	64	Octanols[d]	14
					Octanones[e]	86
Ethylbenzene	96	110	11	222	1-Phenylethanol	22
					Acetophenone	78

Data from Ref. [30]; reaction conditions 1 mmol substrate, **1** (0.5 μmol), isobutyl acetate (3 ml), O_2 atmosphere. [a] Total yield of mentioned products. [b] TON is the sum of oxidation products (mole)/1 (mole). [c] 18.5 mmol cyclohexane was used. [d] 2-ol:3-ol:4-ol=42:33:25. [e] 2-one:3-one:4-one=51:27:22.

$$2\ \text{POM-Ru}^{II} + O_2 \longrightarrow \text{POM-Ru}^{III}\text{-O-O-Ru}^{III}\text{-POM} \longrightarrow 2\ \text{POM-Ru}^{IV}\text{=O}$$

POM = polyoxometalate

Fig. 7 Postulated mechanism for formation of active Ru^{IV}-oxo species in polyoxometalate/O_2 systems

This catalyst functions as a heterogeneous catalyst at the reaction temperature of 110 °C, in contrast to most POMs which operate under homogeneous conditions. The reaction mechanism is unknown but the C^3-H/C^2-H value can be calculated for adamantane as 20 [31], which is in the range of 1–20 encountered for radical reactions [32, 33], whereas representative Ru-oxo catalysts, generated in situ with single oxygen donors, showed C^3-H/C^2-H values of over 100 [11, 14].[1] In the case of the homogeneous catalyst $[WZnRu_2(OH)(H_2O)(ZnW_9O_{34})_2]^{11-}$, Neumann postulated that an active Ru=O species is produced by cleavage of Ru–O–O–Ru species resulting from reaction of ruthenium with O_2 directly (Fig. 7).

However, this catalytic system did not suppress the overoxidation of cyclohexane. Prolonging the reaction time over 48 h resulted in decreasing selectivity towards cyclohexanol and cyclohexanone owing to competing oxidative cleavage of cyclohexanone to adipic acid and glutaric acid. This catalytic system was also applicable to the oxidation of alcohols. These results suggest that nonradical, selective oxidation of hydrocarbons to alcohols and/or ketones, with molecular oxygen in the absence of a reducing agent such as aldehyde, may be feasible.

Another oxidation system using tri-ruthenium-containing POM was published by Bonchio et al. [31]. $[Ru(C_6Me_6)]_3Mo_5O_{18}$ or $[Ru(C_6Me_6)]_3W_5O_{18}$ was used as the oxidation catalyst with 2,6-dichloropyridine-*N*-oxide under acidic conditions (Eq. 3).

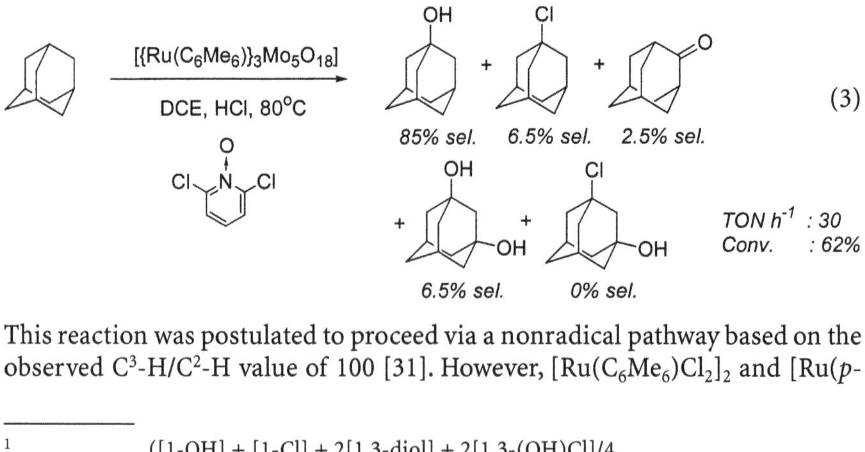

[{Ru(C6Me6)}3Mo5O18]
DCE, HCl, 80°C

OH + Cl + O (3)

85% sel. 6.5% sel. 2.5% sel.

+ OH, OH + Cl, OH

6.5% sel. 0% sel.

TON h⁻¹ : 30
Conv. : 62%

This reaction was postulated to proceed via a nonradical pathway based on the observed C^3-H/C^2-H value of 100 [31]. However, $[Ru(C_6Me_6)Cl_2]_2$ and [Ru(*p*-

[1] $$C^3\text{-H}/C^2\text{-H} = \frac{([1\text{-OH}] + [1\text{-Cl}] + 2[1,3\text{-diol}] + 2[1,3\text{-(OH)Cl}]/4}{([2\text{-OH}] + [2\text{-none}] + [2\text{-Cl}])/12}$$

cymene)$Cl_2]_2$ displayed similar activity compared to the Ru^{III}-containing POM, which makes it doubtful that the POM environment imposes any substantial difference on the reactivity of the ruthenium centre.

Recently, a ruthenium-catalysed oxidation in water was published by d'Alessandro et al. [34]. Water can be regarded as an environmentally friendly solvent which, because it is inert, reduces the risk of explosions. The oxidation of cyclohexane directly to adipic acid was performed using ruthenium catalysts bearing water-soluble phthalocyanine ligands: RuPcS (where PcS is tetrasodium 2,3-tetrasulfophthalocyaninato) with $KHSO_5$ (Eq. 4). However we note that very low TONs were observed and the use of $KHSO_5$ as a primary oxidant is not viable for industrial-scale oxidations.

Ru-cat, $KHSO_5$ 0.5 N; water, 20°C, 72 h

2 TON; trace; 0.08 TON

water-soluble Ru-cat (5 mM) (4)

2.2 Ruthenium-Catalysed Oxidation of Benzene

Direct oxidation of benzene to phenol is of great interest not only for its industrial importance, but also from a purely scientific point of view. Apart from many earlier reports [35] on the oxidation of benzene to phenol by hydroxyl radicals generated by the reaction of Fe^{2+} salt (Fenton reagent) with H_2O_2 not much is known about the homogeneously catalysed oxyfunctionalization of aromatic C–H bonds. The lack of studies is largely attributable to the fact that the activation of the C–H bond in benzene is difficult owing to its resonance stability and the reactivity of phenol, which is consecutively oxidized to quinones and other by-products.

Carbonyl (5,10,15,20-tetrapentafluorophenylporphyrinato)ruthenium(II) displayed activity with 2,6-dichloropyridine-*N*-oxide as an oxidant for the oxidation of benzene to 1,4-benzoquinone [11] (Eq. 5).

$[Ru^{II}(TPFPP)(CO)]$, $pyCl_2NO$; CH_2Cl_2, 65°C, 12h

Conv: 13 %
5320 TON (5)

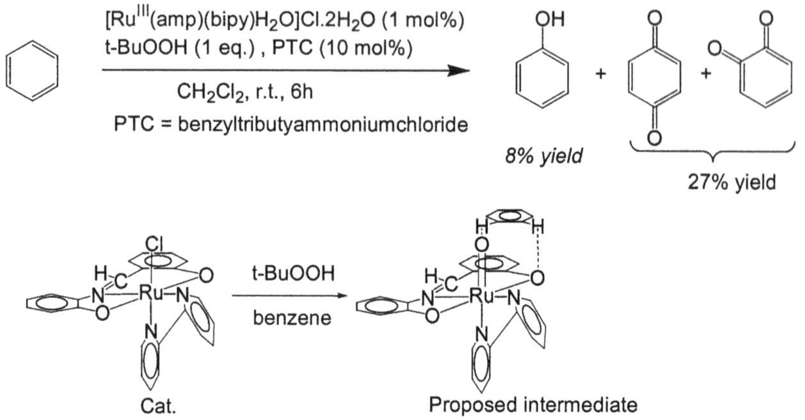

Fig. 8 Oxidation of benzene to phenol using a ruthenium–bipyridyl–Schiff's base type complex as a catalyst and *tert*-BuOOH as the oxidant

This catalyst is known as a highly active catalytic system for the nonradical hydroxylation reaction of adamantane, in which turnovers per hour of 48,000 are observed and only C^3-H is hydroxylated selectively. Its high activity makes the oxidation of the stable C–H bond of benzene possible. However, it was impossible to terminate the oxidation at phenol, and quinone was the final product observed.

Recently, $[Ru^{III}(amp)(bipy)H_2O]Cl{\cdot}2H_2O$, where H_2amp is *N*-(hydroxyphenyl) salicyldiimine and bipy is 2,2′-bipyridyl, was reported to show remarkably high catalytic activity, using *tert*-BuOOH as the oxidant, for the oxidation of benzene to phenol and benzoquinones [36–38] (Fig. 8).

The active oxidant was proposed to be a Ru(V)=O species and access of benzene towards the Ru=O bond is facilitated by the flat structure of the salicyldiimine ligand (see Fig. 8). This catalytic system was also applied to the epoxidation of stilbene, C–H bond activation of cyclohexane or cyclohexene and the oxidation of tetrahydrofuran to γ-butyrolactone [37]. We conclude however, that a suitable and catalytic system for the selective oxidation of benzene to phenol has not yet been forthcoming.

2.3 Ruthenium-Catalysed Oxidation of Alkylaromatics

Methylbenzenes can readily be transformed into benzoic acids at room temperature upon exposure to aqueous sodium hypochlorite at pH 9–10 in the presence of $RuCl_3$ and tetra-*n*-butylammonium bromide (TBAB) catalysts in a two-phase system (Eq. 6) [39]. The proposed mechanism for the Ru–NaOCl phase transfer catalyst system is a hydride abstraction from the substrate by RuO_4 to form a carbonium ion which is promptly hydrolysed. A major advantage of this system is that 100% conversion can be attained with high selectivity.

$$\text{C}_6\text{H}_4(\text{X})\text{CH}_3 + 3\,\text{NaOCl} + \text{NaOH} \xrightarrow[\text{2 h, pH 9}]{\text{RuCl}_3\text{ (1 mol\%), 5 mol\% TBAB}} \text{C}_6\text{H}_4(\text{X})\text{CO}_2\text{Na} + 3\,\text{NaCl} + 2\,\text{H}_2\text{O} \quad (6)$$

yield 92-98%

X = H, Cl, Br, NO_2, CN

TBAB = tetra-n-butyl ammonium bromide

This system was recently extended to the highly selective synthesis of substituted toluic acids. Instantaneous extraction of the benzoic acid product (as a sodium salt) into the aqueous phase prevents oxidation of the second benzylic group [40]. The system is limited to oxidizing xylenes bearing an electron-withdrawing group, since xylenes bearing an electron-donating group result in ring chlorination as the major reaction.

3 Ruthenium-Catalysed Oxidation of Alkenes

Alkenes are useful raw materials for both commodities and fine chemicals because of the high reactivity of the double bond. However an inherent difficulty in the oxidation of olefins is caused by different competing modes of oxidation, for example, epoxidation, allylic oxidation and double-bond cleavage. An overview of the different oxidation reactions of alkenes catalysed by ruthenium is given in Fig. 9. The selectivity of the epoxidation reaction versus allylic oxidation, for example, depends heavily on the nature of the alkene studied. For example, in the case of the oxidation of cyclohexene, the allylic C–H bond is more easily oxidized than the C=C bond as the allylic C–H bond in the relatively small ring structure exhibits enhanced reactivity. On the other hand, in the case of cyclooctene or norbornene, the allylic C–H bonds are much less reactive and epoxidation is highly favoured. Hence, we emphasize that cyclooctene and norbornene are poor model substrates for demonstrating epoxidation with metal catalyst/dioxygen combinations. In the case of the oxidation of styrene or stilbene, both epoxidation and C=C bond cleavage reaction are observed as well as isomerization around the double bond.

When the oxidation of alkenes was performed with low-valent ruthenium (generated in situ from peracetic acid) in an aqueous medium, the formation of α-ketols was observed. For this interesting transformation, for which ruthenium was found to be the best catalyst, the reader is referred to the reviews of Naota et al. [5] and Murahashi and Komiya [6].

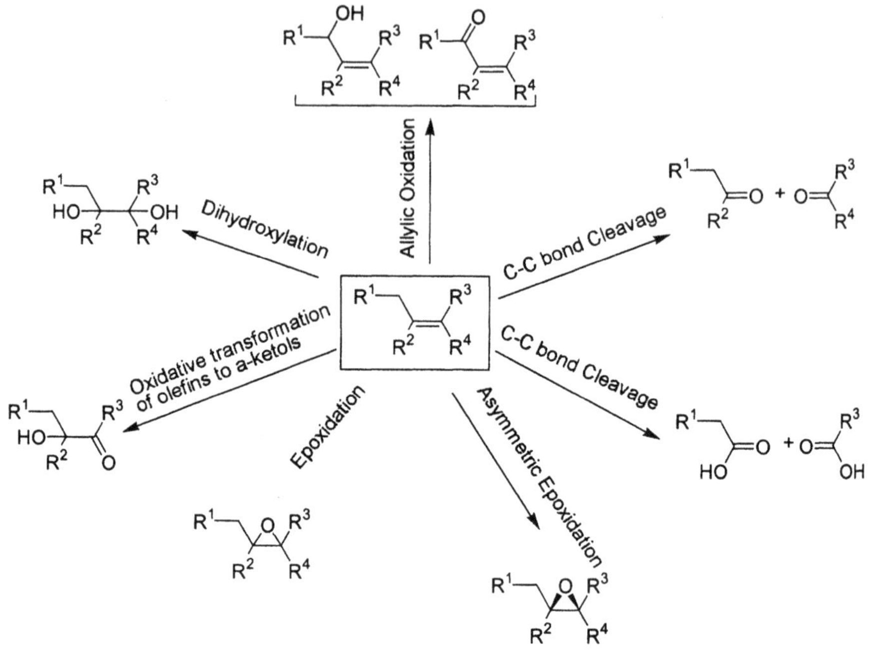

Fig. 9 Overview of different transformations of alkenes catalysed by ruthenium

3.1 Ruthenium-Catalysed Epoxidation of Alkenes

The field of selective epoxidation of alkenes is largely dominated by transition metals in combination with an alkyl hydroperoxide or hydrogen peroxide. Catalysts such as tungsten, manganese, titanium, organometallic methyltrioxorhenium and also selenium, can form electrophilic oxidants by reaction with hydrogen peroxide, which are very selective and active in epoxidizing a range of alkenes, including terminal and cyclic alkenes [41]. All these metals, with the exception of manganese, act via a peroxometal mechanism [1].

Ruthenium, together with, for example, Mn and Cr, belongs to the class of oxo-metal oxidants (see earlier). Ruthenium is a potentially interesting epoxidation catalyst because of the ability of ruthenium mono-oxo and dioxo species to selectively transfer an oxygen atom to the double bond [42]; They exhibit high TONs and can, in principle, be generated using molecular oxygen. The problem, however, is to generate this Ru-oxo species in a catalytic system under conditions which are compatible with the coexistence of often labile epoxides and without side reactions of the alkenes present. Almost all examples with ruthenium catalysts in alkene epoxidation, therefore, involve alkenes which are highly reactive, such as styrenes, stilbenes and cyclohexene (and which thus can react under mild conditions) or which produce epoxides that are highly stable, such as cyclooctene and norbornene. Cyclooctene epoxide, for example, is one of the most stable epoxides known because every reaction of the epoxide group is retarded by the steric hindrance of the ring. On the other hand, cyclohexene epoxide is a very reactive epoxide, since the steric influence of the cyclohexane ring promotes ring opening of the epoxide. Terminal epoxides are generally quite stable towards ring opening, but styrene oxide is highly reactive, as a result of the electronic influence of the aromatic ring.

In practice in the literature of the past 20 years the important results with ruthenium in epoxidation are those where ruthenium was demonstrated to afford epoxides with molecular oxygen as the terminal oxidant. Some examples are presented (see later). Also ruthenium complexes, because of their rich chemistry, are promising candidates for the asymmetric epoxidation of alkenes. The state of the art in the epoxidation of nonfunctionalized alkenes is namely still governed by the Jacobsen–Katsuki Mn-based system, which requires oxidants such as NaOCl and PhIO [43,44]. Most examples in ruthenium-catalysed asymmetric epoxidation known until now still require the use of expensive oxidants, such as bulky amine oxides (see later).

Most ruthenium catalysts used in epoxidation reactions are based on bulky porphyrins or other amine ligands and require the use of PhIO and Cl_2PyNO as oxidants. For examples see the reviews in Refs. [5,6,45] and some recent examples by Liu and coworkers [46,47] and Jitsukawa et al. [48]. Examples for the aerobic epoxidation of alkenes are the ruthenium mesityl porphyrin complex $Ru(TMP)(O)_2$, where TMP is 5,10,15,20-tetramesitylporphyrinato, of Groves and Quinn [12] in 1985 (Eq. 7), the ruthenium dimethylphenanthroline complex, *cis*-$[Ru(2,9\text{-dimethyl-1,10-phenanthroline})(CH_3CN)_2]^{2+}$ published by Goldstein et al. [23] in 1994 (Eq. 8), and the ruthenium POM catalyst $\{[WZnRu_2(OH)(H_2O)](ZnW_9O_{34})_2\}^{11-}$ of Neumann and Dahan [49] in 1997 (Eq. 9).

Ru-1,4,7-trimethyl-1,4,7-triazocyclononane complexes *cis*-$[Ru(1,4,7\text{-trimethyl-1,4,7-triazacyclononane})(O_2)(CF_3CO_2)]^+$ with *tert*-BuOOH [20, 21] also afforded reasonable results and recently this complex was heterogenized on silica [50]. In contrast to the homogeneous catalyst, this heterogeneous ruthenium complex in epoxidation of cyclohexene produced cyclohexene oxide in 75% yield with cyclohexen-1-one in 14% yield, while the homogeneous catalyst produced cyclohexen-1-ol and cyclohexen-1-one mainly. Examples of

$Ru(TMP)(O)_2$

1 atm O_2, 25°C
benzene, 24 h

43 TON

(7)

R = mesityl

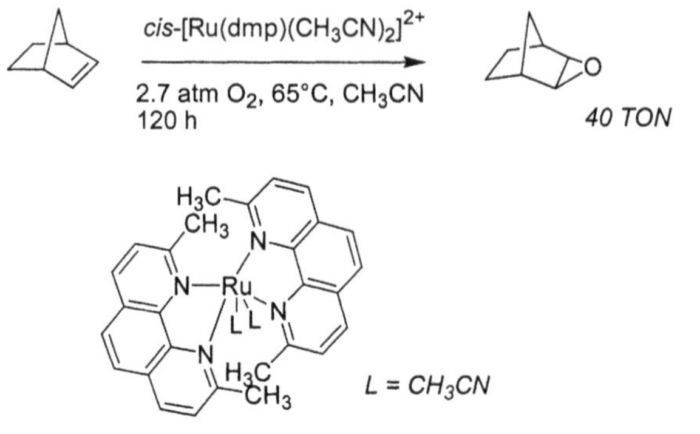

cis-$[Ru(dmp)(CH_3CN)_2]^{2+}$

2.7 atm O_2, 65°C, CH_3CN
120 h

40 TON

H_3C CH_3 N N—Ru L L N N H_3C CH_3

L = CH_3CN (8)

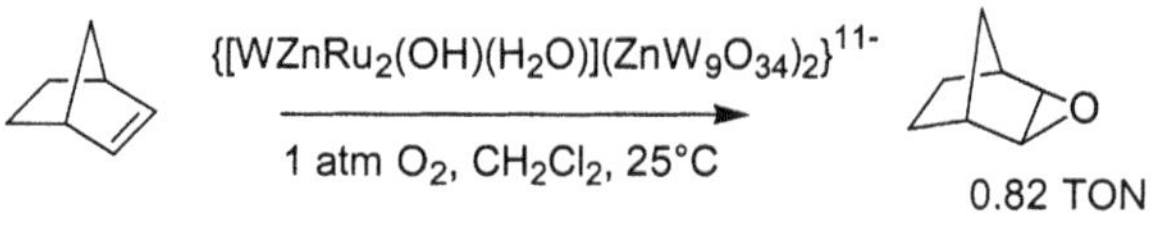

$\{[WZnRu_2(OH)(H_2O)](ZnW_9O_{34})_2\}^{11-}$

1 atm O_2, CH_2Cl_2, 25°C

0.82 TON (9)

the use of a combination of aldehyde and oxygen as oxidant (generating peracids in situ) involve the relatively simple ruthenium(II) perfluorinated 1,3-diketone complex (Eq. 10) [51] and ruthenium complexes with naphthyridine ligands (Eq. 11) [52].

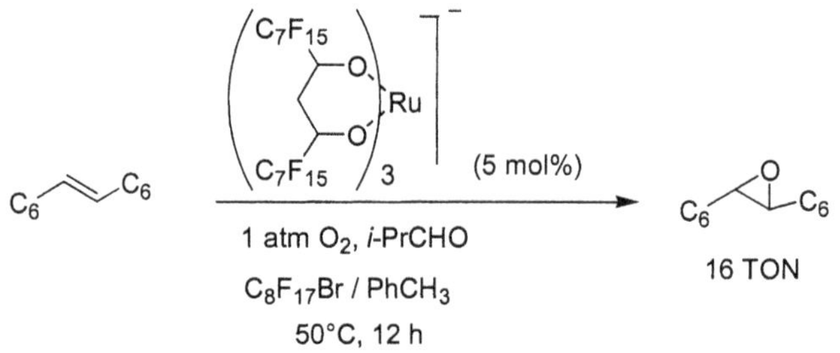

C_7F_{15} O Ru O C_7F_{15} 3 (5 mol%)

C_6 C_6

1 atm O_2, *i*-PrCHO
$C_8F_{17}Br$ / $PhCH_3$
50°C, 12 h

C_6 O C_6

16 TON (10)

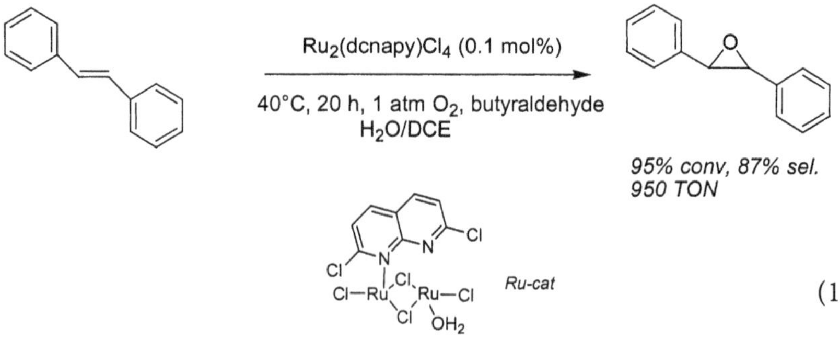

$Ru_2(dcnapy)Cl_4$ (0.1 mol%)

40°C, 20 h, 1 atm O_2, butyraldehyde
H_2O/DCE

95% conv, 87% sel.
950 TON

Cl N N Cl Cl—Ru Cl Cl Ru—Cl OH_2 Cl

Ru-cat (11)

The best result using aldehyde/oxygen was reported recently by Qi et al. [53] using novel Ru(H*L*)(*L*)Cl_2 (H*L* is *N*-2′chlorophenyl-2-pyridine-carboxamide) complexes and isobutyraldehyde/oxygen for the epoxidation of cyclic alkenes. The turnover frequencies (TOFs) in this system were as high as 350 h^{-1} for cyclohexene, with a selectivity towards the epoxide of 87% (see Eq. 12).

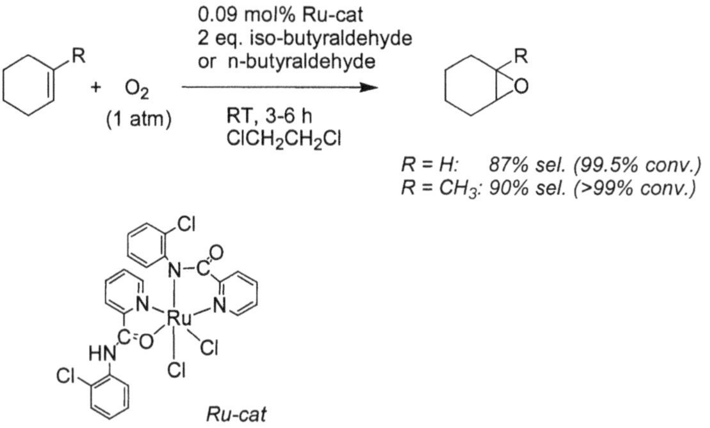

0.09 mol% Ru-cat
2 eq. iso-butyraldehyde
or n-butyraldehyde
+ O_2 (1 atm)
RT, 3-6 h
$ClCH_2CH_2Cl$

(12)

R = H: 87% sel. (99.5% conv.)
R = CH_3: 90% sel. (>99% conv.)

Ru-cat

Another noteworthy example is the use of nitrous oxide as an oxidant to generate Ru-oxo species in situ. Nitrous oxide circumvents the occurrence of competing radical reactions (such as in the case of *tert*-BuOOH and H_2O_2). Moreover, as the by-product of adipic acid production, it is, in principle, available for a reasonable price [54]. Following the publication of Groves and Roman in 1995 [55], Yamada et al. [56] reported that 5 mol % Ru(TMP)$(O)_2$ can catalyse the epoxidation of cholesteryl benzoate at 100 °C under 10 atm N_2O (Eq. 13). Although no other examples of alkenes using this interesting methodology were given, alcohol oxidation could also be performed with reasonable results [57], and thus it seems to demonstrate the principle of Ru-oxo generation by N_2O as shown in Fig. 10.

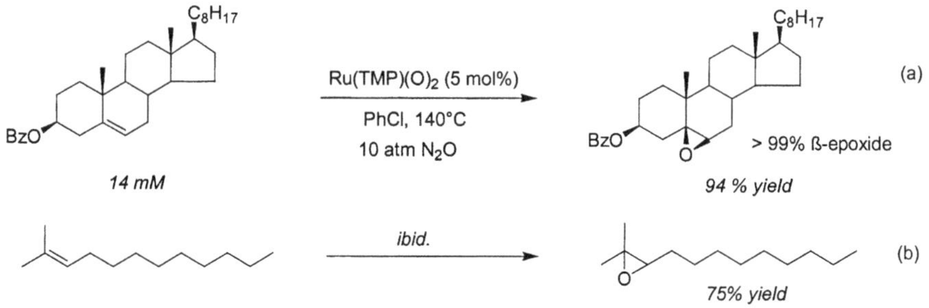

(a) Ru(TMP)$(O)_2$ (5 mol%), PhCl, 140°C, 10 atm N_2O

14 mM

> 99% ß-epoxide

94 % yield

(b) *ibid.*

75% yield

(13)

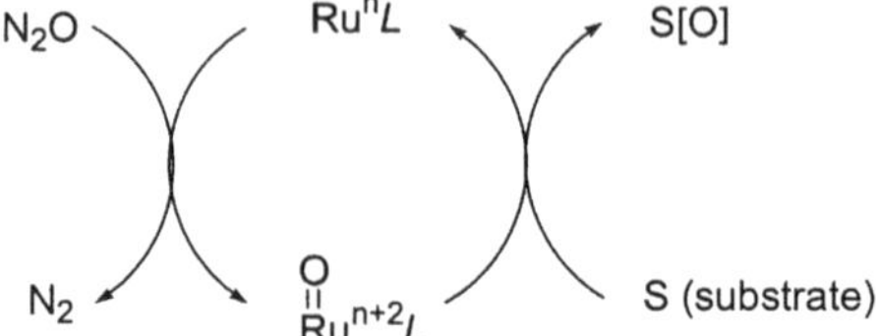

Fig. 10 Generation of active Ru=O intermediates using N_2O as the oxygen donor

3.2 Asymmetric Ruthenium-Catalysed Epoxidation

Some examples of asymmetric epoxidations of alkenes using chiral ruthenium porphyrins have been reported; for example, the previously reported D_4-symmetrical chiral ruthenium porphyrin complex $Ru^{II}(D_4\text{-Por*})(CO)(MeOH)$ [58], which produced (*R*)-styrene oxide in 57% ee with Cl_2PyNO as a donor, was readily converted into the dichloro derivative **A** [59] (Fig. 11). This dichlororuthenium porphyrin gave (*R*)-styrene oxide in 69% ee using Cl_2PyNO and was highly active (875 TON in 1.5 h). The use of unsubstituted pyridine *N*-oxide or NMO as oxidants resulted in low substrate conversions as well as

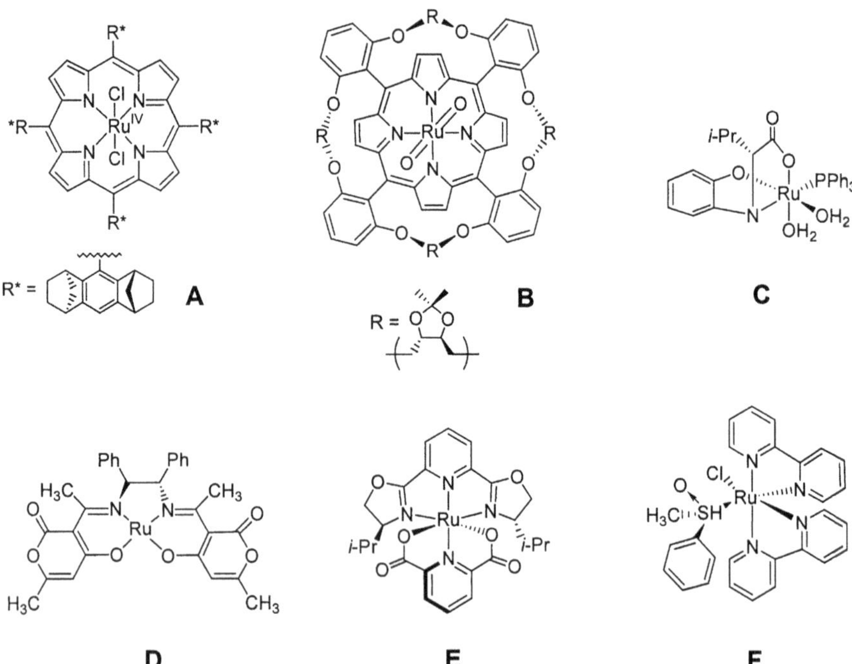

Fig. 11 Structures of chiral ruthenium complexes used as catalysts in epoxidations using a variety of oxygen donors

reduced enantiomeric excesses. The latter catalyst was also heterogenized by sol–gel techniques and thereby maintained its selectivity. The chiral ruthenium porphyrin **B** produced (*S*)-styrene oxide in 70% ee also using Cl_2PyNO as the terminal oxidant [60].

Several ruthenium complexes bearing chiral Schiff's base ligands have been published. Ru*L*(PPh_3)(H_2O)$_2$], complex **C** (Fig. 11), with PhIO produced (*S*)-styrene oxide in 80% ee [61]. Chiral Schiff's base complex **D** was examined using molecular oxygen with aldehyde, with or without 2,6-dichloropyridine *N*-oxide as an axial ligand. Styrene oxide was produced in up to 24% ee[62]. A chiral bis(oxazolinyl)pyridine ruthenium complex **E** with iodosylbenzene diacetate PhI(OAc)$_2$ produced (1*S*,2*S*)-*trans*-stilbene oxide in 74% ee [63]. Similarly, chiral ruthenium bis(bipyridine) sulfoxide complex **F** [64] was effective in combination with PhI(OAc)$_2$ as an oxidant and resulted in in 33% ee for (*R*,*R*) *trans*-stilbene oxide and 94% ee for (*R*,*R*) trans-β-Me-styrene (after 75 h at 25 °C).

Recently, a cationic five-coordinate Ru^{II} complex of the type RuCl(PNNP)]PF_6 (where PNNP is a tetradentate chiral ligand with a P_2N_2 donor set, see Eq. 14) was examined as a chiral epoxidation catalyst for styrene using the cheap and environmentally friendly hydrogen peroxide as an oxidant [65]. Styrene was converted into (*S*)-styrene oxide with 35% ee (Eq. 14). In this reaction 7 eq. H_2O_2 were needed, however, owing to the high catalase activity of ruthenium [66].

1 mol% Ru-cat
7 eq. H_2O_2
CH_2Cl_2, 6 h, 22°C

35% conv.; 81% sel.
37% ee (S) (14)

N N Ru P P Cl Ph Ph Ph Ph PF_6^- +

3.3
Ruthenium-Catalysed Oxidative C=C Bond Cleavage

The mechanism of oxidative cleavage of double bonds is commonly supposed to be the result of coordination of Ru-dioxo species to the double bond (Fig. 12).

Regarding the ruthenium-catalysed C=C bond cleavage reaction, generally "naked" high-valent ruthenium oxo-ruthenium species have been used as catalysts, for example, $RuCl_3$/NaOCl [67], RuO_4/$NaIO_4$ [68], RuO_2/$NaIO_4$ [69], and

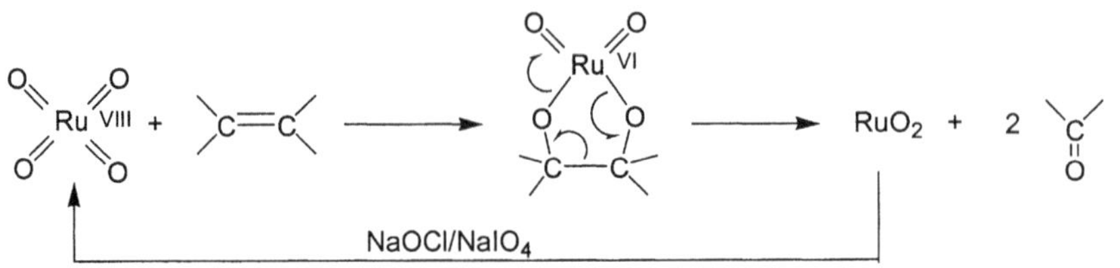

RuO_4 (Ru VIII) + C=C → cyclic Ru(VI) ester → RuO_2 + 2 C=O; NaOCl/$NaIO_4$ regenerates RuO_4

Fig. 12 Mechanism of RuO_4-mediated oxidative cleavage of olefins

$RuCl_3/NaIO_4$ [70]. However these catalytic C=C cleavage reactions lead to carboxylic acids or ketones, not to the aldehydes. Recently, examples of selective C=C cleavage reactions to aldehydes using ruthenium catalysts and oxone or $NaIO_4$ as the terminal oxidant were reported [71]. These are shown in Eqs. (15–17).

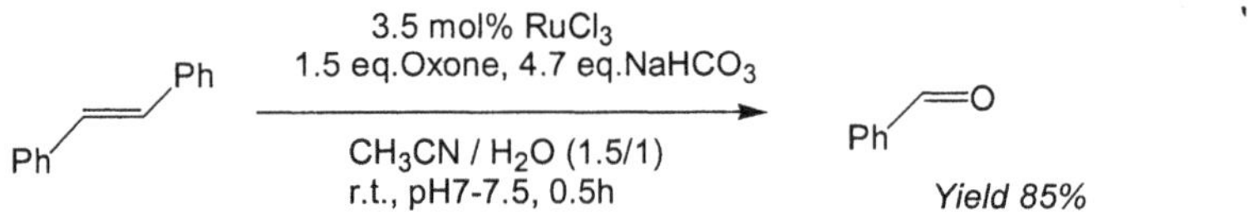

Ph–CH=CH–Ph —[3.5 mol% $RuCl_3$, 1.5 eq. Oxone, 4.7 eq. $NaHCO_3$; CH_3CN / H_2O (1.5/1), r.t., pH7-7.5, 0.5h]→ PhCHO, *Yield 85%* (15)

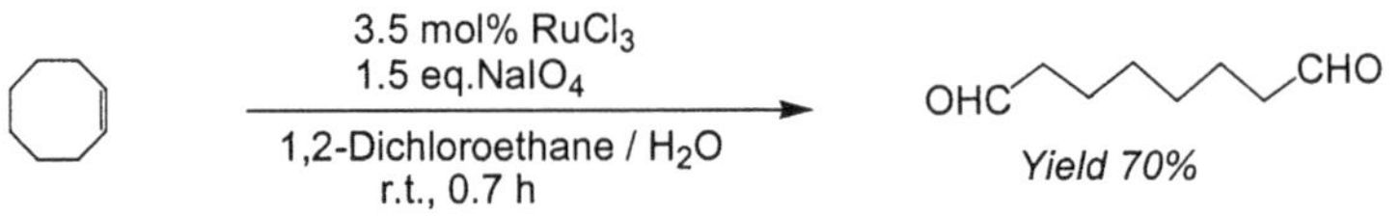

cyclooctene —[3.5 mol% $RuCl_3$, 1.5 eq. $NaIO_4$; 1,2-Dichloroethane / H_2O, r.t., 0.7 h]→ OHC–$(CH_2)_6$–CHO, *Yield 70%* (16)

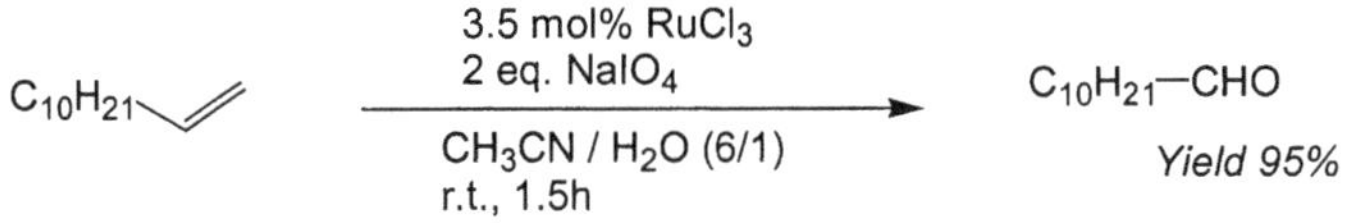

$C_{10}H_{21}$–CH=CH_2 —[3.5 mol% $RuCl_3$, 2 eq. $NaIO_4$; CH_3CN / H_2O (6/1), r.t., 1.5h]→ $C_{10}H_{21}$–CHO, *Yield 95%* (17)

These protocols can be regarded as promising because of their simplicity and broad scope. The conventional method for the production of aldehydes from alkenes consists of ozonolysis followed by workup under reducing conditions [72]. The ruthenium-based method, using either oxone or $NaIO_4$ as the oxidant, is an interesting alternative. The high selectivity to aldehyde versus carboxylic acid was reached by manipulation of the amount of terminal oxidant (2 Eq.) and the reaction time.

A variety of ruthenium complexes have been used in conjunction with $NaIO_4$ [73], PhIO [74], *tert*-BuOOH [38] and O_2 [75] in the oxidation of styrene and/or stilbene. The major reaction was oxidative cleavage rather than epoxidation (Eqs. 18, 19b). Notably in the case of $RuCl_2$[PNNP] complexes [74] as catalysts, *cis*-diols were formed when treating 1-hexene at elevated temperatures and pressures with oxygen (Eq. 19a).

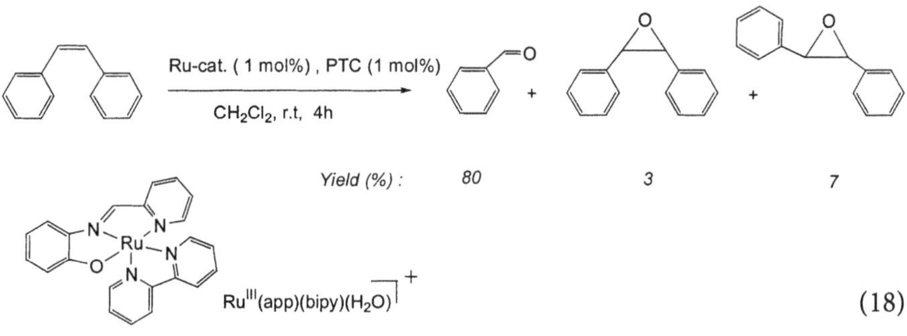

Ru-cat. (1 mol%) , PTC (1 mol%)
CH_2Cl_2, r.t, 4h

Yield (%) : 80 3 7

$Ru^{III}(app)(bipy)(H_2O)^+$

(18)

PTC = Benzyltributylammoniumchloride (phase transfer catalyst)

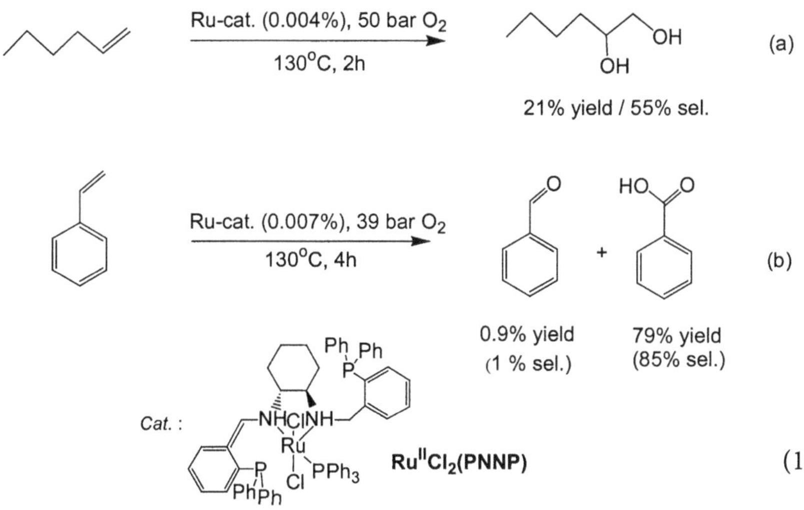

Ru-cat. (0.004%), 50 bar O_2
130°C, 2h

OH
OH

21% yield / 55% sel.

(a)

Ru-cat. (0.007%), 39 bar O_2
130°C, 4h

HO O

0.9% yield (1 % sel.)

79% yield (85% sel.)

(b)

Cat. : $Ru^{II}Cl_2(PNNP)$

(19)

4 Ruthenium-Catalysed Oxidation of Oxygen-Containing Compounds

In the field of alcohol oxidation, ruthenium-based systems belong to the top five catalysts ever reported using O_2 as an oxidant [76]. Therefore the following section is completely devoted to the aerobic ruthenium-based systems. From the viewpoint of green oxidation, the use of H_2O_2 as an oxidant would also be interesting, but owing to the high catalase activity of ruthenium, examples are scarce. One example is the $RuCl_3 \cdot 3H_2O$/didecyl dimethylammonium bromide system reported by Barak et al. [77]. This system catalyses the selective oxidation of a variety of alcohols at high substrate-to-$RuCl_3$ (625:1) ratio in an aqueous/organic biphasic system. However 3–6 eq. H_2O_2 were required, reflecting the propensity of ruthenium for catalysing H_2O_2 decomposition. Alternatively, the use of the oxotriruthenium compound $[Ru_3O(OAc)_6(CH_3OH)_3](OAc)$ was reported by Wynne et al. [78]. In combination with a catalytic amount of TBAB and 3 Eq. H_2O_2, primary aliphatic and benzylic alcohols were converted to their corresponding aldehydes.

4.1 Ruthenium-Catalysed Aerobic Oxidation of Alcohols

Ruthenium compounds are widely used as catalysts in organic synthesis and have been extensively studied as catalysts for the aerobic oxidation of alcohols [79]. In 1978, Tang et al. [80] reported that $RuCl_3 \cdot nH_2O$ catalyses the aerobic oxidation of secondary alcohols into the corresponding ketones, albeit in modest yields. In 1981, Matsumoto and Ito [81] showed that $RuCl_3$ and $RuCl_2(Ph_3P)_3$ catalyse the aerobic oxidation of activated allylic and benzylic alcohols under mild conditions, for example, the oxidation of retinol to retinal could be performed at 25 °C (57% yield was obtained after 48 h). Aliphatic primary and secondary alcohols were more efficiently oxidized using trinuclear ruthenium carboxylates, $Ru_3O(O_2CR)_6L_n$ (L is H_2O or Ph_3P), as the catalysts [82]. With lower aliphatic alcohols, for example, 1-propanol, 2-propanol and 1-butanol, activities were around 10 times higher than with $RuCl_3$ and $RuCl_2(Ph_3P)_3$. Recently somewhat higher activities were reached using $RuCl_2(PPh_3)_2$ as the catalyst with ionic liquids as solvents (Fig. 13). These solvents have been tested as environmentally friendly solvents for a large variety of reactions [83]. In this particular case tetramethylammonium hydroxide and aliquat 336 (tricaprylylmethyl ammonium chloride) were used as solvents and rapid conversion of benzyl alcohol was observed [84]. Moreover the tetramethylammonium hydroxide/$RuCl_2(PPh_3)_3$ could be reused after extraction of the product.

Ruthenium compounds are widely used as catalysts for hydrogen-transfer reactions. These systems can be readily adapted to the aerobic oxidation of alcohols by employing dioxygen, in combination with a hydrogen acceptor as a cocatalyst, in a multistep process. For example, Bäckvall and coworkers [85] used low-valent ruthenium complexes in combination with a benzoquinone and a cobalt Schiff's base complex. The proposed mechanism is shown in Fig. 14. A low-valent ruthenium complex reacts with the alcohol to afford the aldehyde or ketone product and a ruthenium dihydride. The latter undergoes hydrogen transfer to the benzoquinone to give hydroquinone with concomitant

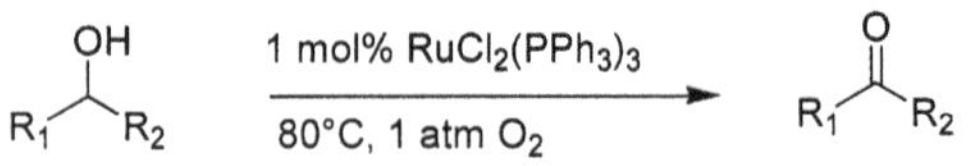

$R_1CH(OH)R_2 \xrightarrow[80°C,\ 1\ atm\ O_2]{1\ mol\%\ RuCl_2(PPh_3)_3} R_1C(O)R_2$

solvent:	tetramethylammonium hydroxide	aliquat
R_1=$PhCH_2$, R_2=H	*91% conv. (5h)*	*58% conv. (5h)*
R_1, R_2, = c-C_7H_{14}	*61% conv. (11h)*	*92% conv. (11h)*
R_1=C_6H_{13}, R_2=CH_3	*43% conv. (25h)*	*81% conv. (25h)*

Fig. 13 Aerobic ruthenium-catalysed oxidation in ionic liquids

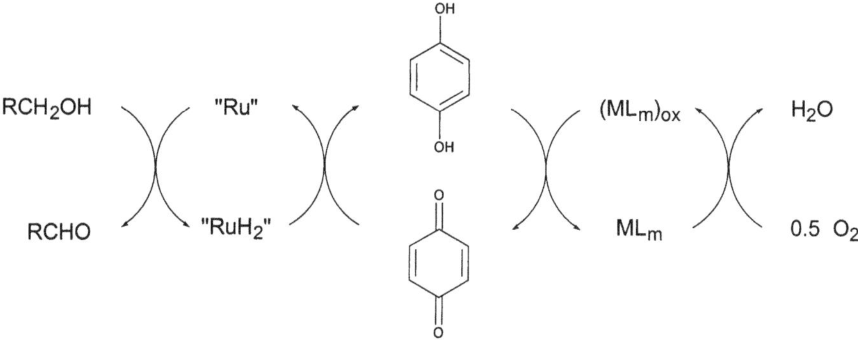

Fig. 14 Ruthenium catalyst in combination with a hydrogen acceptor for aerobic oxidation

regeneration of the ruthenium catalyst. The cobalt Schiff's base complex catalyses the subsequent aerobic oxidation of the hydroquinone to benzoquinone to complete the catalytic cycle.

The regeneration of the benzoquinone can also be achieved with dioxygen in the absence of the cobalt cocatalyst. Thus, Hanyu et al. [86] showed that a combination of $RuCl_2(Ph_3P)_3$, hydroquinone and dioxygen, in $PhCF_3$ as solvent, oxidized primary aliphatic, allylic and benzylic alcohols to the corresponding aldehydes in quantitative yields (Eq. 20).

OH → (1 atm O_2, 60°C, 20 h; $RuCl_2(Ph_3P)_3$ (10 mol%), hydroquinone (10 mol%), K_2CO_3, $PhCF_3$) → O, H

90% conv.
>99% sel.

(20)

A combination of $RuCl_2(Ph_3P)_3$ and the stable nitroxyl radical, 2,2′,6,6′-tetramethylpiperidine-*N*-oxyl (TEMPO) is a remarkably effective catalyst for the aerobic oxidation of a variety of primary and secondary alcohols, giving the corresponding aldehydes and ketones, respectively, with above 99% selectivity [87]. The best results were obtained using 1 mol % of $RuCl_2(Ph_3P)_3$ and 3 mol % of TEMPO (Eq. 21).

OH → ($RuCl_2(PhP)_3$ (1 mol%), TEMPO (3 mol%); 8% O_2/N_2 (10 atm), 100°C, PhCl, 7h) → O (21)

95% conv.
>99% sel.

The results obtained in the oxidation of representative primary and secondary aliphatic alcohols and allylic and benzylic alcohols using this system are shown in Table 2.

Table 2 Ruthenium/2,2′,6,6′-tetramethylpiperidinyl-*N*-oxyl (*TEMPO*) catalysed oxidation of primary and secondary alcohols to the corresponding aldehyde using molecular oxygen. 15 mmol substrate, 30 ml chlorobenzene, $RuCl_2(PPh_3)_3$/TEMPO ratio of 1/3, 10 ml min^{-1} O_2/N_2 (8/92; v/v), P=10 bar, T=100 °C

Substrate	S/C ratio[a]	Time (h)	Conv. (%)[b]
Primary alcohols			
1-Octanol	50	7	85
3-Methyl-2-butenol	67	7	96
Geraniol	67	7	91
Benzyl alcohol[c]	200	2.5	>99
p-Nitrobenzyl alcohol[c]	200	6	97
Secondary alcohols			
2-Octanol	100	7	98
2-Adamantanol	100	7	92
Cyclooctanol	100	7	92
2-Phenyl ethanol	100	4	>99

[a] Substrate-to-Ru ratio.
[b] Conversion of substrate, selectivity to aldehyde or ketone above 99%.
[c] 1 atm O_2.

Primary alcohols give the corresponding aldehydes in high selectivity, for example, 1-octanol affords 1-octanal with more than 99% selectivity. Over-oxidation to the corresponding carboxylic acid, normally a rather facile process, is completely suppressed in the presence of a catalytic amount of TEMPO. For example, attempted oxidation of octanal under the reaction conditions, in the presence of 3 mol % of TEMPO, gave no reaction in 1 week. In contrast, in the absence of TEMPO octanal was completely converted to octanoic acid within 1 h under the same conditions. These results are consistent with overoxidation of aldehydes occurring via a free-radical autoxidation mechanism. TEMPO suppresses this reaction by efficiently scavenging free-radical intermediates, resulting in the termination of free-radical chains, i.e. it acts as an antioxidant. Allylic alcohols were selectively converted to the corresponding unsaturated aldehydes in high yields. No formation of the isomeric saturated ketones via intramolecular hydrogen transfer, which is known to be promoted by ruthenium phosphine complexes [88], was observed.

Although, in separate experiments, secondary alcohols are oxidized faster than primary ones, in competition experiments the ruthenium/TEMPO system displayed a preference for primary over secondary alcohols. This can be explained by assuming that initial complex formation between the alcohol and the ruthenium precedes rate-limiting hydrogen transfer and determines substrate specificity, i.e. complex formation with a primary alcohol is favoured over a secondary one.

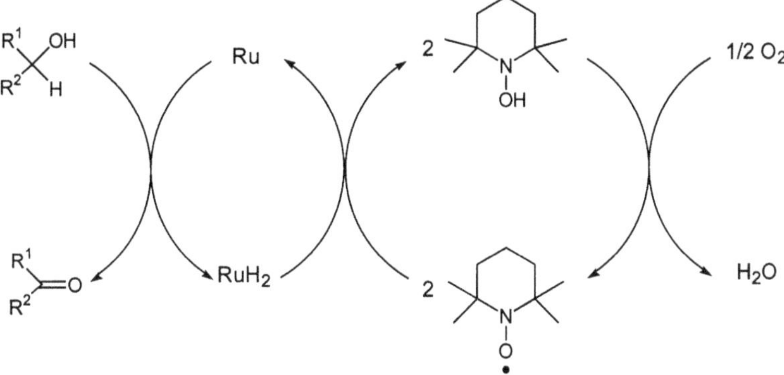

Fig. 15 Ruthenium/2,2′,6,6′-tetramethylpiperidinyl-*N*-oxyl catalysed aerobic oxidation of alcohols

An oxidative hydrogenation mechanism, analogous to that proposed by Bäckvall for the ruthenium/quinone system (see earlier), can be envisaged for the ruthenium/TEMPO system (Fig. 15). The intermediate hydridoruthenium species is most probably $RuH_2(Ph_3P)_3$ as was observed in $RuCl_2(Ph_3P)_3$-catalysed hydrogen-transfer reactions [89]. The observation that $RuH_2(Ph_3P)_4$ exhibits the same activity as $RuCl_2(Ph_3P)_3$ in the ruthenium/TEMPO-catalysed aerobic oxidation of 2-octanol is consistent with this notion. The TEMPO acts as a hydrogen-transfer mediator by promoting the regeneration of the ruthenium catalyst, via oxidation of the ruthenium hydride, resulting in the concomitant formation of the corresponding hydroxylamine, TEMPOH. The latter then undergoes rapid reoxidation to TEMPO, by molecular oxygen, to complete the catalytic cycle (Fig. 15).

A linear increase in the rate of 2-octanol oxidation was observed with increasing TEMPO concentration in the range 0–4 mol % but above 4 mol % further addition of TEMPO had a negligible effect on the rate. Analogous results were observed by Karlson et al. [90] in the ruthenium/benzoquinone system and were attributed to a change in the rate-limiting step. Hence, by analogy, we propose that at relatively low TEMPO-to-ruthenium ratios (up to 4:1) reoxidation of the ruthenium hydride species is the slowest step, while at high ratios dehydrogenation of the alcohol becomes rate-limiting.

Under an inert atmosphere $RuCl_2(Ph_3P)_3$ catalyses the stoichiometric oxidation of 2-octanol by TEMPO to give 2-octanone and the corresponding piperidine, TEMPH, in a stoichiometry of 3:2 (Eq. 22).

3 $R_1R_2CH(OH)$ + 2 TEMPO (N–O•) $\xrightarrow[\text{PhCl, } N_2\text{, 100°C}]{RuCl_2(Ph_3P)_3}$ 3 $R_1R_2C{=}O$ + 2 TEMPH (N–H) + 2 H_2O (22)

This result can be explained by assuming that the initially formed TEMPOH (see earlier) undergoes disproportionation to TEMPH and the oxoammonium cation (Eq. 23). Reduction of the latter by the alcohol affords another molecule of TEMPOH and this leads, ultimately, to the formation of the ketone and TEMPH in the observed stoichiometry of 3:2. The observation that attempts to prepare TEMPOH [91] under an inert atmosphere always resulted in the formation of TEMPH is consistent with this hypothesis.

2 OH → HO^- + H TEMPH (23)

On the basis of the results discussed earlier the detailed catalytic cycle depicted in Fig. 16 is proposed for the ruthenium/TEMPO-catalysed aerobic oxidation of alcohols.

The alcohol oxidations discussed earlier involve as a key step the oxidative dehydrogenation of the alcohol to form low-valent hydridoruthenium inter-

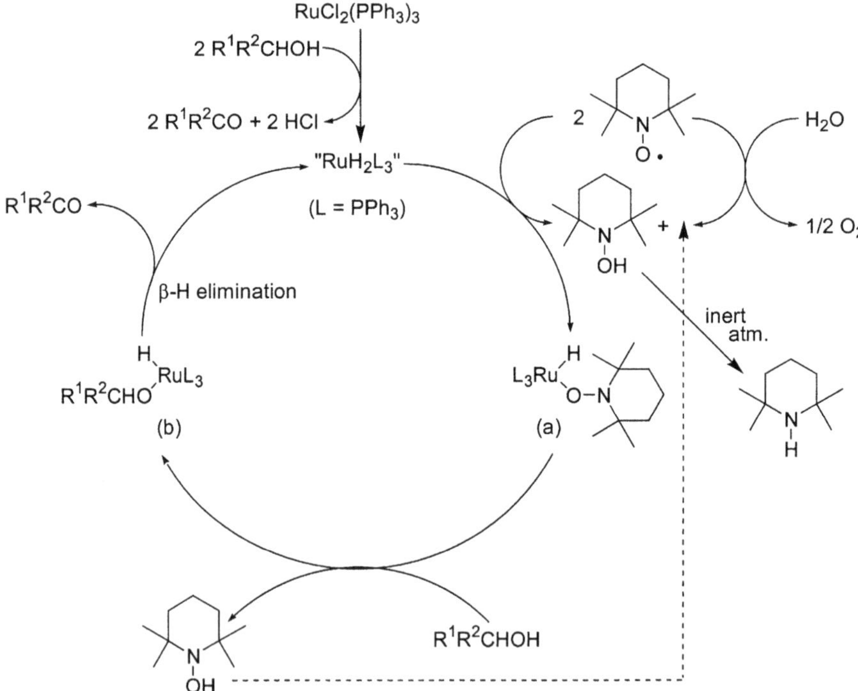

Fig. 16 Proposed mechanism for the ruthenium/2,2′,6,6′-tetramethylpiperidinyl-*N*-oxyl catalysed oxidation of alcohols

mediates. On the other hand, high-valent oxoruthenium species are also able to dehydrogenate alcohols via an oxometal mechanism (see earlier). It has long been known that ruthenium tetroxide, generated by reaction of ruthenium dioxide with periodate, smoothly oxidizes a variety of alcohols to the corresponding carbonyl compounds [92].

Dengel et al. [93] reported the synthesis of the organic soluble TBAP, n-$Bu_4N^+RuO_4^-$, in 1985. It was later found that TPAP, n-$Pr_4N^+RuO_4^-$, is even easier to prepare from RuO_4 and n-Pr_4NOH in water [94, 95]. TBAP and TPAP are air-stable, nonvolatile and soluble in a wide range of organic solvents. Griffith and coworkers [96, 97] subsequently showed that TPAP is an excellent catalyst for the selective oxidation of a wide variety of alcohols using NMO as the stoichiometric oxidant (Eq. 24).

$$\mathrm{R{\frown}OH} \xrightarrow[\text{NMO (1.5 eq), RT, Ar, < 1h}]{\mathrm{Pr_4N^+RuO_4^-}\text{ (5 mol\%), 4A MS}} \mathrm{R{\frown}{=}O} \qquad (24)$$

More recently, the groups of Ley [98] and Marko [99] independently showed that TPAP is able to catalyse the oxidation of alcohols using dioxygen as the stoichiometric oxidant. In particular, polymer-supported perruthenate (PSP), prepared by anion exchange of $KRuO_4$ with a basic anion exchange resin (Amberlyst A-26), has emerged as a versatile catalyst for the aerobic oxidation (Eq. 25) of alcohols [100]. However the activity was around 4 times lower than homogeneous TPAP, and this catalyst could not be recycled, which was attributed to oxidative degradation of the polystyrene support. PSP displays a marked preference for primary versus secondary alcohol functionalities [100]. The problem of deactivation was also prominent for the homogeneous TPAP oxidation, which explains the high (10 mol %) loading of catalyst required.

$$\mathrm{R{\frown}OH} \xrightarrow[\mathrm{O_2},\ 75\text{-}85^\circ\mathrm{C},\ \text{toluene, 0.5-0.8 h}]{\bullet{\frown}\overset{+}{\mathrm{N}}\mathrm{Me_3\ RuO_4^-}\text{ (10 mol\%)}} \mathrm{R{\frown}{=}O} \quad \textit{yield 56-95\%} \qquad (25)$$

Examples illustrating the scope of TPAP-catalysed aerobic oxidation of primary and secondary alcohols to the corresponding aldehydes are shown in Table 3.

Recently two heterogeneous TPAP catalysts were developed which could be recycled successfully and displayed no leaching: In the first example the tetraalkylammonium perruthenate was tethered to the internal surface of mesoporous silica (MCM-41) and was shown [101] to catalyse the selective aerobic oxidation of primary and secondary allylic and benzylic alcohols (Fig. 17). Surprisingly, both cyclohexanol and cyclohexenol were unreactive although these substrates can easily be accommodated in the pores of MCM-41. No mechanistic interpretation for this surprising observation was offered by the authors.

Table 3 Perruthenate-catalysed oxidation of primary and secondary alcohols to aldehydes using molecular oxygen

Substrate	Carbonyl yield[a]		
	Toluene, 75–85 °C, 10 mol % polymer supported perruthenate[b]	Toluene, 70–80 °C, 5 mol % tetra-propylammonium perruthenate, 4-Å molecular sieve[c]	Toluene, 75 °C, 10 mol % tetra-propylammonium perruthenate doped sol–gel ormosil[d]
1-Octanol	91% (8 h)		70% (7 h)
1-Decanol		73% (0.5 h)[e]	
Benzyl alcohol	>95% (0.5 h)		100% (0.75 h)
4-Chlorobenzyl alcohol		81% (0.5 h)	
Cinnamyl alcohol	>95% (1 h)	70% (0.5 h)	90% (5 h)
2-Decanol		88% (0.5 h)	

[a] Yields at 100% conversion; [b] Hinzen et al. [100]; [c] Marko et al. [99]; [d] Pagliaro and Ciriminna [102]; [e] 94% conversion, no molecular sieves were used.

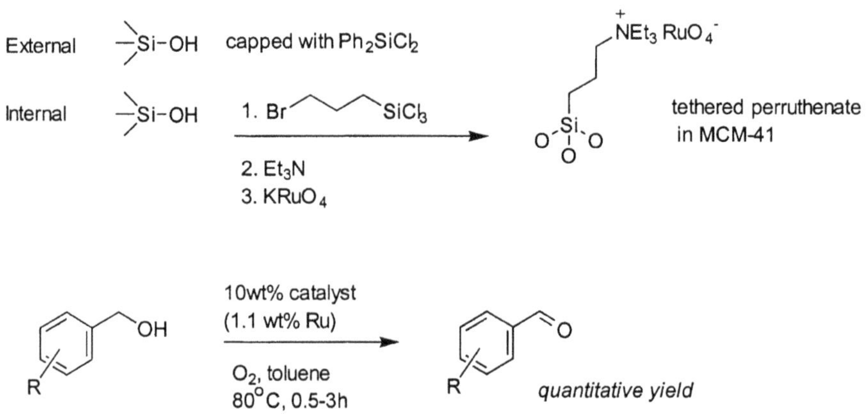

Fig. 17 Aerobic alcohol oxidation catalysed by perruthenate tethered to the internal surface of MCM-41

The second example involves straightforward doping of methyl modified silica, denoted as ormosil, with tetrapropylammonium perruthenate via the sol–gel process [102] (Table 3). A serious disadvantage of this system are the low TOFs of 1.0 and 1.8 h^{-1} observed for a primary aliphatic alcohol and allylic alcohol, respectively.

Sparse attention has been paid to the mechanism of perruthenate-catalysed alcohol oxidations [103]. Although TPAP can act as a three-electron oxidant

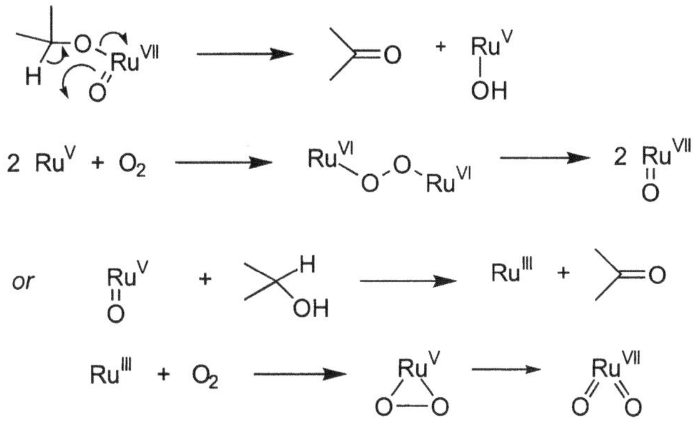

Fig. 18 Proposed catalytic cycle for reoxidation of perruthenate in the oxidation of alcohols

($Ru^{VII} \rightarrow Ru^{IV}$) the fact that it selectively oxidizes cyclobutanol to cyclobutanone and *tert*-Bu(Ph)CHOH to the corresponding ketone, militates against free-radical intermediates and is consistent with a heterolytic, two-electron oxidation [103, 104]. Presumably, the key step involved β-hydride elimination from a high-valent, for example, alkoxyruthenium(VII), intermediate followed by reoxidation of the lower-valent ruthenium by dioxygen. However, as shown in Fig. 18, if this involved the Ru(VII)/Ru(V) couple the reoxidation would require the close proximity of two ruthenium centres, which would seem unlikely in a polymer-supported catalyst. A plausible alternative, which can occur at an isolated ruthenium centre, involves the oxidation of a second molecule of alcohol, resulting in the reduction of ruthenium(V) to ruthenium(III), followed by reoxidation of the latter to ruthenium(VII) by dioxygen (Fig. 18).

More detailed mechanistic studies are obviously necessary in order to elucidate the details of this fascinating reaction. It is worth noting, in this context, that the reaction of TPAP with 2-propanol was found to be autocatalytic, possibly owing to the formation of colloidal RuO_2 [105]. Another possible alternative is one involving the initial formation of oxoruthenium(VI), followed by cycling between ruthenium(VI), ruthenium(IV) and possibly ruthenium(II).

We note, in this context, that the group of James [106] showed that a *trans*-dioxoruthenium(VI) complex of TMP dianion oxidizes 2-propanol, in a stoichiometric reaction, with concomitant formation of a dialkoxyruthenium(IV) TMP complex (Eq. 26).

$$Ru^{VI}(TMP)O_2 + 3\ Me_2CHOH \longrightarrow Ru^{IV}(TMP)(OCHMe_2)_2 + Me_2CO + 2\ H_2O \qquad (26)$$

The oxoruthenium(VI) complex was prepared by exposing a benzene solution of trans-Ru^{II} (tmp)$(MeCN)_2$ to air at 20°C. Addition of 2-propanol to the resulting solution, in the absence of air, afforded the dialkoxyruthenium(IV) complex, in quantitative yield, within 24 hours. In the presence of air, benzene

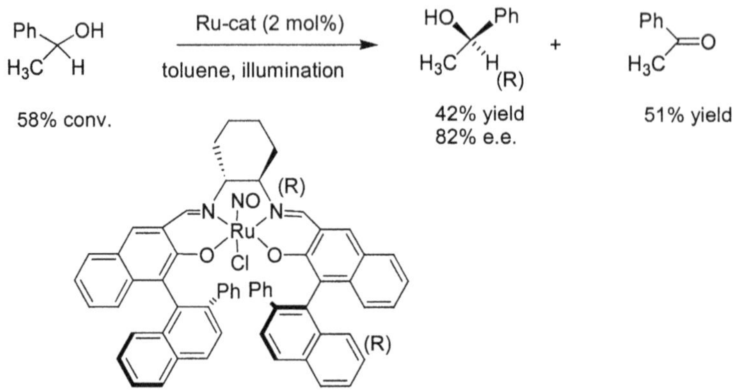

Fig. 19 Ruthenium salen type complexes active as catalysts in alcohol oxidation under illumination

solutions of the dioxoruthenium(VI) or the dialkoxyruthenium(IV) complex effected catalytic oxidation of 2-propanol at room temperature, albeit with a modest rate (1.5 catalytic turnovers per day). Interestingly, with the dialkoxyruthenium(IV) complex, catalytic oxidation was observed with air but not with dry oxygen, suggesting that hydrolysis to an oxoruthenium(IV) complex is necessary for a catalytic cycle.

Other ruthenium-based catalysts for the aerobic oxidation of alcohols have been described where it is not clear if they involve oxidative dehydrogenation by low-valent ruthenium, to give hydridoruthenium intermediates, or by high-valent oxoruthenium. Masutani et al. [107] described (nitrosyl)Ru(salen) complexes, which can be activated by illumination to release the NO ligand. These complexes demonstrated selectivity for oxidation of the alcoholic group versus epoxidation, which was regarded as evidence for the intermediacy of Ru-oxo moieties. Their excellent alcohol coordination properties led to a good enantiomer differentation in the aerobic oxidation of racemic secondary alcohols (Fig. 19) and to a selective oxidation of primary alcohols in the presence of secondary alcohols [108].

Both RuO_2 and 5% ruthenium-on-charcoal catalyse the aerobic oxidation of activated alcohols such as allylic alcohols [109] and α-ketols [110] (Eq. 27).

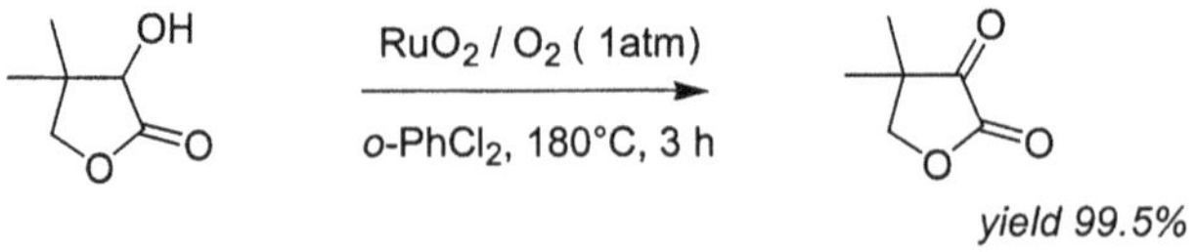

(27)

Vocanson et al. [111] have described the use of ruthenium supported on ceria, CeO_2, as a catalyst for the aerobic oxidation of alcohols. Primary and secondary alcohols are oxidized to the corresponding aldehydes (carboxylic acids) and ketones, respectively, at elevated temperatures (above 140 °C). Surprisingly, allylic alcohols, such as geraniol, and some cyclic alcohols, for example, menthol, are unreactive. The former result suggests that low-valent ruthenium species are possibly involved and that coordination of ruthenium to the double bond inhibits alcohol oxidation. Recently, RuO_2 nanoparticles were incorporated in the supercages of Faujasite zeolite [112]. This material showed good catalytic properties in the oxidation of a variety of activated and unactivated primary and secondary alcohols at 80 °C (Eq. 28). These results clearly indicate that zeolite-confined RuO_2 is much more active than the bulk RuO_2; moreover, it can be easily recycled.

However no functionalized alcohols were tested; high amounts of catalyst are required (7.8 mol %) and the reaction proceeds rather slowly (TOF for 1-octanol is approximately 3 h^{-1}).

1-octanol → octanal (93% yield); RuO_2-FAU (7.8 mol% Ru), toluene, 80°C, 20 h, air (28)

Recently Yamaguchi and Mizuno[113] reported ruthenium on alumina to be a powerful and recyclable catalyst for selective alcohol oxidation. This method displayed a large substrate scope (Eq. 29, Table 4) and tolerates the presence of sulfur and nitrogen groups. Only primary aliphatic alcohols required the addition of hydroquinone. TOFs in the range from 4 h^{-1} (for secondary allylic alcohols) to 18 h^{-1} (for 2-octanol) were obtained in trifluorotoluene, while in the solvent-free oxidation at 150 °C a TOF of 300 h^{-1} was observed for 2-octanol.

$$R_1R_2CH(OH) + 0.5\ O_2 \xrightarrow[\text{PhCF}_3,\ 83^\circ\text{C},\ 1\ \text{atm}\ O_2]{\text{Ru(OH)}_3\ \text{on}\ \text{Al}_2\text{O}_3\ (2.5\ \text{mol\%})} R_1R_2C{=}O + H_2O \qquad (29)$$

The catalyst consists of highly dispersed $Ru(OH)_3$ on the surface of γ-Al_2O_3. On the basis, inter alia, of the fact that this catalyst is also capable of performing a transfer hydrogenation using 2-propanol as the hydrogen donor, it was concluded that the mechanism of this reaction proceeds via a hydridometal pathway.

Ruthenium-exchanged hydrotalcites were shown by Kaneda et al. [114] to be heterogeneous catalysts for the aerobic oxidation of reactive allylic and benzylic alcohols. Hydrotalcites are layered anionic clays consisting of a cationic Brucite layer with anions (hydroxide or carbonate) situated in the interlayer region. Various cations can be introduced in the Brucite layer by ion exchange. For example, ruthenium-exchanged hydrotalcite with the formula $Mg_6Al_2Ru_{0.5}(OH)_{16}CO_3$, was prepared by treating an aqueous solution of $RuCl_3{\cdot}3H_2O$, $MgCl_2 \cdot 6H_2O$ and $AlCl_3 \cdot H_2O$ with a solution of NaOH and Na_2CO_3 followed by heating at 60 °C for 18 h [114]. The resulting slurry was cooled to room temperature, filtered, washed with water and dried at 110 °C for 12 h. The

Table 4 $Ru(OH)_3$-Al_2O_3 catalysed oxidation of primary and secondary alcohols to the corresponding aldehydes and ketones using O_2 (according to Ref. [113]; 2.5 mol % Ru/Al_2O_3, $PhCF_3$ as solvent, 83 °C, 1 atm O_2; conversion and yields determined by GLC)

Substrate	Time (h)	Conv (%)	Sel. (%)
Geraniol	6	89	97
Benzyl alcohol	1	>99	>99
4-Nitrobenzyl alcohol	3	97	>99
2-Octanol	2	91	>99
Cyclooctanol	6	81	>99
2-Phenyl ethanol	1	>99	>99
1-Octanol[a]	4	87	98

[a] 5 mol %Ru/Al_2O_3 and 5 mol % hydroquinone (to suppress over oxidation) were used.

resulting ruthenium hydrotalcite showed the highest activity amongst a series of hydrotalcites exchanged with, for example, Fe, Ni, Mn, V and Cr.

Subsequently, in 1999 the same group showed that the activity of the ruthenium hydrotalcite was significantly enhanced by the introduction of cobalt(II), in addition to ruthenium(III), in the Brucite layer [115]. For example, cinnamyl alcohol underwent complete conversion in 40 min in toluene at 60 °C, in the presence of ruthenium/cobalt hydrotalcite, compared with 31% conversion under the same conditions with ruthenium hydrotalcite. A secondary aliphatic alcohol, 2-octanol, was smoothly converted into the corresponding ketone but primary aliphatic alcohols, for example, 1-octanol, exhibited extremely low activity. The authors suggested that the introduction of cobalt induced the formation of higher oxidation states of ruthenium, for example, Ru(IV) to Ru(VI), leading to a more active oxidation catalyst. However, on the basis of the reported results it is not possible to rule out low-valent ruthenium species as the active catalyst in a hydridometal pathway. The results obtained in the oxidation of representative alcohols with ruthenium hydrotalcite and ruthenium–cobalt–hydrotalcite are compared in Table 5.

In 2000, Yamaguchi et al. [116] synthesized a ruthenium-based hydroxyapatite catalyst, with the formula $(RuCl)_{10}(PO_4)_6(OH)_2$. This catalyst could also be recycled and displayed a reasonable substrate scope in the aerobic alcohol oxidations (Eq. 30). TOFs reported in this case were generally somewhat lower, on the order of 1 h^{-1} for 2-octanol to 12 h^{-1} for benzyl alcohol. The fact that distinct Ru–Cl species are present at the surface points in the direction of a hydridometal mechanism.

RuCl-hydroxyapatite (17 mol%)

OH → O

toluene, 80°C, 6 h, 1 atm O_2 (30)

The same group recently reported the use of a ferrite spinel catalyst ($MnFe_2O_4$), where the iron was partially substituted with ruthenium and cop-

Table 5 Oxidation of various alcohols to their corresponding aldehydes or ketones with ruthenium hydrotalcites (*HTs*) as catalysts using molecular oxygen. 2 mmol substrate, 0.3 g HT (around 14 mol %), in toluene, 60 °C, 1 bar O_2. Conversion 100%

Substrate	$Ru\text{-}Mg\text{-}Al\text{-}CO_3\text{-}HT^a$		$Ru\text{-}Co\text{-}Al\text{-}CO_3\text{-}HT^b$	
	Time (h)	Yield (%)	Time	Yield (%)
Cinnamyl alcohol	8	95[c]	40 min	94
Benzyl alcohol	8	95[c]	1 h	96
4-Chlorobenzyl alcohol	8	61[d]	1.5 h	95
2-Phenyl ethanol	18	100	1.5 h	100
2-Octanol	–	–	2 h	97
Geraniol	–	–	12 h	71[e]

[a] See Ref. [114]; [b] See Ref. [115]; [c] Conversion 98%; [d] Conversion 64%; [e] Conversion 89%.

per, i.e. $MnFe_{1.5}Ru_{0.35}Cu_{0.15}O_4$ for the room temperature oxidation of alcohols [117]. However, 20 mol % of catalyst (based on ruthenium) was necessary to accomplish even the oxidation of benzyl alcohol. For primary and secondary aliphatic alcohols, TOFs of 2 and 3.5 h^{-1}, respectively, were the maximum rates achieved.

Another class of ruthenium catalysts which has attracted considerable interest owing to their inherent stability under oxidative conditions are the POMs [118]. Recently, Yamaguchi and Mizuno [30] reported that a mono-ruthenium-substituted silicotungstate, synthesized by the reaction of the lacunary POM $[SiW_{11}O_{39}]^{8-}$ with Ru^{3+} in an organic solvent, acts as an efficient heterogeneous catalyst with high TOFs for the aerobic oxidation of alcohols (Table 6). Among the solvents used 2-butyl acetate was the most effective and this ruthenium heteropolyanion (HPA) could be recycled. The low loading used resulted in very long reaction times of more than 2 days (Table 6).

An example of a homogeneous ruthenium HPA derivative is also shown in Table 6. In this case the same lacunary silicotungstate was overexchanged with a basic $RuCl_3$ solution. The resulting solution was precipitated in organic solution using $[(C_6H_{13})_4N]HSO_4$ [119] and elemental analysis showed that seven ruthenium molecules were present per molecule of $[SiW_{11}O_{39}]^{8-}$. This led us to postulate a structure comprising ruthenium oxide clusters stabilized by the HPA. This material displayed better results than ruthenium HPAs which were prepared according to previous publications and subjected to the conditions in Table 6 [120, 121]

In contrast to the previously mentioned reactions, which involve either oxoruthenium or ruthenium hydride species as intermediates, free-radical reactions can also be promoted by ruthenium. The aerobic oxidation of alcohols proceeds smoothly at room temperature in the presence of 4 eq. of an aldehyde, for example, acetaldehyde, and a catalyst comprising a 1:1 mixture of $RuCl_3 \cdot nH_2O$ and $Co(OAc)_2$, in ethyl acetate (Eq. 31) [122].

Table 6 Ruthenium-substituted polyoxometalates as catalysts for the oxidation of alcohols

Substrate	Mono Ru silicotungstate[a]		Over exchanged $Ru/SiW_{11}O_{39}$/tetrahexyl-ammonium sulfate[b]	
	Time	Conv. (%) (sel. %)[c]	Time	Conv. (%) (sel. %)
Cinnamyl alcohol			2	100 (96)
Benzyl alcohol	120	36 (65)[d]		100 (96)
Cyclohexanol	48	67 (81)	48	44 (38)
2-Octanol	48	14 (44)[e]		99 (92)

[a] 0.05 mol % (tetrabutylammonium)$_4$H[$SiW_{11}Ru(H_2O)O_{39}$] · $2H_2O$, isobutyl acetate as solvent, 110 °C, 1 atm O_2, see Ref. [30]: [b] 5.8 mol % (on Ru) (tetrahexylammoniumsulfate)$_x$(Ru_7O_y) $SiW_{11}O_{39}$, a homogeneous catalyst, prepared by exchanging lacunary $K_8[SiW_{11}O_{39}]$ with basic $RuCl_3$, followed by precipitation in organic solution. 80 °C, solvent PhCl, see Ref. [119]; [c] Selectivity towards aldehyde or ketone; [d] 10% benzoic acid was also formed; [e] 30% acid was also formed.

$$R_1CH(OH)R_2 \xrightarrow[\text{RCHO (4 eq), } O_2\text{, 20°C, ethylacetate}]{RuCl_3\text{-}Co(OAc)_2\text{ (1mol\%)}} R_1C(=O)R_2 \tag{31}$$

The results were rationalized by assuming that the corresponding percarboxylic acid is formed by cobalt-mediated free-radical autoxidation of the aldehyde. Subsequent reaction of ruthenium(III) with the peracid affords oxoruthenium(V) carboxylate, which is the active oxidant. Compared with the aerobic oxidations discussed earlier the method suffers from the drawback that 1 eq of a carboxylic acid is formed as a coproduct.

4.2 Ruthenium-Catalysed Oxidation of Ethers

The selective α-oxidation of ethers to esters can be applied in the fine chemical industry for the production of complex natural products and in carbohydrate chemistry, where it enables efficient oxidative deprotection of hydroxyl groups. The most efficient oxidant for this transformation is RuO_4 [123]. Classical use involves stoichiometric amounts of this reagent, but RuO_4 can be conveniently generated in situ via a combination of ruthenium precursors, usually $RuCl_3 \cdot H_2O$ or $RuO_2 \cdot H_2O$, with an inorganic oxidant such as $NaIO_4$ [124], $NaBrO_3$ [125] or NaOCl [126]. The Ru/$NaIO_4$ method was substantially improved by Carlsen et al. [127] by addition of MeCN to the traditional CCl_4/H_2O system. The higher catalytic activity was explained by the coordi-

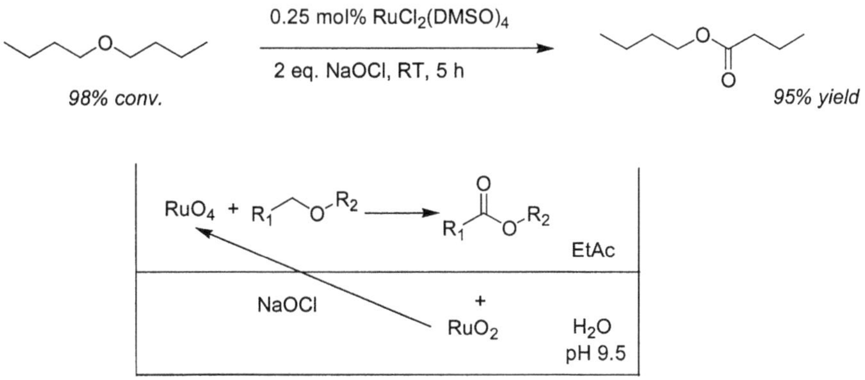

Fig. 20 Simple pH control in ruthenium-catalysed oxidation of ethers leading to higher activities and selectivities

nating properties of MeCN, thus preventing the formation of insoluble ruthenium species which caused catalyst deactivation. We showed that ruthenium-catalysed oxidation of ethers could be dramatically improved, both in activity and selectivity to esters, by simple pH control [128]. This allows for (1) low concentrations of ruthenium precursors, (2) use of a stoichiometric quantity (2 eq) of oxidant, i.e. NaOCl, (3) easy product recovery by simple phase separation and (4) efficient catalyst recycling. The system is exemplified in Fig. 20.

Nearly complete conversion and highly selective ester formation was achieved for a series of symmetrical and unsymmetrical aliphatic, cyclic and benzylic ethers. Efficient use of NaOCl (i.e. 2:1 ratio versus substrate for ethers) could be achieved by optimizing the reoxidation of the catalyst from its reduced form, thus avoiding decomposition to inactive insoluble species. In contrast with earlier observations by Carlsen et al. [127], the presence of a chlorinated solvent is not mandatory. The use of ethyl acetate and methyl *tert*-butyl ether as solvents gave excellent results. The best catalyst precursors under the pH-stat conditions were found to be *cis*-$Ru(DMSO)_4Cl_2$, *trans*-$Ru(dppp)_2Cl_2$, where dppp is 1,3-bis(diphenylphosphino)propane, and TPAP. The pH-control method could be replaced by simple adequate buffer solutions or by adding $NaHCO_3$ to the NaOCl solution (pH 9–10).

The protocol was also successfully applied to the oxidation of alcohols, in particular giving spectacular results for the troublesome selective conversion of 1,2:4,5-di-*O*-isopropylidene-*β*-D-fructopyranose into 1,2:4,5-di-*O*-isopropylidene-*β*-D-*erythro*-2,3-hexadiulo-2,6-pyranose [129].

5 Ruthenium-Catalysed Oxidation of N-Containing Compounds

5.1 Oxidation of Primary and Secondary Amines

The catalytic oxidation of amines results in imines and nitriles which are versatile synthetic intermediates. Few catalytic procedures are known, but most of these are based on ruthenium [130]. In 1985 Murahashi et al. [131] reported that low-valent ruthenium complexes exhibit specific activity towards oxidations of nitrogen compounds with peroxides, sharply differing from RuO_4, which simply converts amides to the corresponding imides. It was proposed that Ru^{IV}=O species generated from $Ru^{II}Cl_2(PPh_3)_3$ and *tert*-BuOOH abstract the α-hydrogen, thus generating an iminium ion complex (Fig. 21). This product decomposes to the imine product, the Ru(II) species and water to complete the catalytic cycle. Various examples of the oxidation of secondary amines [131] and a primary amine [6] are shown in Eqs. (32–34). Molecular sieves (4 Å) were needed to prevent the hydrolysis of product imines in some cases. In the case

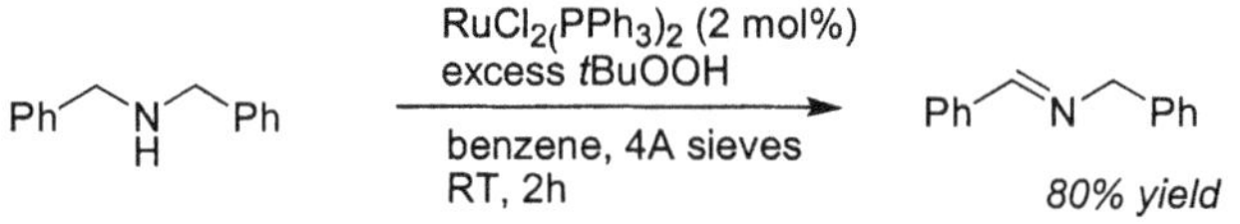

(32)

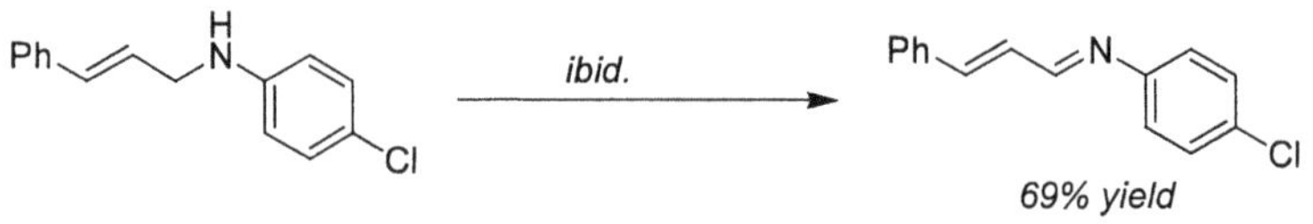

(33)

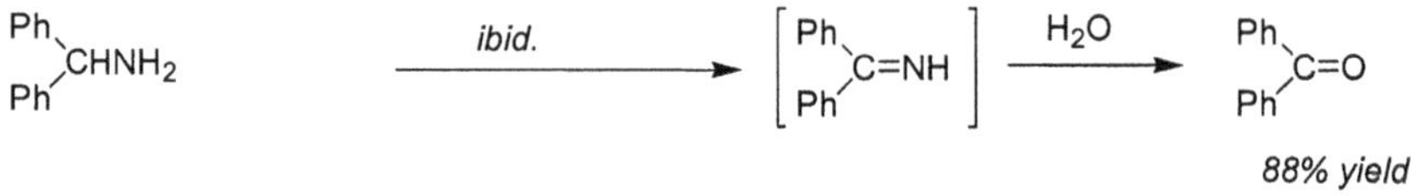

(34)

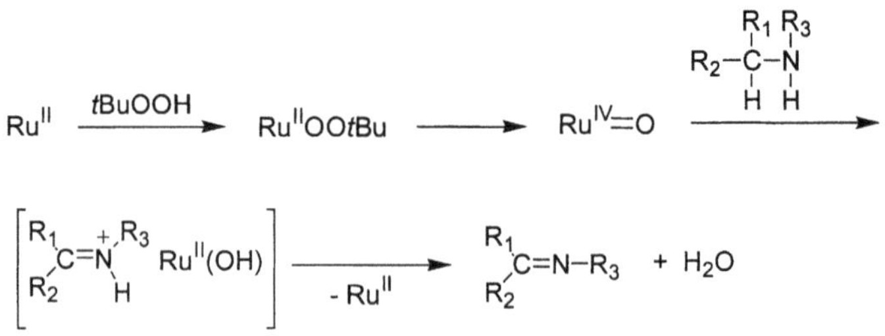

Fig. 21 Oxidation of amines to imines using low-valent ruthenium

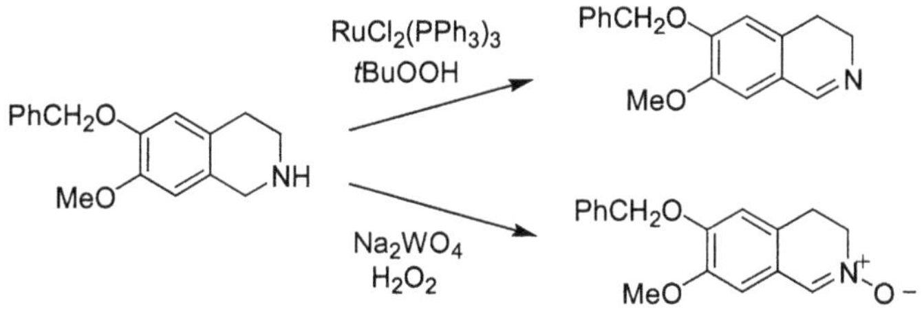

Fig. 22 Reactivity of oxometal (Ru) versus peroxometal (W) species in the oxidation of amines

of diphenylmethyl amine, benzophenone was formed in 88% yield by hydrolysis of the intermediate imine.

It is worthwhile to comment on the catalytic species. As opposed to oxometal species, which convert amines to imines, hydroperoxymetal complexes (*M*OOH) convert amines to nitrones. Thus the oxidation of amines is a convenient way of distinguishing the active species. The reactivity of oxometal versus peroxometal species is illustrated in Fig. 22. In practice, tungsten is the catalyst of choice to convert amines to nitrones [130].

tert-BuOOH can be conveniently replaced by oxygen in the ruthenium-catalysed oxidation of amines as two recent examples of the groups of Mizuno [132] and Kaneda [133] have shown. Earlier reports using Ru(II) and Ru(III) precursors, as well as ruthenium porphyrins already demonstrated the proof of principle, albeit with low TONs [134–136]. The results using heterogeneous Ru/Al_2O_3 or heterogeneous ruthenium hydroxyapatite are given in Table 7.

Table 7 Oxidation of primary and secondary amines using heterogeneous ruthenium catalysts with molecular oxygen

Substrate	Ru/Al_2O_3[a]		Ru hydroxyapatite[b]	
	Time (h)	Yield (%) Nitrile	Time (h)	Yield (%) Nitrile
Benzyl amine	1	82	12	90
4-Methylbenzyl amine	1	93	12	96
	1	40[c]		
1-Octyl amine	2	96	24	>99
		Imine		Imine
Dibenzyl amine	16	85	24	98[d]
Indoline	2	99[e]		

[a] Mizuno system, see Ref. [132]. 2.8 mol % Ru, solvent $PhCF_3$, 100 °C, 1 atmO_2; [b] Kaneda system, see Ref. [133]. 17 mol % Ru, solvent $PhCH_3$, 110 °C, 1 atm O_2: [c] Homogeneous *n*-Pr_4NRuO_4 was used as the catalyst instead of Ru/Al_2O_3; [d] 130 °C; [e] Product indole.

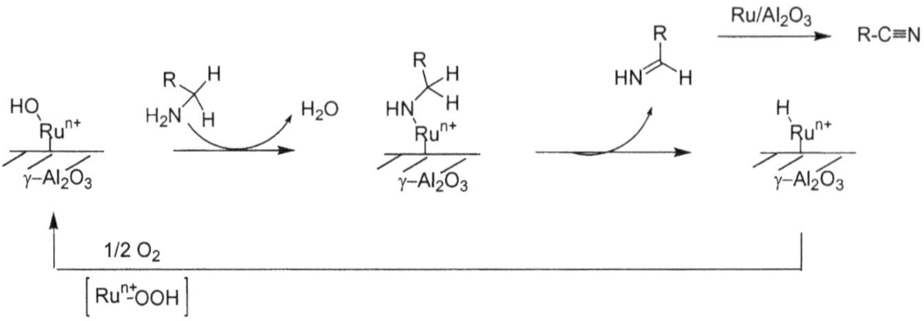

Fig. 23 Proposed mechanism of the Ru/Al_2O_3 catalysed oxidation of primary amines. Adapted from Ref. [132]

Especially the Ru/Al_2O_3 catalyst, which performed well in alcohol oxidation, gave very high activities and selectivities. Primary amines were effectively converted into the corresponding nitriles in high yields, and highly activated secondary amines could be converted to imines. No cleavage or isomerization of double bonds took place. Possible by-products in primary amine oxidation result from condensation of the starting amines and aldehydes through imine hydrolysis. Such findings indicate that imines are the intermediate products which are rapidly dehydrogenated to nitriles. The Ru/Al_2O_3 system was unable to oxidize tertiary amines. The proposed mechanism is illustrated in Fig. 23.

5.2 Oxidation of Tertiary Amines

The Ru(II)/*R*OOH system can also be used to oxidize tertiary amines. The intermediate iminium ion is formed, as described earlier for secondary amines, and can be trapped by nucleophiles. Thus, the ruthenium-catalysed oxidation of tertiary amines with hydrogen peroxide in methanol can be performed to give the corresponding α-methoxyamines with high efficiency as illustrated in Fig. 24 [137]. Another example is the selective demethylation of tertiary amines in methanol with a combination of Ru(II) and H_2O_2, followed by hydrolysis of the intermediate α-methoxylated amines. For example, the methoxylation of *N*,*N*-dimethyl-*p*-toluidine followed by treatment with 2 N HCl solution gave *N*-methyl-*p*-toluidine in 75% yield (Eq. 35) [137].

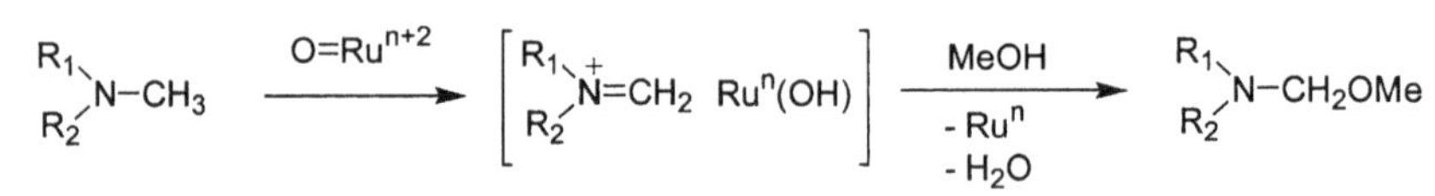

Fig. 24 Ru(II)/*R*OOH system in methanol for oxidation of tertiary amines

$$H_3C\text{-}C_6H_4\text{-}N(CH_3)_2 \xrightarrow[\text{MeOH}]{\text{RuCl}_3\text{ (mol\%), }H_2O_2} \xrightarrow{\text{2 N HCl}} H_3C\text{-}C_6H_4\text{-}NH\text{-}CH_3 + \text{MeOH} + CH_2O \quad (35)$$

5.3 Oxidation of Amides and β-Lactams

An important application of oxidation of a C–H bond adjacent to a nitrogen atom is the selective oxidation of amides. This reaction proceeds in the presence of *tert*-BuOOH as the oxidant and Ru(II) salts. Thus in the example of Eq. (36), the α-*tert*-butylperoxy amide of the isoquinoline was obtained, which is an important synthetic intermediate for natural products [138]. This product can be conveniently reacted with a nucleophile in the presence of a Lewis acid. Direct trapping of the iminium ion complex by a nucleophile was achieved in the presence of trimethylsilyl cyanide, giving α-cyanated amines as shown in Eq. (37) [45]. This ruthenium/peracid oxidation reaction provides an alternative to the Strecker reaction for the synthesis of α-amino acid derivatives since they involve the same α-cyano amine intermediates. In this way *N*-methyl-*N*-(*p*-methoxyphenyl) glycine could be prepared from *N*,*N*-dimethyl-*p*-methoxyaniline in 82% yield.

$$\text{(PhCH}_2\text{O, MeO)-tetrahydroisoquinoline-N-CO}_2\text{Me} \xrightarrow[\text{tBuOOH, benzene}]{\text{RuCl}_2(\text{PPh}_3)_3} \text{(PhCH}_2\text{O, MeO)-1-(OO-}t\text{-Bu)-tetrahydroisoquinoline-N-CO}_2\text{Me} \quad (36)$$

$$\text{MeO-}C_6H_4\text{-}N(CH_3)_2 \xrightarrow[\text{Me}_3\text{SiCN, EtOAc, RT}]{\text{RuCl}_3\text{ (cat.), }CH_3CO_3H} \text{MeO-}C_6H_4\text{-}N(CH_3)\text{-}CH_2CN \xrightarrow[\text{EtOH}]{\text{KOH}} \text{MeO-}C_6H_4\text{-}N(CH_3)\text{-}CH_2COOH \quad \text{82\% yield} \quad (37)$$

One of the most challenging and industrially important reactions is the catalytic oxidation of β-lactams, since they are often unstable under oxidation conditions. In this case a combination of $RuCl_3$ with aldehyde and molecular oxygen was shown to be very effective. Typically, the $RuCl_3$-catalysed oxidation of β-lactam with 1 atm oxygen in the presence of acetaldehyde and sodium carboxylate gave the corresponding 4-acyloxy β-lactam in high yields [139]. Regarding the mechanism, in situ formation of peracids takes place and forms the Ru-oxo species as intermediates (Fig. 25). The remaining steps are analogous to the oxidation of secondary amines and the intermediate iminium ion is

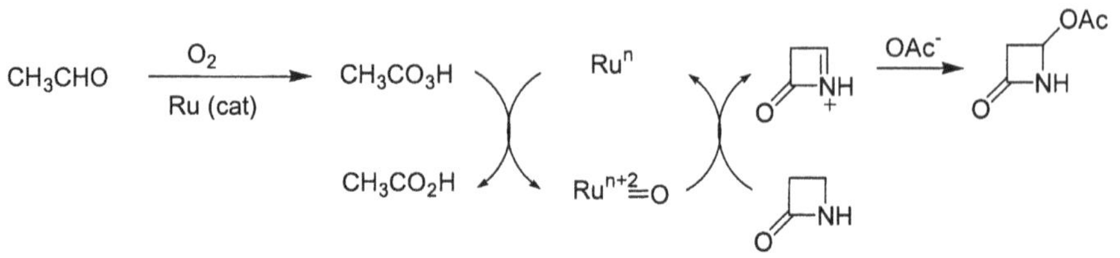

$$CH_3CHO \xrightarrow[\text{Ru (cat)}]{O_2} CH_3CO_3H$$

Fig. 25 Mechanism for the ruthenium/aldehyde aerobic oxidation of β-lactams

trapped by the acetic acid nucleophile. This reaction has found an important industrial application: the oxidation of (1′*R*, 3*S*)-3-[1′-(*tert*-butyldimethylsilyl)oxy]ethyl]azetidin-2-one gives the corresponding 4-acetoxy azetidinone, which is a key intermediate for the synthesis of carbapenem antibiotics, in 91% yield and more than 99% diastereomeric excess (Eq. 38).

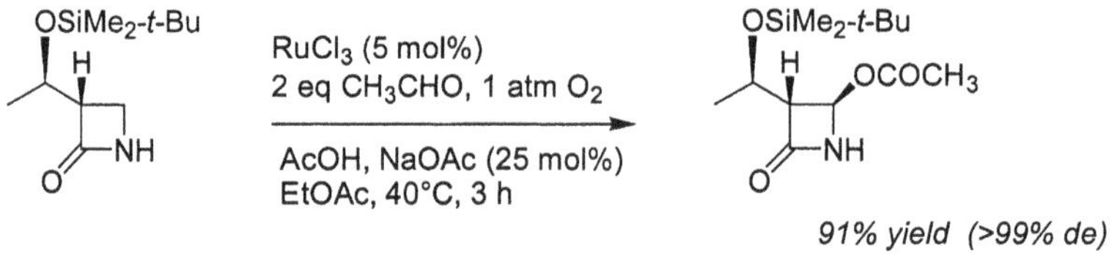

$$\xrightarrow[\text{AcOH, NaOAc (25 mol\%), EtOAc, 40°C, 3 h}]{\text{RuCl}_3\text{ (5 mol\%), 2 eq CH}_3\text{CHO, 1 atm O}_2} \quad (38)$$

91% yield (>99% de)

6
Concluding Remarks

The unique versatility of ruthenium as an oxidation catalyst continues to provide a stimulus for research on a variety of oxidative transformations. Its juxtaposition in the periodic table and close similarity to the biological redox elements, iron and manganese, coupled with the accessibility of various high-valent oxo species by reaction of lower-valent complexes with dioxygen make ruthenium an ideal candidate for suprabiotic catalysis.

Ruthenium-catalysed oxidations with dioxygen or hypochlorite are currently methods of choice for the oxidation of alcohol, ethers, amines and amides. In hydrocarbon oxidations, in contrast, ruthenium has not yet lived up to expectations. The proof of principle with regard to direct oxidation of, for example, olefins, with dioxygen via a nonradical, Mars–van Krevelen pathway has been demonstrated but this has, as yet, not led to practically viable systems with broad scope. The problem is one of rate; although feasible the heterolytic oxygen-transfer pathway cannot compete effectively with the ubiquitous free-radical autoxidation.

With the exception of a few specific examples, the combination of ruthenium catalysts with hydrogen peroxide has not led to viable systems owing to rapid, competing decomposition of the hydrogen peroxide. If the activity of ruthenium with hydrogen peroxide could be tamed this would afford attractive methods for, for example, alkene epoxidation. One would also expect ruthenium to be a good candidate for catalysing asymmetric epoxidations. Although many examples have been described an effective system with broad scope, that utilizes an attractive terminal oxidant, has not yet been forthcoming. However, it would appear to be only a matter of time before this goal will be achieved.

In short, a lot of progress has been made in the development of synthetically useful ruthenium-catalysed oxidations but there are still many goals to be achieved. It remains a fascinating and challenging area of research.

References

1. Sheldon RA, Kochi JK (1981) Metal catalyzed oxidation of organic compounds. Academic, New York
2. Hudlicky M (1990) Oxidations in organic chemistry. American Chemical Society, Washington, DC
3. (a) Clayden J, Greeves N, Warren S, Wothers P (2001) Organic chemistry. Oxford University Press, New York; (b) Bruice PY (2003) Organic chemistry, 4th edn. Prentice Hall, New York
4. Seddon EA, Seddon KR (1984) The chemistry of ruthenium. Elsevier, Amsterdam
5. Naota T, Takaya H, Murahashi SI (1998) Chem Rev 98:2599
6. Murahashi SI, Komiya N (2000) In: Meunier B (ed) Biomimetic oxidations catalyzed by transition metal complexes. Imperial College Press, London, p 563
7. Cainelli G, Cardillo G (1984) Chromium oxidations in organic chemistry. Springer, Berlin Heidelberg New York
8. Friedrich HB (1999) Platinum Met Rev 43:94 and references therein
9. Ortiz de Montellano PR (ed) (1986) Cytochrome P450: structure, mechanism and biochemistry. Plenum, New York
10. Drago RS (1992) Coord Chem Rev 117:185
11. Groves JT, Bonchio M, Carofiglio T, Shelyaev, K (1996) J Am Chem Soc 118:8961
12. Groves JT, Quinn R (1985) J Am Chem Soc 107:5790
13. (a) Leung T, James BR, Dolphin D (1983) Inorg Chim Acta 79:180; (b) Dolphin D, James BR, Leung T (1983) Inorg Chim Acta 79:25
14. Shingaki T, Miura K, Higuchi T, Hirobe M, Nagano T (1997) Chem Commun 861;
15. Ohtake H, Higuchi T, Hirobe M (1992) J Am Chem Soc 114:10660
16. Balkus KJ Jr, Eissa M, Levado R (1995) J Am Chem Soc 117:10753
17. Lau TC, Mak CK (1995) J Chem Soc Chem Commun 943
18. (a) Bressan M, Morvillo A (1988) J Chem Soc Chem Commun 650; (b) Bressan M, Morvillo A (1989) J Chem Soc Chem Commun 421
19. (a) Kojima T (1996) Chem Lett 121; (b) Yamaguchi M, Kousaka H, Yamagishi T (1997) Chem Lett 769
20. (a) Cheng WC, Yu WY, Cheung KK, Che CM (1994) J Chem Soc Chem Commun 1063; (b) Fung WH, Yu WT, Che CM (1998) J Org Chem 63:2873
21. Lau TC, Che CM, Lee WO, Poon CK (1988) J Chem Soc Chem Commun 1406
22. Murahashi SI, Oda Y, Naota T, Kuwabara T (1993) Tetrahedron Lett 34:1299
23. Goldstein AS, Beer RH, Drago RS (1994) J Am Chem Soc 116:2424
24. (a) Neumann R, Abu-Gnim C (1989) J Chem Soc Chem Commun 1324; (b) Matsumoto Y, Asami M, Hashimoto M, Misono M (1996) J Mol Catal A 114:161

25. (a) Murahashi SI, Oda Y, Komiya N, Naota T (1994) Tetrahedron Lett 35:7953; (b) Murahashi SI, Komiya N, Oda Y, Kuwabara T (2000) J Org Chem 65:9186
26. Murahashi SI, Oda Y, Naota T (1992) J Am Chem Soc 114:7913
27. Murahashi SI, Naota T, Komiya N (1995) Tetrahedron Lett 36:8059
28. (a) Taqui Khan MM, Shukla RS (1992) J Mol Catal 77:221; (b) Taqui Khan MM, Bajaj HC, Shukla RS, Mirza SA (1988) J Mol Catal 45:51
29. Neumann R, Khenkin AM, Dahan M (1995) Angew Chem Int Ed Engl 34:1587
30. Yamaguchi K, Mizuno N (2002) New J Chem 26:972
31. Bonchio M, Scorrano G, Toniolo P, Proust A, Artero V, Conte V (2002) Adv Synth Catal 344:841
32. Ruchardt C (1970) Angew Chem Int Ed Engl 9:830
33. Barton DHR, Boivin J, Motherwell WB, Ozbalik N, Schwartzentruber KM (1986) Nouv J Chem 10:387
34. d'Alessandro N, Liberatore L, Tonucci L, Morvillo A, Bressan M (2001) New J Chem 25:1319
35. (a) Walling C, Johnson RA (1975) Acc Chem Res 8:125; (b) Strckhan E, Wellmann J (1976) Angew Chem Int Ed Engl 15:294; (c) Matue T, Fujihira M, Osa T (1981) J Electrochem Soc 128:2656; (d) Brook MA, Castle L, Lindsay Smith JR (1982) J Chem Soc Perkin Trans 687; (e) Kuruta T, Watanabe MK, Sawaki Y (1988) J Am Chem Soc 110:7472
36. Chatterjee D, Mitra A, Mukherjee S (2001) J Mol Catal A 165:295
37. Chatterjee D, Mitra A, Mukherjee S (1999) Polyhedron 18:659
38. Chatterjee D, Mitra A, Roy BC (2000) J Mol Catal A 161:17
39. Sasson Y, Zappi GD, Neumann R (1996) J Org Chem 51:2880
40. Sasson Y, Quntar AEA, Zoran A (1998) Chem Commun 73
41. Arends IWCE, Sheldon RA (1992) Top Catal 19:133
42. Barf G, Sheldon RA (1995) J Mol Catal A 102:23
43. Zhang W, Loebach JL, Wilson SR, Jacobsen EN (1990) J Am Chem Soc 112:2801
44. Irie R, Noda K, Ito Y, Katsuki T (1991) Tetrahedron Lett 32:1055
45. Gore E (1983) Platinum Met Rev 27:111
46. (a) Liu CJ, Yu WY, Peng SM, Mak TCW, Che CM (1998) J Chem Soc Dalton Trans 1805; (b) Liu CJ, Yu WY, Che CM, Yeung CH (1999) J Org Chem 64:7365
47. (a) Liu CJ, Yu WY, Li SG, Che CM (1998) J Org Chem 63:7364; (b) Liu CJ, Li SG, Pang WQ, Che CM (1997) Chem Commun 65
48. Jitsukawa K, Shiozaki H, Matsuda H (2002) Tetrahedron Lett 43:1491
49. Neumann R, Dahan M (1997) Nature 388:353
50. Cheung WH, Yu WY, Yip WP, Zhi NY, Che CM (2002) J Org Chem 67:7716
51. Klement I, Lütjens H, Knochel P (1997) Angew Chem Int Ed Engl 36:1454
52. Boelrijk AEM, Neenan TX, Reedijk J (1997) J Chem Soc Dalton Trans 4561
53. Qi JY, Qiu LQ, Lam KH, Yip CW, Zhou ZY, Chan ASC (2003) Chem Commun 1058
54. McCoy M (2000) Chem Eng News 78:32
55. Groves JT, Roman JS (1995) J Am Chem Soc 117:5594
56. Yamada T, Hashimoto K, Kitaichi Y, Suzuki K, Ikeno T (2001) Chem Lett 268
57. Hashimoto K, Kitaichi Y, Tanaka H, Ikeno T, Yamada T (2001) Chem Lett 922
58. (a) Gross Z, Ini S, Kapon M, Cohen S (1996) Tetrahedron Lett 37:7325; (b) Gross Z, Ini S (1997) J Org Chem 62:5514
59. Zhang R, Yu WY, Wong KY, Che CM (2001) J Org Chem 66:8145
60. Berkessel A, Frauenkron M (1997) J Chem Soc Perkin Trans 12265
61. Kureshy RI, Khan NH, Abdi SHR, Bhatt KN (1993) Tetrahedron Asymmetry 4:1693
62. Kureshy RI, Khan NH, Abdi SHR, Iyer P (1997) J Mol Catal A 124:91
63. Nishiyama H, Shimada T, Ito H, Sugiyama H, Motoyama Y (1997) Chem Commun 1863
64. Pezet F, Aït-Haddoe H, Daran, J, Sasaki I, Balavoine GGA (2002) Chem Commun 510

65. (a) Stoop RM, Mezzetti A (1999) Green Chem 1:39; (b) Stoop RM, Bachmann S, Valentini M, Mezzetti A (2000) Organometallics 19:4117
66. Strukul G (1992) Catalytic oxidation with hydrogen peroxide as oxidant. Kluwer, Dordrecht
67. Wolf S, Hasan SK, Campbell JR (1970) J Chem Soc Chem Commun 1420
68. Stork G, Meisels A, Davies JE (1963) J Am Chem Soc 85:3419
69. Tori S, Inokuchi T, Kondo K (1985) J Org Chem 50:4980
70. Carlsen PHJ, Katsuki T, Martin VS, Sharpless KB (1981) J Org Chem 46:3936
71. Yang D, Zhang C (2001) J Org Chem 66:4814
72. Papps JJ, Keaveney WP, Gancher E, Berger M (1966) Tetrahedron Lett 4273
73. (a) Lin X, Xu J, Liu H, Yue B, Jin S, Xie G (2000) J Mol Catal A 161:163; (b) Zhang X, Chen Q, Duncan DC et al. (1997) Inorg Chem 36:4381; (c) Neumann R, Abu-Grim C (1990) J Am Chem Soc 112:6025
74. Morvillo A, Bressan M (1997) J Mol Catal A 125:119
75. Wong WK, Chen XP, Pan WX, Guo JP, Wong WY (2002) Eur J Inorg Chem 231
76. Sheldon RA, Brink GJ ten, Arends IWCE (2002) Selective oxidation - homogeneous. Encyclopedia of catalysis. Wiley, http://www.mrw.interscience.wiley.com
77. Barak G, Dakka J, Sasson Y (1988) J Org Chem 53:3553
78. Wynne JH, Loyd CT, Witsil DR, Mushrush GW, Talick WM (2000) Org Prep Proc Int 32:588
79. Sheldon RA, Arends IWCE, Dijksman A (2000) Catal Today 57:158
80. Tang R, Diamond SE, Neary N, Mares FJ (1978) J Chem Soc Chem Commun 562
81. Matsumoto M, Ito SJ (1981) J Chem Soc Chem Commun 907
82. Bilgrien C, Davis S, Drago RS (1987) J Am Chem Soc 109:3786
83. Sheldon RA (2001) Chem Commun 2399
84. Wolfson A, Wuyts S, de Vos DE, Vancelecom IFJ, Jacobs PA (2002) Tetrahedron Lett 43:8107
85. (a) Bäckvall JE, Chowdhury RL, Karlsson U (1991) J Chem Soc Chem Commun 473; (b) Wang GZ, Andreasson U, Bäckvall JE (1994) J Chem Soc Chem Commun 1037; (c) Csjernyik G, Ell A, Fadini L, Pugin B, Backvall JE (2002) J Org Chem 67:1657
86. Hanyu A, Takezawa E, Sakaguchi S, Ishii Y (1998) Tetrahedron Lett 39:5557
87. (a) Dijksman A, Arends IWCE, Sheldon RA (1999) Chem Commun 1591; (b) Dijksman A, Marino-González A, Mairata i Payeras A, Arends IWCE, Sheldon RA (2001) J Am Chem Soc 123:6826. (c) Inokuchi T, Nakagawa K, Torii S (1995) Tetrahedron Lett 36: 3223
88. (a) Bäckvall JE, Andreasson U (1993) Tetrahedron Lett 34:5459; (b) Trost BM, Kulawiec RJ (1991) Tetrahedron Lett 32:3039
89. Aranyos A, Csjernyik G, Szabo KJ, Bäckvall JE (1999) Chem Commun 351
90. Karlson U, Wang GZ, Bäckvall JE (1994) J Org Chem 59:1196
91. Paleos CM, Dais PJ (1977) Chem Soc Chem Commun 345
92. (a) Beynon PJ, Collins PM, Gardiner D, Overend WG (1968) Carbohydr Res 6:431; (b) Friedrich HB (1999) Platinum Met Rev 43:94
93. Dengel AC, Hudson RA, Griffith WP (1985) Transition Met Chem 10:98
94. Griffith WP, Ley SV, Whitcombe GP, White AD (1987) Chem Commun 1625
95. Dengel AC, El-Hendawy AM, Griffith WP (1989) Transition Met Chem 40:230
96. Griffith WP, Ley SV (1990) Aldrichimica Acta 23:13
97. Ley SV, Norman J, Griffith WP, Marsen SP (1994) Synthesis 639
98. Lenz R, Ley SV (1997) J Chem Soc Perkin Trans 1 3291
99. Marko IE, Giles PR, Tsukazaki M, Chelle-Regnaut I, Urch CJ, Brown SM (1997) J Am Chem Soc 119:12661
100. Hinzen B, Lenz R, Ley SV (1998) Synthesis 977

101. Bleloch A, Johnson BFG, Ley SV, Price AJ, Shepard AS, Thomas AN (1999) Chem Commun 1907
102. Pagliaro M, Ciriminna R (2001) Tetrahedron Lett 42:4511
103. Hasan M, Musawir M, Davey PN, Kozhevnikov I (2002) J Mol Catal A 180:77
104. Rocek J, Ng CS (1974) J Am Chem Soc 96:1522
105. Lee DG, Wang Z, Chandler WD (1992) J Org Chem 57:3276
106. (a) Cheng SYS, Rajapakse N, Rettig SJ, BR James (1994) J Chem Soc Chem Commun 2669; (b) Rajapakse N, James BR, Dolphin D (1990) Stud Surf Sci Catal 55:109
107. Masutani K, Uchida T, Irie R, Katsuki T (2000) Tetrahedron Lett 41:5119
108. Miyata A, Marakami M, Irie R, Katsuki T (2001) Tetrahedron Lett 42:7067
109. Matsumoto M, Watanabe NJ (1984) J Org Chem 49:3435
110. Matsumoto M, Ito S (1984) Synth Commun 14:697
111. Vocanson F, Guo YP, Namy JL, Kagan HB (1998) Synth Commun 28:2577
112. Zhan BZ, White MA, Sham TK, Pincock, JA, Doucet RJ, Rao KVR, Robertson KN, Cameron TS (2003) J Am Chem Soc 125:2195
113. Yamaguchi K, Mizuno N (2002) Angew Chem Int Ed Engl 41:4538
114. Kaneda K, Yamashita T, Matsushita T, Ebitani K (1998) J Org Chem 63:1750
115. Matsushita T, Ebitani K, Kaneda K (1999) Chem Commun 265
116. Yamaguchi K, Mori K, Mizugaki T, Ebitani K, Kaneda K (2000) J Am Chem Soc 122:7144
117. Ji HB, Ebitani K, Mizugaki T, Kaneda K (2002) Catal Commun 3:511
118. Neumann R (1998) Prog Inorg Chem 47:17; (b) Hill CL, Prosser-McCartha CM (1995) Coord Chem Rev 143:407; (c) Pope MT, Müller A (1991) Angew Chem Int Ed Engl 30:34
119. Nishimura T, Arends IWCE, Sheldon RA (to be published)
120. (a) Neumann R, Abu-Gnim C (1989) J Chem Soc Chem Commun 124; (b) Neumann R, Abu-Gnim C (1990) J Am Chem Soc 112:6025
121. Higashijima M (1999) Chem Lett 1093
122. Murahashi SI, Naota T, Hirai N (1993) J Org Chem 58:7318
123. (a) Courtney JL (1986) In: Mijs WJ, de Jonge CHRI (eds) Organic syntheses by oxidation with metal compounds. Plenum, New York; (b) Haines A (1985) Methods for the oxidation of organic compounds. Academic, London
124. Baker DC, Horton D, Tindall CG (1976) Methods Carbohydr Chem 7:3
125. Giddings S, Mills A (1988) J Org Chem 53:1103
126. Stevens CL, Bryant CP (1972) Methods Carbohydr Chem 3:337
127. Carlsen PHJ, Katsuki T, Martin VS, Sharpless KB (1981) J Org Chem 46:3936
128. Gonsalvi L, Arends IWCE, Sheldon RA (2002) Chem Commun 202
129. Gonsalvi L, Arends IWCE, Sheldon RA (2002) Org Lett 4:1659
130. Murahashi SI (1995) Angew Chem Int Ed Engl 34:2443
131. Murahashi SI, Naota T, Taki H (1985) J Chem Soc Chem Commun 613
132. Yamaguchi K, Mizuno N (2003) Angew Chem Int Ed Engl 42:1480
133. Mori K, Yamaguchi K, Mizugaki T, Ebitani K, Kaneda K (2001) 461
134. Tang R, Diamond SE, Neary N, Mares F (1978) J Chem Soc Chem Commun 562
135. Cenini S, Porta F, Pizzotti M (1982) 15:297
136. Bailey AJ, James BR (1996) Chem Commun 2343
137. Murahashi SI, Naota T, Miyaguchi N, Nakato T (1992) Tetrahedron Lett 33:6991
138. Murahashi SI, Naota T, Kuwabara T, Saito T, Kumobayashi H, Akutagawa S (1990) J Am Chem Soc 112:7820
139. Murahashi SI, Saito T, Naota T, Kumobayashi H, Akutagawa S (1991) Tetrahedron Lett 32:5991

Topics Organomet Chem (2004) 11: 321–336
DOI 10.1007/978-3-540-40027-1

Ruthenium-Catalyzed Organic Synthesis in Aqueous Media

Mingwen Wang[1] · Chao-Jun Li[1,2] (✉)

[1] Department of Chemistry, Tulane University, New Orleans, LA70118, USA
[2] Department of Chemistry, McGill University, Montreal, PQH3A 2K6, Canada
cj.li@mcgill.ca; mingwenwang@yahoo.com

Abstract Ruthenium-catalyzed organic synthesis in aqueous media has attracted much attention recently. The applications of ruthenium in aqueous media are generally in atom-economical and environmentally benign reactions. In this review, our focus is on ruthenium-catalyzed C–C bond formations. Ruthenium-catalyzed activation of the C–H bond of allyl alcohol and tandem isomerization/aldol reaction are discussed. Also, through activation of the C–H bond of terminal alkynes, direct additions of alkynes to aldehydes and imines to give Grignard-type reaction products were realized in aqueous media. Various ruthenium-catalyzed nonmetathesis C–C formations in aqueous media, most of which were on the addition to alkynes, provided useful methodology for organic synthesis. In addition, ruthenium-catalyzed ring-closing metathesis and ring-opening metathesis polymerization reactions in water are briefly overviewed.

Keywords Ruthenium · Catalysis · C–C formation · C–H activation · Aqueous

1 Introduction

There has been growing interest in the development of organic synthesis in aqueous media recently [1–2]. From an environmental perspective, water as an obviously benign and inexpensive solvent could yield significant "green chemistry" benefits. From a synthetic point of view, one of the biggest advantages of using water as a solvent is the potential simplification of protection and deprotection sequences for functional molecules such as alcohols, amines, and acids. Recent studies on organic reactions in water have shed light on the possibility that many more reactions could be carried out in water, although most organic compounds are not soluble in water. The fact is that reactions can occur very well under emulsion without the need of being completely soluble. Also, in an aqueous environment, not all organic intermediates are reactive to water molecules, which leads to the hydrolysis of substrates. In fact, many reactions can still proceed if the intermediate reacts with the desired species faster than with the water molecules. On the other hand, the catalytic actions of transition metals in water have played a key role in various enzymatic reactions, including biocatalysis, biodegradation, photosynthesis, nitrogen fixation, digestions, and the evolution of bioorganisms [3–4]. All of these "natural" catalytic reactions occur in aqueous conditions, which is in sharp contrast to most transition-metal-catalyzed reactions commonly used in the laboratory.

Within the past few decades, the use of late transition metals to catalyze reactions has made a great contribution to modern organic chemistry, and a variety of highly selective and atom-economical reactions have been discovered using group 8 transition metals [5–7]. Ruthenium has a wide range of oxidation states (from –2 to +8) and various coordination geometries. Because of this, it has unique characteristics: high electron-transfer ability, high Lewis acidity, low redox potentials, and stabilities of reactive metallic intermediates such as oxometals, metallacycles, and ruthenium carbene complex. The complexity and diversity of ruthenium's characteristics offers great potential for the exploitation of novel ruthenium-catalyzed methodologies. For ruthenium catalysis, some excellent reviews have been published recently [8–9]. There is no doubt that many more interesting catalytic reactions using ruthenium in aqueous media will be discovered. However, compared with Lewis-acid-catalyzed and some other metal-catalyzed organic syntheses in aqueous media [1–2, 10], aqueous ruthenium catalysis is a relatively unexplored field, mostly limited to hydrogenation or reduction [11]. This review will mainly focus on ruthenium-catalyzed C–C bond formation reactions in aqueous media.

2
Ruthenium-Catalyzed Reactions Involving C–H Activation in Aqueous Media

An ideal organic synthesis is both environmentally benign and atom-economical [12]. The path to reach this ultimate goal is to develop catalytic C–C bond formation reactions through efficient activation of a C–H bond in water. The study on ruthenium-catalyzed activation of a C–H bond in an organic solvent already provided us with the foundation for this endeavor [9]. The first investigation is the activation of sp^3 allylic C–H bond and its tandem aldol and Mannich reactions in aqueous media. The second investigation is on a ruthenium catalytic system giving Grignard-type nucleophilic addition products in water via activation of the *sp* C–H bond.

2.1
Isomerization of Homoallyl Alcohols Through Activation of the sp^3 C–H Bond

Isomerization of allylic alcohol to ketone has been extensively studied [13], and two different pathways have been established, including π-allyl metal hydride and the metal hydride addition–elimination mechanisms [5, 14]. McGrath and Grubbs [15] investigated the ruthenium-catalyzed isomerization of allyl alcohol in water and proposed a modified metal hydride addition–elimination mechanism through an oxygen-functionality-directed Markovnikov addition to the double bond.

Li and coworkers [16] discovered that in the presence of $RuCl_2(PPh_3)_3$, which is compatible with water and air, the allylic C–H bond was activated and the functional groups of homoallyl alcohols were repositioned to give allyl alcohols (Eq. 1). The experimental procedure is very simple: stirring a mixture of homoallyl alcohol **1** with a catalytic amount of $RuCl_2(PPh_3)_3$ in water and air at 90–100 °C for 1–3 h led to the product **2**.

$$\text{R-CH(OH)-CH}_2\text{-CH=CH}_2\ (\mathbf{1}) \xrightarrow[\text{90-100°C, H}_2\text{O, air}]{\text{4 mol\% RuCl}_2\text{(PPh}_3\text{)}_3} \text{R-CH=CH-CH(OH)-CH}_3\ (\mathbf{2}) \quad (1)$$

1a, R=Ph / 2h — **2a**, 63%
1b, R=4-Cl-Ph / 2h — **2b**, 65%
1c, R=3-Br-Ph / 2h — **2c**, 59%
1d, R=4-Me-Ph / 1.5h — **2d**, 75%
1e, R=4-MeO-Ph / 1.5h — **2e**, 67%
1f, R=4-HOCH$_2$-Ph / 3h — **2f**, 49%

This reaction showed unusual selectivity. The solvent has an important effect on the reaction. If water was switched to dimethylformamide (DMF), tetrahydrofuran, dimethylsulfoxide, or toluene, no isomerization product was observed and the starting materials were recovered or an ether was formed instead of isomerization [17]. However, the substrates were limited to benzylic-type homoallyl alcohols and other allyl alcohols demonstrated a lack of regioselectivity. In the case of compound **1g**, where both an allyl and a homoallyl functional group were involved, the reaction occurred exclusively by rearrangement of the homoallyl group to give the conjugated dienol product **2g** (Eq. 2).

1g → 4 mol% $RuCl_2(PPh_3)_3$, 90-100°C, H_2O, air, 2.5 h → **2g**, 61% (2)

During this process, corresponding ketones were found as side products. The reaction showed a marked dependence on the Ru(II)-to-substrate ratio. Increasing the amount of catalyst resulted in increased formation of the phenyl ketone **3** (Eq. 3).

1d (p-MeO-Ph) → $RuCl_2(PPh_3)_3$, 90-100°C, H_2O, air, 2.5 h → **2d** + **3** (3)

cat (mol%)	**2d** : **3**	conversion (%)
2.2	12: 1	87
4.5	4.5: 1	94
7.0	1.7: 1	90
10.0	1.2: 1	92

Under the same conditions, allyl alcohols **4** underwent isomerization to form allyl alcohols **5** (Eq. 4).

4 → 4 mol% $RuCl_2(PPh_3)_3$, 90-100°C, H_2O, air → **5** (4)

R=Ph, R'=Me (90%)
R=p-MePh, R'=n-C_3H_7 (74%)
R=p-ClPh, R'=n-C_3H_7 (65%)
R=Ph, R'=n-C_3H_7 (70%)
R=p-MeOPh, R'=n-C_3H_7 (89%)
R=m-BrPh, R'=n-C_3H_7 (72%)

A proposed mechanism for the isomerization is illustrated in Fig. 1. The ruthenium complex first coordinates to the olefin and transfers it from a terminal position to an internal position, providing an allyl alcohol [17, 18]. The allyl alcohol is then converted to either another allyl alcohol through C–O cleavage (route a) or a ketone through C–H cleavage (route b).

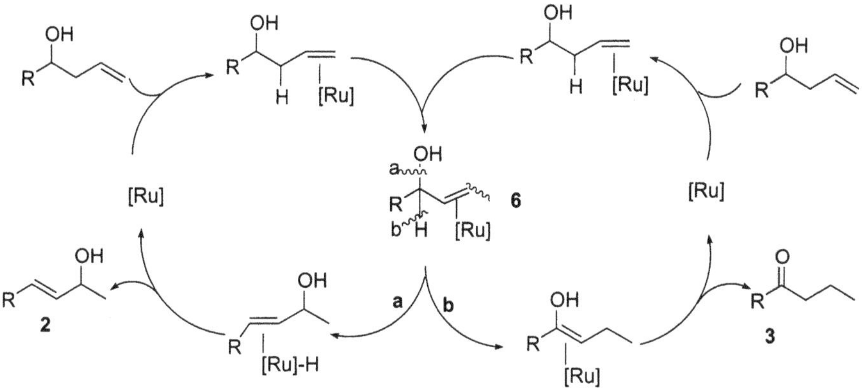

Fig. 1

2.2 Tandem Olefin-Migration/Aldol- and Mannich-Type Reactions

From Fig. 1, there exist two possibilities from the intermediate **6**: one is to form **7**, and the other is to form **8**. The proposal is that such a ruthenium enol **8** could be captured by electrophiles to form new C–C bonds in water (Eq. 5).

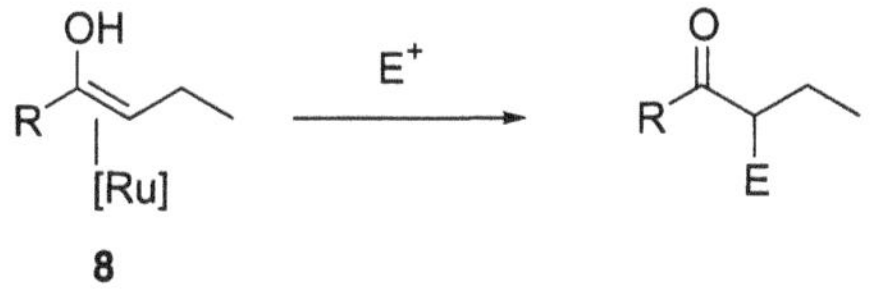

(5)

Indeed, a ruthenium-catalyzed tandem olefin-migration/aldol-type reaction has been realized when an aldehyde is present in aqueous media [18, 19]. For 3-butene-2-ol (**9**), the tandem isomerization/aldol-type reaction was examined. The mixture of **9**, aldehyde (**10**), and a catalytic amount of $RuCl_2(PPh_3)_3$ in H_2O/toluene(4/1) (Eq. 6) or H_2O alone (Eq. 7) was stirred for 5 h at 110 °C (oil bath temperature) and afforded the aldol adduct **11**.

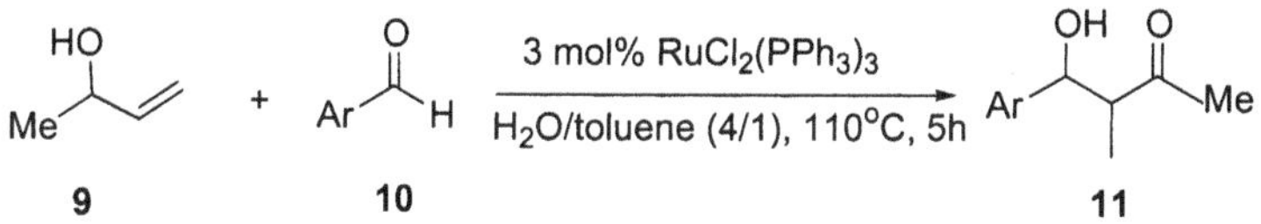

(6)

11a, Ar = Ph, 76% yield, syn:anti = 73:27
11b, Ar = p-ClPh, 70% yield, syn:anti = 74:26
11c, Ar = p-BrPh, 73% yield, syn:anti = 79:21
11d, Ar = p-MeOPh, 68% yield, sym:anti = 51: 49
11e, Ar =p-PhPh, 27% yield, syn:anti =67:33
11f, Ar = 2-Naphthyl, 44% yield, syn:anti =63:37

HO / Me (9) + Ar–CHO (10) $\xrightarrow[H_2O]{3\ mol\%\ RuCl_2(PPh_3)_3}$ Ar–CH(OH)–CH(Me)–C(O)–Me (11) (7)

11b, Ar = p-ClPh, 35% yield, syn:anti = 60:40
11g, Ar = m-F-Ph, 72% yield, syn:anti = 66:34

Under the same conditions, the yield of the corresponding aldol product was very low (10%) when α-vinylbenzyl alcohol (**12**) was used instead of **9** to react with benzaldehyde (**10a**). The allyl alcohol **12** was mainly converted into propiophenone, which was attributed to the olefin migration by path b described in Fig. 1. By adding a Lewis acid, $In(OAc)_3$, as a cocatalyst, the aldol reaction was dramatically improved and the yield of **13a** was increased from 10% to 80% (Eq. 8).

HO / Ph (12) + Ar–CHO (10) $\xrightarrow[H_2O/toluene\ (4:1),\ 110^{\circ}C,\ 7h]{3\ mol\%\ RuCl_2(PPh_3)_3,\ 8\ mol\%\ In(OAc)_3}$ Ar–CH(OH)–CH(Me)–C(O)–Ph (13) (8)

13a, Ar = Ph, 80% yield, syn:anti = 68:32
13b, Ar = p-ClPh, 79% yield, syn:anti = 63:37
13c, Ar = p-BrPh, 82% yield, syn:anti = 61:39
11d, Ar =p-MePh, 54% yield, syn:anti =63:37
11e, Ar = 2-Naphthyl,51% yield, syn:anti =60:40

In addition, the cross-coupling of imines with allyl alcohols to generate Mannich-type reaction products proceeded efficiently under similar conditions in methanol and ionic liquid ([1-*n*-butyl-3-methlyimidazolium]$^+PF_6^-$) [19, 20] (Eq. 9).

HO / Me + Ar–CH=N–Ar' $\xrightarrow[MeOH\ or\ ionic\ liquid]{RuCl_2(PPh_3)_3}$ Ar–CH(NHAr')–CH(Me)–C(O)–Me (9)

2.3 Grignard-Type Reactions

The conventional Grignard reaction (Fig. 2, route I) would generate both a stoichiometric amount of halide waste and a stoichiometric amount of metal waste. It also requires multistep synthesis of the halides. On the other hand, an alternative Grignard-type reaction via catalytic C–H activation in water (Fig. 2, route II) would preclude the use of flammable organic solvents and also avoid the wasteful process of drying them. Obviously, it would provide a cleaner solution for organic synthesis and provide a theoretical 100% atom-efficiency.

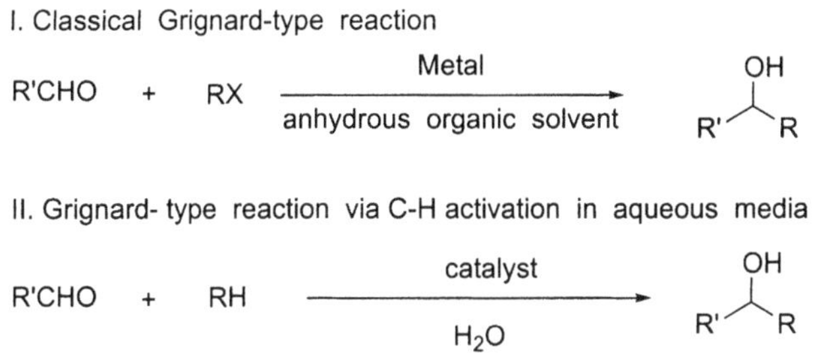

I. Classical Grignard-type reaction

$$R'CHO + RX \xrightarrow[\text{anhydrous organic solvent}]{\text{Metal}} R'CH(OH)R$$

II. Grignard- type reaction via C-H activation in aqueous media

$$R'CHO + RH \xrightarrow[H_2O]{\text{catalyst}} R'CH(OH)R$$

Fig. 2

2.3.1
Addition of Terminal Alkynes to Aldehydes

By using a bimetallic Ru–In catalytic system [21], Wei and Li [22] added phenylacetylene to aldehydes to give Grignard-type nucleophilic addition products via C–H activation in water (Eq. 10). The idea behind the Ru–In system is to use $RuCl_3$ to catalyze the overall reaction and $In(OAc)_3$ to activate the carbonyl group. Although it is not essential for the reaction to proceed, the presence of an organic base morpholine increases conversion of the addition reaction considerably. And the use of 5% aqueous K_2CO_3 instead of water alone further improved the reaction.

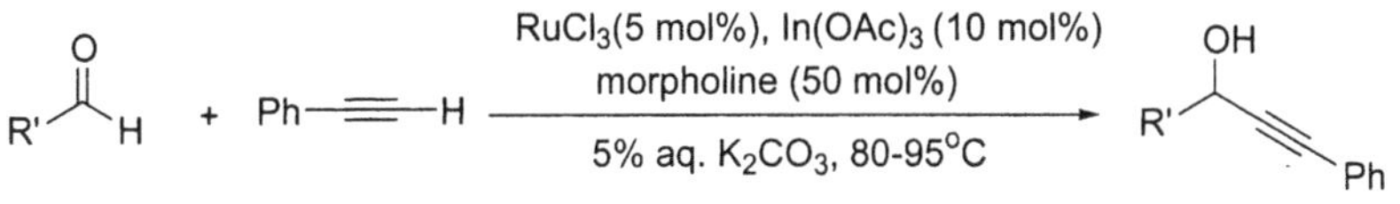

$$R'CHO + Ph{-}C{\equiv}C{-}H \xrightarrow[\text{5\% aq. } K_2CO_3\text{, 80-95°C}]{RuCl_3\text{(5 mol\%), } In(OAc)_3\text{ (10 mol\%), morpholine (50 mol\%)}} R'CH(OH){-}C{\equiv}C{-}Ph \quad (10)$$

R' = Ph, 57% yield
R' = p-ClPh, 52% yield
R' = m-BrPh, 60% yield
R' = p-MePh,46% yield
R' = p-CF_3Ph, 94% yield

R' = p-NCPh, 27% yield
R' = p-PhPh, 42% yield
R' = t-Bu, 38% yield
R' = 1-Naphthyl, 62% yield
R' = 2-Naphthyl, 47% yield

Unlike previous alkyne–aldehyde additions [23], the generation of an alkynyl carbanion is unlikely owing to the large pK_a difference between the terminal acetylene and the solvent water [24]. A mechanism was proposed involving the simultaneous activation of the C–H bond of alkyne by the ruthenium catalyst and the aldehyde carbonyl by the indium ion. The ruthenium intermediate then underwent Grignard-type addition followed by an in situ hydrolysis in water to give the desired carbonyl addition product and regenerated the ruthenium and indium catalysts to catalyze further reactions (Fig. 3).

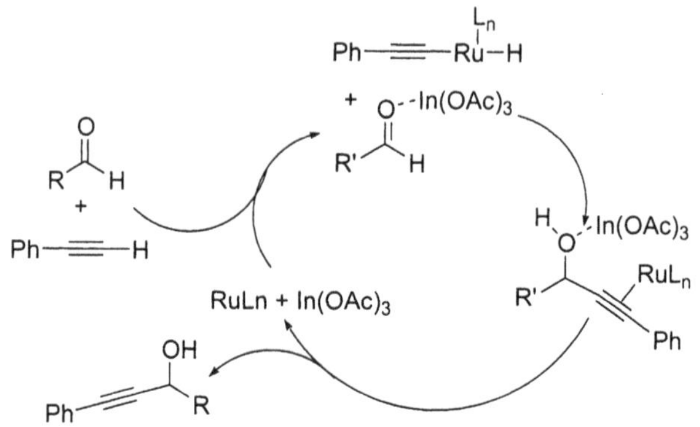

Fig. 3

2.3.2 Addition of Terminal Alkynes to Imines

A reaction related to C=O addition is C=N addition. Direct addition of acetylene to various imines to generate propargyl amines [25–28] via C–H activation in water was investigated (Eq. 11) [29]. The process is simple and generated a diverse range of propargylic amines in excellent yields. The same reaction was done under solvent-free conditions. The key to the reaction is the activation of imines by ions.

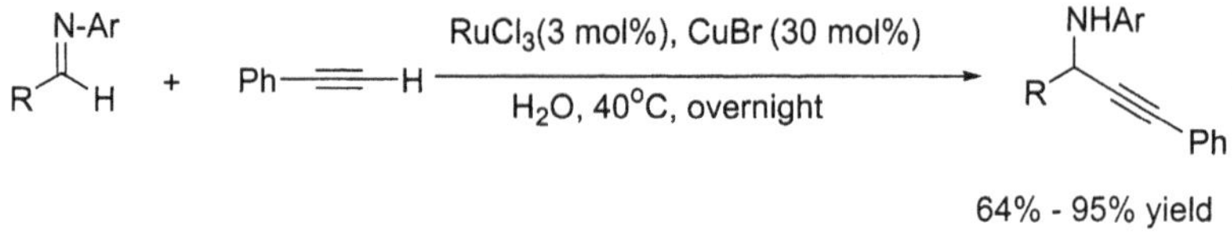

$$\mathrm{RCH{=}NAr} + \mathrm{Ph{-}C{\equiv}C{-}H} \xrightarrow[\mathrm{H_2O,\ 40^{\circ}C,\ overnight}]{\mathrm{RuCl_3(3\ mol\%),\ CuBr\ (30\ mol\%)}} \mathrm{RCH(NHAr){-}C{\equiv}C{-}Ph} \quad \text{64\% - 95\% yield} \tag{11}$$

3 Ruthenium-Catalyzed Nonmetathesis C–C Formation in Aqueous Media

Many ruthenium-catalyzed nonmetathesis C–C bond formation reactions in aqueous media are of high efficiency and exemplify the concept of atom-economy [30]. The vast majority of the reactions are on alkynes.

3.1 Reactions Involving Addition of Water to Alkynes

Trost et al. [31] discovered that by using the RuCp(COD)Cl/In(OCF_3SO_2)$_3$/NH_4PF_6 catalyst system, where Cp is cyclopentadienyl and COD is cyclooctadiene, the reaction of terminal alkyne, water, and α-vinyl ketone afforded the 1,5-diketone in DMF–H_2O. The reaction is highly selective and showed a tolerance to alkynes with various functional groups (Eq. 12).

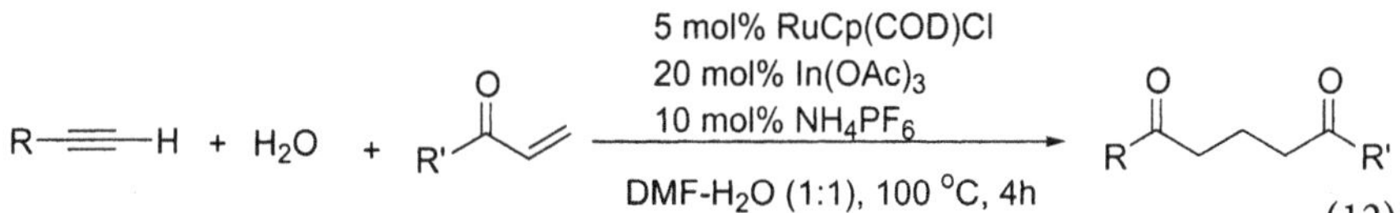

R—≡—H + H_2O + R'C(O)CH=CH$_2$ → (5 mol% RuCp(COD)Cl, 20 mol% $In(OAc)_3$, 10 mol% NH_4PF_6; DMF-H_2O (1:1), 100 °C, 4h) → RC(O)CH$_2$CH$_2$CH$_2$C(O)R' (12)

The reaction was rationalized by a ruthenium enolate mechanism (Fig. 4). Water served as a nucleophile and added to alkynes; then the intermediate isomerized to give a ruthenium enolate, which then underwent addition to α-vinyl ketone followed by protonation to afford the 1,5-diketone. During the reaction, no ketone resulting from the hydration of the alkynes was found, which showed that the conjugate addition is faster than protonation of the ruthenium enolate in this aqueous reaction.

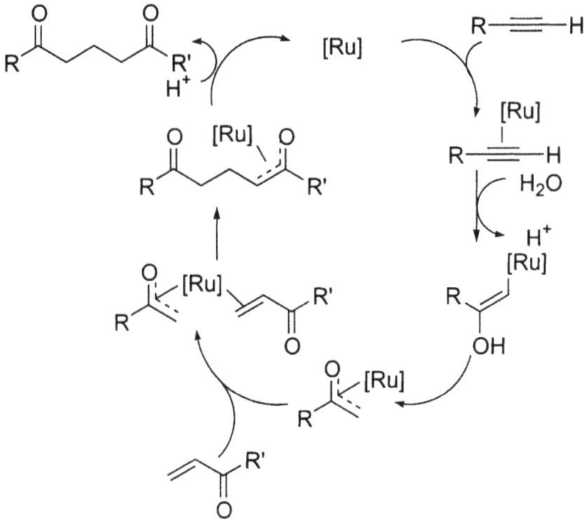

[Ru]
R—≡—H
H_2O
H^+
OH
R'

Fig. 4

An intramolecular version of the 1,5-diketone forming reaction was also realized using 10 mol % of $RuCp(NCCH_3)_3PF_6$ and 10 mol % of camphorsulfonic acid (CSA) as the catalyst system in acetone. It is interesting to note that if no CSA was present in anhydrous acetone then pyran was formed

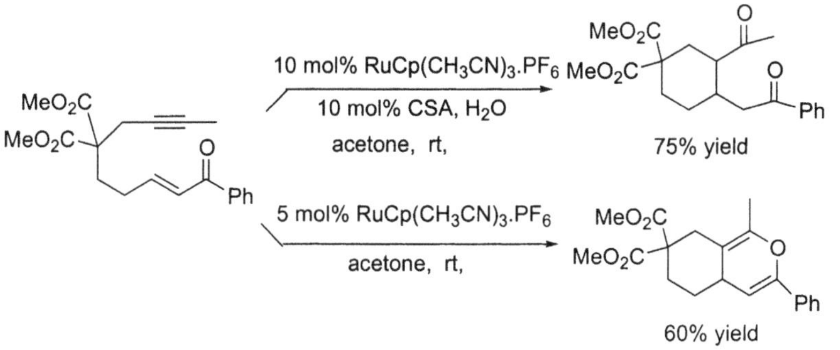

10 mol% $RuCp(CH_3CN)_3.PF_6$, 10 mol% CSA, H_2O, acetone, rt, → 75% yield

5 mol% $RuCp(CH_3CN)_3.PF_6$, acetone, rt, → 60% yield (13)

MeO_2C, MeO_2C, Ph

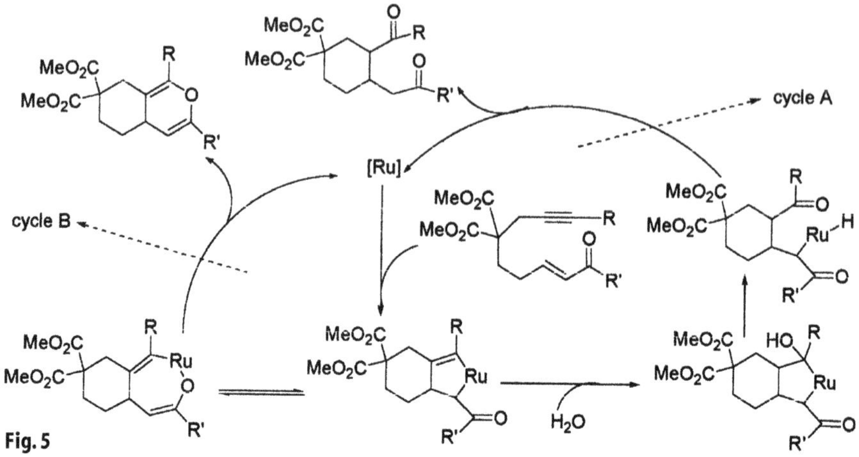

Fig. 5

(Eq. 13), which could be viewed as a [4+2] cycloaddition [32]. Water played an important role in the formation of different products. The formation of the 1,5-diketone and pyran could be explained by two different mechanisms (cycles A and B) (Fig. 5).

3.2 Reactions Involving Addition of Water to Propargyl Alcohol

When propargyl alcohols were used instead of alkynes in the reaction described in Eq. (12), enones were formed (Eq. 14) [33]. The reaction was proposed through a similar mechanism as outlined in Fig. 4.

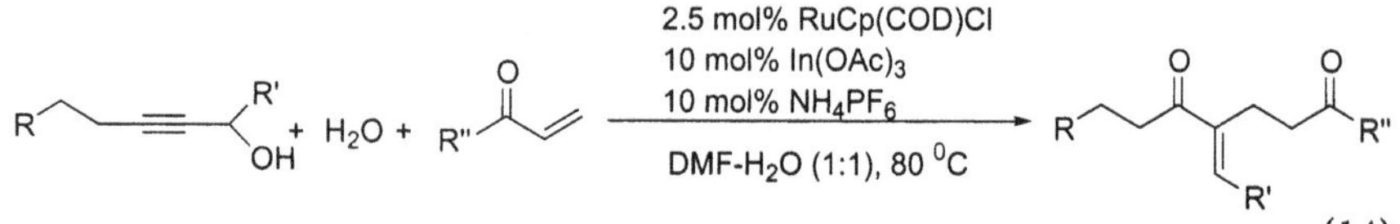

2.5 mol% RuCp(COD)Cl, 10 mol% $In(OAc)_3$, 10 mol% NH_4PF_6, DMF-H_2O (1:1), 80 °C (14)

Ruthenium-catalyzed cycloisomerization of diyn-ols to diene-ones or diene-als was discovered by Trost and Rudd [34], and provided the potential for the intramolecular aldol condensation. In the reaction, water acts as a reactant (Eqs. 15, 16). The reaction was proposed to proceed via a ruthenacyclopentadiene intermediate.

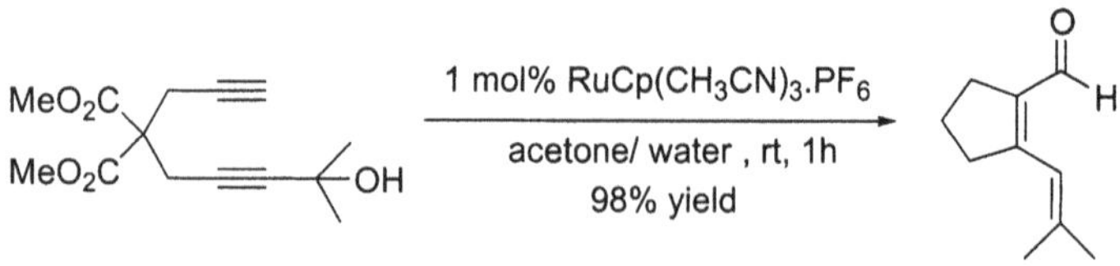

1 mol% $RuCp(CH_3CN)_3.PF_6$, acetone/ water , rt, 1h, 98% yield (15)

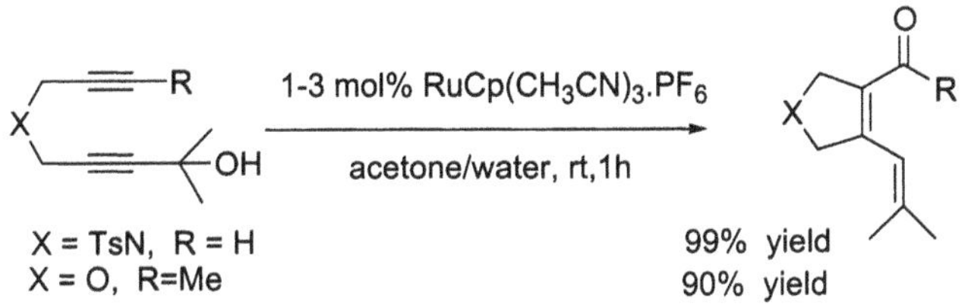

1-3 mol% $RuCp(CH_3CN)_3.PF_6$; acetone/water, rt, 1h

X = TsN, R = H — 99% yield
X = O, R=Me — 90% yield

(16)

3.3 Reactions Involving Addition of Halides to Alkynes

As described in the section Reactions Involving Addition of Water to Alkynes, the reaction of terminal alkynes, water, and α-vinyl ketones afforded 1,5-diketones in DMF–H_2O (Eq. 12). Under similar conditions, in the presence of halide, ruthenium-catalyzed three-component coupling of alkyne, an enone, and halide ion formed vinyl halide (Eq. 17) [35].

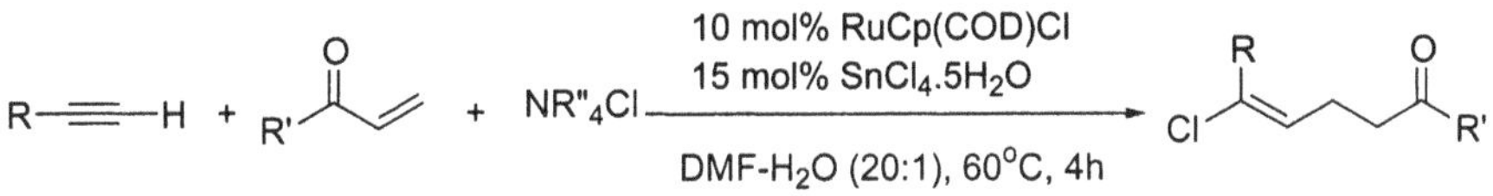

R—≡—H + R'C(O)CH=CH$_2$ + NR"$_4$Cl; 10 mol% RuCp(COD)Cl, 15 mol% $SnCl_4.5H_2O$; DMF-H_2O (20:1), 60°C, 4h

(17)

3.4 Reactions Involving Addition of Alkenes to Alkynes

In aqueous media, addition of unactivated alkynes to unactivated alkenes to form Alder-ene products has been realized by using a ruthenium catalyst (Eq. 18) [36]. A polar medium (DMF-to-H_2O ratio of 1:1) favors the reaction and benefits the selectivity. The reaction was suggested to proceed via a ruthenacycle intermediate.

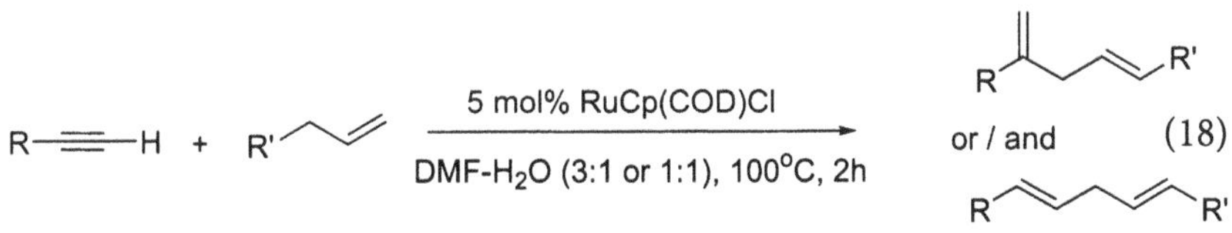

R—≡—H + R'CH$_2$CH=CH$_2$; 5 mol% RuCp(COD)Cl; DMF-H_2O (3:1 or 1:1), 100°C, 2h; or / and

(18)

RuCp(COD)Cl-catalyzed addition of allyl alcohol to alkynes to form γ,δ-unsaturated ketones was developed by Trost at al. [37] in DMF–H_2O. Different from Trost's catalyst, Dérien et al. [38] used $[RuCl(C_5Me_5)]_4$, $RuCl_2$(methallyl)(C_5Me_5) and RuCl(COD)(C_5Me_5) as catalysts to regioselectively form γ,δ-unsaturated aldehydes in aqueous media with the branched aldehydes as the major products (Eq. 19).

$$R-\equiv-H + R'CH(OH)CH=CH_2 \xrightarrow[\text{aqueous media}]{[Ru]} RCH=CHCH_2CH_2C(O)R' + RC(=CH_2)CH_2CH_2C(O)R' \quad (19)$$

López et al. [39] developed a synthetic method to 1,5-oxygen-bridged medium-sized carbocycles through a sequential ruthenium-catalyzed alkyne–alkene coupling and a Lewis-acid-catalyzed Prins-type reaction. The ruthenium-catalyzed reaction can be carried out in aqueous media (DMF-to-H_2O ratio of 10:1) [39].

4 Ruthenium-Catalyzed Olefin-Metathesis Reactions in Aqueous Media

Olefin metathesis is a useful tool for the formation of unsaturated C–C bonds in organic synthesis, and the reaction has been generally accepted to proceed through a series of metallacyclobutanes and carbene complexe intermediates [40–43]. For this type of reaction, the most widely used catalysts include an alkoxyl imido molybdenum complex (Schrock catalyst) [44] and a benzylidene ruthenium complex (Grubbs catalyst) [43]. The former is air- and moisture-sensitive and has some other drawbacks such as intolerance to many functional groups and impurities; the latter has increased tolerance to water and many reactions have been used in aqueous solution without any loss of catalytic efficiency.

4.1 Ring-Opening Metathesis Polymerization

The ring-opening metathesis polymerization (ROMP) of 7-oxanorbornene derivatives initiated by $Ru(H_2O)_6$(4-toluenesulfonyl)$_2$ in aqueous media was reported by Novak and Grubbs [45] (Eq. 20). Compared with the same reaction carried out in organic solvent, the initiation time was greatly decreased. After the polymerization, the aqueous catalyst solution was not only reused but also became more active in subsequent polymerizations.

$$n \text{ (7-oxanorbornene bis(methoxymethyl))} \xrightarrow[H_2O]{Ru(H_2O)_6\,(tos)_2} \text{poly(MeO/OMe-substituted tetrahydrofuran vinylene)}_n \quad (20)$$

Some well-defined ruthenium carbene complexes have been used in the living ROMP in aqueous media using a cationic surfactant to yield polymer latex [46].

Recent developments include the synthesis of new water-soluble ruthenium alkylidenes and their application to olefin metathesis in water [47, 48]. It is interesting to note that the addition of acid made the polymerization rate up to 10 times faster than without acid (Eq. 21).

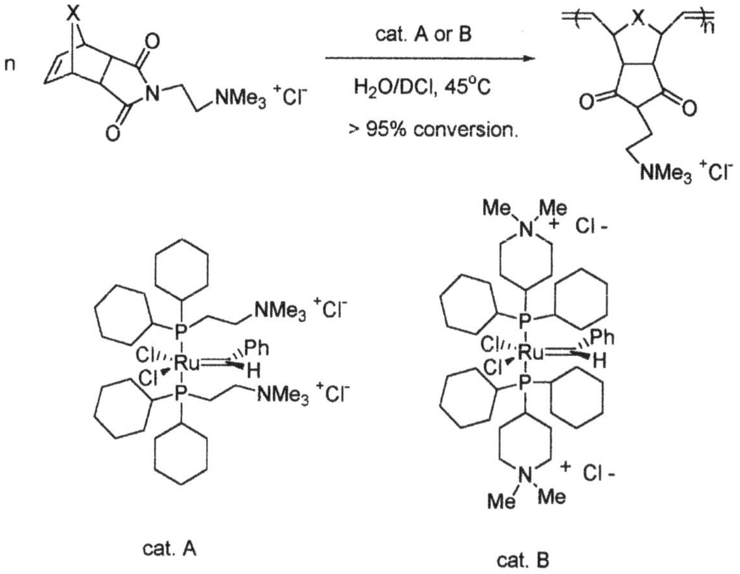

cat. A or B

H_2O/DCl, 45°C

> 95% conversion.

(21)

cat. A

cat. B

The group of Kiessling [49–52] has extended the use of ruthenium alkylidene catalyzed ROMP in aqueous media to give new, biologically active neoglycopolymers (Eq. 22).

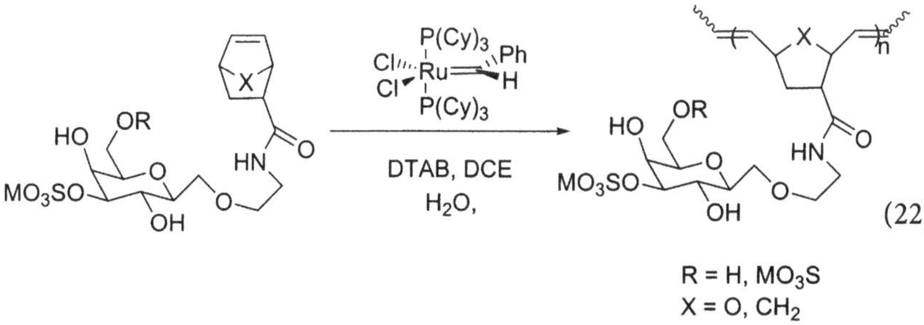

DTAB, DCE

H_2O,

(22)

R = H, MO_3S

X = O, CH_2

4.2 Ring-Closing Metathesis

Ring-closing metathesis (RCM) is an important method for construction of medium- and macro-cycle compounds that has been widely used in organic synthesis [43]. For many biologically related substrates, in order to keep their important higher-order structures, application of RCM must be done in aqueous media [53]. In contrast to ROMP, aqueous RCM has many limits in terms of the substrate and that has greatly retarded its application. For example, RCM of σ, ω-dienes in aqueous media was not successful, owing to the instability of the resulting active ruthenium species. However, through a simple substrate modification (incorporation of an olefin substitute), RCM of α,ω-dienes in aqueous media became highly efficient (Eq. 23) [53].

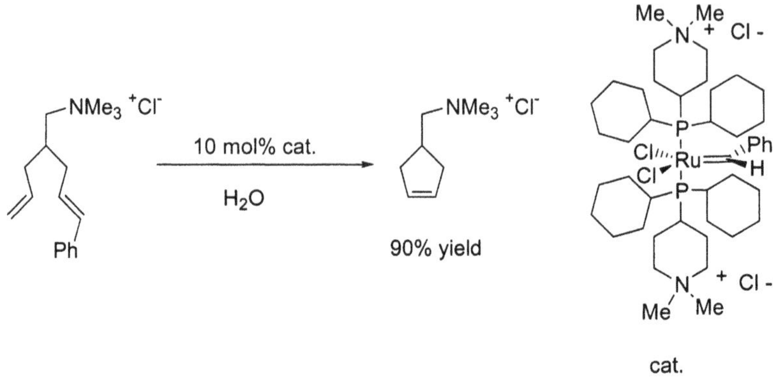

NMe$_3$ $^+$Cl$^-$ — 10 mol% cat., H_2O → NMe$_3$ $^+$Cl$^-$, 90% yield (23)

Ph

cat.

Furthermore, a new metathesis-active ruthenium alkylidene with a sterically bulky and electron-rich phosphine ligand has been synthesized and applied to RCM in aqueous media (Eq. 24) [54].

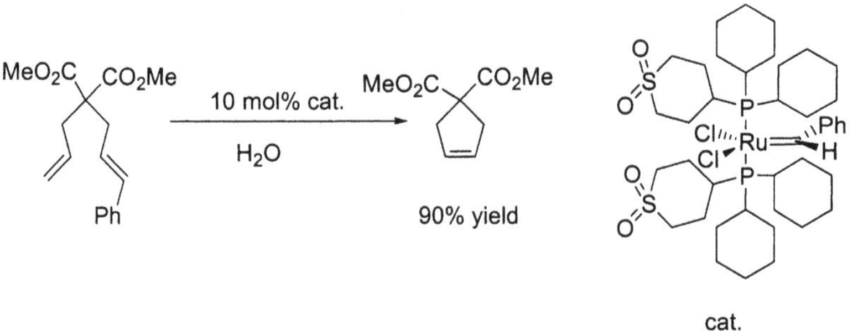

MeO$_2$C CO$_2$Me — 10 mol% cat., H_2O → MeO$_2$C CO$_2$Me, 90% yield (24)

Ph

cat.

5
Concluding Remarks

Organic synthesis in aqueous media has attracted much attention. Ruthenium catalysis in aqueous media is still relatively unexplored. This review briefly discussed the development of this area with representative examples. Many other important contributions could not be covered owing to the space limit.

Acknowledgements We thank NSF and the NSF–EPA joint program of Technology for a Sustainable Environment for support of our research.

References

1. Li CJ, Chan TH (1997) Organic reactions in water. Wiley, New York
2. Grieco PA (1998) (ed) Organic synthesis in water. Black, Glasgow
3. Silverman RB, Silverman RJ (1999) The organic chemistry of enzyme-catalyzed reactions. Academic, New York
4. Jencks WP (1917) In: Prescott SC (ed) Biochemical catalysts in life and industry; proteolytic enzymes. Wiley, New York, pp
5. Crabtree RH (2001) The organometallic chemistry of the transition metals, 3rd edn. Wiley-Interscience, New York
6. Anderson, JR, Boudart M (1997) (eds) Catalysis: science and technology. Springer, Berlin Heidelberg New York
7. McQuillin FJ, Parker DG, Stephenson G R (1991) Transition metal organometallics for organic synthesis. Cambridge University Press, New York
8. (a) Naota T, Takaya H, Murahashi S (1998) Chem Rev 98:2599; (b) Trost BM, Toste D, Pinkerton AB (2001) Chem Rev 101:2067
9. (a) Ritleng V, Sirlin C, Pfeffer M (2002) Chem Rev 102:1731; (b) Jun CH, Moon CW, Lee DY (2002) Chem Eur J 8:2423; (c) Kakiuchi F, Murai S (2002) Acc Chem Res 35:826
10. (a) Li CJ (2002) Acc Chem Res 35:533; (b) Li CJ (1993) Chem Rev 93:2023; (c) Li CJ (1996) Tetrahedron 52:5643; (d) Kobayashi S (1999) In: Kobayashi S (ed) Lanthanides: chemistry and use in organic synthesis. Springer, Berlin Heidelberg New York, p 63
11. (a) Nomura K (1998) J Mol Catal A 130:1; (b) Oehme G, Dwars T (2002) Adv Synth Catal 344:239; (c) Sinou D (2002) Adv Synth Catal 344:221
12. (a) Trost BM (1991) Science 254:1471; (b) Sheldon RA (1994) Chemtech 24:38; (c) Wender PA, Miller BL (1993) In: Hudlicky T (ed) Organic synthesis: theory and applications, vol 2. JAI, New York, pp
13. Uma R, Crévisy C, Grée R (2003) Chem Rev 103:27
14. Parshall GW (1980) Homogeneous catalysis. Wiley, New York
15. McGrath DV, Grubbs RH (1994) Organometallics 13:224
16. (a) Li CJ, Wand D, Chen D (1995) J Am Chem Soc 117:12867; (b) Wang D, Chen D, Haberman JX, Li CJ (1998) Tetrahedron 54:5129
17. Wang D, Li CJ (1998) Synth Commun 28:507
18. Wang M, Li CJ (2002) Tetrahedron Lett 43:3589
19. Wang M, Yang XF, Li CJ (2003) Eur J Org Chem 998
20. Yang XF, Wang M, Varma R, Li CJ (2003) Org Lett 5:657
21. Trost BM, Krause L, Portnoy M (1997) J Am Chem Soc 119:11319
22. Wei C, Li CJ (2002) Green Chem 4:39
23. Yamaguchi M, Hayashi A, Minami T (1991) J Org Chem 56:4091
24. Matthews WS, Bares JE, Bartmess JE, Bordwell FG, Cornforth FJ, Drucker GE, Margolin Z, McCallum RJ, McCallum GJ, Vanier NR (1975) J Am Chem Soc 97:7007
25. Bloch R (1998) Chem Rev 98:1407
26. Katherine BA, Mark DW, David BC (2000) J Am Chem Soc 122:11084
27. Miura M, Enna M, Okuro K, Nomura M (1995) J Org Chem 60:4999
28. Frantz DE, Fâssler R, Carreira EM (1999) J Am Chem Soc 121:11245
29. Li CJ, Wei C (2002) Chem Commun 268
30. Trost BM (2002) Acc Chem Res 35:695
31. Trost BM, Portnoy M, Kurihara H (1997) J Am Chem Soc 119:836
32. Trost BM, Brown RE, Toste FD (2000) J Am Chem Soc 122:5877
33. Trost BM, Krause L, Portnoy M (1997) J Am Chem Soc 119:11319
34. Trost BM, Rudd M (2002) J Am Chem Soc 124:4178
35. Trost BM, Pinkerton AB (2002) J Am Chem Soc 124:7376

36. Trost BM, Indolese AF, Müller TJJ, Treptow B (1995) J Am Chem Soc 117:615
37. Trost BM, Martine JA, Kulawiec RJ, Indolese AF (1993) J Am Chem Soc 115:10402
38. Dérien S, Jan D, Dixneuf PH (1996) Tetrahedron 52:5511
39. López F, Castedo L, Mascareñas JL (2002) J Am Chem Soc 124:4218
40. Grubbs RH, Pine SH (1991) In: Trost BM, Fleming I, Paquette LA (eds) Comprehensive organic synthesis. Pergamon, New York, chap 9.3
41. Schrock PR (1992) In: The strem chemiker, vol XIV, no 1. Strem Chemicals, Newburgport, p 1
42. (a) Ivin KJ, Mol JC (1997) Olefin metathesis and metathesis polymerization. Academic, San Diego; (b) Connon SJ, Blechert S (2003) Angew Chem Int Ed Engl 42:1900
43. Grubbs RH, Chang S (1998) Tetrahedron 54:4413
44. Bezan GC, Oskam JH, Cho HN, Park LY, Schrock RR (1991) J Am Chem Soc 113:6899
45. Novak B, Grubbs RH (1988) J Am Chem Soc 110:7542
46. Lynn DM, Kanaoka S, Grubbs RH (1996) J Am Chem Soc 118:784
47. Lynn DM, Mohr B, Grubbs RH (1998) J Am Chem Soc 120:1627
48. Lynn DM, Mohr B, Grubbs RH, Henling LM, Day MW (2000) J Am Chem Soc 122:6601
49. Mortell KH, Weatherman RV, Kiessling LL (1996) J Am Chem Soc 118:2297
50. Kanai M, Mortell KH, Kiessling LL (1997) J Am Chem Soc 119:9931
51. Manning DD, Hu X, Beck P, Kiessling LL (1997) J Am Chem Soc 119:3161
52. Manning DD, Strong LE, Hu X, Beck P, Kiessling LL (1997) Tetrahedron 53:11937
53. Kirkland TA, Lynn DM, Grubbs RH (1998) J Org Chem 63:9904
54. Rölle T, Grubbs RH (2002) Chem Commun 1070

Author Index Volumes 1–7 and 11

Volumes 8–10 are already in planning and are announced and will be published when the manuscripts are submitted to the publisher.
The volume numbers are printed in italics.

Subject Index

www.ingramcontent.com/pod-product-compliance
Ingram Content Group UK Ltd.
Pitfield, Milton Keynes, MK11 3LW, UK
UKHW021333070726
13610UKWH00011B/52

* 9 7 8 3 6 6 2 1 4 5 5 0 0 *